HANDBOOK OF RESEARCH ON NANO-DRUG DELIVERY AND TISSUE ENGINEERING

Guide to Strengthening Healthcare Systems

HANDBOOK OF RESEARCH ON NANO-DRUG DELIVERY AND TISSUE ENGINEERING

Guide to Strengthening Healthcare Systems

Edited by

Rajakumari Rajendran

Hanna J. Maria, PhD

Sabu Thomas, PhD

Nandakumar Kalarikkal, PhD

First edition published 2022

Apple Academic Press Inc.
1265 Goldenrod Circle, NE,
Palm Bay, FL 32905 USA

4164 Lakeshore Road, Burlington,
ON, L7L 1A4 Canada

CRC Press
6000 Broken Sound Parkway NW,
Suite 300, Boca Raton, FL 33487-2742 USA

4 Park Square, Milton Park,
Abingdon, Oxon, OX14 4RN UK

Library and Archives Canada Cataloguing in Publication

Title: Handbook of research on nano-drug delivery and tissue engineering : guide to strengthening healthcare systems / edited by Rajakumari Rajendran, Hanna J. Maria, PhD, Sabu Thomas, PhD, Nandakumar Kalarikkal, PhD.

Names: Rajendran, Rajakumari, edtior. | Maria, Hanna J., editor. | Thomas, Sabu, editor. | Kalarikkal, Nandakumar, editor.

Description: First edition. | Includes bibliographical references and index.

Identifiers: Canadiana (print) 20210286164 | Canadiana (ebook) 20210286172 | ISBN 9781771889841 (hardcover) | ISBN 9781774639412 (softcover) | ISBN 9781003161196 (ebook)

Subjects: LCSH: Drug delivery systems. | LCSH: Tissue engineering. | LCSH: Nanotechnology. | LCSH: Nanomedicine.

Classification: LCC RS199.5 .H36 2022 | DDC 615/.6—dc23

Library of Congress Cataloging-in-Publication Data

Names: Rajendran, Rajakumari, editor. | Maria, Hanna J., editor. | Thomas, Sabu, editor. | Kalarikkal, Nandakumar, editor.

Title: Handbook of research on nano-drug delivery and tissue engineering : guide to strengthening healthcare systems / edited by Rajakumari Rajendran, Hanna J. Maria, Sabu Thomas, Nandakumar Kalarikkal.

Description: Palm Bay, FL : Apple Academic Press, [2022] | Includes bibliographical references and index. | Summary: "Handbook of Research on Nano-Drug Delivery and Tissue Engineering: Guide to Strengthening Healthcare Systems provides an important and valuable collection of research accomplishments in nanomedicine, drug delivery, tissue engineering, processing, formulations, and their applications. With contributions from leading researchers in the nanomedicine field from industry, academia, and government and private research institutions across the globe, the volume provides an up-to-date record on the major findings and observations in the field of nanomedicine and tissue engineering. Divided into two parts, the book addresses topical issues in nano-drug delivery and nanotechnological approaches to tissue engineering. The first section offers research on a variety of diverse nano-based drug delivery systems, along with discussions of their efficacy, safety, toxicology, and applications for different purposes. Focusing on nanotechnology approaches to tissue engineering, part two of the volume considers the use of hydrogel systems, nanoceria and micro- and nano-structured biomaterials for bone tissue engineering, mesenchymal stem cells, and more. This volume is a systematic scientific reference of the novel research developments specifically in this area. The editors give special emphasis on the new trends and developments in the field of nanomedicine that will be very helpful for pharmaceutical and medical researchers, scientists, faculty, and students"-- Provided by publisher.

Identifiers: LCCN 2021037689 (print) | LCCN 2021037690 (ebook) | ISBN 9781771889841 (hardback) | ISBN 9781774639412 (paperback) | ISBN 9781003161196 (ebook)

Subjects: MESH: Nanotechnology--methods | Drug Delivery Systems--methods | Tissue Engineering--methods | Nanomedicine--methods

Classification: LCC R857.N34 (print) | LCC R857.N34 (ebook) | NLM QT 36.5 | DDC 610.28--dc23

LC record available at https://lccn.loc.gov/2021037689

LC ebook record available at https://lccn.loc.gov/2021037690

ISBN: 978-1-77188-984-1 (hbk)
ISBN: 978-1-77463-941-2 (pbk)
ISBN: 978-1-00316-119-6 (ebk)

About the Editors

Rajakumari Rajendran
Research Scholar, International and Inter University Centre for Nanoscience and Nanotechnology, Mahatma Gandhi University, Kottayam, Kerala, India 686560

Rajakumari Rajendran is a research fellow working at the International and Inter University Centre for Nanoscience and Nanotechnology, Mahatma Gandhi University, Kottayam, Kerala, India. She received a bachelor's degree in Pharmacy from the Tamil Nadu Dr MGR Medical University, India, and a master's degree in Pharmacy (Pharmaceutics), Sastra University, India. She has industrial experience in dietary supplements and medicinal formulations and is currently engaged in developing dietary supplements and nutraceuticals at the International and Inter University Centre for Nanoscience and Nanotechnology. She has awarded a prestigious Inspire Fellowship (DST) and has eight research publications, six book chapters, and two books to her credit.

Hanna J. Maria, PhD

Dr. Hanna J. Maria is a Senior Researcher at the International and Inter University Centre for Nanoscience and Nanotechnology (IIUCNN), Mahatma Gandhi University, Kottayam, Kerala, India. She finished her PhD in the year 2015 from Mahatma Gandhi University. She works to improve the adhesion between the fibers and low-density polyethylen the effect of plasma modification. She has 15 publications, 10 book chapters, and four co-edited books to her credit. She has experience in working with natural rubber composites, their blends, thermoplastic composites, lignin, nanocellulose, bionanocomposites, nanocellulose, rubber-based composites and nanocomposites, and hybrid nanocomposites.

Sabu Thomas, PhD
Professor Sabu Thomas is currently Vice Chancellor of Mahatma Gandhi University and the Founder Director and Professor of the International and Inter University Centre for Nanoscience and Nanotechnology. He is also a

full Professor of Polymer Science and Engineering at the School of Chemical Sciences of Mahatma Gandhi University, Kottayam, Kerala, India. Prof. Thomas is an outstanding leader with sustained international acclaim for his work in nanoscience, polymer science and engineering, polymer nanocomposites, elastomers, polymer blends, interpenetrating polymer networks, polymer membranes, green composites and nanocomposites, nanomedicine, and green nanotechnology. Dr. Thomas's ground-breaking inventions in polymer nanocomposites, polymer blends, green bionanotechnological, and nano-biomedical sciences have made transformative differences in the development of new materials for automotive, space, housing, and biomedical fields. In collaboration with India's premier tire company, Apollo Tyres, Professor Thomas's group invented new high-performance barrier rubber nanocomposite membranes for inner tubes and inner liners for tires. Professor Thomas has received a number of national and international awards which include: Fellowship of the Royal Society of Chemistry, London; FRSC, Distinguished Professorship from Josef Stefan Institute, Slovenia; MRSI Medal, Nano Tech Medal, CRSI Medal, Distinguished Faculty Award, Dr APJ Abdul Kalam Award for Scientific Excellence—2016, Mahatma Gandhi University—Award for Outstanding Contribution—November 2016; Lifetime Achievement Award of the Malaysian Polymer Group; Indian Nano Biologists Award 2017; and Sukumar Maithy Award for the best polymer researcher in the country. He is fifth on the list of most productive researchers in India. Because of the outstanding contributions to the field of nanoscience and polymer science and engineering, Prof. Thomas has been conferred an Honoris Causa (DSc) Doctorate by the University of South Brittany, Lorient, France, and the University of Lorraine, Nancy, France. Very recently, Prof. Thomas has been awarded a Senior Fulbright Fellowship to visit 20 universities in the United States and has received the most productive faculty award in the domain of materials sciences. He was also awarded with a National Education Leadership Award—2017 for Excellence in Education. Prof. Thomas also won he 6th contest of mega-grants in the grant competition of the Government of the Russian Federation (Ministry of Education and Science of the Russian Federation) designed to support research projects implemented under the supervision of the world's leading scientists. He has been honored with Faculty Research Award of India's brightest minds in the field of academic research in May 2018. Professor Thomas was awarded with Trila—Academician of the Year in June 2018, acknowledging his contribution to tire industry. Prof. Thomas was also awarded with the H.G. Puthenkavu Mar Philoxenos Memorial Best Scientist Award. For his outstanding contribution in the field of composite materials, he has been selected for the position of Professor Lorraine at University Lorraine in 2018. Prof. Thomas has been selected by the Materials Research Society of India (MRSI) for the CNR Rao Prize Lecture in Advanced Materials for the year 2019. Professor Thomas has published over 800 peer-reviewed research papers, reviews, and book chapters. He has co-edited 135 books. He is the inventor of 15 (granted—1, filed—15) patents. The H index of Prof. Thomas is 100, with more than 48,197 citations. Prof. Thomas has delivered over 300 plenary/inaugural and invited lectures in national/international meetings over 30 countries.

Nandakumar Kalarikkal, PhD
Director and Professor, International and Inter University Centre for Nanoscience and Nanotechnology, Director and Chair, School of Pure and Applied Physics, Mahatma Gandhi University, Kerala, India

Nandakumar Kalarikkal, PhD, is Director and Professor of the International and Inter University Centre for Nanoscience and Nanotechnology, Mahatma Gandhi University, Kerala, India. He is also the Director of School of Pure and Applied Physics at the same university. He is a fellow of many professional bodies. He has published more than 270 reviewed research papers, reviews, and book chapters and also co-edited 20 books. He also has a patent on the topic "A Polymer Nanocomposite, Process and Application thereof." He is actively involved in research, and his group works on the synthesis, characterization, and applications of various nanomaterials, nanostructured polymer blends, polymer scaffolds for tissue engineering, nutraceutical formulations, polymeric materials for dental applications, carbon hybrid structures, core–shell quantum dots, nanomultiferroics, and nanoferrites, metal nanostructures, nanophosphors, metal organic frameworks (MOFs), polymer nanocomposites, tissue engineering, laser plasma, nonlinear optics, thin films, biosensors, nano devices, ion beam modifications of materials, and phase transitions. He has delivered plenary/inaugural and invited lectures at national/international meetings.

Contents

Contributors

Annie Abraham
Advanced Centre for Tissue Engineering, Department of Biochemistry, University of Kerala, Thiruvananthapuram, India

Rahul Agrawal
Department of Metallurgical Engineering and Materials Science, IIT Bombay, Mumbai, Maharashtra, India

Satar Jabbar Rahi Algraittee
Stem Cell & Immunity Research Group, Immunology Laboratory, Department of Pathology, Faculty of Medicine and Health Sciences, Universiti Putra Malaysia, 43400 Serdang, Selangor, Malaysia
Department of Medical Microbiology, College of Medicine, University of Kerbala, 46001 Karbala, Iraq

P. B. Anjali
Department of Pharmaceutics, JSS College of Pharmacy (JSS Academy of Higher Education and Research), Rocklands, Udhagamandalam, Tamil Nadu 643001, India

Poornima Balan
Biological Materials Laboratory, Central Leather Research Institute (CSIR), Adyar, Chennai, Tamil Nadu 600020, India

Ena Ray Banerjee
Immunobiology and Regenerative Medicine Research Laboratory, Department of Zoology, University of Calcutta, Kolkata, West Bengal, India

R. Bansal
Faculty of Dental Science, Banaras Hindu University, Varanasi, Uttar Pradesh, India

Jaydeep Bhattacharya
School of Biotechnology, Jawaharlal Nehru University, New Delhi, India

Sourish Bhattacharya
Division of Biotechnology and Phycology, CSIR-Central Salt and Marine Chemicals Research Institute (CSIR-CSMCRI), Bhavnagar, Gujarat, India

Sakchi Bhushan
Department of Polymer and Process Engineering, Indian Institute of Technology Roorkee, Saharanpur Campus, Uttar Pradesh, 247001, India

Ranita Bose
Immunobiology and Regenerative Medicine Research Laboratory, Department of Zoology, University of Calcutta, Kolkata, West Bengal, India

Swaroop Chakraborty
Biological Engineering, Indian Institute of Technology Gandhinagar, Gandhinagar 382355, Gujarat, India

V. J. Chakravarthy
P G Department of Computer Science, The New College, Chennai, India

K. Chattopadhyay
Department of Metallurgical Engineering, IIT (BHU), Varanasi, Uttar Pradesh, India

Sonam Dubey
Division of Biotechnology and Phycology, CSIR-Central Salt and Marine Chemicals Research Institute (CSIR-CSMCRI), Bhavnagar, Gujarat, India

Sharida Fakurazi
Department of Human Anatomy, Faculty of Medicine and Health Sciences
Laboratory of Vaccine and Immunotherapeutics, Institute of Bioscience, Selangor, Malaysia

Nandita Ghosh
Immunobiology and Regenerative Medicine Research Laboratory, Department of Zoology, University of Calcutta, Kolkata, West Bengal, India

Lireni C. Humtsoe
Department of Pharmaceutics, JSS College of Pharmacy (JSS Academy of Higher Education and Research), Rocklands, Udhagamandalam, Tamil Nadu 643001, India

Mohd Zobir Hussein
Material Synthesis and Characterisation Laboratory, Institute of Advance Technology, Universiti Putra Malaysia, 43400 UPM, Serdang, Selangor Darul Ehsan, Malaysia

Janani Indrakumar
Biological Materials Laboratory, Central Leather Research Institute (CSIR), Adyar, Chennai, Tamil Nadu 600020, India

N. Jawahar
Department of Pharmaceutics, JSS College of Pharmacy (JSS Academy of Higher Education and Research), Rocklands, Udhagamandalam, Tamil Nadu 643001, India

Ajita Jindal
School of Biotechnology, Jawaharlal Nehru University, New Delhi, India

Annie John
Department of Biochemistry, University of Kerala, Thiruvananthapuram 695581, India

Josna Joseph
Advanced Centre for Tissue Engineering, Department of Biochemistry, University of Kerala, Thiruvananthapuram, India

Md. Asad Khan
Department of Biochemistry, Faculty of Dentistry, Jamia Millia Islamia, New Delhi 110025, India

Purna Sai Korrapati
Biological Materials Laboratory, Central Leather Research Institute (CSIR), Adyar, Chennai, Tamil Nadu 600020, India

P. R. Anil Kumar
Biomedical Technology Wing, Sree Chitra Tirunal Institute for Biomedical Sciences and Technology, Thiruvananthapuram, Kerala 695 012, India

Mas Jafri Masarudin
Department of Cell & Molecular Biology, Faculty of Biotechnology and Biomolecular Sciences, Selangor, Malaysia

Alka Mishra
Department of Microbiology, Parul Institute of Applied Sciences, Parul University, Vadodara, Gujarat, India

Narayan Chandra Mishra
Department of Polymer and Process Engineering, Indian Institute of Technology Roorkee, Saharanpur Campus, Uttar Pradesh, 247001, India

Sandhya Mishra
Division of Biotechnology and Phycology, CSIR-Central Salt and Marine Chemicals Research Institute (CSIR-CSMCRI), Bhavnagar, Gujarat, India

Superb K. Misra
Material Science & Engineering, Indian Institute of Technology Gandhinagar, Gandhinagar 382355, Gujarat, India

Shinjini Mitra
Immunobiology and Regenerative Medicine Research Laboratory, Department of Zoology, University of Calcutta, Kolkata, West Bengal, India

Ranjita Ghosh Moulick
Amity Institute of Integrative Sciences and Health, Amity University, Haryana, India

Suleiman Alhaji Muhammad
Laboratory of Vaccine and Immunotherapeutics, Institute of Bioscience, Universiti Putra Malaysia, 43400 UPM, Serdang, Selangor Darul Ehsan, Malaysia;
Department of Biochemistry, Usmanu Danfodiyo University, Sokoto, Nigeria

Padmaja Murali
Biological Materials Laboratory, Central Leather Research Institute (CSIR), Adyar, Chennai, Tamil Nadu 600020, India

Senthilkumar Muthusamy
Biomedical Technology Wing, Sree Chitra Tirunal Institute for Biomedical Sciences and Technology, Thiruvananthapuram, Kerala 695 012, India

P. Hakkim Devan Mydeen
P G Department of Computer Science, The New College, Chennai, India

Payal Pal
Immunobiology and Regenerative Medicine Research Laboratory, Department of Zoology, University of Calcutta, Kolkata, West Bengal, India

Pramathadhip Paul
Immunobiology and Regenerative Medicine Research Laboratory, Department of Zoology, University of Calcutta, Kolkata, West Bengal, India

Shiv Dutt Purohit
Department of Polymer and Process Engineering, Indian Institute of Technology Roorkee, Saharanpur Campus, Uttar Pradesh, 247001, India

Nancy Raj
Faculty of Dental Science, Banaras Hindu University, Varanasi, Uttar Pradesh, India

Rajesh Ramasamy
Stem Cell & Immunity Research Group, Immunology Laboratory, Department of Pathology, Faculty of Medicine and Health Sciences, Universiti Putra Malaysia, 43400 Serdang, Selangor, Malaysia

M. Moshahid A Rizvi
Department of Biosciences, Jamia Millia Islamia, New Delhi 110025, India

Debasish Saha
Solid State Physics Division, Bhaba Atomic Research Centre, Mumbai, India

C. Sanjeev
Department of Pharmaceutics, JSS College of Pharmacy (JSS Academy of Higher Education and Research, Mysuru), Rocklands, Udhagamandalam 643001, Tamil Nadu, India

Aishwarya Satish
Biological Materials Laboratory, CSIR-Central Leather Research Institute, Adyar, Chennai, Tamil Nadu 600020, India

M. Seenivasan
Department of Mathematics, Annamalai University, Tamil Nadu, India

Adeti Munmun Sengupta
Immunobiology and Regenerative Medicine Research Laboratory, Department of Zoology, University of Calcutta, Kolkata, West Bengal, India

Anupama Shrivastava
Department of Microbiology, Parul Institute of Applied Sciences, Parul University, Vadodara, Gujarat, India

Hemant Singh
Department of Polymer and Process Engineering, Indian Institute of Technology Roorkee, Saharanpur Campus, Uttar Pradesh, 247001, India

V. Singh
Department of Metallurgical Engineering, IIT (BHU), Varanasi, Uttar Pradesh, India

Anbuthiruselvan Solaimuthu
Biological Materials Laboratory, Central Leather Research Institute (CSIR), Adyar, Chennai, Tamil Nadu 600020, India

Myron Spector
Tissue Engineering/Regenerative Medicine, VA Boston Healthcare System, Boston, MA, USA
Department of Orthopedics, Brigham and Women's Hospital, Harvard Medical School, Boston, MA, USA

Gayathri Sundar
Department of Biochemistry, University of Kerala, Thiruvananthapuram 695581, India
Department of Biotechnology, CEPCI Laboratory & Research Institute, Kollam, India

Rebu Sundar
Department of Biochemistry, University of Kerala, Thiruvananthapuram 695581, India

Roshni Thapa
School of Biotechnology, Jawaharlal Nehru University, New Delhi, India

Shiny Velayudhan
Biomedical Technology Wing, Sree Chitra Tirunal Institute for Biomedical Sciences and Technology, Thiruvananthapuram, Kerala 695 012, India

Indu Yadav
Department of Polymer and Process Engineering, Indian Institute of Technology Roorkee, Saharanpur Campus, Uttar Pradesh, 247001, India

Abbreviations

3D	three-dimensional
5-Flu	5-fluorouracil
AA	α-amylase
AC	aceclofenac
AC-SUSP	aceclofenac suspension
AD	atopic dermatitis
AE	alveolar epithelial
AFM	atomic force microscopy
Ag NP	silver nanoparticles
ALD	atomic layer deposition
AMP	antimicrobial peptide
AO	acridine orange
AP	alkaline phosphatase
ATR	attenuated total reflection
AUC	area under the curve
AuNP	gold nanoparticle
AV	aloe vera
BALF	broncho-alveolar lavage fluid
BBB	blood–brain barrier
BCS	biopharmaceutical order framework
BDNF	brain-derived neurotrophic factor
BET	Brunauer–Emmett–Teller
bFGF	basic fibroblast growth factor
BIC	bone–implant contact
Bleo	bleomycin
CC	calibration curve
CCK-8	Cell Counting Kit-8
CIL	cilostazol
CMP	collagen mimic peptide
CNF	cellulose nanofiber
CNS	central nervous system
Cs/PEO	chitosan/polyethylene oxide
DCs	dendritic cells
DDW	double-distilled water

DLS	dynamic light scattering
DMOG	dimethyloxalylglycine
DNA	deoxyribonucleic acid
Dox	doxorubicin
DRG	dorsal root ganglia
DSC	differential scanning calorimetry
DSC	differential scanning calorimetry
EBL	electron-beam lithography
EC	ethyl cellulose
ECM	extracellular matrix
EGF	epidermal growth factor
EPR	enhanced permeability and retention
ESEM	environmental scanning electron microscopy
FESEM	field-emission scanning electron microscope
FGF	fibroblast growth factor
FSC	forward scattering
FTIR	Fourier-transform infrared
GC	gas chromatography
GDNF	glial cell-derived neurotrophic factor
GFs	growth factors
GG	guar gum
GN	guar gum nanoparticle
GNF	gelatin nanofiber
GNP	gold nanoparticle
HA	hyaluronic acid
HA	hydroxyapatite
HA	hyaluronic acid
hAGP	human α1-acid glycoprotein
HB	hydrxybutyrate
HDF	human dermal fibroblast
HF	hydrofluoric acid
HTC	hydrotalcite
HV	hydroxyvalerate
ICATM	International Co-operation on Alternative Test Methods
IL	interleukin
iNOS	inducible nitric oxide synthase
IoBNT	Internet of Bio-NanoTechnology
IoNT	Internet of NanoThings
IPF	idiopathic pulmonary fibrosis

JaCVAM	Japanese Center for the Validation of Alternative Methods
LaBP	laser-assisted bioprinting
LD50	lethal dose 50%
LDH	lactate dehydrogenase
LIF	leukemia inhibitory factor
LMT	lipid microtubule
LPS	lipopolysaccharide
mAB	monoclonal antibody
mAbs	monoclonal antibodies
MALS	multiangle light scattering
MAPK	mitogen-activated protein kinase (MAPK)
MCN	mesoporous carbon nanoparticle
MDR	multidrug-resistant
MI	myocardial infarction
MMPs	matrix metalloproteases
MMT	montmorillonite
MPE	metallocene polyethylene
MRSA	multidrug resistance *Staphylococcus aureus*
MS	mass spectrometry
MSCs	mesenchymal stem cells
NF	nanofiber
NGF	nerve growth factor
NMR	nuclear magnetic resonance
NPs	nanoparticles
NSC	neural stem cells
OA	osteoarthritis
OL	olanzapine
Ova	ovalbumin
Oxa	oxazolone
PAA	polyacrylic acid
PANI	polyaniline
PCL	polycaprolactone
PCL	poly(ε-caprolactone)
PCLF	poly(ε-caprolactone) fiber
PDGF	platelet-derived growth factor
PDI	polydispersity index
PEDOT	poly(3,4-ethylenedioxythiophene)
PEG	poly(ethylene glycol)
PEI	polyethyleneimine

PEO	poly(ethylene oxide)
PF	peritoneal fluid
PHA	polyhydroxyalkanoates
PHB	poly-β-hydroxybutyrate
PHBV	poly(3-hyroxybutyrate-co-3-hydroxyvalerate)
PI	polydispersity index
PLA	poly(L-lactide)
PLACL	poly(L-lactic acid)-*co*-poly (ε-caprolactone)
PLGA	polylactic-*co*-glycolic acid
PLLA	poly(L-lactic) acid
PNCS	phyto-nanocomposite incorporated scaffolds
PNS	peripheral nervous system
PPF	polypropylene fumarate
PPy	polymer polypyrrole
PVA	polyvinyl alcohol
PVD	physical vapor deposition
PVP	polyvinyl pyrrolidone
QC	quality control
RA	retinoic acid
RES	reticuloendothelial system
RGD	arginine–glycine–aspartic acid
RNA	ribonucleic acid
ROS	reactive oxygen species
RSM	response surface methodology
SANS	small-angle neutron scattering
SAXS	small-angle X-ray scattering
SBF	simulated body fluid
SC	stratum corneum
SCI	spinal cord injury
SCL	short chain length
SDF	stromal cell derived factor
SE	secondary electron
SEM	scanning electron microscopy
SF	silk fibroin
SFE	supercritical fluid extraction
SGZ	subgranular zone
SLN	solid lipid nanoparticles
SLN	solid lipid nanoparticles
SLNs	strong lipid nanoparticles

SMAT	surface mechanical attrition treatment
SNC	surface nanocrystallization
SNF	silica nanofiber
SSC	side scattering
SVZ	subventricular zone
TAT	transactivator of transcription
TBI	traumatic brain injury
TEM	transmission electron microscope
TG	thioglycolate
TGA	thermogravimetric analysis
TGFs	transforming growth factors
TLRs	toll-like receptors
TNFs	tumor necrosis factors
TPU	thermoplastic polyurethane
USSP	ultrasonic shot peening
VEGF	vascular endothelial growth factor
XPS	X-ray photoelectron spectroscopy
XRD	X-ray diffraction
YSZ	yttria-stabilized zirconia
ZLT	zeolite

Preface

As we all know, nano-drug delivery and tissue engineering are the application of nanotechnology to achieve breakthroughs in healthcare. They exploit the improved and often novel physical, chemical, and biological properties of materials at the nanometre scale. It has the potential to enable early detection and prevention and to essentially improve diagnosis, treatment, and follow-up of diseases. Nano-drug delivery and tissue engineering are a very special area of nanotechnology because they are an extremely large field ranging from in-vivo and in-vitro diagnostics to therapy including targeted delivery and regenerative medicine; it interfaces nanomaterials (surfaces, particles, etc.) or analytical instruments with "living" human material (cells, tissue, body fluids) and creates new tools and methods that impact significantly existing conservative practices. The field of nanomedicine, drug delivery, and tissue engineering are growing rapidly and its development is making tremendous impacts in life sciences and public health. The importance and significance can be evaluated by the fact that it has made huge improvements over the course of time and is continuing to influence various sectors.

The *ICNT—2018 Fifth International Conference on Nanomedicine and Tissue Engineering* was held on October 12, 13, and 14, 2018, Kottayam, Kerala, India. It was jointly organized by the International and Inter University Centre for Nanoscience and Nanotechnology (IIUCNN), Ayurveda-und Venen-Klinik, Institute for Holistic Medical Sciences (IHMS), and Institute of Macromolecular Science and Engineering (IMSE).

The conference was intended to give emphasis to various aspects of targeted delivery, diagnostics, regenerative medicine, design of nanodrugs, synthesis of nanoparticles for drug delivery, connectivity between traditional and nanomedicine, green nanotechnology in drug discovery, nano-delivery systems, nanoparticles and nanomaterials in therapy, nano-delivery systems, nanoparticles and nanomaterials in therapy, nanomedicine approaches in molecular imaging, nanomedicine in theranostics, novel synthetic approaches in nanomedicine, regulatory aspects toward approval of nanomedicine, toxicology considerations in nanomedicine and nano-delivery systems, cartilage tissue engineering, dental tissue engineering, biomaterials tissue engineering, bioreactors tissue engineering, stem cells tissue engineering, and genetic tissue engineering

This book, titled *Handbook of Research on Nano-Drug Delivery and Tissue Engineering: Guide to Strengthening Healthcare System,* consists of two parts: Part I deals with "New Insights in Nano Drug Delivery" and Part II deals with "Nanotechnology Approaches to Tissue Engineering," which contains collection of chapters from the delegates who presented their papers during the conference. This book includes a wide variety of topics in nanomedicine, drug delivery, tissue engineering, processing, formulations, and their applications.

We appreciate the efforts and enthusiasm of the contributing authors and acknowledge those who were prepared to contribute but unable to do at time. The guest editors are R. Rajakumari, Research Scholar (Inspire Fellow), International and Inter University Centre for Nanoscience and Nanotechnology, Mahatma Gandhi University, Priyadarshini Hills, Kottayam, India; Dr. Hanna J. Maria, Post-Doctoral Fellow, International and Inter University Centre for Nanoscience and Nanotechnology, Mahatma Gandhi University, Priyadarshini Hills, Kottayam, India; Dr. Nandakumar Kalarikkal, Director, International and Inter University Centre for Nanoscience and Nanotechnology, Mahatma Gandhi University, Priyadarshini Hills, Kottayam, India, and the School of Pure and Applied Physics, Mahatma Gandhi University, Kottayam, India; and Prof Sabu Thomas, Vice-Chancellor, Mahatma Gandhi University, Professor, International and Inter University Centre for Nanoscience and Nanotechnology, and Professor, School of Chemical Sciences, Mahatma Gandhi University, Kottayam, India

This book is intended to serve as a "one-stop healthcare book" for important research accomplishments in the area drug delivery, nanomedicine, and tissue engineering. In the present book, we have given special importance to the new trends and developments in the field of nanomedicine and tissue engineering, which will be very helpful to medical, biomedical, and pharmacy students and scientists. We would like to thank all who kindly contributed chapters for this book. We are also very thankful Apple Academic for their kind help and assistance in preparation and publication of this book.

–Prof. Sabu Thomas

PART I
New Insights in Nanodrug Delivery

CHAPTER 1

Recent Advancements on Polymer-Based Drug Delivery System

ROSHNI THAPA[1], DEBASISH SAHA[2], RANJITA GHOSH MOULICK[3], and JAYDEEP BHATTACHARYA[1,*]

[1]*School of Biotechnology, Jawaharlal Nehru University, New Delhi, India*

[2]*Solid State Physics Division, Bhaba Atomic Research Centre, Mumbai, India*

[3]*Amity Institute of Integrative Sciences and Health, Amity University, Haryana, India*

**Corresponding author. E-mail: jaydpb@gmail.com; jaydeep@jnu.ac.in.*

ABSTRACT

The technological advancements and vast scientific explorations have led to emergence of various novel approaches for efficient delivery of biomolecules. Among different strategies employed to circumvent challenges associated with conventional drug delivery system, nanocarriers have shown promising results in various in-vivo models as well as clinical set up. Nanocarriers such as polymer-based nanoparticles and lipid-based liposomes in particular have successfully emerged as potential carrier system with many formulations approved for clinical use. In this chapter, we focus on polymer nanoparticles and various techniques for its preparation and characterization. We have also included elaborate discussion on applications of polymer nanoparticles in drug delivery as well as in achieving targeted drug delivery for improved therapeutic efficacy at a reduced dose and minimized adverse effects.

1.1 INTRODUCTION

Over the last few decades, researchers have exploited lipid and polymer nanoparticles (NPs) as a potential novel drug delivery system with an efficient in-vivo therapeutic effect. Since its advent, numerous synthetic and natural polymers have been approved by the FDA, which resulted in successful and promising clinical results for different life-threatening conditions like cancer, hypertension, etc. These lipid and polymer NPs have shown ideal properties required for the design of smart drug delivery systems like biocompatibility, biodegradability, low cost of production, and simple preparation. These NPs were found to show increased blood circulation time and controlled drug release. Several variables can be manipulated during the design of polymer NPs in order to deliver active ingredients to a specific site at a predetermined rate and time. Conjugation of ligand/antibodies by surface modification of polymer NPs can lead to enhanced localization of biomolecules at the target sites and thus, can render better therapeutic efficacy. Also, the combination of biocompatible polymer NPs with inorganic materials-based components has shown good theranostic properties leading to improved understanding of a medical condition as well as drug delivery. In this chapter, we have addressed the techniques for synthesis, characterization, surface functionalization, and application of FDA-approved polymers in various biomedical fields, as well as different experimental strategies for encapsulation of drugs into these NPs.

Organic materials like lipids have been explored vigorously over the last two decades as a carrier and delivery system of various biomolecules due to its biomimicking properties. These lipid-based nanoparticles (NPs) have shown multiple advantages in the drug delivery system due to their lipophilic property that enhances the entrapment of hydrophobic drug molecule and also enhances absorption. Due to the similarity of lipid molecules with the cell membrane, the immunogenic response is comparatively low and hence they show less toxic [1]. Polymer NPs, on the other hand, have relatively better stability, economical, and easy to prepare which facilitates the possibility of scale-up. Use of biodegradable polymers which breaks down into oligomers and monomers in vivo can also be used further to minimize immunological response. Moreover, hydrophilic polymers like polyethylene glycol (PEG) avoid opsonization of the NPs, increasing circulation time in-vivo and hence enhancing the bioavailability of the biomolecule in the target sites.

There is also significant advancement in the use of conjugated polymer NPs for many biomedical applications including sensing and bioimaging. They have enabled us to better understanding the biological processes and also improved image-guided therapies and surgeries [2].

Co-encapsulation of multiple active compounds, surface functionalization of the NPs has been done. Several preclinical studies on cocktail chemotherapy by co-encapsulation and co-delivery of chemotherapeutics and peptide drugs have proven polymer NPs based combination therapy to be a potential therapeutic approach for efficient anticancer regimens [3, 4].

Novel multifunctional polymeric NPs in combination with inorganic materials, such as various metals and metal oxide, have drawn attention from the researchers to achieve a theranostic approach [5–7]. Chemotherapy in combination with thermotherapy has been reported by encapsulating supramagnetic iron oxide NPs coated with oleylamine into PLGA NPs along with two antitumor agents [8]. The multimodal cancer therapy showed enhanced therapeutic effect as well as localization of chemotherapeutics in the tumor region. PEGylated borate-coordination-polymer-coated polydopamine NPs were designed to achieve synergistic chemo-photothermal tumor-targeted nanoplatform. The dynamically designed NPs showed significantly reduced systematic toxicity and efficiently improved tumor targeting via PEG-directed passive targeting [9].

Polymer NPs have also been used for targeted drug delivery by active or passive targeting. Active targeting includes surface modification of the polymer NPs by antibody conjugation leading to ligand-binding interactions at the target sites. Whereas, passive targeting include modification of physiochemical properties of the polymer with respect to the physiological and histological characteristics [10]. As shown in Figure 1.1, recent years have witnessed polymer NPs playing a key role in many fields apart from biomedical sciences. In this chapter, different types of polymer NPs, as well as various techniques of preparation have been discussed briefly. Furthermore, different characterization techniques, as well as application in drug delivery have been highlighted.

1.1 BIODEGRADABLE POLYMER NPS

One of the fundamental requirements for a biomaterial to be used in drug delivery is its biocompatibility releasing nontoxic metabolites in vivo. Various polymers of natural and synthetic origin have been found to be biocompatible and biodegradable making them potential drug carriers. Table 1.1 shows a list of the most widely used polymers from different sources with the advantages and limitations. These polymers undergo degradation into oligomers and monomers by various enzymatic or nonenzymatic routes leading to the formation of metabolites that are eliminated by various metabolic pathways. The FDA approved

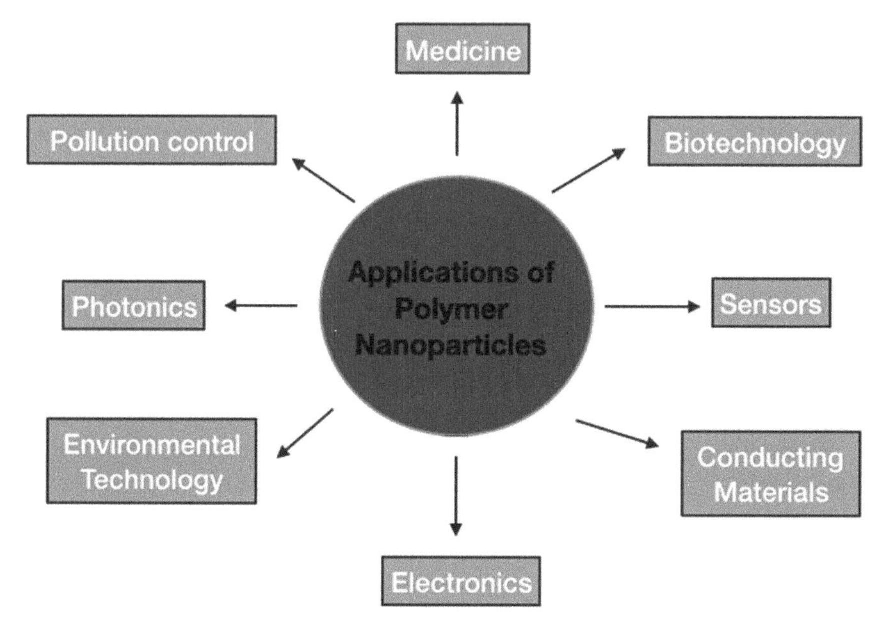

FIGURE 1.1 Application of polymer NPs in different areas.

polymers, poly(lactide) (PLA), PCL, and PLGA are among efficiently employed biodegradable synthetic polymers that are also used as copolymers with PEG, poly-glutamic acid, poly(butylcyanoacrylate) (PBCA), etc. [11–13]. PLGA is one of the most commonly used polymer for drug delivery as it is biodegradable and hydrolyzes to lactide and glycolide that are eliminated from the body via the citric acid cycle [14, 15].

These polymers have been used in the encapsulation of hydrophobic drug molecules and successful delivery to the target sites. On the other hand, polymer-based NPs are not the ideal candidates for delivery of hydrophilic molecules due to unsatisfactory entrapment efficiency and partition of drug in the aqueous phase. One of the major disadvantages associated with polymer-based NPs is the short circulation time in-vivo, limiting the bioavailability of the active ingredient. However, synthetic polymers like PEG with hydrophilic properties are used to develop long-circulating stealth NPs by avoiding opsonization. The enhanced circulating time of PEG NPs confers to the low cell adhesion and low protein absorption due to the formation of a highly water-bound barrier as a result of hydrophilicity of the polymer [16].

TABLE 1.1 Most Widely Used Polymers for Polymer-based NPs for Drug Delivery

	Advantages	**Disadvantages**
Natural polymers Chitosan, dextrins, cyclodextrins, gelatin, albumin, starch, alginates, cellulose, pectins, collage	• Biocompatible • Biodegradable • Easy availability • Less toxic • Presence of reactive sites that facilitates cross-linking, ligand conjugation, and other modifications • Ideal carriers for protein	• High degree of variability complex structure • High cost extraction process
Synthetic polymers Homopolymers Poly(lactide) (PLA) Poly(lactide-*co*-glycolide) (PLGA) Poly(epsilon-caprolactone) (PCL) Poly(isobutylcyanoacrylate) Poly(ethylene glycol) (PEG) Poly(*n*-butylcyanoacrylate) PBCA Poly(acrylate) and poly(methacrylate) Copolymers Poly(lactide)-poly(ethylene glycol) (PLA-PEG) Poly(lactide-*co*-glycolide)-poly(ethylene glycol) (PLGA-PEG) Poly(epsilon-caprolactone)-poly(ethylene glycol) (PCL-PEG)	• Biocompatible • Physical and chemical properties can be tailored, can be modified • Targeted drug delivery • Higher batch reproducibility	• High cost of production • Can induce immune response in some cases

Considering the possible toxicity associated with the polymer NP, the adverse effects could be physiological, physiochemical, and molecular factors. Physiochemical properties of the polymers influence the biodistribution of drug as well as the interaction between the polymer NP and the biological system including the target site [17]. Apart from opsonization, phagocytosis and endocytosis also may cause the potential toxicity of polymer [18]. On the introduction of polymer NPs, opsonization can take place which in turn can potentiate innate as well as an antigen-specific immune response [19].

Interaction of polymer NPs with vascular components of blood can also lead to local inflammation. Interaction of polymeric drug carriers with mitochondria can promote the production of reactive oxygen species leading to toxic cellular events. Moreover, hemotoxicity and tissue accumulation are the major complication associated with potential polymer NP toxicity. Thus, the use of biodegradable and biocompatible polymers becomes an important factor considering the possible toxicity associated with the polymers otherwise.

1.2 TYPES OF POLYMER NPS

Depending upon the properties of polymer as well as the method of preparation, polymer NP can be engineered into polymeric micelles, solid NPs, bilayer, and core–shell structures, as shown in Figure 1.2. These particles are formed either by the self-assembly or emulsification process. The self-assembly of polymer chains into spherical drug-loaded NPs results from inter- and intra-molecular interactions between the drug molecule and the carrier. Amphiphilic polymers spontaneously assemble into micellar structures on reaching critical micellar concentration and form NPs due to hydrophobic interaction whereas electrostatic interaction between cationic polymer and anionic nuclear acids leads to the formation of polyplexes [20]. On the other hand, in the emulsification technique where an organic solution of polymer is dispersed into the continuous aqueous phase in presence of surfactant and continues stirring can lead to the formation core–shell polymer structures.

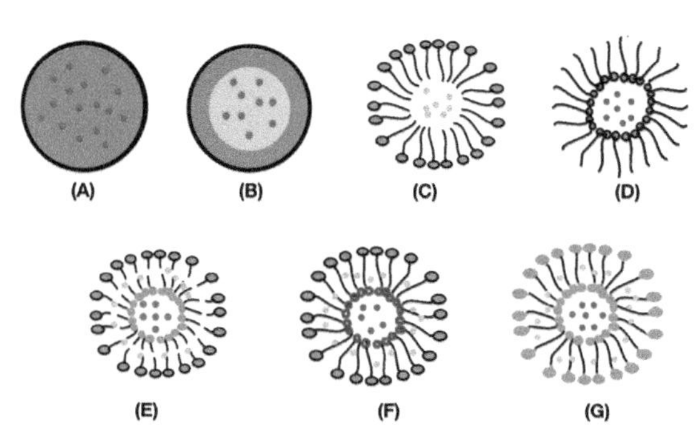

FIGURE 1.2 Biodegradable polymeric NPs: (A) nanospheres with a polymeric matrix with drug molecules dispersed, (B) nanocapsules with polymeric membrane and inner oil/aqueous core polymeric micelles formed by amphiphilic polymers containing a hydrophilic head and hydrophobic tails (C) in hydrophilic solvents (D) in hydrophobic solvents (E) AB-type diblock copolymer forming lipid bilayer like structure in polar solvents and drug encapsulation, (F) polymerosome formed by ABC type block copolymer in polar solvents, (G) polymerosome formed by ABA type block copolymer *n* polar solvents.

Diblock and triblock copolymers have been used to develop an interesting bilayer structures known as polymerosomes. These artificial vesicles are composed of amphiphilic block copolymers and resemble the bilayer structure of the lipid vesicles (liposomes). Similarly, they have bilayer membranes surrounding the aqueous core. The hydrophobic molecules can be entrapped by integrating with the hydrophobic region in the bilayer structures whereas water-soluble drugs, peptides, enzymes, RNA fragments, and DNA can be entrapped in the aqueous core. The bilayer assemblies formed is dependent on the type of copolymer used such as AB (e.g., PEG-PEE: poly(ethylene glycol)-*b*-poly(ethylene ethylene), ABA (e.g., PMOXA-PDMS-PMOXA: poly(2-methyl-2-oxazoline)-*b*-poly(di-methylsiloxane)-*b*-poly(2-methyl-2-oxazoline), or ABC triblock (e.g., PEG-PAA-PNIPAM: Poly(ethyleneglycol)-*b*-poly(acrylic acid)-*b*-poly(*N*-isopropylacrylamide) [21, 22]. Polymerosomes have successfully gained immense attention from researchers as compared to liposome due to their high stability, robust structure, and improved mechanical strength.

Apart from the basic structural differences, different stimuli-responsive polymer NPs can be designed for controlled as well as targeted drug delivery. The external stimuli controlling the drug release could be physical, such as electric field, electromagnetic radiations, thermal energy, or magnetic field or chemical stimuli such as ionic strength, pH, supramolecular chemistry, covalent bonding, etc. [23]. The different approaches to develop stimuli-responsive polymer NPs include the use of stimuli-responsive monomers for copolymerization into the polymeric backbone or chemically incorporating the responsive components into the NPs. An example of the latter is Azobenzene units copolymerized by reversible isomerization from cis-to-trans conformation in response to pH and/or UV radiation [24, 25].

The core of the NPs can also be used for entrapment of stimuli-responsive component. Smart drug delivery using polymeric NPs can be achieved using various stimuli-responsive polymer NPs based on physical stimuli, chemical stimuli, biological stimuli, environmental stimuli, dual stimuli, or multiple stimuli. Physical factors such as osmotic pressure hydrodynamic pressure, vapor pressure, mechanical force, etc., are extensively explored to trigger drug release at the site of action at a predetermined rate that is based on gradients of osmotic pressure, hydrodynamic pressure, and vapor pressure, respectively. The polymeric NP system with magnetic stimuli responsiveness has also been reported that uses electromagnetic field triggering drug release from the carrier. Ultrasonic energy and electrical current can also be utilized to achieve a desired drug release pattern and are called sonophoresis and iontophoresis-stimuli-responsive systems, respectively. Various polymers with hydration-induced-swelling properties and electrically erodible matrix have been utilized too.

Different chemical stimuli employed for smart polymeric NPs are pH, salt concentration, and hydrolysis. In the case of pH-responsive systems, the polymers degrade or erode at a particular c pH which can be used to trigger drug release at a specific region of a gastrointestinal system or at the target tissues, which has altered pH as compared to the healthy tissues. Ionizable drugs can be prepared using ion-exchange resin in which drug release is dependent on the salt concentration. Also, hydrolysis induced polymer degradation can be used to activate the release of the drug from such systems.

Biochemical and enzymatic reactions have also been reported as a stimuli for designing a controlled drug delivery system. Some of the biopolymers can undergo enzymatic degradation leading to the release of encapsulated active ingredient. Feedback-regulated delivery concept and biochemical reactions leading to biodegradation PF polymers have also been applied to design novel drug delivery systems.

Polymers are found to undergo isothermal phase transitions by photon and reversible phase separations through photo-irradiation. In such cases, light can be used as an environmental stimuli to achieve a desired drug release profile. Polymers such as carboxymethyl dextran have been found to bioerode under hypoxic conditions making it applicable for use in the design of stimuli-responsive drug delivery system. More than one stimuli responsiveness has also been successfully used for achieving desired release kinetics. Some of micelles are reported to be thermo as well as pH-responsive [26].

As shown in Figure 1.2, the morphology could vary from hollow, spherical, and concentric and also to more complex cockleblur, jannus, gibbous, and inverse gibbous outer appearance. Spherical polymer NPs have been extensively used whereas chemically and structurally heterogeneous core–shell, hollow, and concentric NPs offer remarkable advantages attributed to their layered structures and cavities [27]. In the case of core–shell assembly, application of external stimuli can lead to disassembly of shell or volume and diffusivity change, leading to the release of entrapped biomolecule. Dual responsive core–shell polymer NPs can be designed employing earth Eu(III) doped with PNIPAM-so-PS copolymer as well as quaternary ammonium tetrapheylethylene derivative (d-TPE)-doped PNIPAM-*co*-PAA shell shows the dual response to photoluminescence and temperature and has been successfully employed as bioprobes [28].

Hollow NPs can be formed using selective monomer to polymerize/copolymerize within the hydrophobic region of the bilayer vesicular structure with liposome as a template [29]. For example, the hydrophobic region of 1,2-dilauroylphosphocholine phospholipid can be utilized for stabilizing the hydrophobic monomers (e.g., styrene). On exposure to UV radiation of wavelength 254 nm, monomer photo-polymerization takes place in presence of photoinitiator, such as 2-hydroxy-2-methyl-propiophenon, leading to the formation of stimuli-responsive hollow vesicle [30].

In the case of concentric NPs, the segregation and entrapment of the molecules are influenced by the composition of outer layers as well as space between the core and shell [31, 32]. Various optical properties like surface plasmon resonance and refractive index can be manipulated using this approach [33, 34]. Surface responsive shells can be used to modulate the reaction by selective permeation of the reactants. This class of polymer NPs can be depicted by a system with stimuli-responsive pMMA outer shell and silver core that enables catalytic reduction of *p*-nitrophenol by sodium borohydride as they diffuse into the interior of the concentric NPs. As a result of reduction, *p*-aminophenol product diffuses out of the pMMA shell [35].

Amphiphilic Janus NPs are formed by the incorporation of two components with different physical and chemical properties that are conventionally prepared by sequential ring-opening metathesis polymerization of polystyrene and poly(ethyleneoxide) (PEO) [36–38]. JNPs are considered multifunctional nanomaterials due to phase-separated morphologies and with the possibility of incorporation of stimuli sensitivity [39].

The bulges and dimples present on the gibbous and inverse gibbous NPs have gained attention from the researchers as it offers localized stimuli responsiveness and molecular recognition. Similar sizes of bulges and dimples can lead to the formation of lock and key structure. Localized copolymerization of monomers swelled into selected seed can be used to develop phase-separated and stimuli-responsive gibbous NPs. The swelling of selected seeds such as pMMA/pMMA, pMMA, pMMA/nBA, pSt, or SiO_2-pMMA with pH and temperature-responsive monomers leads to formation of bulges upon copolymerization making the NP stimuli-responsive [40]. In the case of inverse Gibbous, hydrophobic monomers are using a similar technique [41]. NPs with spikes on the surface resemble the cocklebur and urchins. The high surface area facilitates localized stimuli responsiveness to such particles. The spikes can act as a molecular recognition site and as receptors when small molecules are entrapped [42]. During the synthesis of cocklebur, a stabilized solution of monomer and radical initiator is extruded and the size of the resultant extruded NPs is influenced by the diameter of nanoextruder used. The tubule-forming phospholipids such as 1,2-bis(10,12-tricosadinoyl)-sn-glycero-3-phosphocholine undergo cross-linking upon UV exposure leading to formation of mechanically stable stimuli-responsive spikes [43].

1.3 TECHNIQUES OF PREPARATION

The properties of polymer NPs are significantly influenced by the method of preparation which can be optimized depending upon the intended application of the final product. Depending upon the starting component, the mode of synthesis can be categorized into two major approaches, namely: (1) top-down approach with polymer as a starting material, and (2) bottom-up approach in which monomers are used. In both cases, various synthetic/natural polymers are used for synthesizing nanocapsules and nanospheres. For the preparation of nanocapsules in a specific, interfacial deposition is employed in a top-down approach whereas monomers as starting material, molecular inclusion, and interfacial polycondensation are also used. For the formulation of nano-spheres using a top-down approach, nanoprecipitation emulsification as well as coacervation are the most commonly used techniques. On the other hand, for a bottom-up approach, emulsion polymerization is used to obtain nanospheres. The choice of a particular technique depends on the desired properties of the final product and intended use. Of all the mentioned techniques, nanoprecipitation and emulsification followed by solvent evaporation have been extensively used. A brief discussion of the most commonly used techniques will be discussed in the following section.

1.3.1 EMULSIFICATION-SOLVENT EVAPORATION

This technique was one of the first reported techniques for the synthesis of polymer NPs starting with polymer [44]. Since its advent, researchers have been exploiting this technique to develop drug-loaded biodegradable polymer NPs in the field of pharmaceutical science. Briefly, in this technique, the organic solution of polymer is prepared in a volatile organic solvent, such as chloroform, dichloromehane, etc., depending upon the solubility profile of the polymer as well as drug in case of hydrophobic biomolecule. Introduction of the polymer solution to the aqueous phase in presence of high shear causes diffusion of the solvent into the continuous aqueous phase and hence nanosuspension is formed. The organic solvent is allowed to evaporate from the open system under vacuum or overnight stirring. The different types of emulsions such as oil-in-water (o/w) or water-in-oil-in-water (w/o/w) are formed depending on the properties of the active ingredient to be encapsulated (Figure 1.3). In the case of hydrophobic molecules, a single emulsion technique can give satisfactory entrapment efficiency whereas a double emulsion technique is mostly used for the water-soluble molecules.

Due to the high affinity of hydrophilic molecules to the aqueous phase, very less amount of these biomolecules are entrapped into the polymeric nanocapsules. Thus, double emulsion (w/o/w) has been shown to improve the encapsulation of water-soluble drugs into the polymer NPs. Various techniques like high-pressure homogenization and ultrasonication can be utilized for further size reduction. The major parameters affecting the final particle size are a concentration of the stabilizer, stirring speed, homogenization speed, and polymer concentration. NPs based on polymers like PLGA, PCL PLA, cellulose acetate phthalate, poly(β-hydroxy-butyrate) (PHB) have been formulated using this technique.

1.3.2 *NANOPRECIPITATION*

Nanoprecipitation is one of the most successfully used methods for polymer NP preparation. Also known as the solvent displacement method, this technique involves the introduction of polymer solution in an aqueous phase with continuous stirring in the presence/absence of surfactant. The polar organic solvent with hydrophobic polymer solubilized is injected into the continuous phase. The organic solvent quickly diffuses into the aqueous phase leading to polymer deposition on the interface of organic solvent and water. The resultant colloidal suspension, in most cases, is found to be a homogenous population [45].

A water-miscible nonsolvent for polymer can be used for phase separation which in turn will facilitate the formation of colloidal polymer NPs. Due to the use of polar organic solvent, this method is suitable for the entrapment of hydrophobic drug. However, poor results have been reported for water-soluble drugs. This method has been used for various polymers including PLA, PLGA, and PCL for the incorporation of hydrophobic drugs with varying pharmacological activity. Schematic illustration of nanoprecipitation and solvent evaporation technique has been shown in Figure 1.3.

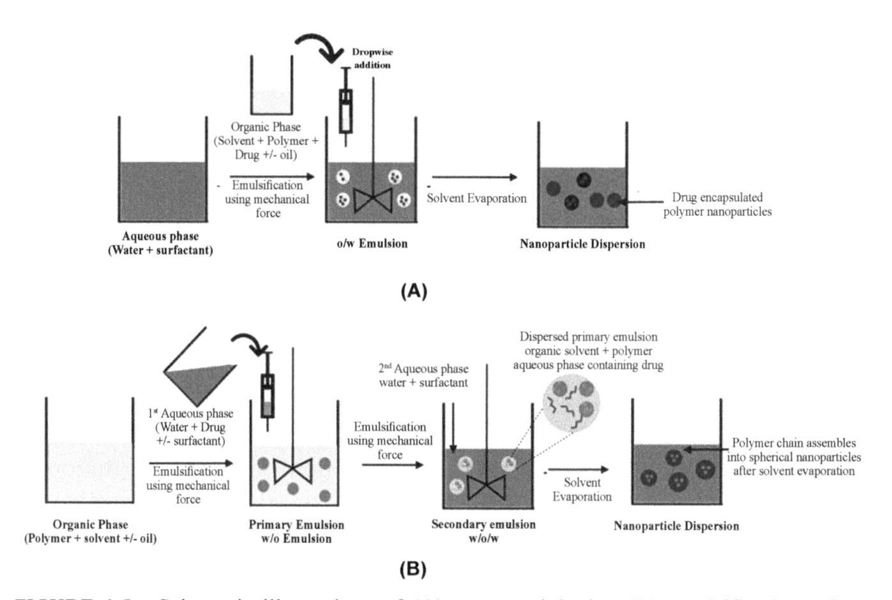

FIGURE 1.3 Schematic illustrations of (A) nanoprecipitation (B) emulsification-solvent evaporation techniques of polymer NPs preparation.

1.3.3 COACERVATION OR ION GELATION METHOD

The need for biocompatible drug carriers has led to an exploration of various biodegradable polymers ignorer to minimize the toxicity while yielding stable NPs. Unlike solvent evaporation and nanoprecipitation, this technique is based on hydrophilic polymers. Calvo and coworkers developed a method for the synthesis of chitosan NPs using the ion gelation technique. This method involves two aqueous phases, one with chitosan, a diblock copolymer of ethylene and propylene oxide (PEO=PPO) while the other with polyanion sodium tripolyphosphate. Coacervates are formed as a result of an interaction between the positively charged amino group of chitosan and the negatively charged tripolyphosphate. While coacervates results from electrostatic interaction, ionic gelation involves phase transition from liquid to gel due to ionic interaction at room temperature [46].

1.3.4 SALTING OUT

This technique is based on the separation of a water-miscible solvent by the salting-out effect. In this method, an organic solvent with polymer and drug is added to the aqueous phase of the system that contains salting agents as well as colloidal stabilizers [47]. The different classes of salting agents used are nonelectrolytes such as sugar and electrolytes such as calcium chloride and magnesium chloride. The resultant oil-in-water emulsion is further diluted to facilitate diffusion of the aqueous solvent leading to the formation of nanospheres. These techniques have been reported to be highly efficient for the preparation of PLA, ethyl cellulose (EC), and poly(methacrylic) acids nanospheres [48].

1.3.5 INTERFACIAL POLYMERIZATION

This technique involves polymerization of the monomer and eventually the formation of polymer NPs. This technique involves two reactive monomers dissolved in two different phases, namely dispersed and continuous phase. Interfacial cross-linking reaction such as polyaddition and polycondensation or radical polymerization takes place leading to the formation of polymer NPs [49–51]. The drug to be encapsulated can be added to the polymerization medium or can be adsorbed onto the NPs after polymerization [52]. Based on the choice of solvent, nanocapsule, as well as nanospheres can be formed. Use of aprotic solvents such as acetone and acetonitrile promotes the formation of nanocapsule whereas using protic solvents like ethanol can lead to the formation of nanospheres along with nanocapsules [53]. Oil containing nanocapsule can be obtained by promoting polymerization of monomers at the oil/water interface of the oil-in-water microemulsion. On the contrary, water containing nanocapsules can be formed by interfacial polymerization of monomers at the interface of water-in-oil microemulsions [54, 55]. Polymers such as poly(alkyl-cyanoacrylate) or polybutylcyanoacrylate have been used for developing NPs using this technique.

1.4 CHARACTERIZATION OF POLYMER NPS

The biodegradable polymeric NPs hold great promise in controlled drug delivery. The pharmacological activity and controlled drug release of polymeric NPs strongly depend on the particle formation and their size [56–58]. For example, NPs of 30–150 nm can be found in the heart kidney, and stomach while NPs with 150–300 nm are mainly going to the liver and the spleen [59]. Therefore, the necessity of controlling the size and polydispersity of polymeric NPs is very high for using them as nanocarriers for drug delivery. Polymer NPs have been prepared using various methods including solvent evaporation, emulsification-diffusion, salting-out, and nanoprecipitation. Among these methods, nanoprecipitation is a simple, facile, mild, and low energy input process to prepare the NPs [60]. It consists of the addition of a hydrophobic polymer dissolved in an organic solvent to form the diffusing phase. The diffusing phase is then added into an aqueous phase (nonsolvent) in the next step. Under these conditions, the precipitation of the polymer NPs occurred spontaneously. The organic solvent is evaporated from the final solution in the second step to obtain the final state of the NPs. The final size and properties of polymer NPs are controlled by a large number of parameters [61]. The initial solute concentration in the organic solvent and the final solute concentration in the nonsolvent are the two most important parameters that control the size of NPs. The selection of organic solvent can also shift the size of the final solute concentration in the non-NPs [62]. The solvent to nonsolvent ratio [63] and the rate of maxing are other factors that can influence the final size of polymer NPs [64, 65].

The physiochemical properties of polymer NPs can be characterized by using advanced technologies like diffraction laser scattering, transmission electron microscopy (TEM), scanning electron microscopy (SEM), and atomic force microscopy (AFM). The characterization based on size, surface charge, surface morphology, and size distribution can have a direct impact on the physical stability as well as the in-vivo fate of the drug carrier. The other physiochemical parameters like drug stability, NP dispersion stability, and release profile also play an important role in the efficiency of polymeric NPs. Drug stability can be assessed using various bioassays of drugs extracted from NPs as well as chemical analysis. However, for investigating the NP dispersion stability, critical flocculation temperature, and AFM are useful techniques. The chemical analysis of the drug and the surface of polymeric nanocolloid can be studied using static secondary ion mass spectroscopy, Sorpto meter, and Fourier-transform infrared spectroscopy. The release profile of the biomolecule from the polymeric drug carrier has a profound effect on the bioavailability and hence, can be assessed by in-vitro experiments under simulated physiological conditions. Some of the important characterization techniques for polymer NPs have been discussed briefly.

1.4.1 PARTICLE SIZE, POLYDISPERSITY INDEX, AND SIZE DISTRIBUTION

The particle size of a polymer NP has a profound effect on stability, biodrug entrapment, biodistribution, cellular uptake, and cellular internalization. The large surface area offered by the smaller particles favors release of entrapped active ingredient. The polymer NPs with particle size closed to 1 μm have shown to be cleared rapidly by macrophage–phagocyte system. Unlike larger particles, NPs within the 150 nm size range have reportedly delayed hepatic clearance enhanced protein interactions and increased transepithelial transport [66, 67]. The above-mentioned factors make particle size a critical parameter for designing an efficient and successful polymer NP system.

A homogenous population of NPs of a particular size ensures stability as well as uniform drug distribution leading to an efficient nanosystem. The homogeneity in the size distribution of a population is given by polydispersity index (PDI), also known as heterogeneity index is used to describe the degree of nonuniformity of size in a given population [67]. PDI is a dimensionless parameter and is calculated from a two-parameter fit to the correlation data. The PDI value of 0.05 or smaller is identified as highly monodisperse population whereas PDI values higher than 0.7 show a population with high heterogeneity and very broad size distribution. The NPs with PDI index higher than 0.7 are not suitable for analysis using dynamic light scattering (DLS) [68]. The PDI values of 0.3 and smaller are considered acceptable. ISO standard documents 13321:1996 E and ISO 22412:2008 defines the calculations for the determination of size and PDI [69].

The various technologies available for particle size determination of an NP are discussed below.

1.4.1.1 PHOTON-CORRELATION SPECTROSCOPY OR DLS

This technique is one of the most extensively used and reliable techniques for the size determination of the NPs. This technique can also be used for the determination of PDI that gives a better understanding of the size distribution in the given population. In this technique, the Brownian motion of the spherical NPs in a nanocolloidal system causes Doppler shift on exposure to the monochromatic laser light. The extent of change in the wavelength of the incoming monochromatic light that hits the particles is based on the size of the suspended particles. This change in the wavelength can help to determine size distribution and the particle's motion which further assists in the measurement of the diffusion coefficient of the particle using the autocorrelation function [46]. As compared to microscopic techniques, DLS can be used to obtain a statistically meaningful analysis of a large population of particles in an aqueous sample, thus making it one of the most efficient techniques for the determination of particle size and size distribution. Figure 1.4 shows DLS spectra of a homogenous population of drug-loaded PCL NPs as well as

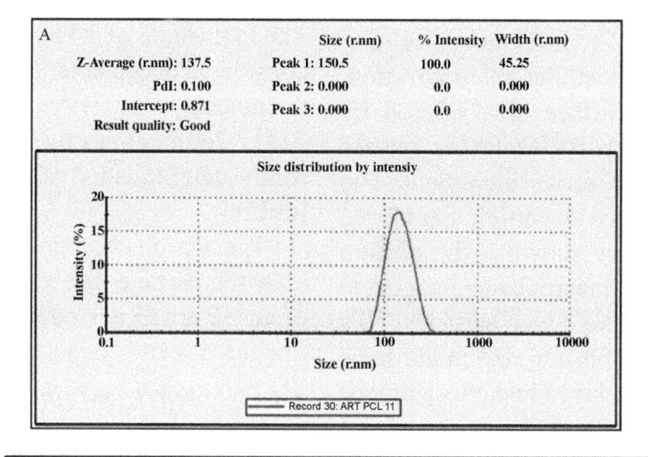

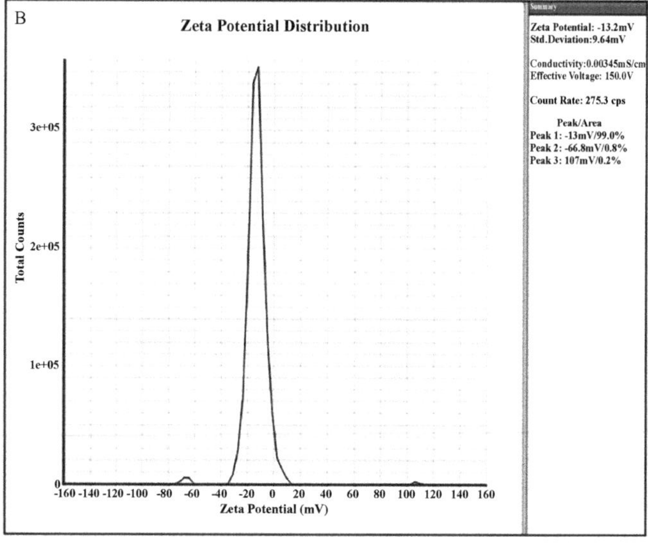

FIGURE 1.4 Dynamic laser scattering spectra of drug-loaded PCL NPs: (A) DLS spectra of a monodisperse PCL NP with radii 137.5 nm and PDI of 0.1 showing homogenous population, (B) PCL NPs with zeta potential of −13 mV showing particles to be stable.

measurement of zeta potential using this technique.

1.4.1.2 ELECTRON MICROSCOPY

Traditionally, electron microscopy was used mainly for the study of surface morphology of the NPs but the advancement in the technology has aided size determination using the microscopic techniques. Electron microscopy can be used to determine shape size and surface morphology by direct visualization of the NPs. In the case of SEM, the powdered sample is coated with a conductive metal, for example, gold, using a sputter coater and analysed by scanning with a focused fine beam of an electron. The surface characteristics of the sample are determined by the secondary electrons emitted from the surface of the sample [70]. In many cases, the high energy electron beam leads to distortion of the physical properties of the polymer NPs. On the other hand, TEM can be used for imaging well has diffraction and spectroscopic information with an atomic or subnanometer spatial resolution. A few limitations in the study of nanostructures pertaining to their small size can be solved by TEM. TEM imaging in combination with nanodiffraction, atomic resolution electron energy-loss spectroscopy, and nanometer resolution X-ray energy dispersive spectroscopy techniques is of significant importance in the fundamental understanding of nanoscience and nanotechnology [46].

SEM and TEM have been found to be of critical importance in the study of surface morphology of polymeric as well as inorganic nanomaterials. However, the sophistication of the instrument, the high cost, and complex sample preparation make the DLS method of choice for size study.

1.4.1.3 ATOMIC FORCE MICROSCOPY

Also known as scanning force microscopy, this technique utilizes a high resolution (fraction of nanometers) scanning probe to scan the specimen at a submicron level and form the images. The use of atomic-scale probe tip, atomic force microscopy (AFM) offers ultra-high resolution size measurement [71]. Samples can be scanned wither in contact or noncontact mode. In the former, tapping the probe on the surface results in the generation of topographical map whereas, in noncontact mode, probe is moved over the conducting surface. Unlike electron microscopy, the nonconducting samples can also be characterized by AFM without any treatment. Additionally, no mathematical calculations are used as it provides the most accurate details of size, size distribution, and surface [72].

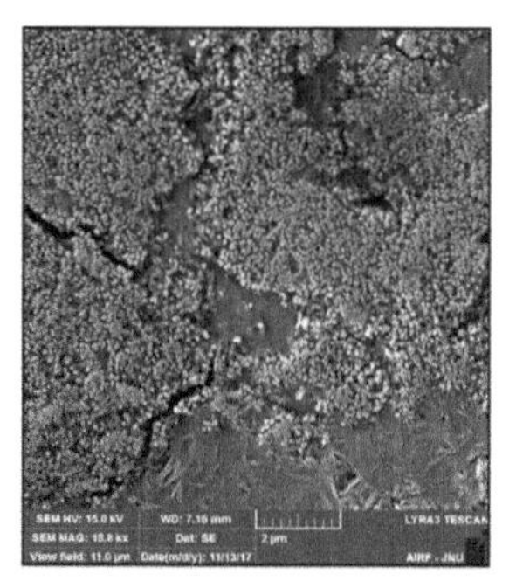

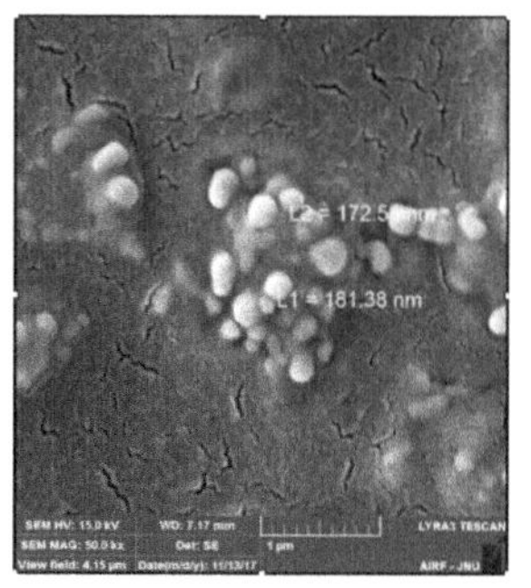

FIGURE 1.5 Scanning electron micrographs of drug-loaded PCL NPs show spherical particles with a size range close to 200 nm.

1.4.2 SURFACE CHARGE

The interaction between the colloidal particles and the surrounding biological components is greatly influenced by the electrostatic charge of the particles. Also, the stability of the colloidal suspension is analyzed by the zeta potential of the NPs. Zeta potential of an NP is the potential difference between the outer Helmholtz plane and the surface of shear and hence is the indirect measure of surface charge [46]. The surface charge is also found to affect cellular uptake and intracellular localization. Also, the balance between van der Waals and the electric double layer greatly affects the stability of a colloidal suspension. High values of zeta potential (positive or negative) cause repulsion between the particles leading to a relatively stable system while low zeta potential can lead to aggregation of particles and thus low stability. Figure 1.6B shows zeta potential measurement using DLS. Also, Figure 1.7 shows the in-vivo fate of the NPs with different sizes

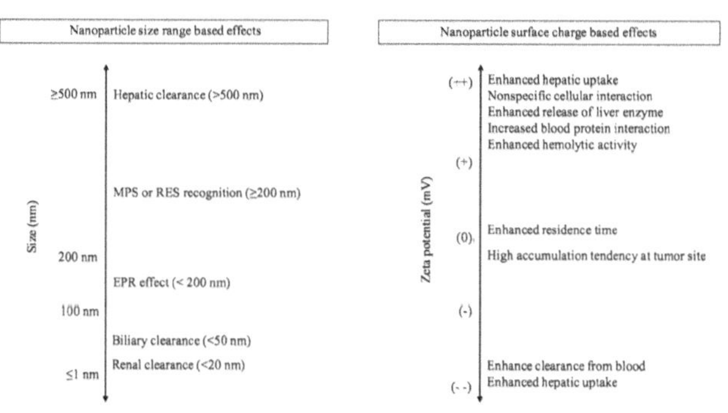

FIGURE 1.6 Relative in-vivo fate of polymeric NPs based on size and surface charge. MPS—monocyte phagocyte system, RES—reticuloendithelial system, and EPR—enhanced permeability and retention.

Source: Reprinted from Ref. [73]. Creative Commons Attribution 4.0 International License.

and charges and different elimination routes for the clearance of the NPs.

1.4.3 SCATTERING TECHNIQUES

As mentioned in the earlier section, different techniques have been used to study the formation and size control of polymeric NPs, such as electron microscopy techniques like SEM and TEM, DLS, and NMR. But, these techniques only give the idea of the overall structure of NPs. Unlike microscopy, there are other techniques, for example, small-angle X-ray scattering (SAXS) and small-angle neutron scattering (SANS) techniques are available, where the measurements are performed in the reciprocal space (Fourier space). The poor electron contrast in SAXS for polymer NPs rules out the use of this technique for polymeric NPs study. On the other hand, SANS offers sufficiently high contrast for such systems with D_2O solvent. The contrast matching properties by SANS also helps to study the intrinsic behavior of polymeric NPs. Riley et al. [74] introduced SANS as a tool to study the structure of PLA-PEG NPs obtained using the nanoprecipitation method.

The scattering intensity of polymeric NPs of radius r and the polydispersity σ can be represented as follows:

$$I(q) = N\int f(r)P(q,r)dr + B \tag{1.1}$$

where N = number of particles per unit volume, $\Delta\rho$ = scattering length density difference of NP and solvent, $f(r)$ = size distribution, and V = volume of the particle. $F(q,r)$ = form factor and B = incoherent background coming from the hydrogenated content.

The interparticle structure factor can be disregarded for such a diluted system. Considering the particles as monodisperse hard sphere, the form factor is defined as follows:

$$P(q,r) = \Delta\rho^2 V^2 \left[\frac{3\{\sin(qr) - qr\cos(qr)\}}{(qr)^3}\right]^2 \tag{1.2}$$

In reality, the polymeric NPs follow a size distribution instead of following one specific size. The size distribution is taken as the log-normal distribution which can be represented as follows:

$$f(r) = \frac{1}{r\sigma\sqrt{2\pi}}\exp\left[-\frac{(\ln(r/r_0))^2}{2\sigma^2}\right] \tag{1.3}$$

where r_0 is the median value. This expression is valid for polymeric NPs obtained from simple polymers, for example, Poly(lactic acid) (PLA), Poly(lactic-*co*-glycolic acid) (PLGA), polycaprolactone (PCL), etc. On the other hand, NPs obtained from conjugated polymers such as PEG conjugated PLA (PLA-PEG) or PEG conjugated PLGA (PLGA-PEG) can be treated by the core–shell model [65]. In the core–shell model, the form factor can be described as follows:

$$P(q,r_c,r_s)=\left[(\rho_c-\rho_{\text{shell}})V_c\left\{\frac{3j_1(qr_c)}{qr_c}\right\}+(\rho_{\text{shell}}-\rho_s)V_s\left\{\frac{3j_1(qr_s)}{qr_s}\right\}\right]^2 \quad (1.4)$$

where ρ_c, ρ_{shell}, and ρ_s are the scattering length densities of the particle core, shell, and solvent, respectively. V_c and V_s are the volumes of the core and the core along with the shell, respectively, and $j_1(x)$ is a first-order spherical Bessel function.

According to the core–shell model, it is assumed that the core is made off biodegradable PLA with a uniform scattering length density, and the shell is made off PEG with nonuniform scattering length density. In principle, the scattering length density of PEG shell is the combination of mostly polymer near the core and turning to be equal to the solvent near the edge of the shell [75]. However, it is found by Yang et al. [75] that the SANS results obtained from PLGA-PEG NPs cannot be fitted by assuming the simple core–shell structure. In addition that the model of a solid PLGA core surrounded by PEG polymer chains unable to fit the data and yields realistic values of the fitting parameters [76]. The introduction of the fractal model by assuming a fractal structure formed by the individual building block of polydisperse PLGA solves this discrepancy to some extent [77]. The scattering from PLGA building blocks in such a fractal-like structure can be expressed as follows:

$$S(q)=1+\frac{\sin\left[(D_f-1)\tan^{-1}(q\xi)\right]}{(qr)^{D_f}}\frac{D_f\Gamma(D_f-1)}{\left[1+1/(q^2\xi^2)\right]^{(D_f-1)/2}} \quad (1.5)$$

where ξ represents correlation length which is related to the overall size of the cluster and D_f is the self-similarity dimension of the fractal structure.

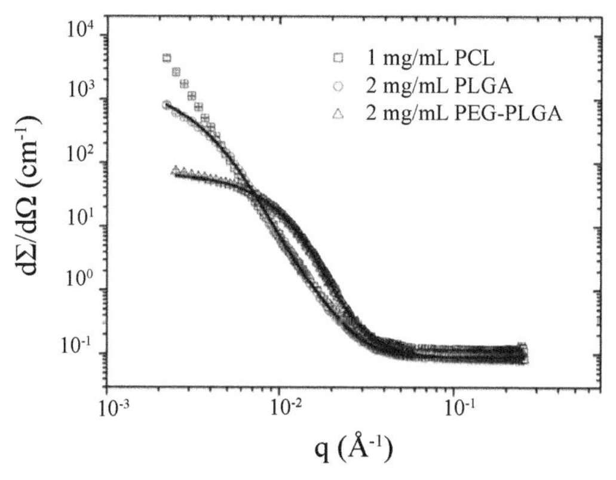

FIGURE 1.7 SANS spectra of polymeric NPs obtained from PCL, PLGA, and PEG-PLGA polymer.

We have recently performed SANS measurements of polymeric NPs obtained from different biodegradable polymers [78]. PCL (Mw = 14,000), PLGA, Resomer® 752 H, Poly(D,L-lactide-*co*-glycolide; lactide:glycolide 75:25, Mw 4000–15,000) and Poly(ethylene glycol) methyl ether-block-poly(L-lactide-*co*-glycolide) (PEG-PLGA; PEG average Mn 5,000, PLGA Mn 7,000) were used for this study. The SANS measurement shows a different kind of assembly of these NPs. The NPs obtained at 1 mg/mL PCL do not show any Guinier regime while the Guinier regime is clearly visible for PLGA and PEG-PLGA NPs even at 2.0 mg/mL concentration. The disappearance of Guinier regime in the scattering curve of PCL NPs depicts the larger assembly of these NPs. On the other hand, the lower forward scattering of PEG-PLGA compared to PLGA polymeric NPs indicates smaller particles formation by this conjugated polymer. In the medium-q range, the scattering follows a trend of q^{-4} (Porod Scattering) reveals abrupt interfaces between the polymer NPs and the solvent [78]. All the scattering data are fitted by assuming a polydisperse sphere (Equations (1.1) and (1.2)). The detailed analysis of SANS data was performed in the same way as proposed for squalene derivated NPs [62]. The use of the spherical model to fit SANS data yields realistic size and number density distribution of PEG-PLGA NPs. The SANS analysis provides size and the number density change of NPs at each step of NPs formation. By knowing these parameters, it is possible to propose the nucleation-growth mechanism of polymeric NPs [78].

1.5 TARGETED DRUG DELIVERY

Drug targeting to the site of action has significantly improved the in-vivo activity making it a potential area in the field of drug delivery. The fate of the active ingredient can be manipulated by the targeting system. With the better molecular understanding of diseases and the availability of targeting ligands, polymer NPs have emerged as a potential targeted drug delivery system. A coherent design of polymer NPs based on tailoring its properties, such as molecular weight, composition, hydrophobicity, composition, crystallinity, and solubility, and to encapsulate certain drugs to the specific site of action, can help in achieving localization of drug in the target cells. Polymer NPs have been successfully used for targeted delivery of anticancer agents making use of the phenotypic diversities of tumor cells as well as targeting ligands.

Passive targeting can be achieved by targeting leaky vasculature exhibited by the solid tumors and metastatic nodules in some of the advanced tumors. The enhanced permeability and retention (EPR) effect and compromised lymphatic drainage due to increased vascularization can lead to the accumulation of polymer NPs with high molecular weight as well as in the size range of 20–500 nm [13]. For efficient targeting using this approach, multiple factors need to be considered such as optimum size, surface charge, and miscibility of polymer nanocolloids with the physiological fluid and its plasma circulation time. PNPs too small in size can escape from the target tumor tissues while the entry of NPs >500 nm will be restricted by the impaired endothelial barrier [79]. Accumulation in the tumor tissues by EPR has been reported for various NPs, such as liposomes, polymer-drug conjugates, polymer micelles, plasma proteins, etc.

Some natural polymers that are commonly used in drug delivery are chitosan, dextrins, cyclodextrins, gelatin, albumin, starch, alginates, cellulose, pectins, and collagen possess highly reactive sites, which makes them potent agents for conjugation. Passive targeting has been explored efficiently using these polymers. Chitosan which is a biological cationic polysaccharide undergoes modification mainly to increase its solubility so that it gets directly absorbed. The basic modification on chitosan is done by *N*-acylation, *O*-alkylation, *N*-alkylation, and oxidation Schiff-base reaction [80]. Cellulose is grafted with polyethylethylene phosphate by the ring-opening polymerization and Cu(I)-catalyzed azide-alkyne cycloaddition to target cancer cells [57]. EC is copolymerized with poly(2-(diethylamino) ethyl methacrylate) and has been used in rifampicin micelles for controlled drug released. Starch is grafted with carboxy-terminal polyethylene glycol (Starch-g-PEG) to form micelles. This micelle is made to deliver the anticancer drug doxorubicin. Engineered glycol-sca-old has multiple multivalent binding site of lectins where the highly branched glucose polymer was activated with lactose and is explored for targeting prostate cancer cells [57]. Surface modification of pectins by acetylation of their free hydroxyl groups protects Ibuprofen a weakly acidic drug throughout the gastrointestinal tract for 10–12 h [81]. Surface modified gelatin with polyethylenimine has high transfection efficiency and low cell toxicity and hence can be an effective gene vector in gene therapy [82]. Alkaline hydrolysis of PLA makes the surface of PLA hydrophilic thereby facilitating the delivery of soluble drugs [83]. Modification of biodegradable PLGA by poly-lactide-PEG improves the hydrophilicity and decreases the surface charge of nanospheres thereby making the particles stable. This decreases protein absorption and increases blood circulation time [84]. Surface modification of MPEG-*b*-PCL-based NPs via oxidative self-polymerization of dopamine is used for malignant melanoma therapy [85]. Surface modification of PCL membrane via aminolysis and biomacromolecule immobilization is used for promoting cytocompatibility of human endothelial cells [86]. Hydroxypartite NPs modified by PEG and FA can be used to successfully target tumor cells expressing folate receptors [87].

On the other hand, efficient "active targeting" has been achieved in numerous cases by surface functionalization of the polymer NPs by attaching various targeting ligands specific to the receptors on target cells. The prerequisite for active targeting is to identify surface receptors, such as lipoproteins, proteins, and glycoproteins, overexpressed by the diseased cells as compared to their healthy counterparts. Next challenge is to identify and design ligands specific to these receptors present on target cells, such as peptides, lectins, enzymes, antibodies, enzyme inhibitors, saccharides, and also antibody fragments. With recent developments, large-scale availability of specific peptides or monoclonal antibodies for extensive in-vivo studies has helped to design various targeting strategies using nanomedicine. Apart from mAb, the other approaches used for targeted drug delivery are based on aptamers, folate receptors, integrin, transferrin, human epidermal growth factors, phosphoinositide-3-kinase inhibitors, and human α1-acid glycoprotein (hAGP) [13, 88]. Various standard reactions of peptide chemistry can be used for conjugation of the targeting ligands to polymer NPs, such as disulfide exchange, aminolysis, the addition of sulfhydryl or amino groups, catalytic acylation using carbodiimide reaction. "Click" chemistry is also a chemoselective covalent technique for peptide conjugation with polymer NPs which is based on azide-alkyne cycloaddition reaction where a terminal group of unprotected peptide attaches to the polymer without need for any reactive functional groups [89]. Alternatively, noncovalent conjugation reaction includes the use of oligopeptides such as coiled-coil peptide which has an ability to self-assemble or stable complexes formation such as biotin-avidin. Numerous reports on monoclonal antibodies derived single-chain fragment variable, antigen-binding fusion protein, have shown promising results in targeting polymer-based NPs [13].

1.6 APPLICATION OF POLYMER NPS IN DRUG DELIVERY

With an emerging need for a smart drug delivery system, nanomedicine has been successfully used for enhancing the bioavailability of therapeutic agents by different approaches. Unlike conventional drug delivery system, an advanced nanomedicine systems can be a combination of targeted delivery with a theranostic approach. Among various types of NPs employed for biomedical applications, biodegradable polymer NPs, in specific, have exhibited significant therapeutic potential for drug delivery as well as diagnostic applications. The physiochemical properties such as particle size and surface properties of polymer NPs can be manipulated in order to achieve controlled drug delivery at the target site. Polymer NPs can be used for encapsulation of drug molecules as well for protein, nucleic acid, or genetic material delivery. For gene delivery, specific genetic materials (DNA or siRNA) are delivered to the target site resulting in the elimination or regulation of defective genes. Achieving genetic therapy using polymer NPs could be the possible solution of next-generation advanced therapeutics [90]. Polymer NPs have also been used in novel therapy development. Few studies showing the use of polymer NPs to attain the enhanced therapeutic effects of different classes of drugs have been briefly discussed below.

There are multiple drawbacks associated with conventional chemotherapy. Thus, nanomedicine has been explored extensively due to its capability of targeting tumor site and enhancing the bioavailability of chemotherapeutics. Consequently, the adverse effects associated with chemotherapy can be reduced significantly. The tumor microenvironment can be used to target biodegradable polymer NPs using various approaches, such as surface modification of NPs using an antibody specific to receptors expressed by the tumor cells or stimuli-responsive such as pH or temperature. A study conducted by Thu et al. [91] showed design of folate conjugated paclitaxel (PTX)-loaded PLA-tocopheryl PEG 1000 succinate (Fol/PTX/PLA-TPGS). The targeted delivery of PTX using folate conjugated nanoparticles showed improved cellular uptake as compared to PTX/PLA-TPGS. The in-vivo targeting effects of Fol/PTX/PLA-TPGS and free PTX/PLA-TPGS were also investigated in tumor-bearing mouse and as a result, the latter demonstrated the best tumor inhibition after the treatment [91]. Polymer NPs have also been employed in combination chemotherapy where more than one antitumor agents are loaded. An intriguing work by Muntimadugu et al. [92] showed successful targeted delivery of PTX and salinomycin (SLM) using PLGA NPs. The particles displayed a positive charge due to dioodecyltrimethylammonium bromide, and hyaluronic acid ligand-modified surface. High entrapment efficiency was observed for both the drugs conferring to their hydrophobic nature; however, coencapsulation of both the drug leads to a significant reduction in PTX load. As a result, the combination of PTX NPs and HA-targeted SLM NPs were administered in-vivo for breast cancer treatment instead of dual-drug loaded polymer NPs. High synergistic effect accompanied by sustained release of the drugs was observed [92]. Another work reported the design of polymeric micelle with multiple drugs encapsulated where paclitaxel, 17-AAG, and rapamycin were conjugated to a PEG-*b*- PLA copolymer. On treatment in an animal model, the mice showed high tolerance of the drug as well as effective localization at the tumor site [93]. The potential combination therapy using polymer NPs has been illustrated by many other research groups.

In a sensing application, it was showed platinum-porphyrin encapsulated PLA NPs could encapsulate the near-infrared flurophore for the detection of physiological concentration of glucose and oxygen with a short-response time [94].

Coencapsulation of betulinic acid and gemcitabine was done in polymer NPs by double emulsion technique. The study showed significant improvement in antitumor efficacy against solid tumor model by dual-drug coencapsulation as compared to the single native gemcitabine and betulinic acid [95].

Polymer NPs have also been successfully used as a carrier for antihypertension and antimicrobial drugs. De Carvalho et al. [56] designed amphotericin-loaded PLGA and dimercaptosuccinic acid NPs and investigated its activity against leishmaniasis. The NPs were decorated with maghemite and the in-vivo antimicrobial activity was assessed on C57BL/5 mice infected intradermally with promastigotes of *Leishmania amazonesis*. On the application of AC magnetic field, mangneto-hypothermia effect was produced leading to a controlled release of the drug. Also, the maghemite decorated NPs were found to have significantly improved activity as compared to the free form of drug, inferred by greater reduction of *Leishmania amazonesis* number and cell viability [56]. In an interesting study conducted by Phan and his team, the scientists investigated natamycin loaded dextran and poly(D,L-lactide) (Dex-*b*-PLA) decorated on contact lenses. As a result, the drug release was found to be extended till 12 h and could be used for targeting the active ingredient to cornea for various fungal infections [96]. Another study by Chaudhary and team has reported results showing antibacterial activity of cefixime-loaded PLGA NPs against intracellular multidrug resistant *Salmonella typhimurium*. The observations were attributed to the sustained release and enhanced permeability of drug across the rat intestine [97].

An antihypertensive drug, carvedilol-loaded poly(ethylene-*co*-vinyl acetate) (PEVA) were designed by Varshosaz and his team [98]. The NPs were coated with chitosan which showed mucoadhesive properties with prolonged release of drug. On spray drying with mannitol, the particles were found to have improved flowability and small aerodynamic diameter making it suitable for pulmonary delivery [89]. Another study reports enhanced bioavailability of felodipine after entrapment in PLGA NPs. The ex-vivo experiment on isolated rat intestine also demonstrated sustained release of the antihypertensive drug [99].

Apart from the above-mentioned examples, the incorporation of biodegradable polymer NPs into scaffolds has shown significant impact in the field of tissue engineering. A study by Nazemi et al. [100] assessed the influence of incorporating PLGA-based polymer NPs in tissue engineering scaffolds. It was reported that by adding PLGA NPs on a chitosan-bioactive glass (CH-BG) scaffold, the swelling property of the scaffold decreased whereas the mechanical strength was enhanced [100]. Also, it was found that such scaffolds can be employed in bone tissue engineering scaffolds for localized delivery of therapeutic agents

Dhas and co-workers [101] designed Folic acid conjugated chitosan (FA-CS) functionalized PLGA NPs. They investigated the activity of FA-CS conjugates against prostate cancer by targeting the folate receptors. The cytotoxicity assay showed a significant improvement in the activity of functionalized PLGA NPs as compared to unfunctioalized though inferring improved efficacy of the delivery system by cell targeting and site-specific delivery [101].

1.7 CONCLUSION

The advancement in polymer science and the extensive research on polymer as a drug delivery vehicle in the field of nanomedicine has led to enormous growth in the field of biomedicine encouraging possibilities of clinical application. The FDA approved polymer has been successfully used for single- and multiple-drug encapsulation with high entrapment efficacy. The smart vesicles have been prepared by the chemical conjugation of the active targeting agent as well as synthesizing metal/ metal oxide polymer hybrid for theranostics. Targeted drug delivery using polymer NPs has helped enormously in the enhancement of therapeutic efficacy of biomolecules and resulted in minimizing dose-related side effects. The biodegradable polymer NPs have also been employed to attain sustained, prolonged, and controlled release of drugs with a prolonged half-life. Consequently, polymer NPs have been successfully employed as a carrier system for not only potent antihypertensive, anticancer drugs but also various antimicrobial, anti-inflammatory agents as well as proteins and nucleic acid.

ACKNOWLEDGMENT

We would like to acknowledge Ms. Ahana Mukherjee for her support in the final formatting.

KEYWORDS

- **drug delivery system**
- **polymer nanoparticles**
- **targeted drug delivery**
- **hydrophobic and hydrophilic drug**

REFERENCES

1. Narvekar M, Xue HY, Wong HL. A novel hybrid delivery system: polymer-oil nanostructured carrier for controlled delivery of highly lipophilic drug all-trans-retinoic acid (ATRA). Int J Pharm 2012, 436, 721–731.
2. Wang M, et al. Engineering multifunctional bioactive citric acid-vectors for intrinsically targeted tumor imaging and specific siRNA gene delivery in vitro/in vivo. Biomaterials 2019, 199, 10–21.
3. Kim MR, et al. Co-encapsulation and co-delivery of peptide drugs via polymeric nanoparticles. Polymers. 2019, 11, 288.
4. Pusuluri A et al. Role of synergy and immunostimulation in design of chemotherapy combination: an analysis of doxorubicin and campothecin. Bioeng Transl Med. 2019, 4(2), e10129.
5. Lyu Y, et al. Enhancing both biodegradability and efficacy of semiconducting polymer nanoparticles for photoacoustic imaging and photothermal therapy. ACS Nano. 2018, 12(2), 1801–1810.
6. Wang Y, Feng L and Wang S. Conjugated nanoparticles for Imaging, cell activity regulation and Therapy. Adv Funct Mater 2019, 29(5), 1806818.
7. Yizhen L, et al. Cell penetrating peptide-modified nanoparticles for tumor targeted imaging and synergistic effect of sonodynamic/HIFU therapy. Int J Nanomed. 2019, 14, 5875–5894.
8. Kandasamy G et al. Multifunctional magnetic-polymeric nanoparticles based ferrofluids for multimodal in vitro cancer treatment using thermotherapy and chemotherapy. J Mol Liquids. 2019, 293, 111549.
9. Liu S, et al. Dynamically PEGylated and borate-coordination-polymer-coated

polydopamine nanoparticles for synergistic tumor-targeted, chemo-photothermal combination therapy. Small. 2018, 14(3), e1703968.
10. Muhammad II, Selvakumaran S and Lazim NA. Designing polymeric nanoparticles for targeted drug delivery system. Nanomedicine. 2014, 287, 287–213.
11. Kreuter J, Drug delivery to the central nervous system by polymeric nanoparticles: What do we know? Adv Drug Deliv Rev 2014, 71, 2–14.
12. Parveen S, Misra, R, Sahoo SK. Nanoparticles: a boon to drug delivery, therapeutics, diagnostics and imaging. Nanomedicine 2012, 8, 147–166.
13. Ulbrich K, Hola K, Subr V, Bakandritsos A, Tucek J, Zboril R. Targeted drug delivery with polymers and magnetic nanoparticles: covalent and noncovalent approaches, release control, and clinical studies. Chem Rev 2016, 116, 5338–5431.
14. Kerimo Lu O, Alarçin E. Poly(lactic-*co*-glycolic acid) based drug delivery devices for tissue engineering and regenerative medicine. ANKEM Derg 2012, 26(2), 86–98.
15. Vroman I, Tighzert L. Biodegradable polymers. Materials 2009, 2, 307–344.
16. Letchford K, Burt H. A review of the formation and classification of amphiphilic block copolymer nanoparticulate structures: Micelles, nanospheres, nanocapsules and polymersomes. Eur J Pharm Biopharm 2007, 65, 259–269.
17. Nel AE, Madler L, Velegol D, Xia T, Hoek EM et al. Understanding biophysiochemical interactions at the nano-bio interface. Nat. Mater. 2009, 8, 543–557.
18. Garnett MC, Kallinteri P. Nanomedicines and nanotoxicology: Some physiological principles, Occup. Med. 2006, 56, 307–311.
19. Akash HSM, Rehman K, Shuqing C. Natural and synthetic polymers as drug carriers for delivery of therapeutic proteins. Polymer Reviews, 2015, 55(3), 371–406.
20. Karlsson J, Vaughan JH and Green JJ. Biodegradable polymeric nanoparticles for therapeutic cancer treatments. Ann Rev Chem Biomed Eng 2018, 9, 105–127.
21. Anajafi T, Mallik S. Polymersome-based drug-delivery strategies for cancer therapeutics. Ther Deliv 2015, 6(4), 521–534.
22. Lee JS, Feijen J. Polymersomes for drug delivery: design, formation and characterization. J Control Release 2012, 161(2), 473–483.
23. Liu S, et al. Dynamically PEGylated and borate-coordination-polymer-coated polydopamine nanoparticles for synergistic tumor-targeted, chemo-photothermal combination therapy. Small 2018, 14(3), e1703968.
24. Ramachandran D, Corten CC, Urban MW. Color- and shape-tunable colloidal nanoparticles capable of nanopatterning. RSC Adv 2013, 3(24), 9357–9364.
25. Tsoi S, Zhou J, Spillmann C, Naciri J, Ikeda T, Ratna B. Liquid crystalline nano-optomechanical actuator. Macromol Chem Phys 2013, 214, 734–741.
26. Bennet D, Kim S. Polymer nanoparticles for smart drug delivery. In: Application in Nanotechnology in Drug Delivery. Ali Demir Sezer, IntechOpen. 2014, 257–310.
27. Arizaga A, Ibarz G, Pinol R. Stimuli-responsive poly(4-vinyl pyridine) hydrogel nanoparticles: synthesis by nanoprecipitation and swelling behavior. J Colloid Interface Sci 2010, 348(2), 668– 672.
28. Zhao Y, Shi C, Yang X, Shen B, Sun Y, Chen Y, Xu X, Sun H, Yu K, Yang

B, Lin Q. pH and temperature sensitive hydrogel nanoparticles with dual photoluminescence for bioprobes. ACS Nano 2016, 10(6), 5856–5863
29. Hotz J, Meier W. Vesicle-templated polymer hollow spheres. Langmuir 1998, 14(5), 1031–1036.
30. Lestage DJ, Urban MW. Stimuli-responsive polymeric films and coatings. Langmuir 2005, 21, 4266–4267.
31. Lee J, Park JC, Song H. A nanoreactor framework of Au@SiO_2 yolk/shell structure for catalytic reduction of p-Nitrophenol. Adv Mater 2008, 20, 1523–1528.
32. W. Zhou, Y. Yu, H. Chen, FJ DiSalvo, H. D. Abruna. Yolk-shell structure of polyaniline-coated sulfur for lithium-sulfur batteries. J Am Chem Soc 2013, 135(44), 16736–16743.
33. Mukherjee S, Sobhani H, Lassiter JB, Bardhan R, Nordlander P, Halas NJ. Fanoshells nanoparticles with built-in fano resonances. Nano Lett 2010, 10, 2694–2701.
34. Wang H, Wu Y, Lassiter B, Nehl CL, Hafner JH, Nordlander P, Halas NJ. Symmetry breaking in individual plasmonic nanoparticles. Proc Natl Acad Sci USA 2006, 103(29), 10856–10860
35. Li GL, Tai CA, Neoh KG, Kang ET, Yang X. Hybrid nanorattles of metal core and stimuli responsive polymer shell for confined catalytic reactions. Polym Chem 2011, 2, 1368–1327
36. Heroguez V, Gnanou Y, Fontanille M. Novel amphiphilic architectures by ring-opening metathesis polymerization of macromonomers. Macromolecules1997, 30(17), 4791–4798
37. Lattuada M, Hatton TA. Synthesis, properties and applications of Janus particles. Nano Today 2011, 6(3), 286–308.
38. Walther A Muller AHE. Janus particle synthesis, self-assembly, physical properties and applications. Chem Rev 2013, 113(7), 5194–5261
39. Tu F, Lee D. Shape-changing and amphiphilic-reverse Janus particles with pH-responsive surfactant properties. J Am Chem Soc 2014, 136(28), 9999–10006.
40. Lu C, Urban MW. Rationally designed Gibbous stimuli responsive colloidal nanoparticles. ACS Nano 2015, 9(3), 3119–3124
41. Lu C, Urban MW, Mater Today Commun 2016, 9, 41
42. Lestage DJ, Urban MW. Cockleblur-shaped colloidal dispersions. Langmuir 2005, 21(23), 10253– 10255
43. Zhu Y, Gao C, Liu X, and Shen J. Surface modification of polycaprolactone membrane via aminolysis and biomacromolecule immobilization for promoting cytocompatibility of human endothelial cell. Biomacromolecules 2002, 3, 1312–1319
44. Vauthier C, Bouchemal K. Methods for the preparation and manufacture of polymeric nanoparticles. Pharm Res 2009, 26(5), 1025–1028.
45. Vanderhoff JW, El Aasser MS, and Ugelstad J. Polymer Emulsification Process. US Patent 4,177,177 (1979).
46. Bhatia S. Nanoparticle types, classification, characterization, fabrication methods and drug delivery applications. In: Natural Polymer Drug Delivery System: Nanoparticles, Plants and Algae. Springer: Berlin, 2016, 33–96.
47. Couvreur P, Dubernet C, Puisieux F. Controlled drug delivery with Nano particles: current possibilities and future trends. Eur J Pharm Biopharm. 1995, 41(1), 2–13.
48. Jung T, Kamm W, Breitenbach A, Kaiserling E, Xiao JK, Kissel T. Biodegradable nano particles for oral

delivery of peptides: is there a role for polymer to affect mucosal uptake? Eur J Pharm Biopharm 2000, 50(1), 147–160.

49. Danicher L, Frere Y, Calve AL. Synthesis by interfacial polycondensation of polyamide capsules with various sizes. Characteristics and properties. Macromol Symp 2000, 151, 387–92.
50. Scott C, Wu D, Ho CC, Co CC. Liquid-core capsules via interfacial polymerization: a free-radical analogy of the nylon rope trick. J Am Chem Soc 2005, 127, 4160–4161.
51. Torini L, Argillier JF, Zydowicz N. Interfacial polycondensation encapsulation in miniemulsion. Macromolecules 2005, 38(8), 3225–3236.
52. Boudad H, Legrand P, Lebas G, Cheron M, Duchene D, Ponchel G. Combined hydroxypro- pyl-[beta]-cyclodextrin and poly(alkylcyanoacrylate) nanoparticles intended for oral administration of saquinavir. Int J Pharm 2001, 218(1/2), 113–124.
53. Puglisi G, Fresta M, Giammona G, Ventura CA. Influence of the preparation conditions in poly(ethylcyanoacrylate) nanocapsule formation. Int J Pharm 1995, 125(2), 283–287.
54. Gasco MR and Trotta M. Nanoparticles from microemulsions. International Journal of Pharmaceutics. 1986, 29(2–3), 267–268
55. Fallouh NAK et al. Development of a new process of manufacture of polyisobutylcyanoacrylate nanocapsules. International Journal of Pharmaceutics. 1986; 28(2–3), 125–132.
56. De Carvalho RF, Ribeiro IF, Miranda-Vilela AL, Filho J de Sousa, Martins OP, Cintra e Silva Dde O, Tedesco AC, Lacava ZGM, Báo SN, Sampaio RNR. Leishmanicidal activity of amphotericin B encapsulated in PLGA–DMSA nanoparticles to treat cutaneous leishmaniasis in C57BL/6 mice. Exp Parasitol 2013, 135, 217–222
57. Gopinatha V, Saravanan S, Al-Maleki AR, Ramesh M, Vadivelu J. A review of natural polysaccharides for drug delivery applications: special focus on cellulose, starch and glycogen. Biomed Pharmacother 2018, 107, 96–108.
58. Jiang W, Kim BYS, Rutka JT, Chan WCW. Nanoparticle-mediated cellular response is size dependent. Nat Nanotechnol 2008, 3(3), 145–150. 27.
59. Moghimi SM. Introduction: targeting of drugs and delivery systems. Adv Drug Deliv Rev 1997, 7, 1–3.
60. Fessi H et al. Nanocapsule formation by interfacial polymer deposition following solvent displacement. Int J Pharm 1989, 55(1), R1–R4.
61. Rodriguez SG et al. Physicochemical parameters associated with nanoparticle formation in the salting-out, emulsification-diffusion, and nanoprecipitation methods. Pharm Res 2004, 21, 1428–1439.
62. Saha D et al. The role of solvent swelling in the self-assembly of squalene-based nanomedicines. Soft Matter 2015, 11(21), 4173–4179.
63. Aubry J, Ganachaud F, Cohen Addad JP, Cabane B. Nanoprecipitation of polymethylmethacrylate by solvent shifting:1. Boundaries. Langmuir 2009, 25, 1970–1979.
64. Zhang C et al. Flash nanoprecipitation of polystyrene nanoparticles. Soft Matter 2012, 8, 86–93.
65. How AH, Tong HH, Chattopadhyay P, Shekunov BY. Particle engineering for pulmonary drug delivery. Pharm Res 2007, 24(3), 411–437
66. Rytting E, Nguyen J, Wang, X, Kissel T. Biodegradable polymeric nano carriers for pulmonary drug delivery. Exp Opin Drug Deliv. 2008, 5(6), 629–639.
67. Bera B. Nanoporous silicon prepared by vapour phase strain etch and

sacrificial technique. In Proceedings of the International Conference on Microelectronic Circuit and System (Micro), Kolkata, India, 2015, 42–45
68. Nobbmann UL. Polydispersity—What Does It Mean for DLS and Chromatography. 2014. Available online: http: //www.materials-talks.com/blog/2014/10/23/polydispersity-what-does-it-mean-for-dlsand-chromatography
69. Worldwide, M.I. Dynamic light scattering, common terms defined. Inform White Paper, Malvern Instruments Limited, Malvern, 2011, pp. 1–6.
70. Jores K et al. Investigation on the stricter of solid lipid nanoparticles and oil-loaded solid nanoparticles by photon correlation spectroscopy, field flow fractionasition and transmission electron microscopy. J Control Release 2004, 95(2), 217–227.
71. Muhlen AZ et al. Atomic force microscopy studies of solid lipid nanoparticles. Pharm Res 1996, 13(9), 1411–1416
72. Polakovic M, Gorner T, Gref R, Dellacherie E. Lidocaine loaded biodegradable nanospheres. II. Modelling of drug release. J Control Release 1999, 60(2–3), 169–77.
73. Sadat MAS, Jahan ST, Hadidi A. Effects of size and surface charge of polymeric nanoparticles on in vitro and in-vitro applications. J Biomater Nanobiotechnology, 2016, 7, 91–108.
74. Riley T et al. Core–shell structure of PLA–PEG nanoparticles used for drug delivery. Langmuir 2003, 19, 8428–8435
75. Yang B, Lowe JP, Schweins R, Edler KJ. Small angle neutron scattering studies on the internal structure of poly(lactide-*co*-glycolide)-block-poly(ethylene glycol) nanoparticles as drug delivery vehicles. Biomacromolecules 2015, 16, 457–464.
76. Pedersen JS, Hamley IW, Ryu CY, Lodge TP. Contrast variation small-angle neutron scattering study of the structure of block copolymer micelles in a slightly selective solvent at semi dilute concentrations, Macromolecules 2000, 33(2), 542–550.
77. Teixeira J. Small-angle scattering by fractal systems. J Appl Crystallography 1988, 21(6), 781–785.
78. Saha D et al., Role of physicochemical parameters associated with the hydrophobic vs. amphiphilic biodegradable polymer nanoparticles formation. J Mol Liquids 2020, 318, 113977.
79. Nakamura H, Jun F, Maeda H. Development of next-generation macromolecular drugs based on the EPR effect: challenges and pitfalls. Expert Opin Drug Deliv 2015, 12, 53−64.
80. Li J Cai C, Li J, et al. Chitosan-based nanomaterials for drug delivery. Molecules 2018, 23, 2661.
81. Bhatia MS et al. Chemical modification of pectins, characterization and evaluation for drug delivery. Sci Pharm 2008, 76, 775–784
82. Brhane Y and Gabriel T. Recent advances in preparation and modification of gelatin nanoparticles for pharmaceutical applications. Int J Pharm Sci Nanotechnol 2018, 11, 3950–3955.
83. Tham CY, Hamid ZAA Ahmad Z, Ismail H, Surface engineered poly (lactic acid) (PLA) microspheres by chemical treatment for drug delivery system. Trans Tech Publications, Switzerland. 2014, 594–595, 214–218.
84. Stolnik S et al. Surface modification of poly(lactide-*co*-glycolide) nanospheres by biodegradable poly(lactide)-poly (ethylene glycol) copolymer. Pharm Res 1994, 11, 1800–1808.

85. Xiong W, Peng L, Chen H, Li Q. Surface modification of MPEG-b-PCL-based nanoparticles via oxidative self-polymerization of dopamine for malignant melanoma therapy. Int J Nanomed 2015, 10, 2985–2996
86. Zhu Y, Li J, Wan M, Jiang L. Superhydrophobic 3D microstructures assembled from 1D nanofibers of polyaniline. Macromol Rapid Commun 2008, 29(3), 239–224
87. Venkatasubbu D et al. Surface modification and paclitaxel drug delivery of folic acid modified polyethylene glycol functionalized hydroxyapatite nanoparticles. Powder Technol 2013, 235, 437– 442.
88. Masood F. Polymeric nanoparticles for targeted drug delivery system for cancer therapy. Mater Sci Eng, C 2016, 60, 569–578.
89. Pola R, Braunova A, Laga R, Pechar M, Ulbrich K. Click chemistry as a powerful and chemoslective tool for attachment of targeting ligands to polymer drug carriers. Poly Chem 2014, 5, 1340–1350.
90. Moritz M, Geszke-Moritz M. Recent developments in the application of polymeric nanoparticles as drug carriers. Adv Clin Exp Med 2015, 24(5), 749–758.
91. Thu et al. In vitro and in-vivo targeting effect of folate decorated paclitaxel loaded PLA-TPGS nanoparticles. Saudi Pharmaceutical Journal. 2015; 23(6): 683–688.
92. Muntimadugu E et al. CD44 targeted chemotherapy for co-eradication of breast cancer stem cells and cancer cells using polymeric nanoparticles of salinomycin and paclitaxel. Colloids Surf B Biointerfaces 2016, 143, 532–546.
93. Shin HC et al. A 3-in-1 polymeric micelle nanocontainer for poorly water-soluble drugs. Mol Pharm 2011, 8, 1257–1265.
94. Pandey G et al. Fluorescent biocompatible platinum-porphyrin-doped polymeric hybrid particles for oxygen and glucose biosensing. Sci Rep 2019, 9, 5029.
95. Saneja A et al. Gemcitabine and betulinic acid co-encapsulated PLGA-PEG polymer nanoparticles for improved efficacy of cancer chemotherapy. Mater Sci Eng. C. 2019;98: 764–771.
96. Phan CM et al. In vitro uptake and release of natamycin Dex-b-PLA nanoparticles from model contact lens materials. J Biomater Sci, Polym Ed 2014, 25(1), 18–31.
97. Chaudhary SH, Kumar V. Taguchi design for optimization and development of antibacterial drug loaded PLGA nanoparticles. Int J Biol Macromol 2014, 64, 99–105
98. Varshosaz J, Taymouri S, Hamishehkar H. Fabrication of polymeric nanoparticles of poly(ethyleneco-vinylacetate) coated with chitosan for pulmonary delivery of carvedilol. J Appl Polym Sci 2014, 131, 39694–39701.
99. Shah U, Joshi G, Sawant K. Improvement in antihypertensive and antianginal effects of felodipine by enhanced absorption from PLGA nanoparticles optimized by factorial design. Mater Sci Eng C 2014, 35, 153–163.
100. Nazemi K et al. Tissue-engineered chitosan/bioactive glass bone scaffolds integrated with PLGA nanoparticles: a therapeutic design for on-demand drug delivery. Mater Lett 2015, 138, 16–20.
101. Dhas NL, Ige PP, Kudarha RR. Design, optimization and in-vitro study of folic acid conjugated chitosan functionalized PLGA nanoparticle for delivery of bicalutamide in prostate cancer. Powder Technol 2015, 283, 234–245.

CHAPTER 2

Biomedical Application of Polyhydroxybutyrate Biopolymer in Drug Delivery

MD. ASAD KHAN[1,*] and M. MOSHAHID A RIZVI[2]

[1]*Department of Biochemistry, Faculty of Dentistry, Jamia Millia Islamia, New Delhi 110025, India*

[2]*Department of Biosciences, Jamia Millia Islamia, New Delhi 110025, India*

[*]*Corresponding author. E-mail: asad1amu@gmail.com.*

ABSTRACT

Poly-β-hydroxybutyrate (PHB) is an environmentally safe polymer which has a remarkable importance in pharmaceutical area. PHB is an effective polymer for controlling drug release after nanoformulations. It is a natural energy reservoir polymer. The recombinant microorganisms have been also used for PHB formation. It is a well-known capacity of biocompatibility and biodegradability activity so that it could be broadly used in biomedical applications. Drug targeting tool is an emerging technology in interdisciplinary science to focus on improving human health. The administered design delivery of active compounds at the specific site for desire dose and therapeutically optimal rate has been the major aim. It is a candidate's polymer used for control and sustains drug release delivery vehicles for therapeutic tools. The desired use of PHB in biomedical is area due to their biocompatible and patient's acquiescence. This chapter emphasizes on various key findings in the biomedical importance of PHB nanoformulation for drug targeting.

2.1 INTRODUCTION

Biopolymers are formed from natural renewable raw materials or organisms (like bacteria) and they can be easily degraded naturally [21].

This is a great example for turning a biopolymer into a native biopolymer [15]. Biocompatible polymers are capable of biological functions for a particular time interval and then subsequently breakdown into useful products in a controlled manner. The waste products are excreted from the body via metabolic pathways after degradation. It also has the capability to deteriorate in biological fluids with the progressive release of drugs from polymers. Biopolymers are biocompatible, biodegradable, and slightly soluble in water but show a wide range of applications. Therefore, polyhydroxybutyrate (PHB) is often referred to as the "sleeping giant" [52]. There are different kinds of PHBs: for example, P(3HB) and P(4HB). Structural and atomic deference's in the biopolymers show different characteristics [33]. Despite the potential, the importance of PHB is in manufacturing and widespread areas. PHB is formed by a linear polymer of D(−)3-hydroxybutyric acid and is the most common commercial synthetic biopolymer form (Figure 2.1).

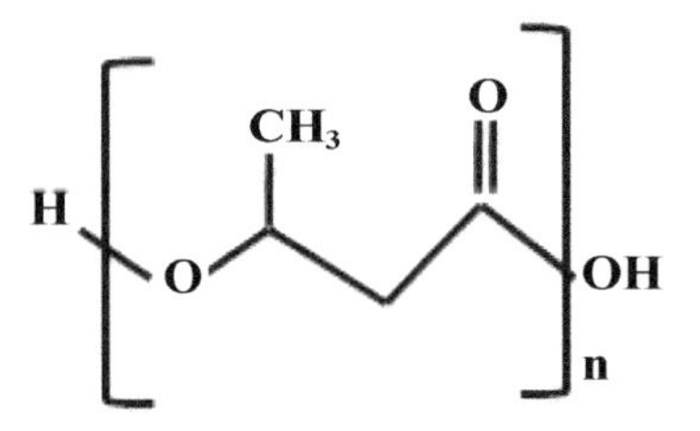

FIGURE 2.1 General structure of polyhydroxybutarate (PHB).

However, formulations for its production via natural raw materials (e.g., cereals or potatoes) have been discovered. Specialized enzymes are fermented starches to PHB [36,37] production. The cost-effective production remains a challenge, and the manufacturing process is not well defined in industrial scale for PHB production. Most studies showed PHB production from bacteria via an enzymatic method. It was discovered by Maurice Lemoigne in 1926 in bacteria (*Bacillus megaterium*). It is present in intracellular granules in a wide variety of Gram-positive and Gram-negative microorganisms [17]. The PHB, a form of polyhydroxyalkanoates (PHA), has a function in intracellular storage, carbon reserve, and facilitates cell survival under stress conditions [3, 18, 65]. PHB is mostly produced by microorganism from excess carbon and other nutrients like nitrogen, phosphorus, magnesium, oxygen, and sulfur present in a lesser amount [65]. PHB is a hydrophobic biomaterial that is biodegradable and biocompatible at high melting temperatures [35]. The physicochemical characteristics of PHB such as thermal stability and other physical properties have been used in the formulation of several types of pharmaceutical drugs. The in vivo breakdown of PHB provides D-3-hydroxybutyric acid to form ketone bodies (acetone and acetoacetate) in blood via ketogenesis. The metabolic pathway of PHB synthesis and regulation is summarized in Figure 2.2.

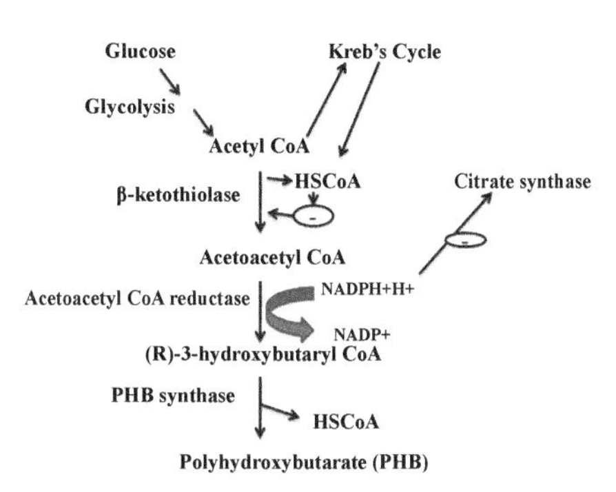

FIGURE 2.2 PHB synthesis and regulatory pathway.

The biophysical characteristics of PHB are more or less similar properties like polymers used in petrochemical such as polypropylene, increasing the number of possible applications [22]. An advantage of making biopolymers like PHB is the lower requirement of temperatures and pressures being cost-effective than other manufacturing methods [77]. Nevertheless, high production costs of PHB have kept it confined to the medical and aerospace fields [85]. However, it could be used for urban-farming because of its biodegradable characteristics [66].

Recently, an emerging area in ideal drug targeting is through nanotechnology. A nanoparticle is an encouraging drug targeting molecule for release in a controlled way. Nanotechnology has become a novel field of research in the area of novel drug delivery. It has the capability to release a drug in several organs/tissues and the biological oxygen demand is retained for a long time. Several drugs can be delivered through a single nanoparticle via a number of routes. Nanomaterials have the tendency to deliver drugs, vaccines, bioactive macromolecules, and so forth [87]. Specific drug targeting systems are mapped to release drugs at a proper dosage for the required time interval to a proper site of the organs/tissues without any toxic side effects.

2.2 METHODS OF PHB PRODUCTION

2.2.1 COLLECTION AND ISOLATION OF PHB PRODUCING BACTERIA

The bacteria were collected from nursery and soils from fodder field for the screening of high PHB producing bacteria and below-mentioned media is used as a nutrient source. Various samples which are collected with serially diluted and 10^{-5} dilution was plated on nutrient media (g/L), with peptone-2, beef extract-2, NaCl-1, agar-4, distilled water in 1 L, the media was then autoclaved [24]. These plates were incubated overnight.

2.2.2 SCREENING FOR PHB PRODUCING BACTERIA

Isolated bacteria were tested by Sudan black B stain for PHB production [68], and also the PHB producing bacterial granules were detected by fluorescent staining methods using acridine orange as described by Senthilkumar and Prabhakaran [69]. The bacterial granules were observed by Carl Zeiss and fluorescent microscope under 100×.

2.2.3 MORPHOLOGICAL CHARACTERIZATIONS OF ISOLATED BACTERIA

Colony and cell morphology of bacteria were characterized on the basis of their color, shape, margin, elevation, surface, and arrangement of bacteria.

2.2.4 BIOCHEMICAL CHARACTERIZATIONS OF ISOLATED BACTERIA

Gram staining, nitrate reduction test, MR, VP test, indole test, citrate utilization test, oxidase test, gelatin liquifaction test, catalase activity, H2S production, starch hydrolysis, D-fructose, sodium malonate, sodium acetate, D-fucose, D-sorbitol, D-glucose, L-alanine, salicin, rhamnose, propinoic acid, valeric acid, trisodium citrate, L-proline, L-rhamnose, D-ribose, inositol, glycogen, capric acid, L-histidine, L-arabinose, hydroxyl butyric acid tests were carried out using standard protocols proposed by Cappuccino and Sherman [62] for the biochemical characterization of isolates.

2.3 CHARACTERIZATION TECHNIQUES

2.3.1 WEIGHT LOSS

Each sample was weighed before the start of the study and after being extracted from the glasses, rinsed with water, and dried for 12 h in a vacuum drying oven at 40 °C. They were weighed with a microbalance ACL-110.4 from Aculab, Milton Keynes, United Kingdom.

2.3.2 VISCOSIMETRY

Viscosity measurements were processed with an Ubbelohde from SI Analytics GmbH, Mainz, Germany. The Ubbelohde capillary was set into a water bath at 25 °C as described earlier [48].

2.3.3 FTIR SPECTROSCOPY

FTIR spectra were recorded with the Avatar 360 FT_IR Nicolet and analyzed with the software OMNIC Spectra 8.3 (Thermo Fischer Scientific, Waltham, USA). Every spectrum, including background and samples, was recorded 32 times. For the measurement, the attenuated total reflection method was used.

2.3.4 CONTACT ANGLE MEASUREMENTS

Water contact angles were measured with OCA 20 from Data Physics, Filderstadt, Germany, and analyzed with the SCA 22 Software as a static contact angle with the Young equation. Each sample was tested seven times with the contact angle of water at different extraction points. Deionized water was used at room temperature.

2.3.5 DIFFERENTIAL SCANNING CALORIMETRY (DSC)

For the DSC measurement, the DSC 823 with the STARe Software (Mettler Toledo, Columbus, USA) was used. All samples were tested with the following program: (i) heating from 25 to 190 °C; heat rate: 20 K/min; (ii) holding phase: 2 min at 190 °C, (iii) cooling from 190 °C to −20 °C; cooling rate: −20 K/min, (iv) holding phase: 2 min at −20 °C, (v) heating from −20 °C to 200 °C; heat rate: 20 K/min, (vi) holding phase: 2 min at 200 °C,(vii) cooling from 200 to 25 °C; cooling rate: −20 K/min. The polymer samples were weighed in an aluminum crucible during measurements in the DSC oven and nitrogen was used as inert gas with a flow rate of 30 mL/min during the measurement. To calculate the degree of crystallinity, a melting enthalpy ΔH_f of 146 J/g was assumed for 100% crystalline PHB; 93.6 J/g for 100% crystalline PLLA, and 106.7 J/g for the blend according to Vogel [83].

3.6 PH-VALUE MEASUREMENTS

The pH value of the degradation media was monitored during the in vitro study with the pH 1970i and pH electrode SenTix 22 Basis pH-electrode (WTW, Dinslaken, Germany).

2.3.7 IN VITRO DRUG RELEASE STUDY

Conditions (medium, temperature, shaking, etc.) for drug release were identical to those of the degradation study. Briefly, samples of each polymer type were placed into clear glass vials and doused with PBS. Samples were kept at 65 °C and shaken constantly at 120–150 rpm. Then, at various timepoints (these had to be varied in comparison to the degradation study due to saturation and instability of the model drug) the solution was completely removed from the vials, and a fresh buffer was added. Of the removed solution, 10 μL was mixed with 90 μL of 1.0 M NaOH in a 96-well plate in replicates. Fluorescence intensity (excitation wavelength: 485 nm, emission wavelength: 520 nm) was then measured with a microplate reader (BMG Labtech, Ortenberg, Germany) to estimate the released drug amount. PHB granules were detected using TEM microscopy, and by monitoring UV spectra of PHB samples (after treatment with concentrated sulfuric acid) by scanning between 220 and 300 nm, and compared with standard PHB which has the highest absorbance at 235 nm.

2.4 SYNTHESIS OF PHB FROM MICROORGANISM

The PHBs are synthesized from cultures of microbes especially *Cupriavidus necator* (also known as *Alcaligenes eutrophus* or *Ralstonia eutropha*) that are used for production on a large scale. Imperial Chemical Industries (ICI plc) is one of the innovators that utilized the bacterial strain for the formation of poly-(3-hydroxybutyrate-*co*-3-hydroxyvalerate) copolymer by the trade name BIOPOL™ [14]. A number of other bacterial strains that have been utilized for the production of PHB are summarized in Table 2.1.

TABLE 2.1 Microbial Species for PHB Production

Alcaligenes latus	*Bacillus spp.*
Burkholderia sacchari sp. nov.	*Burkholderia cepacia*
Caulobacter crescentus	*Escherichia coli mutants*
Halomonas boliviensis	*Legionella pneumophila*
Methylocystis sp.	*Microlunatus phosphovorus*
Pseudomonas aeruginosa	*Pseudomonas oleovorans*
Pseudomonas putida	*Pseudomonas stutzeri*
Rhizobium meliloti, R. viciae	*Rhodopseudomonas palustrism*
Spirulina platensis (cyanobacterium)	*Staphylococcus epidermidis*
Cupriavidus necator	*Cupriavidus necator H16*

2.5 FACTORS RESPONSIBLE FOR INFLUENCING THE PROPERTIES OF PHB PREPARATION

2.5.1 PHYSIOLOGICAL AND ENVIRONMENTAL FACTORS AFFECTING PHB

Various environmental factors such as pH, inoculums size, culture volume, and incubation temperature were tested and screened for maximum PHB production [2]. Optimization of nutritional factor is affecting PHB production (Plackett-Burman).

2.6 PHB USED AS A DRUG TARGETING

Delivery of drugs has become a principal tool in the biomedical area and administered targeting of biologically active molecules to tissues and organs. There are various methods for drug-targeting through biopolymers to effectively release the drug to a specific target site as a therapeutic effect with minimizing toxicity. Due to the pliability of polymers, it facilitates efficient drug delivery, maintaining biocompatibility, superfacial manufacturing, and stable nanoformulation. In addition, it is a principle that the biopolymer should maintain the biophysical properties as well as biocompatibility. Traditionally used silicone polymer has been fancy to cause cancer [10]. Therefore, it is better to use nontoxic, biodegradable, and biocompatible polymers for the drug delivery.

Biopolymer is an alternative for drug delivery because it is biodegradable, environmentally safe, biocompatible, and nonimmunogenic. There is growing interest in the biopolymers known as PHA, a monomer of hydroxyalkanoate (Figure 2.3). It has the capacity to accumulate energy and store carbon materials by granular inclusions in the cytoplasm of several bacterial stains. There are approximately 150 different types of bacterial strains containing hydroxyalkanoate, first discovered by Griebel in 1968 in *Bacillus megateriu.* It is composed of 97.7% polyester, 1.87% protein, and 0.46% lipids [26].

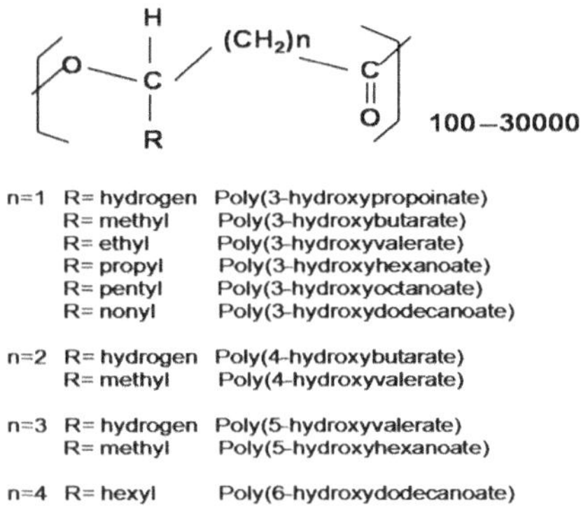

H
O — C — $(CH_2)n$ — C=O
R
100–30000

n=1 R= hydrogen Poly(3-hydroxypropoinate)
R= methyl Poly(3-hydroxybutarate)
R= ethyl Poly(3-hydroxyvalerate)
R= propyl Poly(3-hydroxyhexanoate)
R= pentyl Poly(3-hydroxyoctanoate)
R= nonyl Poly(3-hydroxydodecanoate)

n=2 R= hydrogen Poly(4-hydroxybutarate)
R= methyl Poly(4-hydroxyvalerate)

n=3 R= hydrogen Poly(5-hydroxyvalerate)
R= methyl Poly(5-hydroxyhexanoate)

n=4 R= hexyl Poly(6-hydroxydodecanoate)

FIGURE 2.3 General structure of polyhydroxyalkanoates (PHAs).

2.7 PHA AS A BIODEGRADABLE

PHA is easily biodegradable, environmentally safe, and easily degrades into aqueous solution. Several types of microorganisms exist in soil, sludge, and sea water to release PHA degrading enzymes to decompose PHA material into hydrophilic oligomers, monomers, and simultaneously use the end products as nutrients by the cells. PHB is decomposed faster in in vivo than in vitro at body temperature [60]. PHB have their edges to use in drug targeting for various biomedical and therapeutic applications. It has the characteristic to form nanofabrication to be accountable for conversion into several morphological shapes with features in the pore formation. PHB has the potential to encapsulate several types of active compounds for use in drug targeting. The breakdown of PHA biomaterials is loaded with active compounds to release a compound in the organ/tissues to the host serve as an anticancer drug. PHB releases entrapped drugs in a controlled and extended manner. The drug carrier breaks into a nontoxic and biodegradable polymer which is subsequently metabolized and excreted in the urine through kidneys.

2.7.1 BIOCOMPATIBILITY OF PHB

The PHB used for drug targeting and biomedical applications depends not only on the biodegradation properties but also on biocompatibility and should not elicit an immune response when implanted into soft tissues of the host organism. PHB has also been explored in a variety of biomedical applications. It is used to form several tools like nerve repair gadget, cardiovascular patches and devices, orthopedic pins, adhesion barriers, guided tissue repair/regeneration tools, tendon repair instruments, bone-marrow scaffolds, and wound dressings [17, 82].

2.7.2 BIOCOMPATIBILITY OF PHB IN VITRO AND *IN VIVO*

PHB polymers generally reveal a good response to in vitro and in vivo biocompatibility [16]. In vitro, the PHB biocompatability has been monitored in various types of cell culture mediums of fibroblasts, mesenchymal stem cells, osteoblasts, bone marrow cells, articular cartilage chondrocytes, endothelial cells, smooth muscle cells. The culture PHB isolated from monkey kidney grows into PHB films to form scaffolds within 3–60 days [1]. PHB did not interfere in cell proliferation and viability during cell culture incubation. The cells were not affected by any polymer-loaded compounds, release of active substances (biopolymer degradation), and modification of polymer surface. In addition, cell cultures further grown in laboratory plastic petri dishes did not interfere in proliferation, expansion, and movement of cells. So, PHB is biologically not toxic and biocompatible in vitro isolated from cell cultures medium. PHB biocompatibility has been examined in vivo under subcutaneous implantation of films [1]. The subcutaneous implantation of different molecular weight PHB films was relatively low and did not react between glass plates [32]. In addition, the implanted PHB films indicate better biocompatibility in vivo due to nonimmunogenicity with tissues. Thus, it shows that PHB is useful for biomedical application and drug targeting due to nontoxicity, nonimmunogenicity, and biocompatibility for implantation in human tissues.

2.8 PHB-BASED NANOPARTICLES FOR TARGETED DRUG DELIVERY

Encapsulation of bioactive molecules, like antibiotics or therapeutic drugs to use as the level of micro-and nanosized delivery for performing the transport of bioactive compounds to target tissues should be in the form of spherical, fibrous, or rod-shaped vehicles. These nanoparticles with a typical diameter in the range of 10^2 nm, display high overall surface area for the release of the bioactive molecules and it is possible to profit from surface interactions, which enhance the bioavailability of the drug, controlling the pharmacokinetic properties of the dosage and consequently enhancing the drug's therapeutic value. Further, enhanced tissue selectivity can be obtained by using nanoparticles for drug delivery to increase the uptake of drugs in target tissues [67].

PHA microspheres range from 120 to 200 μm in size, by a solvent evaporation technique for targeting drug release in animal experiments. Renal angiograms and histopathological observations demonstrated that PHB microspheres can readily be used as alternative embolization/chemoembolization agents [39]. Generally, drug release occurred very fast, that means 90% of rifampicin was released within 24 h. However, this releasing rate is strongly dependent on microsphere diameter and drug loading [40].

PHAs are biocompatible, hydrophobic, and thermostable which can be converted into films, matrices, microcapsules, microspheres, and nanoparticles. These modified PHB are capable of entrapping or encapsulating active compounds as biomedical tools, drug targeting, and tissue bioengineering [47]. Nanoparticle-based drug delivery systems have been progressively used for targeting a number of therapeutic drugs (anti-inflammatory and anticancer), anesthetics, antibiotics, hormones, steroids, and vaccines [58]. PHB is exhibited to be useful in implantable nanoformulations for the local targeting of specific antibiotics/drugs in acute/chronic osteomyelitis treatments [33, 81]. The comparisons of in vitro and in vivo release of the active drugs from PHB are possible carriers for drug delivery [6]. The increased efficacy of drug release from PHB nanoparticles to incorporate ethyl or butyl esters of fatty acids [45]. In our chapter, PHB nanoparticles showed delivery of the antitumor drugs to inhibit proliferation of carcinoma [46, 72]. PHB nanoparticles have biphasic nature to release maximally drug loading at initial burst and followed slow release of drug for a longer period of time or even days. The PHB nanoparticles demonstrated no toxicity, nonimmunogenicity, and better compatibility in vivo. A controlled release of P13K inhibitor

(TGX221) encapsulated with PHB nanoparticle was formulated to inhibit the proliferation and growth of cancer cell lines [49].

Nanoparticles targeted delivery may be of importance because it is used for toxic drugs and fragile molecule like immunosupressors, small peptides, and biomolecules. It is reported that the slight modification in nanoparticle with active drugs lead to a significant increase in the conventional therapeutic potential and a decrease in the toxic effects of targeting [5]. In cancer treatment, the targeting of active ligands (drugs) via nanoparticles can greatly increase intracellular uptake and retention through receptor mediated endocytosis [59]. This process can lead to higher retention of intracellular drug concentration and enhance therapeutic activity in the targeted tissues [71]. PHB nanoparticles have been formulated for the targeted release of staphylococcal enterotoxin B vaccine in the gut of lymphoid tissues [20]. However, the PHB nanoparticles exhibited better absorption in vitro and in vivo. Formulation is an ideal targeting drug delivery system using PHB as the drug vector for folic acid in anemic patients and doxorubicin as the anticancer drug model [89]. The average size of drug-loaded PHB nanopaticle was found to be higher than native drugs due to incorporation of the biopolymer. In vitro, the intracellular uptake of drug released from PHB nanoparticles was efficiently taken up by cancer cell lines than native drugs. Similarly, drug loaded nanoparticles showed greater cytotoxicity to cancer cell lines than the normal cell lines. In vivo, antitumor activity of the nanoparticles loaded drugs demonstrated a better therapeutic potential and antiproliferative activity in tumor growth and such nanoparticles are effective for targeting of anticancer drug for the treatment of cancer cells.

Nowadays, the nanodelivery system for targeting drugs to cancer cells is through the catalytic activity of PHA synthase [46]. The PHA synthase enzyme is formulated with a biopolymer (ligand) to prepare a PHA synthase nanoparticle by simple oil in water emulsion method [46]. The hydrophobic surface of PHB nanoparticle couples with the side chain of PHA synthase enzyme nanoparticle in aqueous environment. The functionalized PHB synthase nanoparticles reveal a specific targeting to cancer cells through tumor-specific receptor mediated endocytosis for drug delivery.

2.9 PHARMACEUTICAL AND BIOMEDICAL APPLICATIONS OF PHB

The biopolymers have been utilized in medical applications such as surgical tools, preventive medicine, medical threads, biofilm, and drug targeting [70]. The powerful tool for preparation of medical-grade PHA is supercritical fluid extraction (SFE). Williams and Martin report the high efficiency of pure supercritical CO_2 (sCO_2) in extracting lipophilic compounds from PHA-rich biomass like mcl-PHA shows outstanding solubility in supercritical mixtures. PHHxHO extracted by supercritical solvents reached a purity level of 100% in only one single extraction step; this product contained 25–150 times less LPS than PHHxHO isolated by traditional solvent–nonsolvent extraction and precipitation [84]. SFE for the generation of highly pure scl-PHA was also investigated by Khosravi Darani studied the recovery of PHB by disruption of the Gram-negative bacterium *Ralstonia eutropha* (*Cupriavidus necator*) cells by sCO_2. The impact of applied drying strategy, modifier, physiological stage of cells, repeated release of supercritical CO_2 pressure, operating pressure, and temperature on PHB purity and molecular mass have been evaluated. PHB recovery was studied based on a combination of chemical pretreatments (NaCl or alkaline) and SFE. Cells were exposed for 1 h either to NaCl (140 mM) at 60 °C, 1 h, or to NaOH (0.2–0.8 wt.%). At least 0.4% (wt./wt.) NaOH enabled complete cell disruption, when releasing sCO_2 pressure twice. Pretreatment with NaCl was less effective than alkali pretreatment. Cells in the growth phase were less resistant to disintegration than PHA-rich cells in the later, nutritionally limited, stage of cultivation. Moreover, products obtained from lyophilized biomass were of higher purity than PHB recovered from wet biomass. The method was proposed by the authors as economic and competitive with solvent-based recovery methods in terms of PHB recovery yield and energy consumption; moreover, the method was reported to be environmentally superior to solvent-based techniques [44].

The major advantages of these biopolymers are in human body implantation as a biodegradable material that cannot be removed from the body. PHB has bioavailability and biocompatibility characteristics and it is easily metabolized in cells to produce 3-hydroxybutyric acid monomer released in blood circulation [91]. There is a large application in biomedical areas especially in orthopedics, dental materials, wound management, and urological stents [91].

2.9.1 ANTIBACTERIALS

PHAs are chiral compounds, which have the potential for being used as building blocks for compounds for use in the pharmaceutical industry [4]. PHA can be transformed into different hydroxycarboxylic acids such as 2-alkylated 3HB and β-lactones. These can be employed as oral drugs [13]. Depolymerase enzyme of *Pseudomonas fluorescens* GK13 encoded by gene *pha*ZGK13 can depolymerize PHAs to monomers [19]. These monomeric units can reduce bacterial infection those infected by *Staphylococcus aureus* [55], and those conjugated with D-peptide prove anticancerous [57]. PHB also helps in enhancing angiogenic properties of skin and wound healing [73].

2.9.2 BIOCONTROL AGENTS

Antibiotics are commonly used as feed supplement for animals. At low concentration, these antibiotics have been reported to influence the growth of the animals—livestock and aquaculture [8]. As the incorporation of antibiotic in a consistent manner can be risked affair and the gastrointestinal microflora is likely to develop resistance [54]. One has to ensure that there is a complete elimination of antibiotics from the digestive system of the animal. It has been observed that short-chain fatty acids (SCFAs) are effective in controlling agents against pathogens [53]. As PHAs are biopolymers of beta hydroxy SCFAs, these can be metabolized in the intestinal tract. The metabolites can be exploited as biocontrol agents for giant tiger prawn *Penaeus monodon* [50].

2.9.3 TISSUE ENGINEERING

PHAs available in general were not targeted for use as medical implants and lacking the quality which can get approval of the drug administrators. The need is to produce PHAs of high purity, check their biodegradation in vivo, fabrication of scaffolds, modified their surface [30]. PHAs with necessary modifications hold great potential to contribute to tissue engineering, developing tissue products for medical and therapeutic applications such as (1) vascular grafts, (2) heart valves, (3) nerve tissue engineering [41, 54–65]. PHAs can be used to produce scaffolds, which have higher mechanical strength. These scaffolds promote growth of the cells by supplying nutrition [14]. These products are available as screws, pins, sutures, films, and so forth [25]. PHB has been exploited for fiber meshes by providing support to stem cell growth for proliferation and cell adhesion [11]. It can also be employed as scaffolds for engineering liver tissue [78] and cartilage repair [16]. Further to enhance mechanical strength and flexibility of PHAs, inorganic bioceramics have been combined with PHAs, which produce novel composites for using them for engineering tissues. Composites of PHA and ceramic are employed to form different blends. Hydroxyapatite and PHA are also used in tissue engineering [88].

2.9.4 MEDICAL DEVICES

PHAs have been envisaged to prove useful in making medical devices because they are biocompatible, biodegradable, and have strong mechanical characteristics. Some of the most potential devices like adhesions barriers, articular cartilage repair, cardiovascular patch grafting, meniscus repair device, orthopedic pins, repair patch, rivets and tacks, staples and screws, stents, surgical mesh, sutured fastener [43]. The preparation of cardiovascular patches should have very high quality features, primary being resistance to infection and degradation. In addition, these bioproducts should be durable, lack immunogenicity, and should not be toxic.

2.9.5 BIODEGRADABLE IMPLANTS

Implants are used on a large scale in a very skilful manner. However, invariably the issue of their getting infected with pathogens comes as a major hurdle. Nowadays, biomaterials are also used as implants. Biomaterial-associated infection is a serious health issue. In order to meet the functional demand, materials with desirable physical, chemical, biological, biochemical properties must be selected [43]. Use of biopolymers as biodegradable implants has greatly influenced the modern medicine [80]. The uses of PHA degradation product like PHAs for preparing biodegradable implants, and to fabricate systems to deliver antibiotics in chronic osteomyelitis therapy [13]. PHB and its copolymer P(3HB-3HHx) sheets loaded with lysozyme are being used for biofilm inhibition and in the fabrication of wound dressing [42]. Implant coated with PHB copolymer fastens the degradation process for a stable drug release within a given time period as compared to its copolymer [62]. PHB fiber covered with fibronectin and alginate was used as an implant in spinal cord injury [56].

2.9.6 ANTI-OSTEOPOROSIS EFFECT

Ketoacidosis is induced in human beings by the accumulation of a high concentration of PHB [79]. Oligomers of PHB have properties to act as good energy substrate for patients, where these compounds undergo rapid diffusion within peripheral tissues. It also prevents brain damage as it can enhance cardiac efficiency by regenerating mitochondrial energy. PHB can potentially cure Parkinson's and

Alzheimer's diseases, where they act by reducing the death rate of neuronal cells [38]. It can also enhance osteoblasts growth and anti-osteoporosis activity, by rapidly depositing calcium and its strong serum alkaline phosphatase activity. It helps in preventing the lowering of bone mineral density and reducing the level of serum osteocalcin [12].

2.9.7 MEMORY ENHANCER

Memory loss and related abilities are serious enough to disturb our daily routine. Among the different forms of dementia, Alzheimer's disease is the most common. As a consequence of memory loss, there are problems of thinking and behavior. Derivatives of PHBs such as 3-hydroxybutyrate methyl ester have the potential to act as drug against Alzheimer's disease and protecting mitochondrial damages [90]. During ketogenesis, D-b-HB prevents neuronal death, which is induced by glucose deprivation [9]. PHA monomers derived from PHAs can stimulate the Ca^{2+} channels, which can help in enhancing memory [51].

2.9.8 CHALLENGES FOR INDUSTRIAL SCALE PRODUCTION OF PHAS

PHAs have ideal characteristics, which make them appropriate for commercial production. However, there are still a few challenges, which limit their upscaling [74]. The biophysical properties focused attention on lowering melting point, glass transition temperature, elastic modulus, tensile strength, and elongation. These characteristics depend on the composition and molecular weight of the polymer. High molecular weights' PHAs have the potential to overcome these limitations [74]. In addition, strategies to manipulate composition [51], culture conditions such as independence from nutritional imbalance [64], increased microbial biomass, high expression of polymerase genes, and genetic modification to synchronize termination of PHA biosynthesis with cell lysis [75] will certainly help in economic production of PHAs on an industrial scale.

2.10 CONCLUSION AND FUTURE PROSPECTS

PHB has been established and efficiently used in the area of biomedical sciences and nanoformulation of drug targeting. These formulations may be able to enhance mechanical strength as well as degradation rate of the encapsulated nanopaticles. The PHB biopolymers have been demonstrated for the evolution of new and efficient concept for the delivery of bioactive molecules and prompt biomedical practices and are being used as a novel drug carrier. In addition, PHB-loaded drug delivery is a novel area for the release of anticancer drugs from nanoformulated biopolymers for cancer treatment. PHB have a broad application especially in biomedical field and which are practically very cost effective. PHB is an environmentally safer biopolymer for the next generation research with a broad concept of application mostly in medicine. PHB is useful as a nonimmunogenic and stimuli-responsive polymer in future.

ACKNOWLEDGMENT

The authors would like to acknowledge the University of Grant Commission (UGC-No. F. 359/2017, BSR-Startup Grant), New Delhi, India for providing financial assistance. The authors would like to thank Prof. Haseeb Ahsan for proofreading and improving the language and grammar of the chapter.

KEYWORDS

- **polyhydroxybutyrate (PHB)**
- **polyhydroxyalkanoates (PHA)**
- **biopolymer**
- **biodegradation**
- **drug delivery**
- **nanoparticles**

REFERENCES

1. Akshay SJ, Vyankatesh SB, Vaishali KK, Santosh AP, John ID (2013) Poly-β-hydroxybutyrate: intriguing biopolymer in biomedical applications and pharma formulation trends. Int J Pharm Biol Arch 4(6):1107–1118.
2. Alyaa H, Zakia O, Hanafi H (2013) Microbial production of polyhydroxybutyrate, a biodegradable plastic using agro-industrial waste products. Glob Adv Res J Microbiol 2(3): 054–064.
3. Anderson AJ, Dawes EA (1990) Occurrence, metabolism, metabolic role and industrial uses of bacterial polyhydroxyalkanoates. Microbiol Rev 54:450–472
4. Babel W, Ackermann JU, Breuer U (2001) Physiology, regulation, and limits of the synthesis of poly(3HB). Adv Biochem Eng Biotechnol 71:125–157.
5. Bartlett DW, Su H, Hildebrandt IJ, Weber WA, Davis ME (2007) Impact of tumor-specific targeting on the biodistribution and efficacy of siRNA nanoparticles measured by multimodality in vivo imaging. Pro Nat Acad Sci USA 104(39):15549–15554.
6. Bissery MC, Valeriote FA, Thies C (1985) Therapeutic efficacy of CCNU-loaded microspheres prepared from poly(D,L) lactide (PLA) or poly-*b*-hydroxybutyrate (PHB) against Lewis lung (LL) carcinoma. Pro Am Asso Cancer Res 26:355.
7. Byrom D (1987) Polymer synthesis by microorganisms: technology and economics. Trends Biotech 5(9):246–250. https://doi.org/10.1016/0167-7799(87)90100-4
8. Cabello FC (2006) Heavy use of prophylactic antibiotics in aquaculture: a growing problem for human and

animal health and for the environment. Environ Microbiol 8:1137–1144.
9. Camberos-Luna L, Gerónimo-Olvera C, Montiel T, Rincon Heredia R, Massieu L (2016) The ketone body, b-Hydroxybutyrate stimulates the autophagic flux and prevents neuronal death induced by glucose deprivation in cortical cultured neurons. Neurochem Res 41:600–609.
10. Cammas S, Béar MM, Moine L (1999) Polymers of malic acid and 3-alkylmalic acid as synthetic PHAs in the design of biocompatible hydrolyzeable devices. Int J Biol Macromol 25(1): 273–282.
11. Canadas RF, Cavalheiro JMBT, Guerreiro JDT, de Almeida MCMD, Pollet E, da Silva CL, da Fonseca MMR, Ferreira FC (2014) Polyhydroxyalkanoates: waste glycerol upgrade into electrospun fibrous scaffolds for stem cells culture. Int J Biol Macromol 71:131–140.
12. Chen GQ (2011) Biofunctionalization of polymers and their applications. In: Biofunctionalization of Polymers and Their Applications. Berlin: Springer, pp 29–45.
13. Chen GQ, Wu Q (2005) The application of polyhydroxyalkanoates as tissue engineering materials. Biomaterials 26:6565–6578.
14. Chen W, Tong YW (2012) PHBV microspheres as neural tissue engineering scaffold support neuronal cell growth and axon–dendrite polarization. Acta Biomater 8:540–548.
15. Cheng YT, Jae J, Shi J, et al. (2012) Production of renewable aromatic compounds by catalytic fast pyrolysis of lignocellulosic Biomass with bifunctional Ga/ZSM-5 catalysts. Angew Chem Int Ed 51:1387–1390.
16. Ching KY, Andriotis OG, Li S, Basnett P, Su B, Roy I, Stolz M (2016) Nanofibrous poly (3-hydroxybutyrate)/ poly (3-hydroxyoctanoate) scaffolds provide a functional microenvironment for cartilage repair. J Biomater Appl 31:77–91.
17. Dawes E (1988) Polyhydroxybutyrate an intriguing biopolymer. Biosci Rep 8 (6): 537–538.
18. Dawes EA, Senior PJ (1973) The role and regulation of energy reserve polymers in micro-organisms. Adv Microb Physiol 10:135–266.
19. De Eugenio LI, Escapa IF, Morales V, et al. (2010) The turnover of medium-chainlength polyhydroxyalkanoates in KT2442 and the fundamental role of PhaZ depolymerase for the metabolic balance. Environ Microbiol 12:207–221.
20. Eldridge JH, Hammond CJ, Meulbroek JA, Staas JK, Gilley RM, Tice TR, (1990) Controlled vaccine release in the gut-associated lymphoid tissues. I. Orally administered biodegradable microspheres target the Peyer'spatches. J Controlled Release 11(1–3):205–214.
21. Endres HJ. Biopolymers facts and statistics (2017) Hannover: IfBB—Institute for Bioplastics and Biocomposites.
22. Freier T (2017) Biopolyesters in tissue engineering applications. Polymers for regenerative medicine [Internet]. Berlin, Heidelberg: Springer; p. 1–61. https://link.springer.com/chapter/10.1007/12_073.
23. Grande D, Ramier J, Versace DL, Renard E, Langlois V (2017) Design of functionalized biodegradable PHA-based electrospun scaffolds meant for tissue engineering applications. New Biotechnol 37:129–137.
24. Griebel R, Smith Z, Merrick JM (1968) Metabolism of Polyβ-hydroxybutyrate. I. Purification, composition, and properties of nativepoly-β-hydroxybutyrate granules from bacillus megaterium. Biochemistry 7(10):3676–3681.

25. Gürsel I, Korkusuz F, Türesin F, Gürdal Alaeddinŏglu N. Hasirci V (2001) In vivo application of biodegradable controlled antibiotic release systems for the treatment of implant-related osteomyelitis. Biomaterials 22(1):73–80.
26. Gurubasappa G, Biradar, Shivasharana CT, Basappa BK (2015) Isolation and characterization of polyhydroxybutyrate (PHB) producing Bacillus species from agricultural soil. Eur J Exp Biol 5(3):58–65.
27. Hazer B (2010) Amphiphilic poly(3-hydroxy alkanoate)s: potential candidates for medical applications. Intl J Poly Sci ArticleID423460, http://dx.doi.org/10.1155/2010/423460.
28. Hutchinson F, Furr B (1990) J Controlled Release 13:279.
29. Ikada Y (1998) Tissue engineering research trends at Kyoto University, in tissue engineering for therapeutic use. Am Chem Soc 1–14.
30. Insomphun C, Chuah JA, Kobayashi S, Fujiki T, Numata K (2016) Influence of hydroxyl groups on the cell viability of polyhydroxyalkanoate (PHA) scaffolds for tissue engineering. ACS Biomater Sci Eng. doi:10.1021/acsbiomaterials.6b00279.
31. Iordanskii A, Rogovina S, Berlin A (2013) Current state and developmental prospects for nanopatterned implants containing drugs. Rev J Chem 3(2):117–132.
32. Iordanskii AGB, Yu P, Rogovina KG, Zaikov G, Berlin A (2014) Current status and biomedical application spectrum of poly (3-hydroxybutyrate) as a bacterial biodegradable polymer. Key Eng Mater. 1
33. Isikgor FH, Becer RC (2015) Lignocellulosic biomass: a sustainable platform for the production of bio-based chemicals and polymers. Polym Chem 6:4497–4559.
34. Jing A, Weimin S (2003) The adverse drug reaction of indapamide. Pract J Med Pharm 11:843–844.
35. Kalia S, Kaith BS, Kaur I (Eds.) Cellulose fibers: bio- and nano-polymer composites, Green Chem Technol 2011 ISBN 978-3-642-17370-7.
36. Kamravamanesh D, Kovacs T, Pflügl S, et al. (2018) Increased poly-β-hydroxybutyrate production from carbon dioxide in randomly mutated cells of cyanobacterial strain synechocystis sp. PCC 6714: mutant generation and characterization. Bioresour Technol 266:34–44.
37. Kamravamanesh D, Pflügl S, Nischkauer W, et al. (2017) Photosynthetic poly-β-hydroxybutyrate accumulation in unicellular cyanobacterium synechocystis sp. PCC 6714. AMB Express 7:143.
38. Kashiwaya Y, Takeshima T, Mori N, Nakashima K, Clarke K, Veech RL (2000) d-b-Hydroxybutyrate protects neurons in models of Alzheimer's and Parkinson's disease. Proc Natl Acad Sci USA 97:5440–5444.
39. Kassab AC, Piskin E, Bilgic S, et al. (1999) Embolization with polyhydroxybutyrate (PHB) microspheres: In vivo studies. J Bioact Compat Polym 14: 291–303.
40. Kassab AC, Xu K, Denkbas EB, et al. (1997) Rifampicin carrying polyhydroxybutyrate microspheres as a potential chemoembolization agent. J Biomater Sci Polym Ed 8:947–961.
41. Ke Y, Zhang XY, Ramakrishna S, He LM, Wu G (2017) Reactive blends based on polyhydroxyalkanoates: preparation and biomedical application. Mater Sci Eng C Mater Biol Appl 70:1107–1119.
42. Kehail AA, Brigham CJ (2017) Anti-biofilm activity of solventcast and electrospun polyhydroxyalkanoate membranes treated with lysozyme.

J Polym Environ. doi:10.1007/s10924-0160921-1.
43. Kenar H, Kose GT, Hasirci V (2010) Design of a 3D aligned myocardial tissue construct from biodegradable polyesters. J Mater Sci Mater Med 21:989–997.
44. Khosravi-Darani K, Vasheghani-Farahani E, Shojaosadati SA, Yamini Y (2004) Effect of process variables on supercritical fluid disruption of Ralstonia eutropha cells for poly(R-hydroxybutyrate) recovery. Biotechnol Prog 20:1757–1765.
45. Kubota M, Nakano M, Juni K (1988) Mechanism of enhancement of the release rate of aclarubicin from poly*β*-hydroxybutyric acid microspheres by fatty acid esters. Chem Pharm Bull 36(1):333–337.
46. Lee J, Jung SG, Park CS, Kim HY, Batt CA, Kim YR (2011) Tumor-specific hybrid polyhydroxybutyrate nanoparticle: surface modification of nanoparticle by enzymatically synthesized functional block copolymer. Bioorg Med Chem Lett 21(10):2941–2944.
47. Lenz RW, Marchessault RH (2005) Bacterial polyesters: biosynthesis, biodegradable plastics and biotechnology. Biomacromolecules 6(1):1–8.
48. Li T (2013) Charakterisierung von Biopolymeren. Darmstadt: Technische Universität Darmstadt.
49. Lu XY, Ciraolo E, Stefenia R, Chen GQ, Zhang Y, Hirsch E (2011) Sustained release of PI3K inhibitor from PHA nanoparticles and in vitro growth inhibition of cancer cell lines. Appl Micro Biotech 89(5):1423–1433.
50. Ludevese-Pascual G, Laranja JLQ, Amar EC, Sorgeloos P, Bossier P, De Schryver P (2016) Poly-beta-hydroxybutyrate-enriched Artemia sp. for giant tiger prawn *Penaeus monodon* larviculture. Aquaculture 23:422–429.
51. Magdouli S, Brar SK, Blais JF, Tyagi RD (2015) How to direct the fatty acid biosynthesis towards polyhydroxyalkanoates production? Biomass Bioenerg 74:268–279.
52. Markl E, Grünbichler H, Lackner M (2018) PHB—bio based and biodegradable replacement for PP: A Review. Nov Tech Nutri Food Sci 4:2.
53. Martinez JL (2017) Effect of antibiotics on bacterial populations: a multi-hierarchical selection process. F1000 Res 6:51.
54. Martínez V, de la Peña F, García-Hidalgo J, de la Mata I, García JL, Prieto MA (2012) Identification and biochemical evidence of a medium-chain-length polyhydroxyalkanoate depolymerase in the Bdellovibrio bacteriovorus predatory hydrolytic arsenal. Appl Environ Microbiol 78:6017–6026.
55. Martinez V, Dinjaski N, De Eugenio LI, et al. (2014) Cell system engineering to produce extracellular polyhydroxyalkanoate depolymerase with targeted applications. Int J Biol Macromol 71:28–33.
56. Novikov LN, Novikova LN, Mosahebi A, Wiberg M, Terenghi G, Kellerth JO (2002) A novel biodegradable implant for neuronal rescue and regeneration after spinal cord injury. Biomaterials 23:3369–3376.
57. O'Connor S, Szwej E, Nikodinovic-Runic J, et al. (2013) The anti-cancer activity of a cationic anti-microbial peptide derived from monomers of polyhydroxyalkanoate. Biomaterials 34:2710–2718. doi:10.1016/j.biomaterials.2012.12.032.
58. Orts WJ, Nobes GAR, Kawada J, Nguyen S, Yu GE, Ravenelle F (2008) Poly(hydroxyalkanoates): biorefinery polymers with a whole range of

applications. The work of Robert H. Marchessault. Canad J Chem 86(6): 628–640.
59. Pirollo KF, Chang EH (2008) Does a targeting ligand influence nanoparticle tumor localization or uptake? Trends Biotech 26(10):552–558.
60. Piskin E (1995) Biodegradable polymers as biomaterials. J Biomate Sci Poly 6(9):775–795.
61. Pouton CW, Kennedy JE, Notarianni LJ, Gould PL (1988) Biocompatibility of polyhydroxy-butyrate and related copolymers. Proc Int Symp Controlled Release Bioact Mater 15:179–180.
62. Raoga O, Sima L, Chirioiu M, et al. (2017) Biocomposite coatings based on poly(3-hydroxybutyrate-co-3hydroxyvalerate)/calcium phosphates obtained by MAPLE for bone tissue engineering. Appl Surf Sci. doi:10.1016/j.apsusc.
63. Rawte T, Mavinkurve S (2002) Indian J Exp Biol, 40:924–929.
64. Ray S, Kalia VC (2017) Co-metabolism of substrates by Bacillus thuringiensis regulates polyhydroxyalkanoate co-polymer composition. Bioresour Technol 224:743–747.
65. Reddy SV, Thirumala M, Mahmood SK (2009) Production of PHB and P (3HB-co-3HV) biopolymers by *Bacillus megaterium* strain OU303A isolated from municipal sewage sludge. World J Micro Biotech. 25 (3):391-397.
66. Rettenbacher L (2014) Biokunststoffe Überblick und aktuelle Trends in der Biokunststoffwelt. Wien: OFI.
67. Rezaie Shirmard L, Bahari Javan N, Khoshayand MR, et al. (2017) Nanoparticulate fingolimod delivery system based on biodegradable poly(3-hydroxybutyrate-co-3-hydroxyvalerate) (PHBV): design, optimization, characterization and in vitro evaluation. Pharm Dev Technol 22:860–870.
68. Schlegel HG, Lafferty R, Krauss, I (1970) The isolation of mutants not accumulating poly-beta-hydroxybutyric acid. Arch Microbiol 70:283–294, 70.
69. Senthilkumar B, Prabhakaran G (2006) Isolation and identification of new strains to enhance the production of biopolymers from marine sample in Karankura, Tamil Nadu. Indian J Biotechnol 5:76–79.
70. Shalaby M, Ikada Y, Lander R, Williams J (1994) Polymers of biological and biomedical significance. ACS Symp Ser 540.
71. Shi J, Votruba AR, Farokhzad OC, Langer R (2010) Nanotechnology in drug delivery and tissue engineering: from discovery to applications. Nano Lett 10(9):3223–3230.
72. Shishatskaya EI, Goreva AV, Voinova ON, Inzhevatkin EV, Khlebopros RG, Volova TG (2008) Evaluation of antitumor activity of rubomycin deposited in absorbable polymeric microparticles. Bul Expt Biol Med 145(3):358–361.
73. Shishatskaya EI, Nikolaeva ED, Vinogradova ON, Volova TG (2016) Experimental wound dressings of degradable PHA for skin defect repair. J Mater Sci Mater Med 27:165.
74. Singh M, Kumar P, Ray S, Kalia VC (2015) Challenges and opportunities for the customizing polyhydroxyalkanoates. Indian J Microbiol 55:235–249.
75. Singh M, Patel SKS, Kalia VC (2009) Bacillus subtilis as potential producer for polyhydroxyalkanoates. Microb Cell Fact 8:38.
76. Smith R, Stewart J (1981) Procurement and characterization of standard reference material. 4: 256–258.
77. Somleva MN, Peoples OP, Snell KD (2013) PHA bioplastics, biochemicals,

and energy from crops. Plant Biotechnol J 11:233–252.
78. Su Z, Li P, Wu B, Ma H, Wang Y, Liu G, Wei X (2014) PHBVHHx scaffolds loaded with umbilical cord-derived mesenchymal stem cells or hepatocyte-like cells differentiated from these cells for liver tissue engineering. Mater Sci Eng C 45:374–382.
79. Tokiwa Y, Calabia BP (2007) Biodegradability and biodegradation of polyesters. J Polym Environ 15:259–267.
80. Ulery BD, Nair LS, Laurencin CT (2011) Biomedical applications of biodegradable polymers. J Polym Sci Part B Polym Phys 49:832–864.
81. Uresin FT, G¨ursel I, Hasirci V (2001) Biodegradable polyhydroxyalkanoate implants for osteomyelitis therapy: in vitro antibioticrelease. J Biomater Sci Poly 12(2):195–207.
82. Valappil SP, Misra SK, Boccaccini A, Roy I (2006) Biomedical applications of polyhydroxyalkanoates, an overview of animal testing and in vivo responses. Expert Rev Med Dev 3(6):853–868.
83. Vogel C (2008) Charakterisierung der thermischen und mechanischen Eigenschaften von Polyhydroxyalkanoat (PHA) Homopolymeren, Copolymeren und Polymermischungen. Essen: Duisburg-Essen.
84. Williams SF, Martin DP (2002) Applications of PHAs in medicine and pharmacy. In Biopolymers Polyesters III-Applications and Commercial Products; Doi, Y., Steinbüchel, A., Eds.; Chichester, UK Wiley-VCH, 91–127.
85. Wong JWC, Tyagi RD, Pandey A (2016) Current developments in biotechnology and bioengineering: solid waste management. Amsterdam: Elsevier.
86. Xiong Y, Yao Y, Zhan X, Chen G (2010) Application of polyhydroxyalkanoates nanoparticles as intracellular sustained drug-release vectors. J Biomater Sci 21:127–140.
87. Yamamoto H, Kuno Y, Sugimoto S, Takeuchi H, Kawashima Y (2005) Surface-modified PLGA nanosphere with chitosan improved pulmonary delivery of calcitonin by mucoadhesion and opening of the intercellular tight junctions. J Controlled Release 102(2):373–381.
88. Yucel D, Kose GT, Hasirci V (2010) Polyester based nerve guidance conduit design. Biomaterials 31:1596–1603.
89. Zhang J, Qian C, Shaowu L, et al. (2013) 3-Hydroxybutyrate methyl ester as a potential drug against Alzheimer's disease via mitochondria protection mechanism. Biomaterials 34:7552–7562.
90. Zhang C, Zhao LQ, Dong YF, Zhang XY, Lin J, Chen Z (2010) Folate-mediated poly(3-hydroxybutyrate-co-3hydroxyoctanoate) nanoparticles for targeting drug delivery. Eur J Pharm Biopharm 76(1):10–16.
91. Zinn M, Witholt B, Egii T (2001) Occurrence synthesis and medical application of bacterial polyhydroxyalkanoate. Adv Drug Rev 53:5–21.

APPENDIX

Wound management: Biopolymers are actively used in the sutures, staples, clips, adhesives, and surgical meshes.

Orthopedic devices: Pins, rods, screws, tacks, and ligaments are mostly made by biodegradable polymers to easily degrade in vivo without any toxicity.

Dental applications and maxillofacial: The biodegradable materials are mostly used for the guided tissue membrane regeneration, bone regeneration, and void filler following tooth extraction.

Cardiovascular applications: Cardiovascular stents, vascular grafts, pericardial patch, and heart valve are generally made by biocompatible and biodegradable polymers in nature.

Tissue engineering: Two-dimensional biopolymer matrices are used in the formation of thin films for tissue engineering applications [1]. These biodegradable thin films are successfully used in skin regeneration. However, these products have provided mechanical and physical properties in the skin graft and low risk of immunological rejection. PHB is usually used to improving the skin regeneration [29].

Trans-dermal patches: Transdermal drug delivery is a tool that control drug concentration in blood for therapeutic treatment and also reduce side effects with improve patient compliance [34, 76].

Cardiac stent: The earliest advance in the field of nanobiotechnology is noticeable in the formation of cardiovascular stents. The concept in a blueprint to form the stents that provided prolonged and sustained drug targeting locally. However, the cardiovascular stents are effective degradable and biocompatible in nature. The cardiac implants are formed by nanoformulating of biodegradable polymer with a characteristic of meso and nanoporous layers responsible for controllable drug release, protein adsorption, and adhesion of blood cells [31].

Urology implants: A biodegradable polymer has a characteristic unique functional group to use for the urology implantation. Several analogies of biopolymers have been used for the production of poly(α-oxyacids) and poly(β-oxyalkanoates) for implantation. One of the multifunctional implanted system Zoladex was designed over two decades ago and currently it is widely used in clinical practice for the long term therapy of endometriosis, breast tumors, and prostate tumors [28].

CHAPTER 3

Brain Targeting Efficacy of Novel Solid Lipid Nanoparticles

N. JAWAHAR*, LIRENI C. HUMTSOE, and P. B. ANJALI

Department of Pharmaceutics, JSS College of Pharmacy (JSS Academy of Higher Education and Research), Rocklands, Udhagamandalam, Tamil Nadu 643001, India

**Corresponding author. E-mail: jawahar.n@jssuni.edu.in*

ABSTRACT

This work deals with the investigation carried out on the formulation and evaluation of solid lipid nanoparticles containing olanzapine (OL), for brain targeting for increasing the bioavailability in the brain, maintain prolonged therapeutic levels, and to reduce the dose to achieve the same pharmacological effect. Preformulation studies confirm the compatibility of different excipients with the drug. Strong lipid nanoparticles (SLNs) were prepared by solvent diffusion method and optimized by using Box–Behnken design. Batch prepared with 30 mg lipid concentration, 4 h stirring time, 5000 rpm stirring speed, and 1.5% w/v surfactant concentration showed minimum particle size and was identified as an ideal batch, that is, glyceryl tripalmitate (110.5 nm) and glyceryl monostearate (165.1 nm). SLNs were evaluated for zeta potential, entrapment efficiency of the drug, scanning electron microscopy studies, polydispersity, in vitro release studies, and in vivo biodistribution and pharmacokinetic studies. In vivo biodistribution studies were carried out in albino Wistar rats where two different formulations, namely, GTP-SLN and OL suspension were administered intravenously and pharmacokinetic parameters were calculated. When administered intravenously in the SLNs as compared to the aqueous suspension, SLNs were more effective in enhancing the bioavailability of OL in the brain. In the brain, the concentrations of OL were significantly higher, that is, 23-fold increased for OL-SLN compared to OL-suspension, OL in the brain tissues. Thus, these results indicate that OL SLNs improve brain OL level as well as the therapeutic efficacy of the OL in the treatment of schizophrenia.

3.1 INTRODUCTION

In the most recent decade, a developing interest has been developing toward cerebrum focusing on where the expanding familiarity with the absence of reasonable and normal endeavors among various and reciprocal research zones has brought up the requirement for a more profound understanding and a closer joint effort among differing research specialists of the field. These issues alongside poor information with respect to the physiology of the focal sensory system cCentral Nnervous sSystem (CNS) have been the fundamental constraining elements in the improvement of powerful medications and suitable medication conveyance frameworks dDrug dDelivery Ssystem (DDS) for cerebrum focusing on. Actually, conveying medications to the CNS are weakened by the nearness of the blood–cerebrum boundary blood–cerebrum boundary (BBB) that speaks to the principal impediment for CNS tranquilize advancement [1]. Among the various medications, just 5% of the medications treat the downturn, schizophrenia, and sleep deprivation [2]. Energizing trial results got in vitro are all the time followed by baffling outcomes in vivo. Principal explanations behind treatment disappointment include insufficient drug concentration due to poor absorption, rapid metabolism, and elimination (e.g., peptides and proteins). Drug distribution to other tissues combined high drug toxicity (e.g., cancer drugs):

(1) Poor medicate solvency which rejects i.v. infusion of watery medication arrangement.
(2) High change of plasma levels because of unusual bioavailability after peroral organization remembering the impact of nourishment for plasma levels (e.g., cyclosporine) [3].

To build the particularity toward cells or tissues, to improve the bioavailability of medications, and to secure them against protein inactivation, colloidal bearers, for example, micelles, emulsions, liposomes, and nanoparticles (nanocapsules and nanospheres) are commonly utilized. Furthermore, the colloidal frameworks permit access over the BBB of nontransportable medications by covering their physico-substance qualities through their exemplification in these frameworks [4]. Parenteral medication conveyance took a significant jump after effective improvement of the submicron parenteral fat emulsion (Intralipid) in the 1960s. Brisk commercialization of submicron emulsion-based items, for example, Diazemuls (Diazepam) and Diprivan (Propofol), was the pointer of the enthusiasm of the pharmaceutical industry in colloidal transporters. The main disadvantage related to these submicron emulsions was the low consistency of the beads causing quick discharge and vulnerability of the consolidated actives toward debasement by the watery persistent stage. Liposomes speak to the original of the novel colloidal bearers, which reformed the situation in parenteral medication conveyance. The fruitful commercialization of different injectable liposomal items, for example, AmBisome_ (Amphotericin B), DaunoXome_ (Daunorubicin), and so on, obviously shows the potential preferences of liposomes as novel lipid transporters. However, the multifaceted nature related to the assembling of liposomes, troubles in scale-up, constrained physical solidness, and huge expense of the liposomal detailing are the significant obstructions in the fruitful commercialization of liposomes. Polymeric nanoparticles are likewise remembered for the original novel colloidal bearers created with the goal to improve parenteral conveyance. In any case, because of different inconveniences, for example, challenges in scale-up, significant expense of biodegradable polymers, possibly harmful/unfavorably susceptible finished results of biodegradable polymers, there is no business item dependent on polymeric "nano" particles considerably following 35 years of their revelation and the main portrayal of polymeric nanocapsules by Speiser and Birenbach. All things considered, polymeric "small scale" molecule-based stop details are accessible in the market, for example, Lupron_ containing leuprolide and Parlodel_ containing bromocriptin.

In the 1990s, scientists (Mueller and colleagues and Gasco and associates) began investigating the capability of nanoparticles-based strong lipids or SLNs in the medication conveyance. SLN consolidates points of interest of previously mentioned colloidal medication bearer frameworks like liposomes, polymeric nanoparticles, and emulsions, and yet keep away from or limit the downsides related to them [5].

Along these lines, SLN conveyance can be an inventive method to control atoms into the cerebrum by potentially surviving or mitigating the solvency, penetrability, and harmfulness issues related to the particular medication particles thus additionally it can have a bit of leeway over the regular obtrusive techniques for sedate conveyance to the mind [6].

3.2 CHALLENGES OF CNS DRUG DEVELOPMENT

Powerful medications have not been created for most CNS issues. Lion's share of little particle drugs does not cross the BBB. More than 7000 medications in the far-reaching restorative science (CMC) database (see Table 3.1), just 5% of all medications treat the CNS, and these CNS dynamic medications just treat despondency, schizophrenia, and sleep deprivation [2]. Studies show that solitary 12% of medications were dynamic in the CNS; however, just 1% of all medications were dynamic in the CNS for sicknesses other than full of a feeling issue [7].

TABLE 3.1 The Comprehensive Medicinal Chemistry for Some of the Drugs Used in CNS Disorders

CNS Disorders Treatable with Small Molecule Drug Therapy	CNS Disorders Largely Refractory to Small Molecule Drug Therapy
Depression Schizophrenia Chronic pain Epilepsy	Neurodegenerative diseases (Alzheimer's, Huntington's, Parkinson's disease), Inflammatory diseases (Amytrophic lateral sclerosis, multiplesclerosis, Neuro-AIDS, Brain Cancer, Stroke, Brain or Spinal cord Trauma, Autism, Lysosomal storage disorder, Fragile X syndrome, Inherited ataxias, Blindness), Cerebro vascular Disease

3.3 LIMITATIONS FOR DRUGS TO ENTER INTO CNS

Drugs may be transported through the BBB by passive or active transport (Figure 3.1).

3.3.1 TRANSPORT ROUTES ACROSS THE BBB

1. Leukocyte movement across the BBB.
2. Molecules may diffuse across the BBB by passive diffusion through the cell membranes and across the endothelial cells.
3. Active carrier-mediated efflux transporters to expel a wide range of molecules out of the BBB.
4. Necessary polar metabolites and nutrients are carried into the CNS by transporters in the BBB.

5. Macromolecules, proteins, and peptides are transported by receptor-mediated transcytosis.
6. Some macromolecules like cationic macromolecules induce adsorptive-mediated transcytosis.
7. To allow an increased movement of polar solutes, tight junctions may be modulated through the aqueous paracellular pathway. Adapted from [8].

3.3.2 STRATEGIES FOR DRUG DELIVERY TO THE BRAIN

(1) Invasive approach.
   (a) Disruption of the BBB.
      (i) Osmotic disruption.
      (ii) MRI-guided focused ultrasound BBB disruption technique.
      (iii) Application of bradykinin-analog.
         - Intracerebro-ventricular infusion.
         - Convection-enhanced delivery.
         - Intracerebral injection or use of implants.

(2) Noninvasive approach.
   (b) Physiological approach.
      (i) Transporter-mediated delivery.
      (ii) Receptor-mediated transcytosis.
      (iii) Transferrin receptor.
      (iv) Insulin receptor.
      (v) Colloidal carrier systems.

(3) Vesicular systems—liposomes, niosomes.

(4) Nanocarrier systems—nanoparticles, nanosuspension.

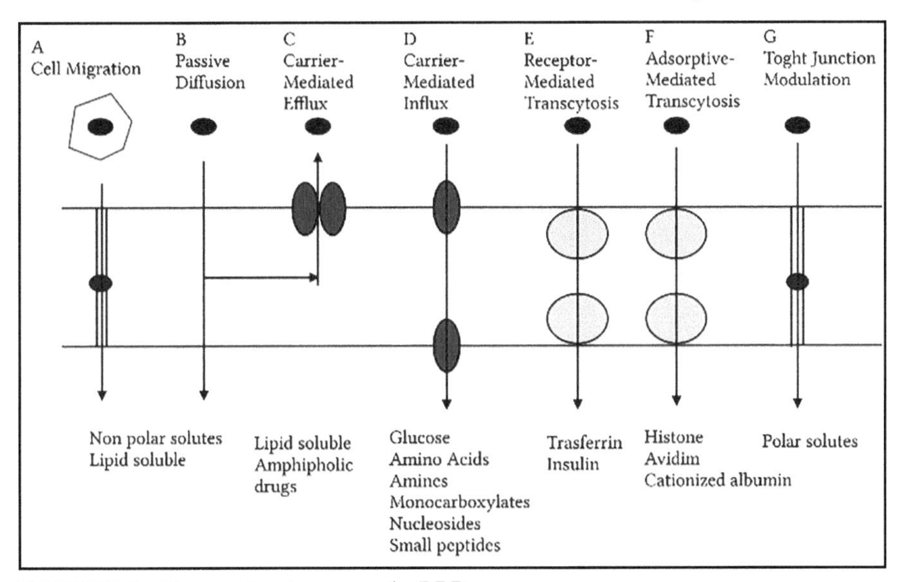

FIGURE 3.1 Transport routes across the BBB.

(5) Lipid-based drug delivery systems.
(6) Self-microemulsifying/nanoemulsifying drug delivery systems (SMEDDS/SNEDDS).
(7) SLNs.
(8) Emulsions—macro/micro/multiple/nanoemulsions.
   (c) Chemical approach (prodrug approach).
   (d) Biological approach (conjugation of drug with antibodies).

Colloidal frameworks effectively enter mind vessels before arriving at the outside of the cerebrum microvascular endothelial cells, when the outside of these colloids are changed in a legitimate manner (for example, by PEG or PS-80). These surface changed colloidal particles improve introduction of the BBB because of delayed blood course, which favors connection and entrance into cerebrum endothelial cells. Colloidal frameworks may additionally be changed with an assortment of specialists on their surface, each with an interesting capacity prompting multifunctional treatment. In this way, when medication is stacked, colloidal transporters might be useful for the treatment of mind infections, since they offer clinical advantages, for example, diminished medication portion, diminished symptoms, expanded medication feasibility, nonobtrusive courses of organization, and improved personal satisfaction to persistent.

3.4 MECHANISM OF SOLID LIPID NANOPARTICLE TRANSPORT ACROSS THE BBB

Out of the few distinct methodologies recently referenced, the utilization of NPs has been viewed as having incredible potential for the conveyance of medications into the CNS. First assumed that the surfactant contact point might be the characteristic in anticipating the body for the adoption of covered NPs. In light of this presumption, in a later work, the creators considered the in vivo body dispersion of surfactant covered and noncovered poly(methyl methacrylate) (PMMA) NPs of 131 ± 30 nm in size. A progression of poloxamers (likewise known with the exclusive name Pluronic®) (poloxamer 188, 407, 184, 338, and poloxamine 908), polysorbates (additionally known with the restrictive name Tween®) (polysorbates 20, 60, and 80), and Brij® 35 were utilized as covering material to examine the conceivable relationship between the NP surface properties and their in vivo biodistribution. The outcomes demonstrated that (1) no relationship exists between the body conveyance design and the contact edge of the NP covering materials; (2) albeit every single utilized surfactant diminished the liver take-up and expanded the take-up in different organs and tissues, poloxamer 338 and poloxamine 908 were the most dominant surfactants in improving NP blood levels and in expanding the spleen take-up while lessening the liver take-up; (3) P80 and poloxamer 184 yielded the most noteworthy gathering in the heart, muscles, kidneys, and the cerebrum. In this manner, it was believed that polysorbates and poloxamers could be helpful for explicit focusing to non-reticuloendothelial system (RES) organs. Specifically, the all-out NP cerebrum sum was expanded with all surfactants up to 24 h, particularly by polysorbates 60, and 80, poloxamers 407, 184, and 338, and poloxamine 908.

A theory was proposed that adsorption of the ApoE or B on P80-covered NPs could be liable for the collaboration with the BBB and the consequent endocytosis.

The component of the medication transport over the blood-cerebrum hindrance with the nanoparticles has all the earmarks of being endocytotic take-up by the mind fine endothelial cells followed either by the arrival of the medications in these cells and dissemination into the cerebrum or by transcytosis. The nanoparticles after infusion, apolipoprotein E (apo E), or apo B adsorb on the molecule surface and afterward communicate with the LDL receptor followed by endocytotic take-up. The nanoparticles subsequently would mirror the take-up of normally happening lipoprotein particles. This speculation was bolstered by the accomplishment of an antinociceptive impact utilizing dalargin-stacked poly(butyl cyanoacrylate) nanoparticles with adsorbed apo E or loperamide-stacked egg whites nanoparticles with covalently bound apo E.

A portion of the conceivable outcomes that clarify the component of medication conveyance by nanoparticles over the blood-cerebrum hindrance are as per the following:

1. Solubilization of endothelial cell layer lipids by surfactant impact prompts film fluidization and upgraded tranquilize penetrability through the blood mind boundary.
2. Retention of nanoparticle in the mind blood vessels joined with an adsorption to the slim dividers makes a higher focus slope that improves the vehicle of medications over the endothelial cell layer and in this manner, brings about conveyance to the cerebrum.
3. The nanoparticles could prompt an opening of the tight intersections between the endothelial cells. The medication could

then saturate through the tight intersections in free structure or together with the nanoparticles in bound structure.

4. The nanoparticles might be endocytosed by the endothelial cells followed by the arrival of the medications inside these cells and conveyance to the cerebrum.
5. The nanoparticles with bound medications could be transcytosed through the endothelial cell layer.
6. Using of covering specialist, for example, polysorbate 80 could restrain the efflux framework, particularly the *p*-glycoprotein that represses the vehicle of medications over the blood mind boundary.

3.5 SOLID LIPID NANOPARTICLES

SLNs are lipid-based submicron particulate colloidal carrier systems ranging in size from 10 to 1000 nm. SLNs combine advantages of other colloidal carrier systems like liposomes, polymeric nanoparticles, and emulsions, but at the same time avoid or minimize the drawbacks associated with them [10]. (Figure 3.2). Comparison of SLN with Oother cColloidal cCarriers were provided in Table 3.2.

3.5.1 ADVANTAGES OF SLNS (FIGURE 3.3)

1. The nanoparticles and SLN particularly those in the range of 120–200 nm are not readily taken up by the cells of RES and bypass liver and spleen filtration [11].
2. Controlled release of the incorporated drug can be achieved up to several weeks [12]. Further by coating with or attaching ligands to SLN, there is an increased scope of drug targeting.
3. SLN formulations stable for even 3 years have been developed this is of paramount importance with respect to other colloidal carrier systems [13].
4. High drug payload.
5. Excellent reproducibility with the cost-effective high-pressure homogenization as a preparation procedure.
6. Feasibility of incorporating both hydrophilic and hydrophobic drugs.
7. The lipids are biodegradable, biocompatible, nontoxic, and safe.
8. Avoidance of organic solvents.
9. Feasible large scale production and sterilization [14].
10. Ability to prevent chemical, photochemical, or oxidative degradation of the drug.

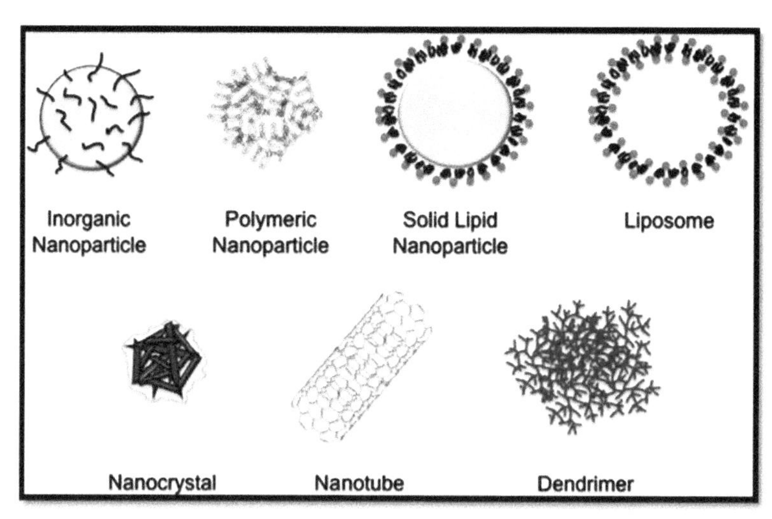

FIGURE 3.2 Various types of nanoparticles used in biomedical research and drug delivery. *Source*: Reprinted with permission from Ref. [xx]. © 2009 Elsevier.

TABLE 3.2 Comparison of SLN With Other Colloidal Carriers (Adapted from [15])

SLN		**Liposome**	**Nanosuspension**	**Polymeric Nanoparticles**	**Emulsions**
Possible		Possible	Possible	Possible	Possible
Parenteral delivery					
Oral delivery	Possible	Not possible	Possible	Possible	Possible
Ability to deliver Hydrophobic and hydrophilic drugs	Yes	Yes	Only Hydrophobic drugs	Yes	Yes
Physical stability	+++	+	+++	+++	++
Biological stability	++	+	++	+++	++
Biocompatibility	+++	+++	++	++	+++
Ease of sterilization	++	+	++	++	++
Drug targeting	++	++	+	++	+
Drug loading	Low to moderate	Low to moderate	High	Moderate	High
Ease of commercialization	++	+	++	+	++
Ability to deliver biotechnological therapeutics	++	++	_	++	+
Regulatory acceptance of excipients	++	++	+++	+	+++

Source: Reprinted/adapted with permission from Ref. [39]. © 2007 Elsevier.

3.5.2 GENERAL INGREDIENTS

General fixings incorporate strong lipid, emulsifiers, and water. The term lipid is utilized in the more extensive sense and incorporates triglycerides (e.g., tristearin), halfway glycerides (for example, imwitor), unsaturated fats (for example, stearic corrosive), steroids (for example, cholesterol), and waxes (for example, cetyl palmitate). All classes of emulsifiers (regarding charge and subatomic weight) have been utilized to settle the lipid scatterings. It has been discovered that a mix of emulsifiers may forestall molecule accumulation all the more proficiently (Table 3.3.).

TABLE 3.3 Lipids Used for Preparation of SLN [16]

Triglycerides	Hard fat	Waxes	Fatty acids
Tricaprin	Witepsol® W 35	Cetyl palmitate	Stearic acid
Trilaurin	Witepsol® H35		Palmitic acid
Trimyristin	Witepsol® H42		Decanoic acid
Tripalmitin Tristearin	Witepsol® E85	Partial glycerides Glyceryl monostearate (Imwitor® 900) Glyceryl behenate (Compritol® 888)	Behenic acid

TABLE 3.4 Emulsifiers/Coemulsifiers Used in Preparation of SLN

Soyabean lecithin (Lipoid® S 75, Lipoid® S 100)	Poloxamers (182,188,407) Poloxamine 908	Taurocholic acid sodium salt Taurodeoxycholic acid sodium salt
Egg Lecithin	Polysorbates (20,60,80)	Butanol
Phosphatidylcholine (Epikuron® 170, Epikuron 200)	Sodium cholate Sodium glycocholate	Dioctyl sodium Sulfosuccinate

3.5.3 PREPARATION OF SLN

3.5.3.1 HIGH SHEAR HOMOGENIZATION AND ULTRASOUND

The SLN was created from lipid microparticles delivered by shower solidifying followed by lipid nanopellets, delivered by rapid mixing or sonication. An incredible bit of leeway of this technique is the way that the gear is regular in each lab and the generation should effortlessly be possible. The issue of rapid mixing was a more extensive molecule size dispersion running into the micrometer extend. This leads to physical insecurities, for example, molecule development upon capacity. This could be improved by higher surfactant fixations, which all together may be corresponded with toxicological issues after parenteral organization. A further disservice is potential metal pollution because of ultrasonication.

For the most part, fast mixing and ultrasonication are consolidated and performed at raised temperatures for quite a while. Very tight and physically stable dispersions can be accomplished, anyway, the lipid focus is low (<1%) and the surfactant fixation is equivalently high [17].

3.5.3.2 HIGH-PRESSURE HOMOGENIZATION

High-weight homogenization (HPH) has developed as a dependable and amazing strategy for the planning of SLN. Homogenizers of various sizes are economically accessible from a few makers. Rather than different systems, scaling up speaks to no issue as a rule. High weight homogenizers push a fluid with high weight (100–2000 bar) through a limited hole (scarcely any microns). The liquid quickens on an extremely short separation to exceptionally high speed (more than 1000 km/h). Exceptionally high shear pressure and cavitation powers disturb the particles down to the submicron run. Commonplace lipid substances are in the range 5%–10% and speak to no issue to the homogenizer.

Two general methodologies of the homogenization step, the hot and cold homogenization can be utilized for the creation of SLN. In the two cases, a preliminary advance includes the medication consolidation into the mass lipid by dissolving or scattering the medication in lipid soften.

Hot homogenization: It is done at temperatures over the liquefying purpose of the lipid and can in this way be viewed as the homogenization of an emulsion. A pre-emulsion of the medication stacked lipid liquefies and the fluid emulsifier stage (same temperature) is acquired by high shear blending gadget. HPH of the pre-emulsion is completed at temperatures over the softening purpose of the lipid. Higher temperature brings about lower molecule measures because of diminished consistency of the internal stage. The homogenization step can be rehashed a few times. Much of the time, 3–5 homogenization cycles at 500–1500 bar are adequate. The essential result of the hot homogenization is a nanoemulsion because of the fluid condition of the lipid. Strong particles are framed by the accompanying cooling of the example to room temperature or temperatures underneath.

Cold homogenization: Cool homogenization is a reasonable procedure for preparing temperature labile or hydrophilic medications. Here, lipid and medication are liquefied together and afterward quickly ground under fluid nitrogen framing strong lipid microparticles. A presuspension is framed by the fast mixing of the particles in a cool watery surfactant arrangement. This presuspension is then homogenized at or underneath room temperature shaping SLN, NLC (nanostructured lipid transporters), or LDC (lipid tranquilize conjugates), the homogenizing conditions are commonly five cycles at 500 bar (Figure 3.3).

3.6 SOLVENT EMULSIFICATION /EVAPORATION

Nanoparticle dispersions are set up by precipitation in o/w emulsions. The lipophilic material is broken down in a water-miscible natural dissolvable (e.g., cyclohexane) that is emulsified in a fluid stage. Endless supply of the dissolvable nanoparticle scattering is framed by precipitation of the lipid in the fluid medium. The mean molecule size relies upon the grouping of the lipid in natural stage. Small particles can be acquired with low-fat burdens (5% w/w) identified with the natural dissolvable. The upside of this technique over

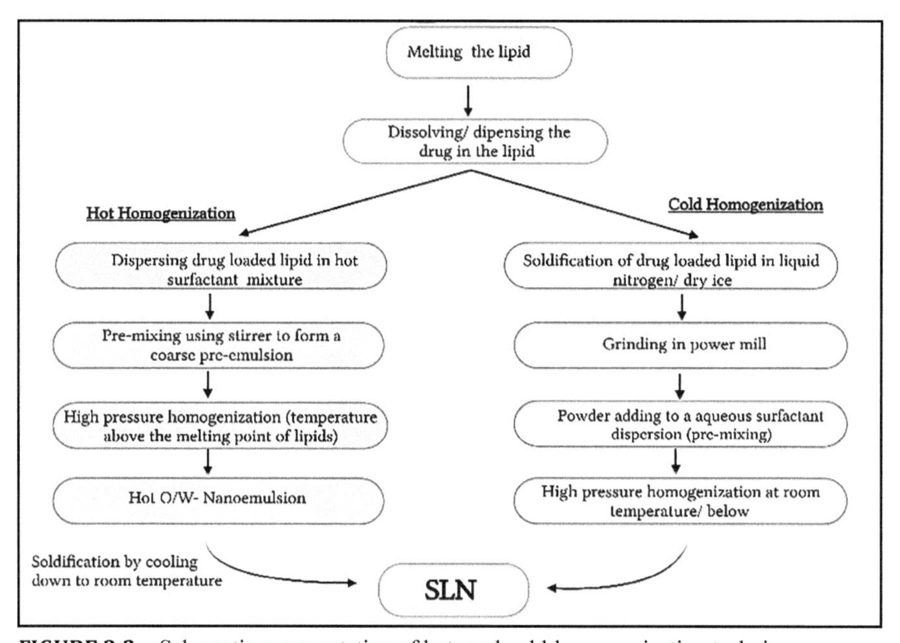

FIGURE 3.3 Schematic representation of hot- and cold-homogenization techniques.

the chilly homogenization process is the evasion of any warm pressure. An unmistakable weakness is the utilization of natural solvents. Additionally, these scatterings are commonly very weakened, on account of the restricted dissolvability of lipid in the natural material. Regularly, lipid fixations in the last SLN scattering range around 0.1 g/L, in this manner, the molecule focus must be expanded by methods for, for example, ultrafiltration or vanishing [17].

3.7 MICROEMULSION METHOD

This method depends on the weakening of microemulsions. They are made by blending an optically straightforward blend at 65–70 °C, which is regularly made out of a low liquefying unsaturated fat (stearic corrosive), an emulsifier (polysorbate 20, polysorbate 60, soy phosphatidylcholine, and sodium taurodeoxycholate), coemulsifiers (sodium monooctylphosphate), and water. The hot microemulsion is scattered in cool water (2–3 °C) under blending. Run of the mill volume proportions of the hot microemulsion to cold water are in the scope of 1:25–1:50. The weakening procedure is basically controlled by the creation of the microemulsion [18].

Test factors, for example, microemulsion creation, scattering gadget, temperature, and lyophilisation on size and structure of the acquired SLN have been examined seriously. It must be commented basically that the expulsion of abundance water from the readied SLN scattering is a troublesome assignment with respect to the molecule size. Likewise, high groupings of surfactants and co-surfactants (for example, butanol) are essential for defining purposes, anyway less attractive regarding administrative purposes, and application [17]. The weakening procedure is basically controlled by the arrangement of microemulsion. As per writing, the bead structure is as of now contained in the microemulsion, and accordingly, no vitality is required to accomplish submicron molecule sizes.

The molecule size is basically dictated by the speed of the circulation forms. Nanoparticles were delivered uniquely with solvents that disseminate quickly into the watery stage (e.g., CH3)2CO), while bigger molecule sizes were gotten with increasingly lipophilic solvents. Because of the weakening advance, reachable lipid substances are impressively lower contrasted and the HPH-based definitions.

3.8 W/O/W DOUBLE EMULSION METHOD

As of late, a novel strategy dependent on dissolvable emulsification–dissipation for the planning of SLN stacked with hydrophilic medications has been acquainted with mainstream researchers. Here, the hydrophilic medication is exemplified, alongside a stabilizer to forestall tranquilize dividing to the outer water stage during dissolvable vanishing, in the inward water period of a w/o/w twofold emulsion. This method has been utilized for the arrangement of sodium cromoglycate-containing SLN, in any case, the normal size was in the micrometer extend so the term "lipospheres" in the sense as a term for nanoparticles is not utilized accurately for these particles [19].

3.9 SLN PREPARATION BY USING SUPERCRITICAL FLUID

This is a generally new method for SLN generation and has the benefit of dissolvable less preparation. There are a few varieties in this stage of innovation for powder and nanoparticle arrangement. SLN can be set up by the fast development of supercritical carbon dioxide arrangements (RESS) strategy. Carbon dioxide (99.99%) was an acceptable decision as a dissolvable for this technique.

3.10 SPRAY DRYING METHOD

It's an elective system to lyophilization so as to change a fluid SLN scattering into a medication item. It's a less expensive strategy than lyophilization. This technique cause molecule total because of high temperature, shear powers, and fractional dissolving of the molecule. Freitas et al. [17] prescribe the utilization of lipid with softening point >70 °C for shower drying. The best outcome was acquired with an SLN grouping of 1% in an answer of trehalose in water or 20% trehalose in ethanol–water blends (10/90 v/v).

3.11 DRUG INCORPORATION AND LOADING CAPACITY OF SLN

The basic elements for SLN are lipids, and a solitary or blend of emulsifiers. Contingent upon the lipid (triglycerides, unsaturated fat, steroids, waxes, and so forth), emulsifier (anionic, cationic, or nonionic), and technique for planning the molecule size, stacking limit, and the size appropriation of SLNs is found to differ.

Factors affecting the loading capacity of a drug in the lipid are as follows:

1. Solubility of the drug in the lipid melt.

2. Miscibility of the lipid melt and the drug melt.
3. Chemical and physical structure of solid lipid matrix.
4. Polymorphic state of lipid material.

Specifically, there is an opposite connection between solvency of the medication and stacking limit. Upgrade in watery solvency of medication drives lower to ensnarement productivity. The essential to acquire adequate stacking limit is an adequately high dissolvability of the medication in the lipid liquefy. Commonly, the solvency ought to be higher in the softened state than that required in the strong state in light of the fact that the dissolvability diminishes when the liquefy cools and may even be lower in the strong lipid. To improve the dissolvability in the lipid soften, one can include solubilizers. Furthermore, the nearness of mono- and di-glycerides in the lipid utilized as the framework material additionally advances sedate solubilization. The substance idea of the lipid is additionally significant in light of the fact that lipids that structure exceptionally crystalline particles with an ideal grid (e.g.,, monoacid triglycerides) lead to medicate removal.

Increasingly intricate lipids being blends of mono-, di-, and tri-glycerides and furthermore containing unsaturated fats of various chain lengths structure less ideal gems with numerous flaws offering space to suit the medications. Polymorphic structure is a significant parameter deciding medication fuse. Crystallization of the lipid in nanoparticles is unique in relation to the mass material; lipid nanoparticles recrystallize at any rate mostly in the α-structure, while mass lipids tend to recrystallize especially in the β′-change and change quickly to the β-structure. With the expanding development of the more steady alterations, the cross-section is getting increasingly great and the quantity of blemishes diminishes, which implies arrangement of β′/β-adjustment advances sedate removal. All in all, the change is slower for long-chain than for short-chain triglycerides. An ideal SLN transporter can deliver in a controlled manner when a specific portion of β′-structure can be made and protected during the capacity time. By doing this typical SLN bearer changes to an astute medication conveyance framework by having a worked-in activating component to start the change from β′- to β-structures and thusly controlled medication discharge.

The consolidation of the helpful specialist in SLN can be depicted by three models, namely, homogenous network model or strong arrangement (in which medication is either molecularly scattered or present as nebulous bunches in the lipid framework), sedate advanced shell model (external lipid shell containing drug with lipid center), and medication improved center model (tranquilize center encompassed by lipid layer or supply type framework) [16].

On account of the strong arrangement model, the medication is molecularly scattered in the lipid lattice when the particles are created by the chilly homogenization system and utilizing no surfactant or no medication solubilizing surfactant. The medication has emphatically articulated associations with the lipid.

As per the medication improved shell model of medication fuse, a strong lipid center structures when the recrystallization temperature of the lipid is come to. On diminishing the temperature of the scattering, the medication gathers in the still fluid external shell of the SLN.

As per the medication improved center model of medication fuse, cooling the nanoemulsion prompts a super immersion of the medication which is broken down in the lipid dissolve at or near its immersion solvency and the medication hastens preceding lipid recrystallization. Further cooling at long last prompts the recrystallization of the lipid encompassing the medication as a layer.

3.11.1 DRUG RELEASE FROM SLNS

The general principles of drug release from lipid nanoparticles are as follows:

1. There is an inverse relationship between drug release and the partition coefficient of the drug.
2. Higher surface area due to smaller particle size in nanometer range gives higher drug release.
3. Slow drug release can be achieved when the drug is homogeneously dispersed in the lipid matrix. It depends on the type and drug entrapment model of SLN.
4. Crystallization behavior of the lipid carrier and high mobility of the drug lead to fast drug release. There is an inverse relationship between crystallization degree and mobility of the drug.

The medication fuse model of SLN is vital to the medication discharge design. It is identified with the arrangement and creation technique for SLN as clarified previously. For example, the medication stacked lipid stage remains fundamentally in the strong state on account of generation by cool homogenization method. The strong arrangement medicates consolidation model shows up here. Medication discharge is drawn out more than a little while since the versatility of the medication molecularly scattered in colloidal particles is constrained.

Quick starting medication discharge (burst impact) exists in the initial 5 min in the medication improved shell model (i.e., about 100% inside <5 min) because of the external layer of the particles because of the huge surface zone of medication statement on the molecule surface. The burst discharge is diminished with expanding molecule size and delayed discharge could be acquired when the particles were adequately enormous, that is, lipid microparticles. The kind of surfactant and its focus, which will interface with the external shell and influence its structure, ought to be noted as the other significant factor, on the grounds that a low surfactant fixation prompts a negligible burst and delayed medication discharge. In the medication improved center model, the medication discharge is film controlled and is represented by the Fick law dissemination since the lipid encompasses the medication as a layer.

The molecule size that influences sedate discharge rate straightforwardly relies upon different parameters, for example, organization of SLN definition, (for example, surfactant/surfactant blend, measure of medication consolidated, auxiliary properties of lipid and medication), generation strategies and conditions, (for example, time, creation temperature, gear, sanitization, and lyophilization). Every one of those parameters has been broadly examined and information has been accounted for in the writing for a considerable length of time. Also, surface modifiers to decrease phagocytic take-up, for example, polyethylene oxide and PEG may change the molecule size. The impact of surface modifiers on molecule size and on sedate discharge rate is examined in up and coming segments.

3.12 CHARACTERIZATION OF SLN

3.12.1 PARTICLE SIZE

1. *Photon correlation spectroscopy* (technique based on dynamic laser light scattering due to Brownian motion of particles in solution/suspension, suitable for measurement of particles in the range of 3 nm to 3 mm. The photon correlation spectroscopy (hydrodynamic diameters) diameters are based on the amount of light scattering from the nanoparticles. The nanoparticles are usually polydisperse in nature and polydispersity index (PI) gives a measure of the size distribution of the nanoparticle population. Theoretically, the PI for a monodisperse system is zero, for polydisperse systems the PI should be <0.5 (PI greater

than 0.5 indicates a very broad size distribution).

2. *Transmission electron microscopy* (uses electron transmitted through the specimen to determine the overall shape and morphology, that is, both particle size as well as distribution.)
3. *Scanning electron microscopy* (uses electron transmitted from the specimen to determine the overall shape and morphology, that is, both particle size as well as distribution).
4. *Scanned probe microscopes.*
5. *Polarization intensity differential scattering* (measures the particle size as low to 40 nm).
6. *Field flow fractionation* (based on the elution of the smaller particles when placed on a parabolic flow profile. All the eluted fractions are analyzed by multiangle light scattering (MALS) where a photometer records the scattering signal of the particles and calculates size weighed radius. MALS allows to measure particle radius from 10 nm to1 μm.)
7. *X-ray diffraction (XRD)* (a useful technique to exclude aggregate of >1 μm and substantial polymorphic β1 transition form to stable; thus help in characterizing the crystalline nature of the compound and determine the polymorphic shifts present).
8. *Freeze-fracture electron microscopy* (solid spherical structure with no internal lamellae).

3.12.2 MOLECULAR WEIGHT

1. Gel chromatography
2. Static secondary-ion mass spectrometry
3. Atomic force microscopy (to determine the original unaltered shape and surface properties of the particles).

3.12.3 SURFACE ELEMENT ANALYSIS

1. X-ray photoelectron spectroscopy for chemical analysis (ESCA)
2. Electrophoresis
3. Laser Doppler anemomometry
4. Amplitude-weighted phase structure determination
5. XRD
6. Differential scanning calorimetry (DSC) (yields information on melting behavior and crystallization behavior of solid and liquid constituents of the particles).

3.12.4 DENSITY

1. Helium compression pychnometry.
2. Contact angle measurement.
3. Hydrophobic interaction chromatography.

3.12.5 MOLECULAR ANALYSIS

1. ^{1}H *NMR* (mobility of molecules inside the SLN).
2. *Infrared analysis* (structural property of lipids).

3.13 APPLICATIONS OF SLN

SLNs have a superior solidness and simplicity of upgradability to the creation scale when contrasted with liposomes. This property might be significant for some methods of focusing on. SLNs structure the premise of colloidal medication conveyance frameworks, which are biodegradable and equipped for being put away for in any event 1 year. They can convey medications to the liver in vivo and in vitro to cells that are effectively phagocytic. There are a few potential utilizations of SLNs some of which are given beneath.

3.13.1 SLNS AS QUALITY VECTOR BEARER

SLN can be utilized in the quality vector definition. In one work, the quality exchange was improved by the fuse of a polar HIV-1 HAT peptide (TAT 2) into an SLN quality vector. There are a few ongoing reports of SLN conveying hereditary/peptide materials, for example, DNA, plasmid DNA, and other nucleic acids [21]. The lipid-nucleic corrosive nanoparticles were set up from a fluid nanophase containing water and a water-miscible natural dissolvable where both lipid and DNA are independently disintegrated by expelling the natural dissolvable, stable, and homogeneously measured lipid-nucleic corrosive nanoparticle (70–100 nm) were framed. It is called genospheres. It is focused on explicit by the inclusion of a counteracting agent lipo polymer conjugated in the molecule.

3.13.2 SLNS FOR TOPICAL USE

SLNs and NLCs have been utilized for topical application for different medications, for example, tropolide, imidazole antifungals, enemies of malignant growths, nutrient An, isotretinoin, ketoconazole, DNA, flurbiprofen, and glucocorticoids. The infiltration of podophyllotoxin-SLN into stratum corneum alongside skin surface leads to the epidermal focusing on. By utilizing glyceryl behenate, nutrient A-stacked nanoparticles can be readied. The techniques are valuable for the improvement of an entrance with supported discharge. The isotretinoin-stacked lipid nanoparticles were detailed for the topical conveyance of medication. The soyabean lecithin and Tween 80 were utilized for the hot homogenization technique for this. The strategy is valuable on account of the expansion of aggregate take-up of isotretinoin in the skin.

3.13.3 SLNS AS COSMECEUTICALS

The SLNs have been applied in the planning of sunscreens and as a functioning bearer operator for subatomic sunscreens and UV blockers. The in vivo examination demonstrated that skin hydration will be expanded by 31% following a month by an expansion of 4% SLN to an ordinary cream. SLN and NLCs have end up being controlled discharge imaginative occlusive topicals. Better restriction has been accomplished for nutrient in upper layers of skin with glyceryl behenate SLNs contrasted with traditional plans.

3.13.4 SLNS FOR POTENTIAL FARMING APPLICATION

Fundamental oil extricated from *Artemisia arboreseens* L. when consolidated in SLN had the option to lessen the quick dissipation contrasted and emulsions and the frameworks have been utilized in agribusiness as an appropriate bearer of naturally safe pesticides. The SLN was set up here by utilizing compritol 888 ATO as lipid.

3.13.5 LNS AS FOCUSED ON TRANSPORTER FOR ANTICANCER MEDICATION TO STRONG TUMORS

SLNs have been accounted for to be valuable as medication transporters to treat neoplasms. Tamoxifen, an anticancer medication joined in SLN to drag out the arrival of medication after i.v. organization in bosom disease and to improve the penetrability and maintenance impact. Tumor focusing on has been accomplished with SLNs stacked with drugs like methotrexate and camptothecin.

3.13.6 SLNS IN BOSOM MALIGNANCY AND LYMPH HUB METASTASES

Mitoxantrone-stacked SLN nearby infusions were defined to decrease the lethality and improve the wellbeing and bioavailability of medication. Adequacy of doxorubicin (Dox) has been accounted for to be improved by consolidation in SLNs. In the strategy, the Dox was complexed with soybean-oil-based anionic polymer and scattered together with a lipid in water to shape Dox-stacked SLNs. The framework is upgraded its viability and decreased bosom malignant growth cells.

3.13.7 ORAL SLNS IN HOSTILE TO TUBERCULAR CHEMOTHERAPY

Against tubercular medications, for example, rifampicin, isonizide, pyrazinamide-stacked SLN frameworks had the option to diminish the dosing recurrence and improve tolerant consistence. By utilizing the emulsion dissolvable dispersion strategy, this enemy of tubercular medication stacked SLNs was readied. The nebulization in the creature by consolidating the above medication in SLN likewise detailed for improving the bioavailability of the medication.

3.14 STEALTH NANOPARTICLES

3.14.1 AIM AND OBJECTIVE OF THIS STUDY

Olanzapine (OL) is an atypical antipsychotic that has a place with Thioenobenzodiazepine class utilized orally in the treatment of schizophrenia, intense blended or hyper scenes in bipolar issue, and intra-solid for control of unsettling and upset conduct in schizophrenia or madness. Conveying this medication to the cerebrum gangs an incredible test in the intersection of the BBB, that is, BBB. In the oral course, it experiences broad first disregard digestion with 40% of the medication being processed before arriving at the foundational dissemination. Additionally, it has poor watery solvency (BCS Class II medicate). It is related to extreme portion symptoms which incorporate medication—incited parkinsonism, intense dystonia response, akathisia, tardive dysakinesia, and tardive dystonia. These reactions are seen at doses that yield a gainful impact on the indications of the illnesses. The seriousness of the unfriendly occasions and additionally absence of adequacy in an impressive number of patients often brings about poor patient consistence or end of treatment. A considerable lot of the patients are not in full control of their intellectual capacities. A dose type of OL having delayed movement and along these lines requiring less regular organization is exceptionally fitting. This is on the grounds that such a dose structure would limit inconveniences caused because of patients absent or neglecting to take a portion. As of late, SLNs have been misused as plausible potential outcomes as transporters for tissue focusing on. Likewise, SLN is made out of a strong lipid network which can give a successful way to control the arrival of medication.

The principle point of the work is to plan and create SLN containing OL utilizing lipids like unsaturated fats glyceryl monostearate and glyceryl tripalmitin as medication bearers over the BBB to improve the medication bioavailability in the cerebrum.

3.15 METHODOLOGY

3.15.1 PREFORMULATION STUDIES

Predetailing might be portrayed as a period of the innovative work process where the definition researcher describes the physical, compound, and mechanical properties of another medication substance, so as to create steady, protected, and powerful dose structures. In a perfect world, the preplan stage starts right off the bat in the disclosure procedure with the end goal that suitable physical substance information is accessible to help in the determination of new synthetic elements that enter the advancement procedure. During this assessment, conceivable cooperation with different inactive fixings planned for use in definite dose structure is likewise considered.

3.15.2 SOLUBILITY STUDIES

The dissolvability of the OL was resolved in different solvents by including an abundance measure of medication to 10 mL of solvents in cone-shaped flagons. The carafes were kept at 25 ± 0.5 °C in an isothermal shaker for 72 h to arrive at harmony. The equilibrated tests were expelled from the shaker and centrifuged at 4000 rpm for 15 min. The supernatant was taken and sifted through a 0.45-µm film channel. The convergence of OL was resolved in the supernatant after reasonable weakening by utilizing HPLC at 258 nm.

3.15.3 COMPATIBILITY STUDY

I.R spectroscopy can be utilized to research and anticipate any physio-chemical communication between various segments in a detailing and in this manner it very well may be applied to the choice of reasonable concoction good excipients while choosing the fixings, we would pick, those which are steady, perfect, and restoratively adequate.

Infrared spectra coordinating methodology was utilized for the location of any conceivable compound cooperation between the medication, lipid, and surfactants. A physical blend of medication, lipid, and surfactants was arranged and blended in with the appropriate amount of potassium bromide. This blend was packed to shape a straight-forward pellet utilizing a water-driven press at 15 tons pressure. It was examined from 4000 to 400 cm^{-1} out of an FTIR spectrophotometer (FTIR 8400 S, Shimadzu). The IR range of the physical blend was contrasted and those of unadulterated medication, lipid and surfactants, and pinnacle coordinating were done to identify any appearance or vanishing of pinnacles.

3.15.3.1 DEVELOPMENT OF CALIBRATION CURVE

A stock arrangement of OL was set up by dissolving 100 mg of medication in 10 mL of 0.1 N HCl and made up to 100 mL (with various cradles, namely, 0.1 N HCl and pH 7.4) to give a stock arrangement of fixation 1 mg/mL. From this stock arrangement

2–10 μg/mL weakenings were readied. The μ_{max} of the medication was controlled by checking one of the weakenings somewhere in the range of 400 and 200 nm utilizing an UV-obvious spectrophotometer. At this wavelength, the absorbance of the various arrangements was estimated against a clear. Standard bend among fixation and absorbance was plotted.

3.15.4 PARTITION COEFFICIENT STUDIES

Parceling conduct of OL was resolved in various lipids, namely, glyceryl tripalmitate and glyceryl monostearate [22]. Ten milligrams of OL were scattered in a blend of softened lipid (1 g) and 1 mL of hot phosphate cushion pH 7.4 (PB) and shaken for 30 min in a water shower shaker kept up at 10 °C over the dissolving purpose of lipid. The fluid period of the above blend was isolated from the lipid by centrifugation at 10,000 rpm for 20 min. The reasonable supernatant acquired was appropriately weakened with 0.1 N HCl and OL content was resolved in UV–visible spectrophotometer at 258 nm against dissolvable clear. The parcel coefficient was determined as

$$PC = (C_{OLI} - C_{OLA})\, C_{OLA}$$

where C_{OLI} = the initial amount of OL added (10 mg) C_{OLA} = the concentration of OL in pH 7.4 PB

3.16 PREPARATION OF SLNS

SLNs were prepared based on the principal of "emulsion solvent diffusion" technique [23]. Briefly, the required quantity of the drug and lipid was put in a mixture of acetone and ethanol and heated to 60–70 °C in a water bath. The resulting solution was poured into 50 mL of 1%w/v aqueous tween 80 under the mechanical stirring drop by drop with a syringe and is stirred for the required time. The SLNs thus formed is a centrifuge and freeze-dried. Stearyl amine was used as a positive charge inducer and added along with the lipid and drug (Figure 3.4).

3.17 STUDY ON THE EFFECT OF FORMULATION PROCESS VARIABLES

The effect of formulation/process variables such as stirring time, stirring speed, surfactant concentration, lipid concentration on the particle size was studied using Box–Behnken design. To investigate the effect of formulation/process variables, a 4-factor, 2-level Box–Behnken design was used to optimize nanoparticle formulation with stirring time, stirring speed, surfactant concentration, lipid concentration as independent variables with low- and high-concentration values each, as described in Table 3.5. The range of concentration was established according to the previous studies for the development

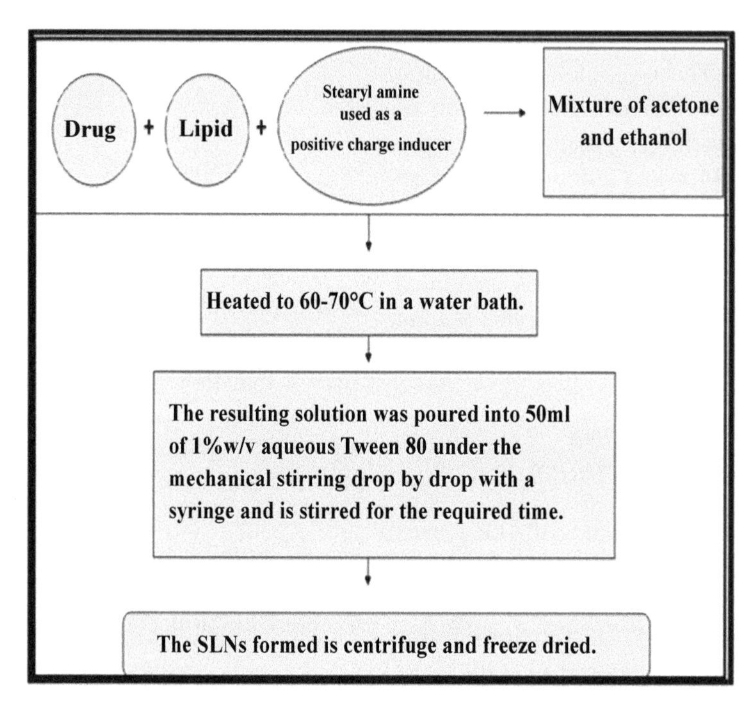

FIGURE 3.4 Preparation of SL.

Source: Reprinted from Ref. [38]. https://creativecommons.org/licenses/by/4.0/

of SLN. The dependent variables were particle size, PI, zeta potential. Design–Expert software (v.7.1.5 Stat-Ease Inc., Minneapolis, USA) was used for the generation and evaluation of the statistical experimental design. The concentrations of the formulation parameters and the corresponding observations for these dependent variables are presented in Table 3.5 and cComposition and Oobserved rResponses in Box–Behnken dDesign for GMS and TP were presented in Table 3.6 and Table 3.7, respectively.

TABLE 3.5 Variables in Box–Behnken Design

Factor Independent Variables	-1	Levels Used	1
Lipid concentration	30		60
Surfactant concentration	1.5% w/v		2%w/v
Stirring speed	2500 rpm		5000 rpm
Stirring time	2 h		4 h
Dependent variables Particle size (nm) Polydispersity index Zeta potential (mV) Entrapment efficiency (%)	Constraints Minimize Minimize >30 mV Maximize		

TABLE 3.6 Composition and Observed Responses in Box–Behnken Design for GMS

STD	RUN	Lipid %w/v	Surfactant Conc. %w/v	Stirring time (rpm)	Stirring Time (h)	Particle size nm	Zeta Potential (mV)	Polydispersity index
6	1	60	1.5	5000	2	349	39.71	0.741
12	2	60	2	2500	4	340.1	36.15	0.730
5	3	30	1.5	5000	2	350.1	44.11	0.801
4	4	60	2	2500	2	540	42.16	0.733
14	5	60	1.5	5000	4	262.5	39.76	0.726
1	6	30	1.5	2500	2	429	37.02	0.801
2	7	60	1.5	2500	2	433.3	38.54	0.910
13	8	30	1.5	5000	4	165.1	35.29	0.742
10	9	60	1.5	2500	4	335	45.09	0.780
11	10	30	2	2500	4	341	38.71	0.90
3	11	30	2	2500	2	541	48.90	0.81
8	12	60	2	5000	2	275	37.91	0.780
15	13	30	2	5000	4	270	38.10	0.770
16	14	60	2	5000	4	268	36.91	0.910
7	15	30	2	5000	2	230.7	35.04	0.618
9	16	30	1.5	2500	4	276	36.19	0.761

TABLE 3.7 Composition and Observed Responses in Box–Behnken Design for TP

STD	RUN	Lipid %w/v	Surfactant Conc. %w/v	Stirring Time (rpm)	Stirring time (h)	Particle size nm	Zeta Potential (MV)	Polydispersity index
6	1	60	1.5	5000	2	321	44.29	0.510
12	2	60	2	2500	4	239	36.09	0.407
5	3	30	1.5	5000	2	308	40.21	0.396
4	4	60	2	2500	2	245	36.99	0.660
14	5	60	1.5	5000	4	186.8	41.50	0.158
1	6	30	1.5	2500	2	344	43.91	0.416
2	7	60	1.5	2500	2	362	44.06	0.780
13	8	30	1.5	5000	4	110.5	66.50	0.340
10	9	60	1.5	2500	4	231	53.09	0.791
11	10	30	2	2500	4	226	37.86	0.364
3	11	30	2	2500	2	230.7	45.04	0.618
8	12	60	2	5000	2	340	44.26	0.540
15	13	30	2	5000	4	361.1	52.21	0.471
16	14	60	2	5000	4	212.3	43.21	0.398
7	15	30	2	5000	2	351	39.50	0.658
9	16	30	1.5	2500	4	241	35.01	0.406

3.18 EVALUATION OF SLPS

3.18.1 PARTICLE SIZE AND ZETA POTENTIAL

Particle size and zeta potential of the SLNs were measured by photon correlation spectroscopy using a Malvern Zetasizer Nano ZS90 (Malvern Instruments, Worcestershire, UK), which works on the Mie theory. All size and zeta potential measurements were carried out at 25 °C using disposable polystyrene cells and disposable plain folded capillary zeta cells, respectively, after appropriate dilution with the original dispersion preparation medium. In order to investigate the effect of stearyl amine on zeta potential two batches were prepared with and without stearylamine and their zeta potential was measured.

3.18.2 POLYDISPERSITY INDEX

Polydispersity was determined according to the following equation:

$$\text{Polydispersity} = D\,(0.9) - D\,(0.1) / D(0.5)$$

where

D(0.9) corresponds to particle size immediately above 90% of the sample. *D*(0.5) corresponds to particle size immediately above 50% of the sample.

D(0.1) corresponds to particle size immediately above 10% of the sample [24].

3.18.3 DIFFERENTIAL SCANNING COLORIMETRY

A differential scanning colorimetry DSC Q200 V24.4 Build 116 was used. The instrument was calibrated with indium for melting point and heat of infusion. A heating rate of 20 °C/min was employed throughout the analysis in the 25 °C–200 °C. Standard aluminum sample pans were used for all the samples; an empty pan was used as reference. The thermal behavior was studied under a nitrogen purge; triplicate run was carried out on each sample to check reproducibility.

3.18.4 ENTRAPMENT EFFICIENCY

SLNs-entrapped OL were separated from the free drug by dialysis method. The prepared SLNs were filled into dialysis bags and the free OL dialyzed for 24 h into 50 mL of phosphate buffer 7.4 saline. The absorbance (*A*) of the dialysate was measured at 258 nm against blank phosphate buffer 7.4 saline, and the absorbance (*A*0) of the corresponding blank phosphate buffer 7.4 saline was measured under the same condition. The concentration of free OL could be obtained from the absorbance difference (_*A*=*A* − *A*0) based on the standard curve. A standard curve was made by measuring absorbance at 258 nm for known concentrations of OL solution. The eEntrapment eEfficiency of SLNs is represented byin Table 3.8. The entrapment efficiency of the drug was defined as the ratio of the mass of SLNs-associated drug to the total mass of drug.

3.19 EXTERNAL MORPHOLOGICAL STUDY (SCANNING ELECTRONMICROSCOPY)

The external morphology of nanoparticles was determined using scanning electron microscopy (SEM). Samples were diluted with ultrapurified water to obtain a suitable concentration. Then the samples were spread on a sample holder and dried using a vacuum. They were subsequently coated with gold (JFC 1200 fine coater, Japan) and examined by an SEM.

3.20 IN VITRO RELEASE STUDIES

The release of OL from the SLNs was studied under sink conditions. Tripalmitate-SLN (TP-SLN) and glyceryl monostearate (GMS-SLN) which showed higher drug content and entrapment efficiency (8th run of Table 3.8) were evaluated for in vitro release. Five milliliters of SLNs equivalent to 1 mg were put in dialysis bags (MWCO 12000, HiMedia). The dialysis bags were placed in 50 mL of dissolution medium and stirred under magnetic stirring at 37 °C. Aliquots of the dissolution medium were withdrawn at each time interval and the same volume of fresh dissolution medium was added to maintain a constant volume. Samples withdrawn from pH 7.4 phosphate buffer saline were analyzed for OL content spectrophotometrically at 258 nm against solvent blank (Hao et al. 2002).

3.21 RELEASE KINETICS

In-vitro dissolution has been recognized as an important element in drug development. Under certain conditions, it can be used as a surrogate for the assessment of bioequivalence. Several theories/ kinetic models describe drug dissolution from immediate and modified release dosage forms. There are several models to represent the drug dissolution profiles where f_t is the function of t (time) related to the amount of drug dissolved from the pharmaceutical dosage system. To compare dissolution profiles between two drug products model dependent (curve fitting), statistical analysis and model-independent methods can be used [25].

In order to elucidate the mode and mechanism of drug release, the in-vitro data were transformed and interpreted at a graphical interface constructed using various kinetic models. The zero-order release equation (3.1) describes the drug

TABLE 3.8 Entrapment Efficiency of SLNs

Std	Run	Lipid %w/v	Surfactant Conc. %w/v	Stirring Time (rpm)	Stirring Time (h)	Particle Size (nm)	Zeta Potential (mV)	Polydispersity Index	Entrapment Efficiency (%)
13	8a	30	1.5	5000	4	165.1	10.1	0.742	67.2
13	8b	30	1.5	5000	4	110.5	13.93	0.340	96.3

dissolution of several types of modified release pharmaceutical dosage forms, as in the case of transdermal systems, matrix tablets with low soluble drugs, coated forms, osmotic systems, etc., where the drug release is independent of concentration.

$$Q_t = Q_o + K_o t \quad (3.1)$$

where Q_t is the amount of drug released in time t, Q_o is the initial amount of the drug in the solution, and K_o is the zero-order release constant

The first-order equation (3.2) describes the release from the system where the release is concentration dependent, for example, pharmaceutical dosage forms containing water-soluble drugs in porous matrices.

$$\log Q_t = \log Q_o + K_1 t/2.303 \quad (3.2)$$

where Q_t is the amount of drug released in time t, Q is the initial amount of drug in the solution, and K is the first s order release constant.

Higuchi described the release of drug from the insoluble matrix as a square root of time

$$Q_t = K_H \sqrt{t} \quad (3.3)$$

where Q_t is the amount of drug released in time t, K_H is Higuchi's dissolution constant. Weibull equation can be applied to almost all kinds of dissolution curves and is given as:

$$\log[-\ln(1 - Q_t)] = b \times \log t - \log a \quad (3.4)$$

where Q_t is the amount of drug released in time t, b is shape parameter, a is scale parameter.

The following plots were made: *cumulative% drug release versus time* (zero order kinetic models); *log cumulative of percentage drug remaining versus time* (first-order kinetic model); cumulative *percentage drug release versus square root of time* (Higuchi model) and Weibull model

$$[\ln\ln\{1/(1-Q_t)\} \text{ vs. } \ln t]$$

3.22 MECHANISM OF DRUG RELEASE

Korsmeyer et al. developed a simple, semiempirical model, relating exponentially the drug release to the elapsed time (t)

$$f_t = K t^n$$

where K is a constant incorporating structural and geometrical characteristic of the drug dosage form, n is the release exponent; f_t is M_t/M_∞ (fractional release of the drug).

Contingent upon the general size of the pace of polymer expanding to the pace of medication dissemination, different discharge profiles might be conceivable. The circumstance where the polymer basic modification happens quickly in light of the expanding dissolvable when contrasted with the pace of medication dissemination by and large prompts Fickian dispersion, or the supposed first request discharge, described by the square base of time reliance in both the sum discharged and the infiltrating dispersion front situation in piece geometry.

In the event of the sorption process is totally represented by the pace of polymer unwinding, the purported Case II transport, described by straight time reliance in both the sum diffused and the infiltrating growing front position, results. In many frameworks, the halfway arrangement, which is frequently named non-Fickian or peculiar dissemination, will win at whatever point the paces of dispersion and polymer unwinding are practically identical.

Dynamic consistent consolidates auxiliary and geometrical characters of the medication/polymer framework. For non-Fickian discharge, the n esteem falls somewhere in the range of 0.5–1.0 ($0.5 < n < 1.0$), though on account of Fickian dispersion, $n = 0.5$; for zero-request discharge (case transport), $n = 1$, and for super case-II transport, $n > 1$. The estimations of n as assessed by the direct relapse of $\log(M_t/M_\infty)$ versus log(t) of various details were determined. Table 3.9 represents the interpretations of diffusion mechanisms from dosage forms.

3.23 IN VIVO BIODISTRIBUTION STUDIES

Albino Wistar rats (male/female) weighing 200 ± 20 g were used for brain bioavailability studies. All animal experiments were approved by Institutional Animal Ethical Committee, J.S.S. College of Pharmacy, Ooty (Proposal no. JSSCP/IAEC/M. PHARM/PH.CEUTICS/03/2010-11). All the rats were fasted for overnight before the experiments but had free access to water.

Groupings: Six animals were taken in each group (male).

1. Group 1—control group.
2. Group 2—drug suspension treated group.
3. Group 3—TP–SLNs treated group.

Drug suspension (0.3% w/v CMC) and SLNs were administered intravenously by causal vein at a dose of 0.18 mg (calculated based on the human dose of 10 mg, conversion factor = 0.018). Blood (0.5 ml) was collected by retro-orbital venous plexus puncture at 0, 0.25, 0.50, 1, 2, 3, 4, 6, 8, 12, and 24 h after administration separately. Blood samples were placed into Eppendorf tubes containing 0.3 mL of anticoagulant (citrate) solution and centrifuged immediately. After centrifugation, the plasma obtained was stored at −20 °C until further analysis

3.24 BIOANALYTICAL METHOD DEVELOPMENT AND ANALYSIS

Reverse-phase HPLC method is the most popular mode for analytical and preparative separations of the compounds in chemical, biological, pharmaceutical, and food samples. In reversed-phase mode, the stationary phase is nonpolar and the mobile phase is polar. The polar compounds get eluted first in this mode and nonpolar compounds are retained for a longer time. In the present study, methods for the estimation of the OL present in the blood plasma samples were developed and validated. For the estimation of OL in blood plasma, the chromatographic variables, namely pH, solvent strength, solvent ratio, flow rate, the addition of peak modifiers in mobile phase, nature of the stationary phase, detection wavelength, and internal standard (IS), were studied and optimized for the separation and retention of the drug. The drug was extracted from the blood plasma by solid-phase extraction method using an SPE cartridge. The following are the optimized chromatographic conditions, preparation of standard and sample solutions, and the methods used for the estimation of OL in plasma.

3.25 OPTIMIZED CHROMATOGRAPHIC CONDITIONS FOR FLUTICASONE PROPIONATE

Stationary phase: Hibar C_{18}(250 × 4.6 mm i.d.,5μ)

Mobile phase: Acetonitrile: Phosphate buffer

Mobile phase ratio: 40:60

Flow rate: 1.0 mL/min

Sample volume : 20 μL using Rheodyne7725iinjector

Detection : 258 nm

pH : 6.5

Buffer strength : 25 mM

Data station : LC-20AD

Internal std : Atorvastatin

Detector : PDA

Pump : Gradient

Drug run time : 6.9 min

IS Runtime : 14.4 min

TABLE 3.9 Interpretations of Diffusion Mechanisms from Dosage Forms

Release Exponent (n)	Drug Transport Mechanism	Rate as a Function of Time
0.5	Fickian diffusion	$t^{-0.5}$
0.5<n<1.0	Anomalous transport	t^{n-1}
1.0	Case-II transport	Zero order release
Higher than 1.0	Super case-II transport	t^{n-1}

3.26 SELECTION OF DETECTION WAVELENGTH FOR OL

A total of 100 μg/mL of OL was prepared in ACN. This solution was scanned in the UV region of 200–400 nm and the UV spectrum was recorded. From the spectra, detection wavelength at 258 nm was selected for OL.

3.27 PREPARATION OF STANDARD AND SAMPLE OL SOLUTIONS

3.27.1 STANDARD STOCK SOLUTION OF OL

Ten milligrams of OL working standard was accurately weighed and transferred into a 10 mL volumetric flask and dissolved in ACN and made up to the volume with the same solvent to produce a 1 mg/mL of OL. The stock solution was stored in a refrigerator at –20 ± 2 °C until analysis.

The stock solution was diluted to suitable concentrations to obtain the calibration curve (CC) standards and quality control (QC) samples.

3.27.2 STANDARD STOCK SOLUTION OF IS

Ten milligrams of Atorvastatin working standard was accurately weighed and transferred into a 10 mL volumetric flask and dissolved in ACN and made up to the volume with the same solvent to produce a 1mg/mL of Atorvastatin. The stock solution was stored in refrigerator at –20 ± 2 °C until analysis.

The stock solution was diluted to suitable concentrations to obtain CC standards and QC samples.

3.27.3 CC STANDARDS AND QC SAMPLES

Working solutions for calibration and controls were prepared from the stock solution by an adequate dilution using ACN: water. Calibration standards curve was prepared for the concentration from 1 to 100 mcg.

3.28 PREPARATION OF BLANK PLASMA

Blank plasma (0.5 mL) was transferred into 2.0 mL Eppendorf tube and 0.1 mL of the mobile phase was added. The resulting solution was vortexed for 5 min. The plasma was extracted using an SPE cartridge with ACN–WATER (50:50) and analyzed.

3.29 PREPARATION OF BIOANALYTICAL CC SAMPLES

A total of 0.1 mL of 5, 10, 15, 20, and 25 μg/mL of OL solutions were transferred to 2.0 mL Eppendorf tube, respectively, to this 0.5 mL of plasma, 0.5 mL of IS were added. The resulting solution was vortexed for 5 min. The plasma was extracted using an SPE cartridge with ACN–WATER (50:50) and analyzed.

3.30 PREPARATION OF PLASMA SAMPLES

Plasma samples (0.5 mL) obtained from study subjects were transferred into 2.0 mL Eppendorf tube and 0.5 mL of IS was added. The resulting solution was vortexed for 5 min. The plasma was extracted using an SPE cartridge with ACN-WATER (50:50) and analyzed.

3.30.1 METHOD OF ANALYSIS

The bioanalytical CC samples and plasma sample solutions were injected with the above chromatographic conditions and the chromatograms were recorded. The quantification of the chromatogram was performed using the peak area.

3.30.2 STATISTICAL ANALYSIS

Data are expressed as mean ± standard error of mean (S.E.M.). For statistical analysis of pharmacokinetic parameters, an unpaired student's *t*-test was applied. A $P < 0.05$ was considered statistically significant. Statistics were carried out using GraphPad Prism software, v.5.01, USA.

3.31 RESULTS AND DISCUSSION

3.31.1 PREFORMULATION STUDIES

3.31.1.1 SOLUBILITY STUDIES

The saturation solubility of OL was determined in different solvents, namely, 0.1 N HCl, phosphate buffer pH 6.8, pH 7.4, and distilled water. The results are given in Table 3.10. OL was found to be highly soluble in 0.1 N HCl. The solubility in phosphate buffer pH 6.8, pH 7.4 and distilled water was found to be 192.7, 182.4, and 55.46 μg/mL, respectively.

3.31.1.2 COMPATIBILITY STUDIES

The spectra obtained from IR studies at a wavelength from 4000 to 400 cm^{-1} are shown in Figures 3.5–3.11 and table Table 3.11.

After translation of the above spectra, it was affirmed that there was no significant moving, misfortune or presence of practical tops between the spectra of medication, lipid, physical blend of medication and lipid. From the spectra, it was presumed that the medication was entangled into the polymer framework with no compound collaboration. From the IR study, it was inferred that the chose lipids glyceryl tripalmitate and glyceryl

TABLE 3.10 Solubility Profile of OL in Different Media

Solvent	Solubility
0.1 N HCl	251.3 ± 6.1 mg/mL
Phosphate buffer pH 6.8	192.7 ± 2.8 μg/mL
Phosphate buffer pH 7.4	182.4 ± 4.1 μg/mL
Distilled water	55.46 ± 0.92 μg/mL

TABLE 3.11 Functional Groups Present in the IR Spectrum

Sl. No	Functional Group	Wave Number in Std Drug Spectra	Wave Number in Mixture Spectra	Observation
1	(N–H) S	3236.66	3236.66	No shifting
2	Thio (C–H) S	3051.49	3051.49	No shifting
3	Ar (N–H)	1587	1589.40	No shifting
4	Ar (C–N)	1419	1419	No shifting
5	Ar (C–H)	758.05	758.05	No shifting

monostearate were seen as good in entangling the chose medication OL.

3.31.1.3 DEVELOPMENT OF CC

CC of the drug was developed to found out the linearity between the concentration of drug in the solution and its optical density. It was concluded that the perfect linearity between the concentration and absorbance was observed when the concentration range was 2–10 μg/mL. Tables 3.12 and 3.13 and Figures 3.17 and 3.18 show the calibration of OL using 0.1 N HCl and phosphate buffer pH 7.4. The "Slope (K)" and "Intercept (β)" value was found to be 0.075 and 0.004 for 0.1 N HCl, and 0.038 and 0.0142 for phosphate buffer pH 7.4 (Figures 3.12 and 3.13).

TABLE 3.12 CC of OL in 0.1NHCl (λ_{max} = 258 nm)

Concentration (μgmL/mL)	Absorbance
2	0.161
4	0.283
6	0.441
8	0.605

TABLE 3.13 CC of OL in phosphate buffer saline pH 7.4 (λ_{max} = 258 nm)

Concentration (μg/mL)	Absorbance
2	0.095
4	0.168
6	0.245
8	0.320
10	0.407

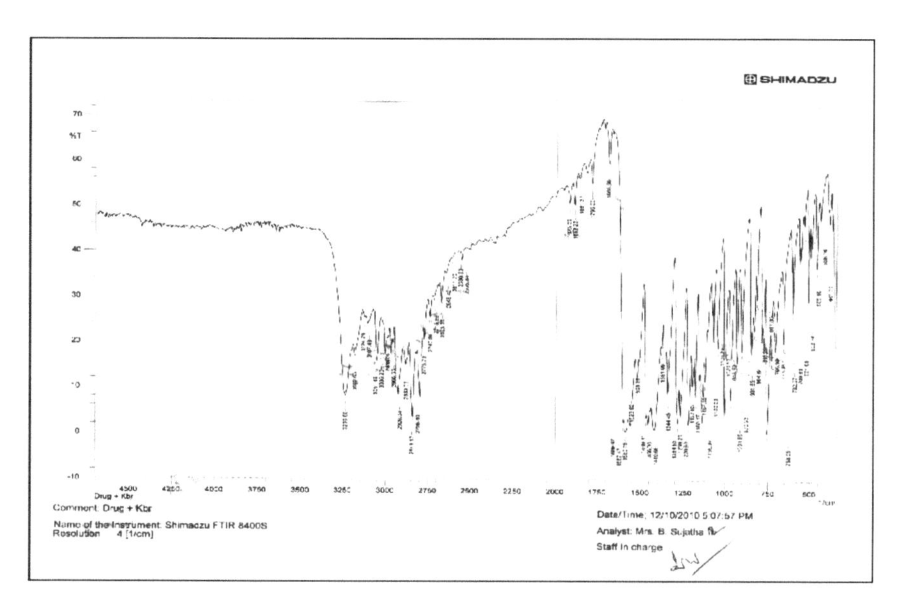

FIGURE 3.5 IR spectra of drug.

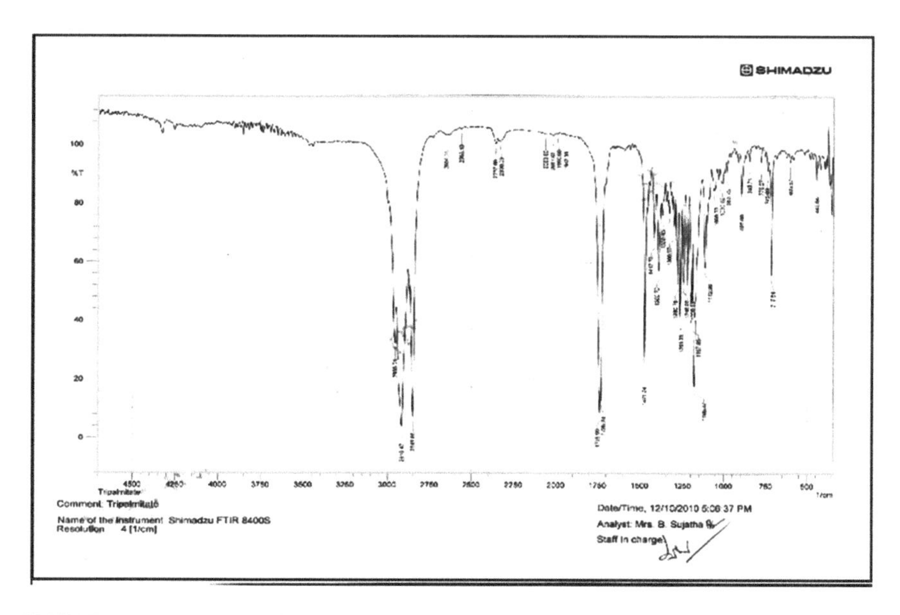

FIGURE 3.6 IR spectra of glyceryl tripalmitate.

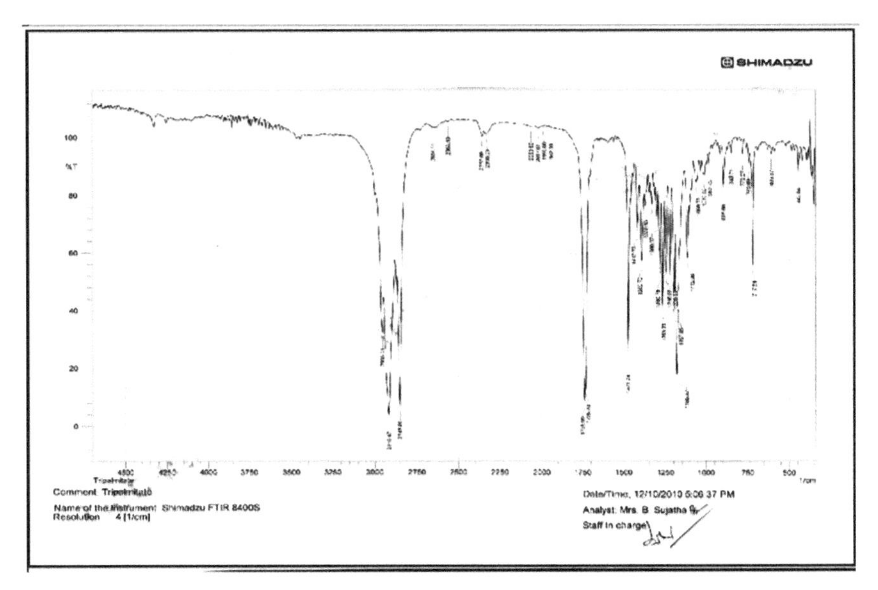

FIGURE 3.7 IR Spectra of glyceryl monostearate.

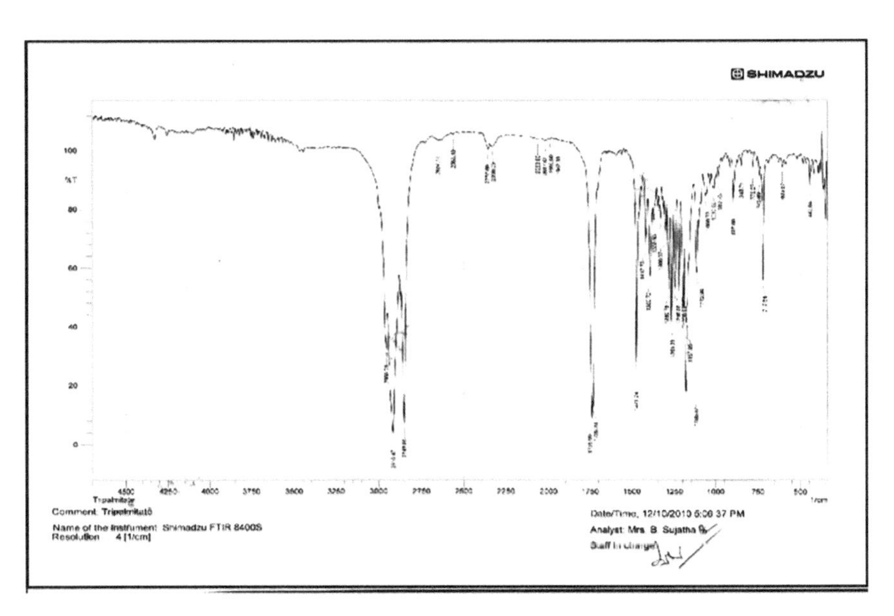

FIGURE 3.8 IR spectra of Tween 80.

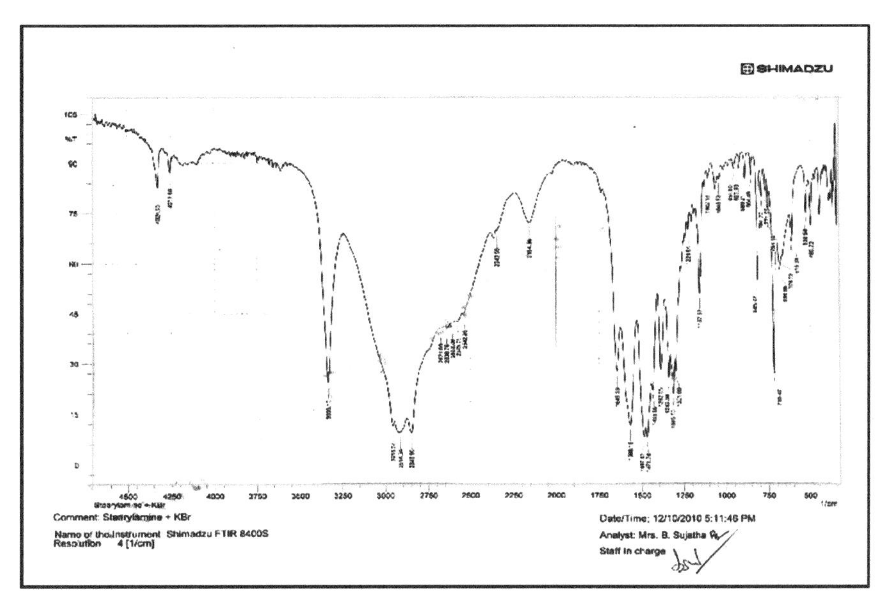

FIGURE 3.9 IR spectra of stearyl amine.

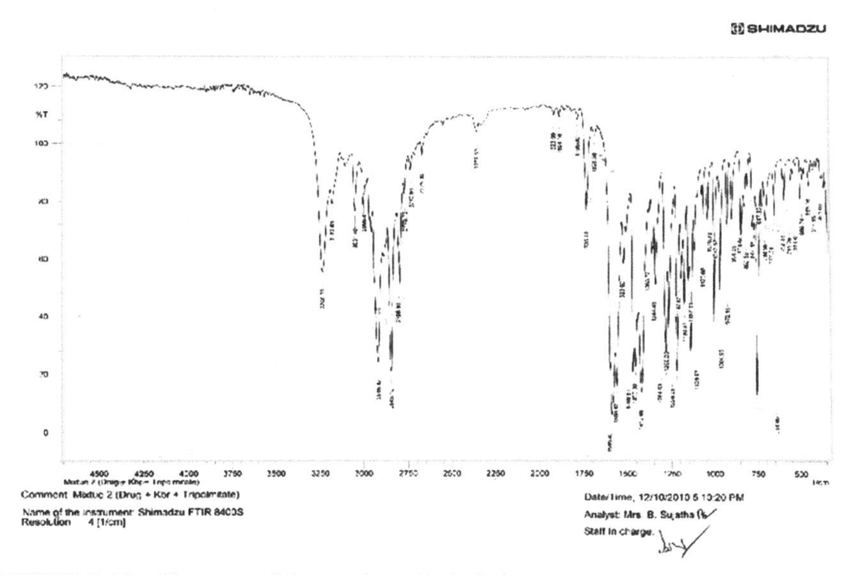

FIGURE 3.10 IR spectra of drug + glyceryl tripalmitate.

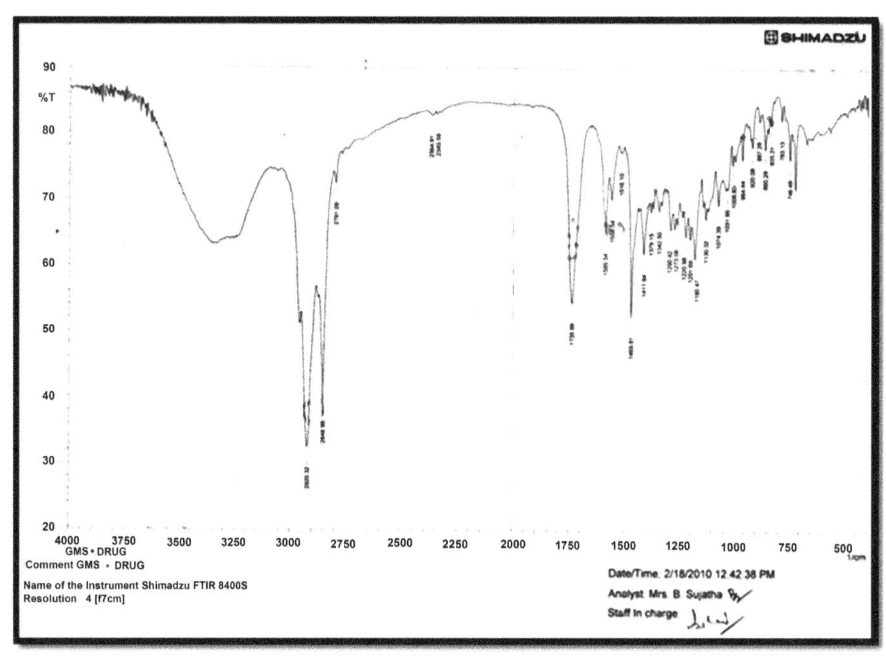

FIGURE 3.11 IR spectra of drug + glyceryl monostearate.

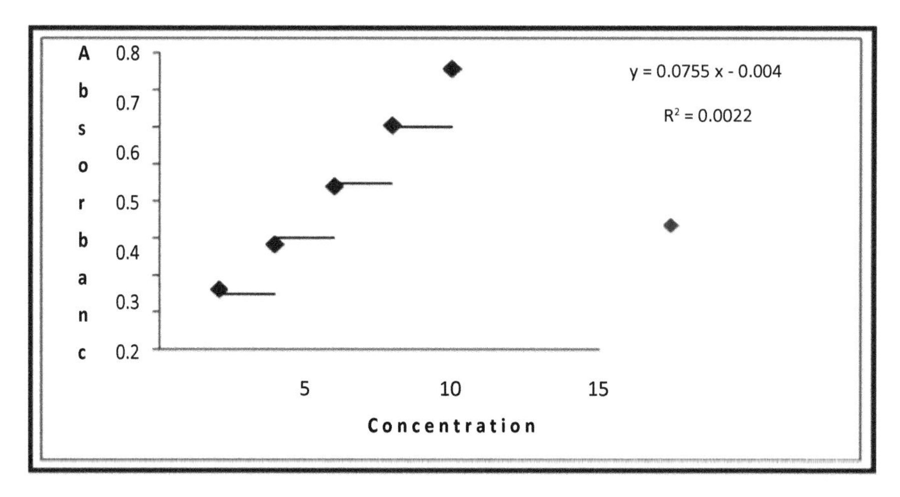

FIGURE 3.12 CC of OL in 0.1 N HCl (λ_{max} = 258 nm).

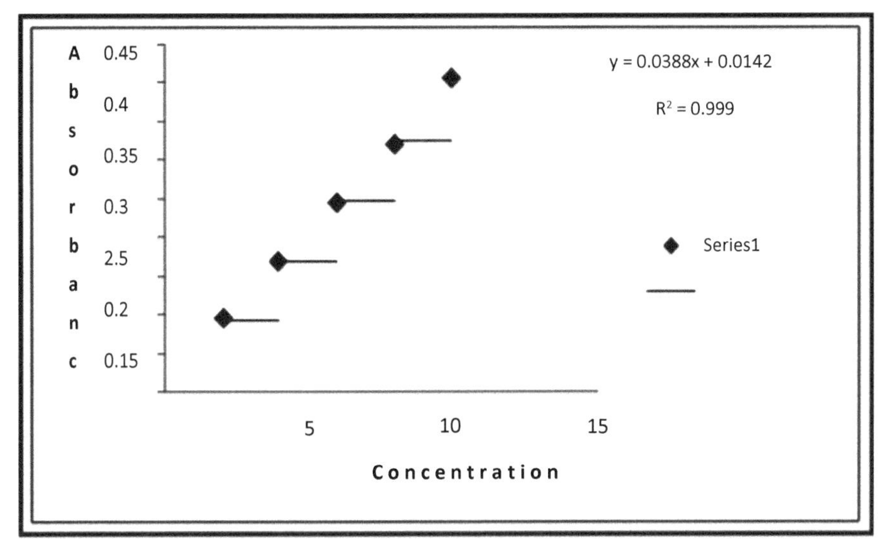

FIGURE 3.13 CC of OL in phosphate buffer saline pH 7.4 (λ_{max} = 258 nm).

3.31.1.4 PARTITION COEFFICIENT STUDIES

OL is a hydrophobic drug with a log *P* value of 2.199. The partition coefficient was found in order of glyceryl tripalmitate (TP) (282.5), glyceryl monostearate (54.86). TP is more lipophilic of the two lipids and had higher affinity for OL. It is suspected to have more imperfections in its lipid matrix to accommodate the drug. Hence, an initial study of the partitioning nature of the drug between the melted lipid and aqueous media can provide some clues about the entrapment in SLN formulation.

3.31.2 EFFECT OF FORMULATION PROCESS VARIABLES

SLNs were prepared by solvent diffusion technique by varying the amount of the lipid. Sixteen different batches were prepared separately for both GMS and TP using as lipids, as shown in Table 3.8. It was observed that at a higher concentration of the lipid 60 mg, there was an increase in the particle size as observed. The possible reason might be that amount of lipid was high compared to surfactants used. Surfactant concentration was not sufficient or enough to effectively cover the lipid nanoparticles; hence, larger particles were observed.

Thus, due to lower particle size in the eigth run with 165.1 nm in GMS was taken as the optimized batch and in the case of TP 13th batch was considered as the optimized batch with a particle size as 110.5 nm.

The effect of stirring time on particle size is given in Table 3.8. It was observed that particle size was in the range of 165–541 nm as in GMS and 110.5–361 nm as in TP. It is obvious that with an increase in stirring time and stirring speed, there was a decrease in particle size up to 4 h in stirring time and 5000 rpm in stirring speed.

There was no significant changes with the increase in surfactant concentration thus, 1.5% is considered as the optimized one.

3.32 EVALUATION OF SLNS

3.32.1 *ZETA POTENTIAL*

Zeta potential is the potential difference between the dispersion medium and the stationary layer of fluid attached to the dispersed particle. The significance of zeta potential is that its value can be related to the stability of colloidal dispersions. The zeta potential indicates the degree of repulsion between adjacent, similarly charged particles in the dispersion. For molecules and particles that are small enough, a high zeta potential will confer stability, that is, the solution or dispersion will resist aggregation. When the potential is low, attraction exceeds repulsion and the dispersion will break and flocculate.

It has been reported that positively charged SLNs have a better uptake by brain than neutral or negatively charged particles [26]. Thus, batches were prepared stearylamine as a positive charge inducer. It was found that batch prepared with stearylamine had zeta potential of 10.1 and 13.93 (Table 3.8 in the 8th run) mV. Stearylamine contains lipophilic hydrocarbon chain (18 carbons), which is accommodated in the lipid core projecting the amine group into the aqueous phase and induces positive charge [27].

3.32.2 *ENTRAPMENT EFFICIENCY*

The lipid core was found to affect the extent of entrapment efficiency (Table 3.11). As observed with TP-SLN and GMS-SLN, the maximum entrapment efficiency was 96.3% in TP-SLN and 67.2% in GMS-SLN, respectively. The entrapment efficiency was higher in the TP-SLN compared to GMS-SLN, when the amount of drug taken for preparation of SLN was more (10 mg), higher entrapment was obtained. TP-SLN showed higher entrapment efficiency compared to GMS-SLN because TP is more lipophilic than GMS and can accommodate more drug in the lipid matrix [28].

3.32.3 POLYDISPERSITY

The PI is the measure of size distribution of the nanoparticle formulation. PI was measured using Malvern Zetasizer. PI values range from 0.000 to 1.000, that is, monodisperse to very broad particle size distribution. PI values of all the formulations indicate that particle size distribution was unimodal. The optimized batch having the least particle size (110.5 nm) had a PI of 0.340 in TP-SLN and GMS-SLN having particle size (165.1 nm) had a PI of 0.726.

3.32.4 DIFFERENTIAL SCANNING COLORIMETRY

The DSC curve of OL (Figure 3.14) showed a melting endotherm at 194.78 °C. The peak intensity corresponding to the melting of OL decreased in thermograms of OL-loaded TP and GMS SLNs. These results indicate that only a small fraction of the drug substance existed in the crystalline state. Reduction in the melting point and the enthalpy of the melting endotherm was observed when the lipid was formulated as SLNs. Incorporation of OL inside the lipid matrix results in an increase in number of defects in the lipid crystal lattice, and hence causes a decrease in the melting point of the lipid in the final SLNs formulations. Freitas and Muller also observed the crystalline behavior of Compritol SLNs differed distinctly from that of the bulk lipid. The small particle size of SLNs leads to high surface energy, which creates an energetically suboptimal state causing a decrease in the melting point.

3.32.4.1 SCANNING ELECTRON MICROSCOPY [29]

The external morphological studies (SEM) revealed that maximum nanoparticles were spherical in shape (Figure 3.19). The nanoparticle size observed by SEM correlated well with the particle size measured by zeta sizer (Malvern instrument) (Figure 3.15).

3.32.4.2 IN VITRO RELEASE STUDIES

In vitro dissolution studies were carried out in phosphate buffer pH 7.4 saline. The release profiles indicate that SLN formulations showed a retarded release of the drug from the lipid matrix when compared with plain OL solution (OL-SOL). The in vitro release data and graph of SLN formulations and OL-SOL in phosphate buffer 7.4 saline are shown in Table 3.14 and Figure 3.20. It was observed that OL-SOL showed 90.24% release in 4 h compared to 6.62% and 16.68% release for TP-SLN and GMS-SLN at the end of the same time. This is due to fact that there is no barrier for diffusion at dialysis membrane

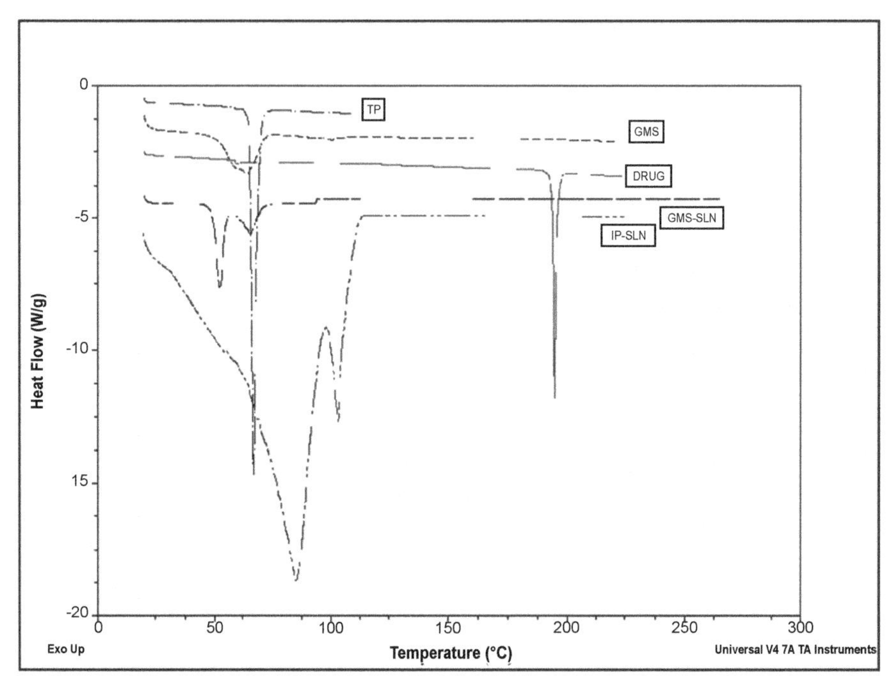

FIGURE 3.14 Differential scanning calorimeter curves of OL and Ol-loaded SLN formulations. OL indicates OL; TP—glyceryl tripalmitate; GMS—glyceryl monostearate; SLNs—solid lipid nanoparticles.

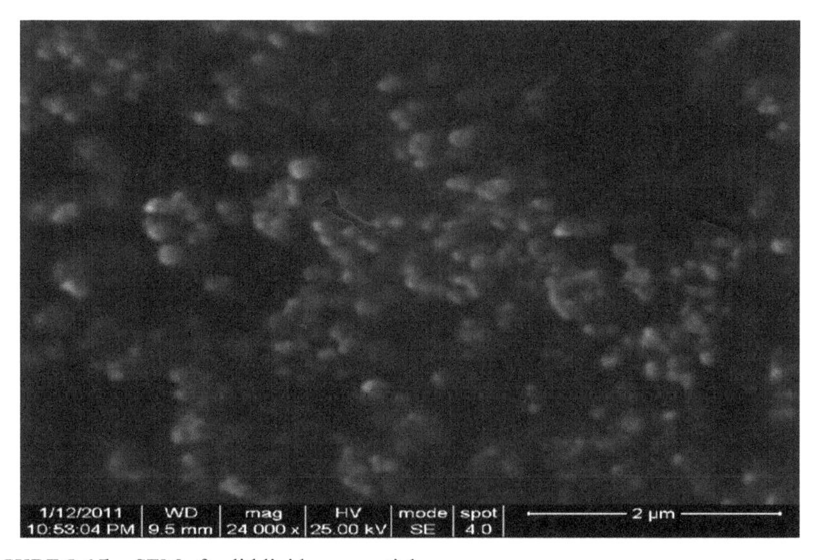

FIGURE 3.15 SEM of solid lipid nanoparticle.

interface for OL molecules. Hence, higher release was observed in the case of OL-SOL. But, TP-SLN and GMS-SLN released the drug in a sustained manner. It is evident that there is an inverse relationship between the percent drug release and the partition coefficient of OL. TP having a higher partition coefficient than GMS released the drug in a sustained manner compared to TP-SLN. GMS-SLN release rate was nearly 50% higher than that of TP-SLN at the end of 48 h.

TP-SLN-glyceryl tripalmitate SLN, GMS-SLN-glyceryl monostearate SLN, OL-SOL-OL solution (Figure 3.17).

3.32.5 RELEASE KINETICS

In order to elucidate mode and mechanism of drug release, the in vitro data were transformed and interpreted at a graphical interface constructed using various kinetic models. The in vitro release data obtained for SLN formulations, in phosphate buffer pH 7.4, were fitted into various kinetic models. The results are shown in Table 3.15.

The best linearity was obtained in Higuchi's plot for both SLN formulations indicating the release from the matrix as a square root of the time-dependent process. Higuchi's plot for TP-SLN and GMS-SLN are shown in Figures 3.18 and 3.19, respectively.

TABLE 3.14 Comparative In Vitro Release Data for SLN Formulations Compared to Pure Drug in Phosphate Buffer pH 7.4

Time (h)	Cumulative Percent Release (%)		
	TP-SLN	GMS-SLN	OL-SOL
0	0	0	0
0.5	0.31 ± 0.89	0.56 ± 0.87	12.3 ± 0.87
0.75	0.93 ± 0.67	1.1 ± 0.65	17.9 ± 1.45
1	1.84 ± 0.65	2.12 ± 0.46	21.14 ± 2.33
2	3.92 ± 0.43	7.74 ± 0.87	49.51 ± 2.19
4	6.62 ± 0.89	16.18 ± 0.99	90.24 ± 2.26
6	8.57 ± 1.19	21.20 ± 1.10	–
12	13.25 ± 1.23	33.35 ± 1.21	–
24	19.75 ± 1.44	48.32 ± 1.88	–
48	29.52 ± 1.33	70.51 ± 1.66	–

TABLE 3.15 Regression Value for Various Kinetic Models

Formulation	R^2		
	Zero Order	First Order	Higuchi's
TP-SLN	0.943	0.964	0.999
GMS-SLN	0.927	0.988	0.998

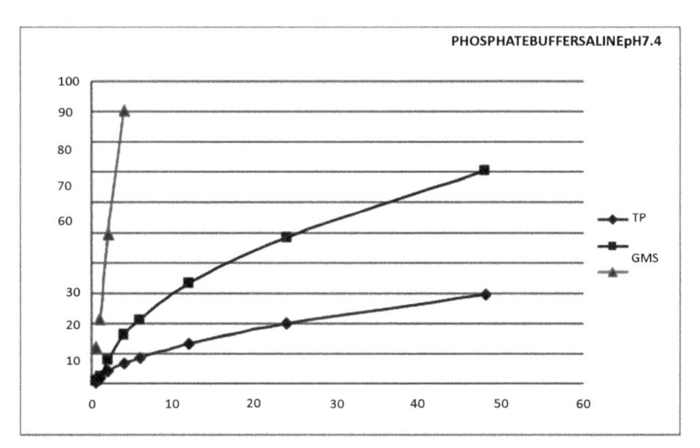

FIGURE 3.16 Comparative in vitro release profile in phosphate buffer saline pH 7.4.

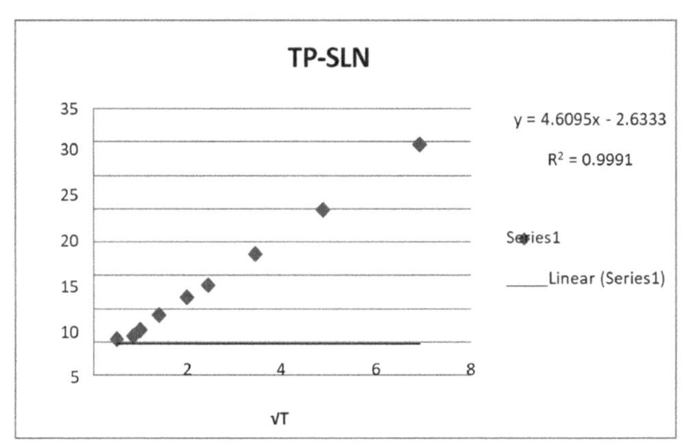

FIGURE 3.17 Higuchi's plot for TP-SLN.

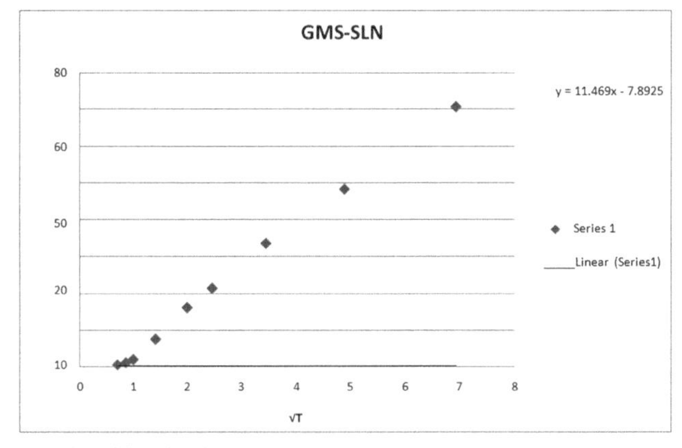

FIGURE 3.18 Higuchi's plot for GMS-SLN.

3.32.6 RELEASE MECHANISM

By incorporating the release data in Korsmeyer–Peppa's equation, the mechanism of the drug release can be indicated according to the value of release exponent $-n$. Korsmeye–Peppa's plot for TP-SLN and GMS-SLN are given in Figures 3.19 and Figure 3.20, respectively.

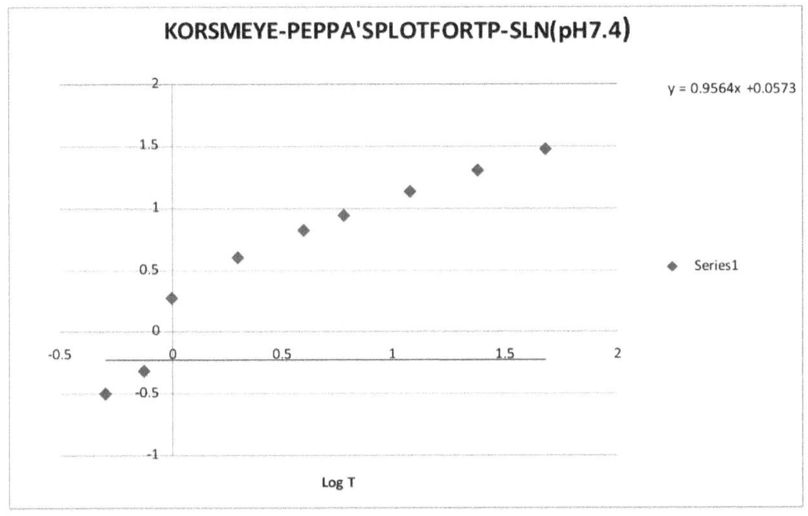

FIGURE 3.19 Korsmeye–Peppa's plot for TP-SLN.

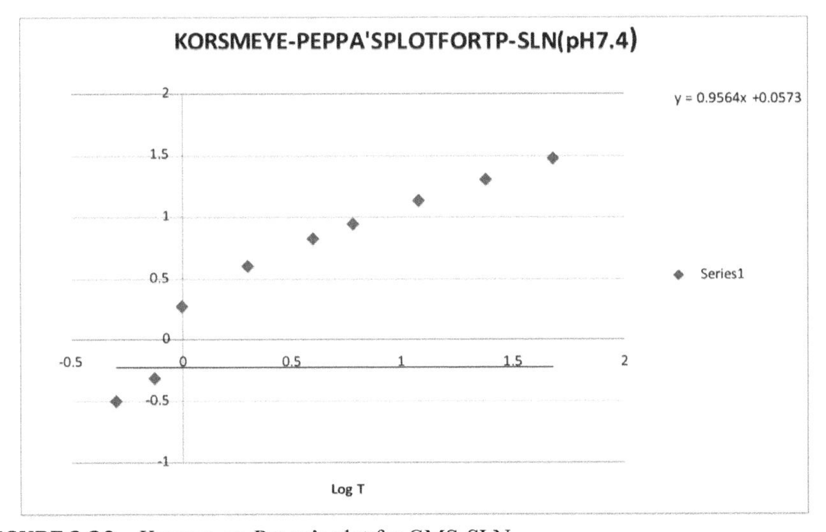

FIGURE 3.20 Korsmeyer–Peppa's plot for GMS-SLN.

The release exponent values "n" for TP-SLN and GMS-SLN was found to be 0.9564 and 0.9414, respectively. Since the release exponent "n" values were between 0.5 and 1, it indicates that the SLN formulations undergo anomalous diffusion.

3.33 BIOANALYTICAL METHOD DEVELOPMENT AND ANALYSIS

Bioanalytical CC of OL was prepared in the mobile phase and peak area was calculated (Figure 3.22) and Bbioanalytical cCalibration of OL is represented byin Table 3.16.

TABLE 3.16 Bioanalytical Calibration of OL

Sr. No	Concentration (ng/mL)	Response Factor
1	11	0.075681
2	15	0.082144
3	20	0.081026
4	30	0.090446
5	110	0.159171
6	130	0.176028
7	180	0.230747
8	220	0.282964

The chromatograms of OL showed a stable baseline. The regression equation in the range of 11–222 ng/mL was as follows: y=0.001x + 0.062, R^2 = 0.992.

The concentration of OL was determined in plasma samples separated at different time intervals by HPLC analysis. The concentration of plasma samples was determined from the area of the chromatographic peak using the calibration graph (Figures 3.21–3.24).

3.34 DATA ANALYSIS [30–34]

Pharmacokinetic parameters of OL after i.v. administration are shown in Table 3.17. Peak concentration (C_{max}) and time of peak concentration (T_{max}) were obtained directly from the individual plasma concentration-time profiles. The area under the concentration-time curve from time zero to time t ($AUC_{0\to t}$) was calculated using the Wagner–Nelson method. The area under the curve (AUC) determines the bioavailability of the drug for the given the same dose in the formulation. The area under the total plasma concentration–time curve from time zero to infinity was calculated by

$$AUC_{0\to\infty} = AUC_{0\to t} + C_t/K_e$$

where C_t is the OL concentration observed at last time, and K_e is the apparent elimination rate constant obtained from the terminal slope of the individual plasma concentration–time curves after logarithmic transformation of the plasma concentration values and application

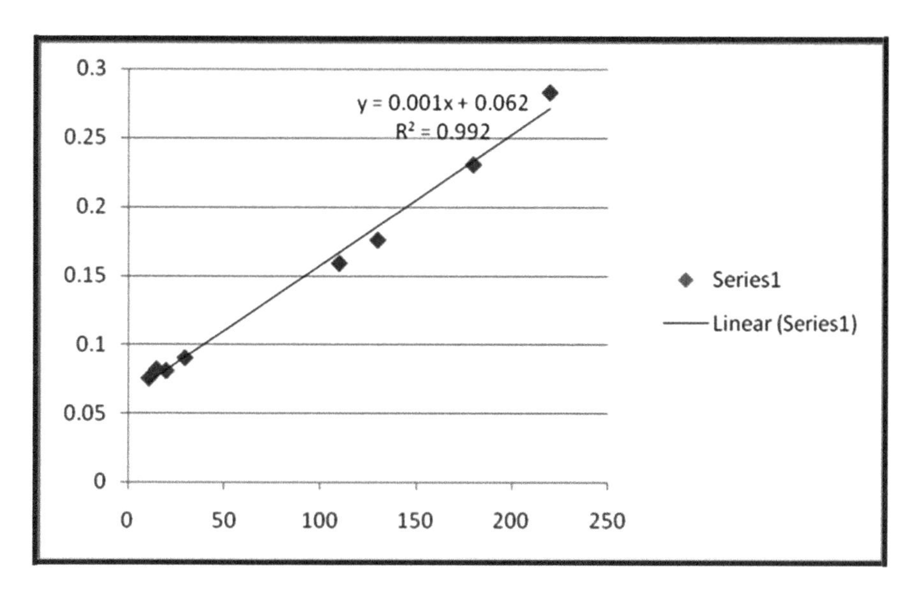

FIGURE 3.21 Bioanalytical CC of OL.

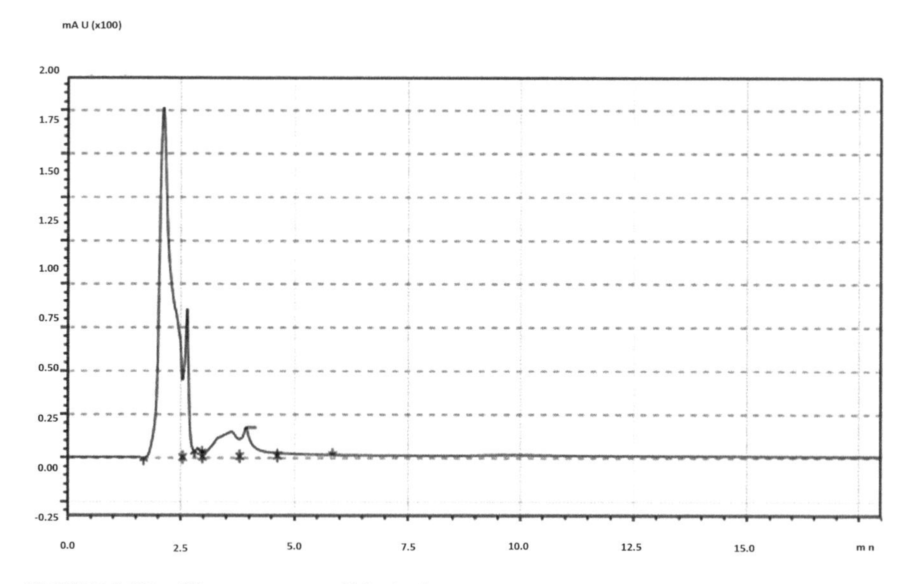

FIGURE 3.22 Chromatogram of blank plasma.

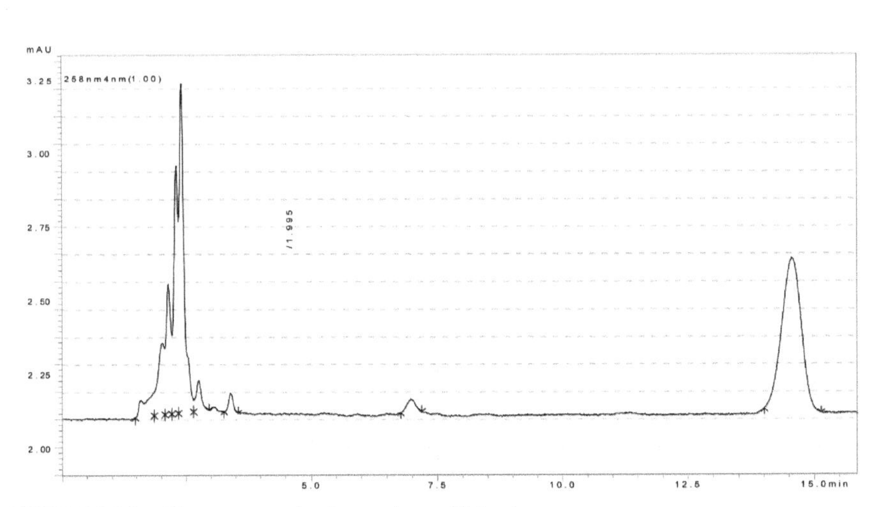

FIGURE 3.23 Chromatograph of pure drug+ IS in plasma.

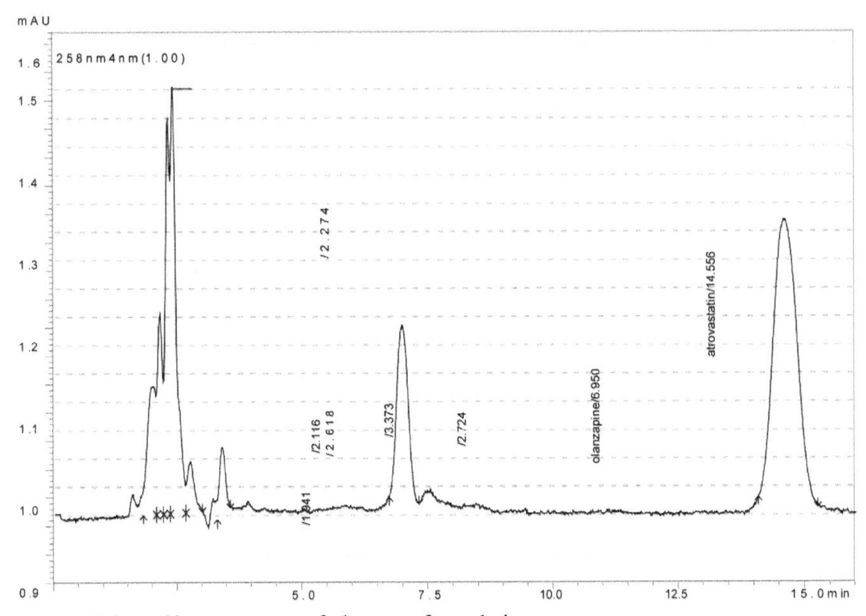

FIGURE 3.24 Chromatogram of plasma + formulation.

of linear regression (Figure 3.25). The relative bioavailability (F_r) at the same dose was calculated as

$$F_r = \text{AUCSLN}, 0 \rightarrow t / \text{AUCSOL}, 0 \rightarrow t$$

TABLE 3.17 Pharmacokinetic Parameters for OL in the Plasma After i.v. Administration

Time	Formulation	Pure Drug
$AUC_{0\text{-}t}$	1723.58914 ± 0.037*	578.409833 ± 0.040
keli (h^{-1})	1.75 ± 0.01	2.57 ± 0.10
$t_{1/2}$ (h)	0.39 ± 0.04	0.26 ± 0.04
$AUCo\text{-}_{\alpha}$	8072.921 ± 2.83*	4303.078 ± 2.99
CL(L/h)	0.41 ± 1.42	0.44 ± 1.50
Vd (L)	0.257 ± 0.3	0.163 ± 0.30
$AUMC_{o\to t}$	8660.67±2.93	1835.017±6.93
$AUMC_{o\to\alpha}$	12722±1.99	2223.222±5.90
F_r	2.6±0.473	1
MRT (h)	2.196±0.17	0.5166±0.29

Data are expressed as mean ± S.E.M.; $n = 6$.
*$P < 0.05$ versus pure drug.
*$P < 0.05$ is considered statistically significant.
Statistics was done using Student's t-test with GraphPad Prism software, v.5.01.

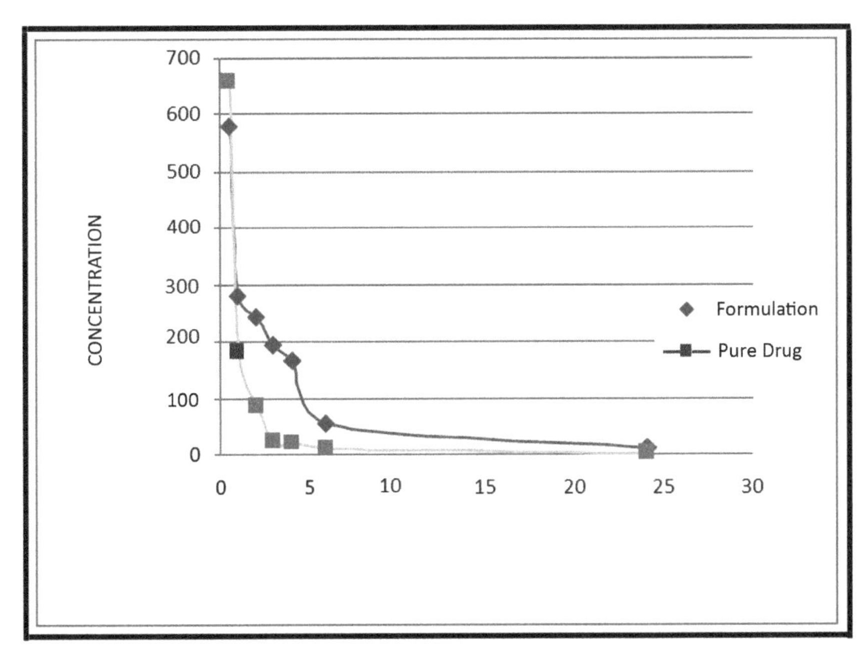

FIGURE 3.25 Concentration–time curve of glyceryl tripalmitate solid lipid nanoparticles (TP-SLN) and olanzapine-suspension (OL-SUSP).

After the injection of OL SLNs, the OL concentration was still present after 6 h, with OL suspension concentration was reached minimum after 3 h. Following i.v. administration of the OL suspension, the mean measured peak serum concentration achieved was 25.83 ng/mL, with OL SLNs it was 55.92 ng/mL.

After intravenous administration of OL-SUSP, free OL is available for solvation and might have solubilized in plasma, subsequently the solution of OL distributed rapidly compared to the distribution of OL entrapped in SLNs. The initial serum concentration (at 30 min) observed is higher for OL-SUSP than OL-SLN, following intravenous administration due to its solubility in plasma and subsequent distribution and elimination. In the case of SLNs of OL, free drug is available for distribution only after its release. In vitro release studies showed that OL released slowly from SLNs for an extended period of time, thus free OL is not readily available for distribution (Figure 3.26).

A difference in OL concentrations between the SLNs and OL-SUSP was observed, OL-TP-SA showed higher concentrations than OL-TP. Possible explanation for higher AUC ($P < 0.05$), elimination $t_{1/2}$ and MRT for OL-TP than OL-SUSP is due to slower release of OL from SLN than the OL-SUSP which leads to lower clearance. Drug release can take place either by diffusion or by the degradation of the lipid matrix which occurs mainly by the enzyme, lipase. The nature of the lipid matrix has influence on biodegradation of SLN. Thus, OL released slowly from OL-SLN might be the reason for higher AUC and lower clearance for OL-SLN than OL-SUSP. Serum kinetics of OL showed in Table 3.18 with a higher AUC and lower rates of clearance are due to slower distribution of OL incorporated in SLN than that of OL suspension. Increased MRT values were due to the slow release of OL from SLN.

3.35 BIODISTRIBUTION

In order to evaluate the enhanced distribution pattern in the brain following i.v. administration, control aqueous suspension and the solid lipid nanoparticle formulations containing OL were administered to conscious male Wistar rats. Figure 3.27 shows the brain drug concentration versus time profile of OL following i.v. administration in aqueous suspension (OL-SUSP) and OL-SLNs.

When administered intravenously in the SLNs as compared to the aqueous suspension, SLNs were more effective in enhancing bioavailability of OL in the brain. In the brain, the concentrations of OL were significantly higher for OL-SLN compared to OL-SUSP. OL was present in the brain until 24 h after administration of OL-SLNs.

AUC(0–*t*), MRT(0–*t*), and F_r values of the brain for OL formulations were given in Table 3.18. When compared with OL-SUSP, the relative bioavailability of OL-TP in the brain was 23.66. AUC(0–*t*) and MRT(0–*t*) values for the brain in SLNs were found higher than OL-SUSP. SLNs of OL release drug at a slow rate. The highest F_r value was observed with OL-TP in the brain. The F_r value was increased about 23 times compared with OL-SUSP which may be due to the presence of Tween 80 on the surface of nanoparticles and slow release of OL from SLN.

In non-RES organ, such as the brain, OL-SLN showed enhanced brain concentrations of OL and maintained high drug levels for 24 h. The AUC of OL-SLNs in the brain were higher than OL-SUSP. OL exhibits pharmacological action by binding to dopamine receptors (D1 and D4) in the brain. These results indicate that OL SLNs improve brain OL level as well as the therapeutic efficacy of the OL in the treatment of schizophrenia.

Recent reviews on the enhancement of brain uptake of various drugs using novel drug delivery systems and their transport mechanisms were outlined [35]. Biodistribution of nanoparticles can be altered by coating these particles with nonionic surfactants such as polysorbates, poloxamers and poloxamines. In the literature, it was reported that drug-loaded nanoparticles coated by polysorbates (especially polysorbate 80) could able to cross-BBB after intravenous administration. Dalargin loaded polybutylcyanoacrylate nanoparticles coated with polysorbate 80 crossed BBB by endocytic uptake and showed significant analgesic effect. After intravenous injection of doxorubicin loaded polybutylcyanoacrylate coated with polyosrbate 80, very high concentration of doxorubicin was found in the brain due to proposed mechanisms either endocytic uptake or inhibition of the P-glycoprotein efflux pump. SLN containing camptothecin and doxorubicin showed higher concentrations of respective drugs in the brain of different species tested after intravenous administration.

Among all the mechanisms proposed endocytosis of nanoparticle-mediated drugs into the brain was proved by many experiments. Overcoating of drug-loaded nanoparticles with polysorbate 80 led to the adsorption of apolipoprotein E of plasma onto the nanoparticle surface. This coated particle behaves as LDL particles and could interact with LDL receptors, initiating their uptake by endothelial cells. Subsequently, either the released drug might diffuse into brain cells or the particles may be transcytosed. Polysorbate 80 appears to act as an anchor for apolipoprotein Gessner et al. [36] showed that the ability of BBB passage by a sufficient amount of diminizene-LDC nanoparticles is not only due to adsorption of

TABLE 3.18 Pharmacokinetic Parameters for Formulation and Pure Drug in Brain

TIME	Formulation	Pure drug
AUC0-t	4804.467 ± 0.75*	03.8751 ± 0.20
keli (h^{-1})	0.093841 ± 0.1132	0.096002 ± 0.0015
$t_{½}$ (h)	7.386417 ± 1.43	7.220105 ± 1.08
AUCo-$_{\alpha}$	16219.18 ± 1.70*	420.2113 ± 1.26
$AUMC_{0\to t}$	32953 ± 1.29	1428 ± 0.32
$AUMC_{0\to\infty}$	33861 ± 2.01	3506 ± 1.99
F_r	23.66 ± 0.421	1
MRT	4.788 ± 0.15	0.8147 ± 0.18

Data are expressed as Mean ± S.E.M.; $n = 6$.
*$P < 0.05$ versus pure drug.
$P < 0.05$ is considered statistically significant.
Statistics was done using Student's t-test with GraphPad Prism software, v.5.01.

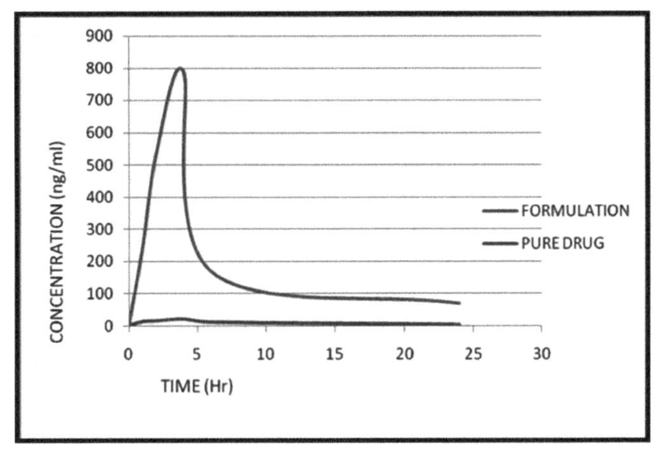

FIGURE 3.26 Brain drug concentration versus time profile of OL following i.v. administration in aqueous suspension (OL-SUSP) and OL-SLNs.

apolipoprotein E and also due to other apolipoproteins like A-I and A-IV.

Therefore, in accordance with the literature, the higher concentration and F_r for SLNs of OL in the brain may be the result of the transport of intact OL nanoparticles through the BBB by endocytosis.

3.36 SUMMARY

1. Based on the literature studies various excipients, namely, lipids (glyceryl monostearate, glyceryl tripalmitate) and surfactants (tween 80) were selected for the preparation of solid lipidnanoparticles.

2. In preformulation studies, saturation solubility studies were carried out in different solvents and the solubility was found to be highest in 0.1 N HCl.
3. Compatibility of the selected lipids, surfactants, and OL were carried by FT-IR peak matching method. It was found that there was no interaction between the drug, lipid, and surfactant, and it was compatible based on the FT-IR peak matching method.
4. Partitioning behavior of OL was determined in selected lipids to determine the affinity of the drug toward selected lipids. Partitioning studies indicates that glyceryl tripalmitate had a higher partition coefficient for OL compared to glyceryl monostearate. Higher partition coefficient indicates better entrapment efficiency and prolonged drug release.
5. In formulation development SLN were prepared by solvent diffusion method and the effect of a certain process and formulation variables such as lipid concentration, stirrings peed, stirring time, and surfactant concentration of particle size was optimized using Box–Behnken design and studied for further studies.
6. The study on various process/formulation variables revealed that all the variables are important in the formulation of SLNs.
7. SLNs were evaluated for zeta potential, entrapment efficiency of the drug, SEM studies, polydispersity, in vitro release studies, and in vivo biodistribution and pharmacokinetic studies. The entrapment efficiency was found to be higher in GTP-SLN (96.3%) compared to GMS-SLN (67.2). Therefore, GTP-SLN was selected for animal studies.
8. In order to elucidate mode and mechanism of drug release the in vitro data were transformed and interpreted at graphical interface constructed using various kinetic models. The best linearity was obtained in Higuchi's plot for both SLN formulations indicating the release from the matrix as a square root of time dependent process. SLN formulations released the drug in sustained manner over a period of 48 h.
9. In vivo biodistribution studies were carried out in albino Wistar rats where two different formulations, namely, GTP-SLN and OL suspension were administered intravenously and pharmacokinetic parameters were calculated.

3.37 CONCLUSION

To conclude, the solvent diffusion technique was suitable for producing SLNs. Lipophilic drugs like OL can be successfully incorporated into the lipids. In vitro release test showed OL-SLN exhibited sustained release. The formulated SLNs showed a significant increase, that is, 23-folds in relative bioavailability in the brain compared to a pure drug suspension. Higher relative bioavailability in the brain may be the result of the transport of intact OL nanoparticles through the BBB by endocytosis. In particular, OL-SLN can effectively target the brain by crossing the BBB. The dose of the OL-SLN needs to be corrected in accordance with increase bioavailability, to minimize its dose-related adverse effects. OL-SLNs provided sustained release of the drugs and have the potential in delivering the drugs across the BBB. Thus, these systems are the preferred drug carriers to achieve higher brain targeting efficiency and therapeutic efficacy in the case of schizophrenia a devastating chronic psychotic diseases by nullifying the common adverse effects.

KEYWORDS

- **solid lipid nanoparticles**
- **olanzapine**
- **brain targeting**
- **Box-Behnken design**

REFERENCES

1. Paolo Blasi, Stefano Giovagnoli, Aurélie Schoubben, Maurizio Ricci, Carlo Rossi, 2007. Solid lipid nanoparticles for targeted brain drug delivery, *Adv Drug Deliv Rev* **59**(6): 454–477.
2. Ghose, A.K., Viswanadhan, V.N., Wendoloski, J.J. 1999. A knowledge-based approach in designing combinatorial or medicinal chemistry libraries for drug discovery. 1. A qualitative and quantitative characterization of known drug databases. *J. Comb. Chem.* **1:** 55–68.
3. Mehnert, W., Mäder, K. 2001. Solid lipid nanoparticles production, characterization and applications. *Adv. Drug Deliv. Rev.* **47**: 165–196.
4. Karanth, K., Kramer, R.A., Qian, S.S., Christensen, N.L. 2008. Examining conservation attitudes, perspectives, and challenges in India, *Biol. Conserv.*, **141**(9): 2357–2367.
5. Joshi, P., Carrington, E.A., Wang, L., Ketel, C.S., Miller, E.L., Jones, R.S., Simon, J.A. 2008. Dominant alleles identify SET domain residues required for histone methyltransferase of polycomb repressive complex 2. *J. Biol. Chem.* **283**(41): 27757–27766.
6. Kaur, I.P., Bhandari, R., Bhandari, S., Kakkar, V. 2008. Potential of solid lipid nanoparticles in brain targeting. *J. Control Rel.* **127**: 97–109.
7. Lipinski, C.A. 2000. Drug-like properties and the causes of poor solubility and poor permeability. *J. Pharmacol. Toxicol. Meth* **44**: 235–249.
8. Begley, D., Brightman, M.W. 2003. Structural and functional aspects of the blood brain barrier. In: L. Prokai, K. Prokai-Tatrai (eds.), Peptide Transport and Delivery into the Central Nervous System. Progress in Drug Research, 61. Birkhauser Verlag: Basel, pp. 39–78.
9. Ricci, M., Blasi, P., Giovagnoli, S., Rossi, C. 2006. Delivering drugs to the central nervous system: a medicinal chemistry or a pharmaceutical

technology issue? *Curr. Med. Chem.* **13**: 1757–1775.

10. Ricci, M., Blasi, P., Giovagnoli, S., Rossi, C. 2006. Delivering drugs to the central nervous system: a medicinal chemistry or a pharmaceutical technology issue? *Curr. Med. Chem.* **13**: 1757–1775.
11. Chen, Y., Dalwadi, G., Benson, H.A.E. 2004. Drug delivery across the blood brain barrier. *Cur. Drug Deliv.* **1**: 361–376.
12. Azur Mühlen , Schwarz, C., Mehnert, W. 1998. Solid lipid nanoparticles (SLN) for controlled drug delivery–drug release and release mechanism, *Eur J Pharm Biopharm*, **45**(2):149-55.
13. Freitas, C., Müller, R.H. 1998. Effects of light and temperature on zeta potential and physical stability in solid lipid nanoparticles (SLN) dispersions. *Int. J. Pharm.* **168**: 221–229.
14. Cavalli, R., Caputo, O., Carlotti, M.E., Trotta, M., Scarnecchia, C., Gasco, M.R. 1997. Sterilization and freeze drying of drug-free and drug-loaded solid lipid nanoparticles. *Int. J. Pharm.* **148**: 47–54.
15. Date, A.A., Joshi, M.D., Patravale, V.B. 2007. Parasitic diseases: liposomes and polymeric nanoparticles versus lipid nanoparticles. *Adv. Drug Deliv. Rev.* **59**: 505–521.
16. Date, A.A., Joshi, M.D., Patravale, V.B. 2007. Parasitic diseases: liposomes and polymeric nanoparticles versus lipid nanoparticles. *Adv. Drug Deliv. Rev.* **59**: 505–521.
17. Wissing, S.A., Kayser, O., Müller, R.H. 2004. Solid lipid nanoparticles for parenteral drug delivery. *Adv. Drug Deliv. Rev.* **56**: 1257–1272.
18. Mukherjee, S., Ray, S., Thakur, R.S. 2009. Solid lipid nanoparticles: a modern formulation approach in drug delivery system. *Indian J. Pharm. Sci.* **71**: 349–358.
19. Cortesi, R., Esposito, E., Luca, G., Nastruzzi, C. 2002. Production of lipospheres as carriers for bioactive compounds. *Biomaterials* **23**: 2283–2294.
20. Rudolph, C., Schillinger, U., Ortiz, A., Tabatt, K., Plank, C., Müller, R.H. 2004. Application of novel solid lipid nanoparticles (SLN)-gene vector formulations based on a diametric HIV-1 VAT-peptide in vitro and in vivo. *Pharm. Res.* **21**: 1662–1669.
21. Chuan-Ming Hao, Reyadh Redha, Jason Morrow and Matthew D Breyer, 2002. PPAR activation promotes cell survival following hypertonic stress,. JBC Papers, in press.
22. Vobalaboina Venkateswarlu, Kopparam Manjunath, 2004. Preparation, characterization and in vitro release kinetics of clozapine solid lipid nanoparticles, *J. Controlled Rel.*, **95**(3), 62--638.
23. Pandey, R., Heidmann, S., Lehner, C.F. 2005. Epithelial re-organization and dynamics of progression through mitosis in Drosophila separase complex mutants. *J. Cell Sci.* 118(4): 733–742.
24. Bunjes, H., Siekmann, B. 2007. Manufacture, characterization and application of solid lipid nanoparticles as drug delivery system, In: M. Deleers, Y. Pathak, D. Thassu (eds.), *Nanoparticulate Drug Delivery Systems, Informa Healthcare*, Apple Academic Press: Palm Bay, FL, USA, pp. 213–268.
25. Benjamin P. Chapman, Duberstein, Paul R., Sörensen, S. and Lyness, Jeffrey M. 2001. Gender differences in five factor model personality traits in an elderly cohort: extension of robust and surprising findings to an older generation, *Pers Individ Dif*, **43**(06): 1594–1603.
26. Kopparam Manjunath, J.S. Reddy, Vobalaboina Venkateswarlu, 2005.

Methods and findings in experimental and clinical pharmacology, *SJR*, **27**(2): 127–144
27. Stéphanie Poullain-Termeau, Sylvie Crauste-Manciet, Denis Brossard,, Saleh Muhamed, Georges Nicolaos, Robert Farinotti, Christine Barthélémy, Hugues Robert & Pascal Odou, 1999. Effect of oil-in-water submicron emulsion surface charge on oral absorption of a poorly water-soluble drug in rats, *Drug Deliv*., **15**(8), 503–514.
28. Gill, P.S., Espina, B.M., Muggia, F., Cabriales, S., Tulpule, A., Esplin, J.A., Liebman, H.A., Forssen, E., Ross, M.E., Levine, A.M. 1995. Phase I/II clinical and pharmacokinetic evaluation of liposomal daunorubicin. *J. Clin. Oncol*. **13**: 996–1003.
29. Peracchia, M.T., Harnisch, S., Pinto-Alphandary, H., Gulik, A., Dedieu, J.C., Desmaele, D., d'Angelo, J., Muller, R.H., Couvreur, P. 1999b. Visualization of in vitro protein-rejecting properties of PEGylated stealth polycyanoacrylate nanoparticles. *Biomaterials* **20:** 1269– 1275.
30. Tröster, S.D., Müller, U., Kreuter, J. 1990. Modification of the body distribution of poly(methyl methacrylate) nanoparticles in rats by coating with surfactants. *Int. J. Pharm*. **61**: 85–100.
31. Göppert, T.M., Müller, R.H. 2005. Adsorption kinetics of plasma proteins on solid lipid nanoparticles for drug targeting. *Int. J. Pharm*. **302**: 172–186.
32. Göppert, T.M., Müller, R.H. 2004. Alternative sample preparation prior to two dimensional electrophoresis protein analysis on solid lipid nanoparticles. *Electrophoresis* **25**: 134–140.
33. Göppert, T.M., Müller, R.H. 2003. Plasma protein adsorption of Tween® 80- and poloxamer 188-stabilized solid lipid nanoparticles. *J. Drug Target*. **11**: 225–231.
34. Göppert, T.M., Müller, R.H. 2005. Polysorbate-stabilized solid lipid nanoparticles as colloidal carriers for intravenous targeting of drugs to the brain: comparison of plasma protein adsorption patterns. *J. Drug Target* **13**: 179–187.
35. Kreuter, J. 2001. Nanoparticulate systems for brain delivery of drugs. *Adv. Drug Deliv. Rev*. **47**: 65–81
36. Gessner, A., Olbrich, C., Schröder, W., Kayser, O., Müller, R.H. 2001. The role of plasma proteins in brain targeting: species dependent protein adsorption patterns on brain-specific lipid drug conjugate (LDC) nanoparticles. *Int. J. Pharm*. **214**: 87.
37. Faraji AH, Wipf P., Nanoparticles in cellular drug delivery. Bioorg Med Chem. 2009 Apr 15;17(8):2950-62. doi: 10.1016/j.bmc.2009.02.043
38. Source: S. Mukherjee, S. Ray, and R. S. Thakur, Solid Lipid Nanoparticles: A Modern Formulation Approach in Drug Delivery System, Indian J Pharm Sci. 2009 Jul-Aug; 71(4): 349–358. doi: 10.4103/0250- 474X.57282
39. Date AA, Nagarsenker MS., Design and evaluation of self-nanoemulsifying drug delivery systems (SNEDDS) for cefpodoxime proxetil., Int J Pharm. 2007 Feb 1;329(1-2 DOI: 10.1016/j. ijpharm.2006.08.038):166-72. Epub 2006 Sep 1.

CHAPTER 4

Nanoclay: A Novel Drug Delivery System for Improved Oral Bioavailability

N. JAWAHAR*, C. SANJEEV C. and P. B. ANJALI

Department of Pharmaceutics, JSS College of Pharmacy (JSS Academy of Higher Education and Research, Mysuru), Rocklands, Udhagamandalam 643001, Tamil Nadu, India

Corresponding author. E-mail: jawahar.n@jssuni.edu.in

ABSTRACT

In this chapter, the huge impact of nanoclay on improving the disintegration rate and bioavailability of Aceclofenac has been exhibited by readiness of nanoclay. The unadulterated medication with various centralization of clay (0.5%–1%) was portrayed regarding solvency, sedate substance, molecule size, warm conduct (differential scanning calorimetry), morphology (scanning electron microscopy), and in vitro execution was surveyed by pharmacokinetic examines. The molecule size of the readied nanoclay was radically decreased during the plan procedure. The differential scanning calorimetry indicated that there is no substance communication between tranquilize, earth, and physical blend. The disintegration showed a plan increment in the disintegration rate in correlation with unadulterated medication. In vivo examinations uncovered that the streamlined detailing gave a fast pharmaceutical reaction in rodents other than displaying improved pharmacokinetics parameters in rodents.

4.1 INTRODUCTION

Oral course is the most enjoyed and palatable technique for calm association because of insignificant exertion and effortlessness of association which prompts high patient consistency. Nevertheless, lipophilic prescriptions show low oral bioavailability as a result of various reasons like poor watery dissolvability and high first-pass assimilation. Oral medicine movement is unendingly examining increasingly current streets on account of affirmation of the factors like poor drug dissolvability just as maintenance, quick processing, high instability in the prescription plasma level, and capriciousness in light of sustenance sway which are expecting noteworthy occupation in astounding in vivo results provoking disillusionment of the ordinary transport structure. Since the last decade, the oral prescription movement has taken another estimation with the extending utilization of lipid as a transporter for the transport of ineffectually water dissolvable, lipophilic drugs.

4.2 BIOPHARMACEUTICAL ORDER FRAMEWORK (BCS)

The Biopharmaceutical Classification (BCS) (Figures 4.1 and 4.2) is a logical structure for grouping a medication substance dependent on its watery dissolvability and porousness. The biopharmaceutical order framework was grown fundamentally with regard to prompt discharge (IR) strong oral measurement structures. It is a medication improvement device that permits estimation of the commitments of three central points, disintegration, dissolvability, and intestinal penetrability that influence oral medication assimilation from prompt discharge strong oral dose structures [1].

4.3 WHAT IS CLAY?

Clay is a normally happening material made fundamentally out of fine-grained minerals, which shows pliancy through a variable scope of water substance and which can be solidified when dried or fried. Dirt

BCS CLASS	SOLUBILITY	PERMEABILITY	ORAL DOSAGE APPROACH	CHANCES OF NON-ORAL DOSAGE FORM
I	High	High	Simple oral dosage form	INCREASES
II	Low	High	➢ Techniques to increase surface area ➢ Solutions using solvents and/ or surfactants	
III	High	Low	Incorporate permeability enhancers , maximize local luminal concentration	
IV	Low	Low	Combine 2 & 3	

FIGURE 4.1 BCS classification.

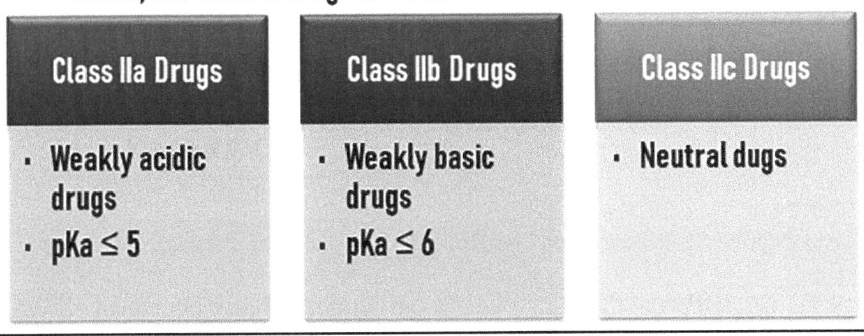

FIGURE 4.2 BCS classification of class 2 drugs.

stores are for the most part made out of earth minerals (*Phyllosilicate minerals*) and variable measure of water caught in the mineral. Clay is a costly material which can be altered by particle trade, metal/metal complex impregnation, pillaring, and corrosive treatment to create impetuses with wanted usefulness.

General structure and classification of clay were provided in Figure 4.3 and Figure 4.4, respectively.

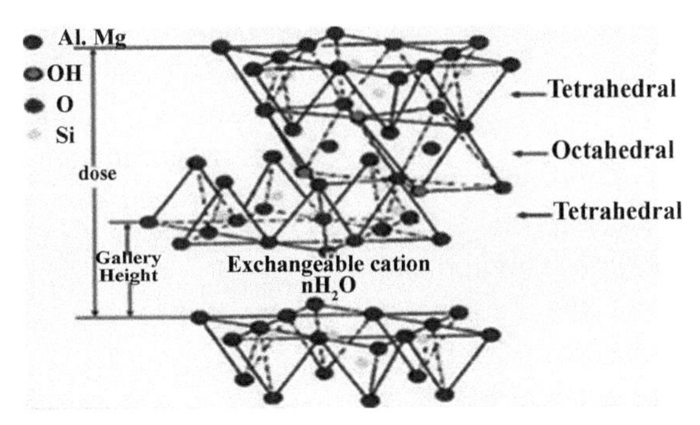

FIGURE 4.3 General structure of clay.
Source: Reprinted with permission from Ref. [31]. © 2016 Springer Nature.

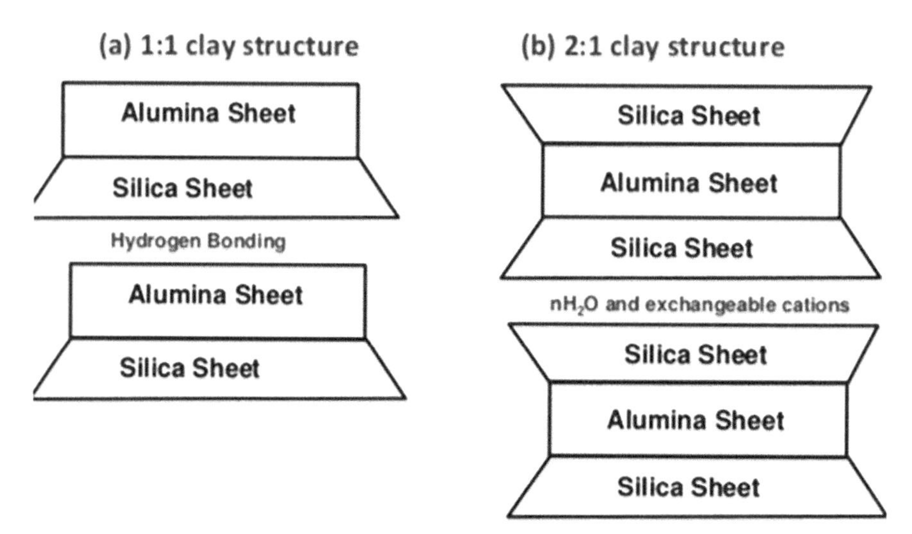

FIGURE 4.4 Classification of clay.
Source: Reprinted with permission from Ref. [31]. © 2016 Springer Nature.

4.3.1 GROUPS OF CLAY

There are four main groups of clay:

1. Kalonite
2. Montmorillonite (MMT)—smectite the essential nanoclay raw material and a 2:1 layered smectite clay mineral with a platy structure
3. Illite
4. Chlorite

4.3.1.1 NANOCLAY

Nanoclay is a mineral that has a high perspective proportion and with at any rate one component of the particles in the nanometer run. The most significant factor is the viewpoint proportion of earth minerals. The muds having a platy structure and a thickness of short of what one nanometer are the muds of decision. The length and width of the decision dirts are in the micron extend. Viewpoint proportions of the decision muds are in the 300:1 to 1500:1 range [2]. Nanoclay has gotten a lot of consideration as fortifying materials for polymer due to its conceivably high perspective proportion and novel intercalation attributes. The modest quantity expansion of nanoclay into polymer grid shows sudden properties including diminishing gas porousness, improved dissolvable opposition, being prevalent in mechanical properties and warm steadiness, and upgraded fire-resistant properties. Earth minerals have been proposed as key constituents of a few adjustment of medication conveyance frameworks, with various purposes and acting through different components. In view of their high maintenance limits just as expanding and colloidal properties, muds have been proposed as valuable materials to change sedate conveyance. In view of their expanding potential, dirt minerals can be viably used to delay (broadened discharge frameworks) medicate discharge or even improve sedate solvency [3–6].

4.3.1.2 PHYSICAL PROPERTIES OF NANOCLAY

Color	Off white
Shape	Fine powder
Bulk density g/cc	1.5–1.7
Weight loss at 1000 °C	37%
D spacing at d001	17.2 Å

The benefits of nanoclay are:

1. Easy dispersion
2. Viscosity stability
3. Increased efficiency
4. Increased mechanical properties such as modulus
5. Scratch resistances
6. Increased heat resistance and dimensional stability

The nanoclays that researchers have concentrated on are listed below:

1. Hydrotalcite (HTC)
2. MMT
3. Mica Fluoride
4. Octasilicate

Types of composites from polymer–clay interactions are shown in Figure 4.5

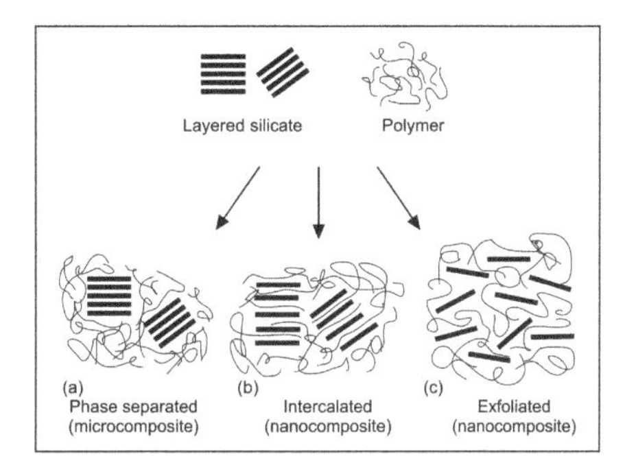

FIGURE 4.5 Types of composites from polymer–clay interactions.

Source: Reprinted from Ref. [32]. Open access.

4.3.1.3 MMT CLAY (FIGURE 4.6)

So as to alter medicate discharge, an extremely fascinating plausibility is to utilize mud mineral–polymer composites. In spite of the fact that earth minerals and polymers were as often as possible utilized in their unadulterated structure as single medication transporters, this sort of drug delivery system (DDS) frequently did not meet the necessities of everyone. The planning of polymer-layered silicate composite offered the probability of improving the properties of every single segment: those of the earth mineral particles alone (steadiness of the dirt mineral scatterings and changes in its particle trade conduct) and, all the more every now and again, those of the polymer (mechanical properties, growing limit, film framing capacities, rheological properties, bio bond or cell take-up) [7]. MMT has a place with the smectite gathering, composed of silica tetrahedral sheets layered between alumina octahedral sheets at a proportion of 2:1, individually [8]. The imperfection of the gem cross-section and the isomorphic substitution induces a net negative character that prompts the adsorption of soluble earth metal particles in the interlayer space. Such blemish was answerable for the movement and trade responses with organic mixes [9]. MMT likewise contains dangling hydroxyl end-bunches on the surfaces [10]. It has huge specific surface territory, displays great absorbance capacity, high feline particle trade limit, champion bio adhesiveness, and medication conveying ability. It was accounted for [11] that medication consolidation into clays took place by adsorption, both by intercalation into the earth structure inside the interlayer space (by supplanting the water particles) and superficially. The most significant cooperation that took place between the two parts of the half-breed framework was ionic [12]. The particle trade procedure may occur by blending particle exchangers with ionic medications in arrangement. In biological fluids, "counterions" could dislodge the medication from the substrate and convey it into the body. The exchanger may be then wiped out or biodegraded MMT nanoclays are interesting muds having a platy structure with a unit thickness of one nanometer or less. This dirt additionally has an angle proportion in the 1000:1 territory in light of the fact that MMT earth is hydrophilic, it is not perfect with most polymers and must be synthetically changed to make its surface increasingly hydrophobic. MMT is a subclass of smectite, is a 2:1 earth mineral, meaning that it has two tetrahedral sheets of silica sandwiching a focal octahedral sheet of alumina. MMT is a characteristic mineral comprised of an insoluble enormous layer and pitifully bound cations to the space between layers. Each layered hydrated aluminum silicate whose unit cell is made out of one Al-octahedral sheet (O) sandwiched between two Si-tetrahedral sheets (T). It has a net negative charge because of isomorphous ionic substitutions in the T–O–T structure. This charge is remunerated by interlayer hydrated cations, which can be traded by an assortment of natural atoms. MMT can grow with the expansion of huge cations into the space between the layers of MMT. Intercalated dirt is a subsidiary of smectite which the cations have been traded by enormous cations and these cations fill in as a mainstay of help between MMT layers. MMT is a smectite bunch of dirt and has an enormous explicit surface territory, great adsorption, and cation trade limit [13–18].

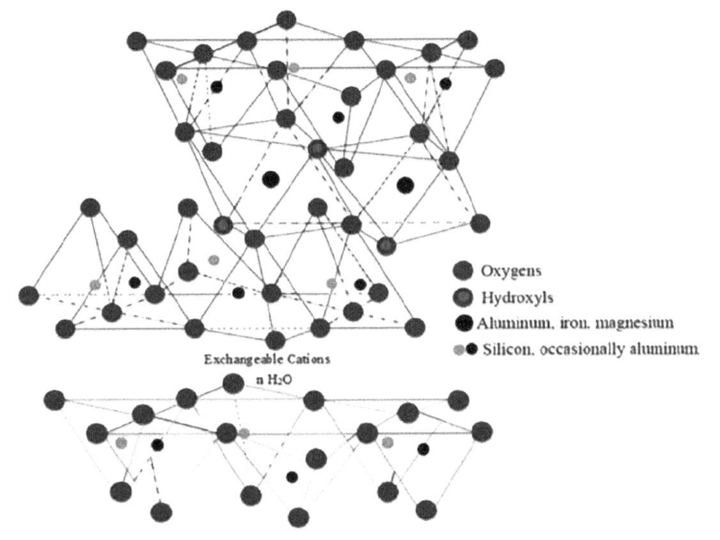

FIGURE 4.6 Structure of Na-MMT (Adapted from Ref. [34]).

4.3.1.4 ADVANTAGES OF NANOCLAY IN THE DRUG DELIVERY SYSTEM

1. Nanoclay on the nanoscale level shows significant improvements in mechanical properties, heat distortion temperatures, thermal stability, flame retardancy, and enhanced barrier properties.
2. Improve the compatibility with the polymer matrix.
3. Low cost.

4.4 MECHANISM OF CLAY-DRUG INTERACTION

For the most part clay minerals are made out of inorganic cations subsequently they promptly experience particle trade with fundamental medications. In contrast with different gatherings in clay minerals, smectites are all the more normally contemplated because of its higher cation trade limit. By and by; there are a few instruments that might be engaged with the collaboration between clay minerals and natural particles [19].

The structure of 2:1 layered silicates treatment was provided in Figure 4.7 and Figure 4.8 represents nanoclay, nanocrystal, and nanotube.

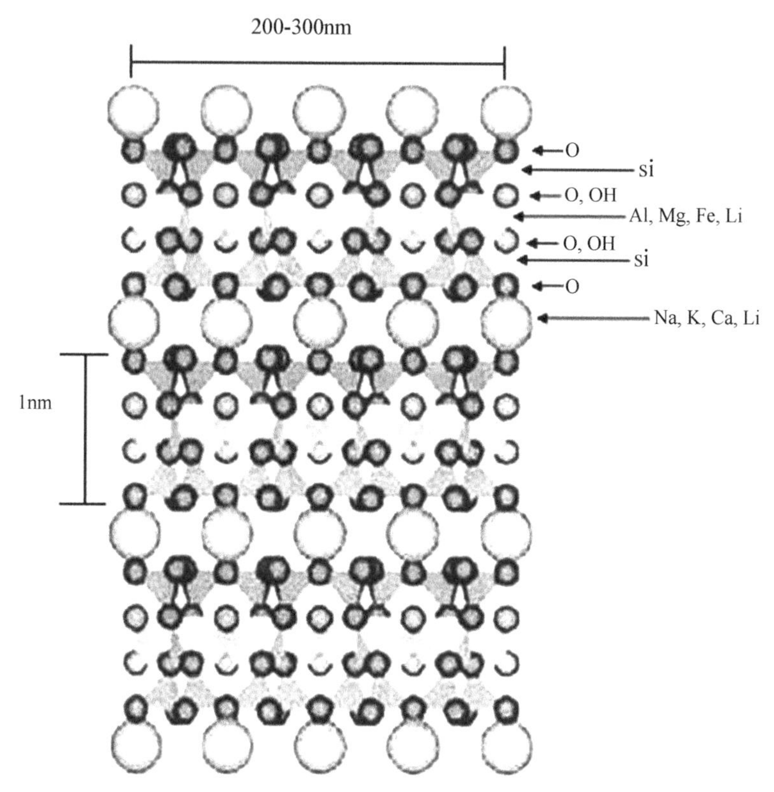

FIGURE 4.7 The structure of 2:1 layered silicates treatment.
Source: Reprinted with permission from Ref. [33].

4.4.1 DISSOLUTION ENHANCEMENT OF NANOCLAY

Improving disintegration of ineffectively water dissolvable medications stays one of the more significant changes for pharmaceutical technologists. Among the few methodologies applied, the surface adsorption of medication is one fascinating methodology, lastly atoms are isolated into strong and afterward significantly build the surface territory accessible to the disintegration medium. Smectites were found to successfully improve the in vitro disintegration pace of nonionic and acidic insoluble medications. At the point when medication discharges from the mud surface and advanced or raise by the feeble holding and because of hydrophilic properties of the earth [20].

TABLE 4.1 Comparison of Nanoclay with Other Colloidal Carrier

Route of Administration	Nanoclay	Nano-oxide	Carbon Nanotubes	Carbon Nanofillers
Oral delivery	Possible	Possible	Not possible	Not possible
Transdermal delivery	Possible	Possible	Not possible	Not possible
Local delivery	Possible	Possible	Not possible	Not possible
Physical stability	++++	+++	+	+++
Biological stability	++	++	+	++
Biocompatibility	+++	+++	+++	++
Ease of sterilization	++	++	+	++

+: Mild
++: Moderate
+++: High
++++: Extremely high

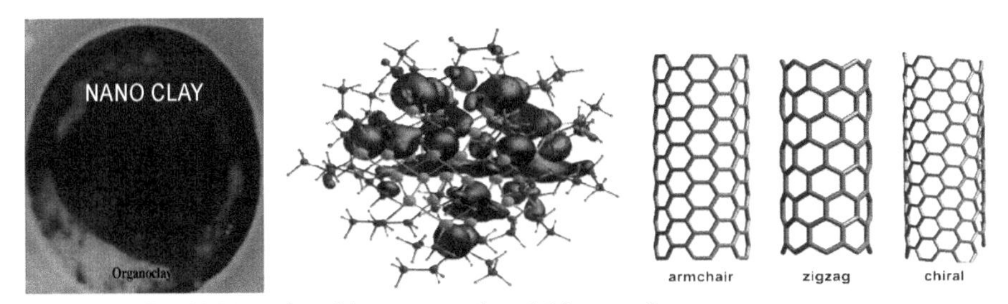

FIGURE 4.8 (a) Nanoclay, (b) nanocrystal, and (c) nanotube.

Source: a) https://www.slideshare.net/taifalawsi1/nano-clay-and-its-applications; b) https://www.labmate-online.com/news/microscopy-and-microtechniques/4/breaking-news/an-introduction-to-nanocrystals/30231; c) https://coecs.ou.edu/Brian.P.Grady/nanotube.html

TABLE 4.2 Application of Nanoclays [21]

Type of Drug-Delivery System	Natural Minerals Employed	Mechanism
Extended release systems	Smectites MMT fibrous minerals, HTC	They can retain large amounts of drug due to their high cation exchange capacity
Targeted delivery systems	Natural, synthetic, nanocomposites clay–polymers, films, and hydrogels composites clay-polymers	Interact with drugs reducing their absorption. Therefore, such interactions can be used to achieve technological and biopharmaceutical advantages, regarding the control of release

TABLE 4.1 (Continued)

Type of Drug-Delivery System	Natural Minerals Employed	Mechanism
Colon delivery systems	MMT	Pharmaceutical natural minerals and drug interactions
Periodontal systems	Laminar minerals	Improved bioadhesion
Hydration-activated extended-release systems	Smectites	Act as disintegrant agents in tablet formulations because of their hydrophilic and swelling properties
Microparticles	Amorphous silica bentonite attapulgite kaolin talc	Encapsulation of surface, precipitation inclusion

4.4.2 NANOCLAY FOR CONTROLLED DRUG DELIVERY SYSTEM

Nanoclay shows more points of interest when contrasted with the polymers and lipids, for example, mechanical and rheological properties, great intercalation limit, huge surface region, and high incorporation limit. These are utilized generally as they are practical and biocompatible in nature. Thus dirt mineral particles have been utilized to grow new controlled discharge frameworks. The layered materials suits polar natural mixes between their layers and structure an assortment of intercalated mixes and the arrival of medications is possibly controllable, subsequently can be utilized in the pharmaceutical field. These shows amplify remedial action and limit reactions. Calcium MMT has likewise been utilized for the treatment of torment, open injuries, colitis, loose bowels, hemorrhoids, stomach ulcers, intestinal issues, skin inflammation, frailty, and an assortment of other medical problems. Not exclusively does MMT fix minor issues, for example, looseness of the bowels and clogging through the nearby application, it follows up on all organs too [12]. Yuancai and Si-Shen have depicted the novel poly(D,L-lactide-*co*-glycolide)/MMT nanoparticle sedate conveyance framework, detailing the medication transporter from a material, which can likewise have remedial impacts, either synergistic with or competent to intervene the symptoms of the epitomized medication.

Zeolite (ZLT) and HTC incorporate swellability, cation/anion trade and intercalation which make them promising applicant being developed of medication nanoclay composites for controlled discharge sedate conveyance frameworks. These have huge pores and medication can be stacked into the nanoporous structure normally. The littler pores than mesoporous material are utilized to accomplish controlled discharge adequately. The medication discharge relies upon the size selectivity and pore size of the grid. Nanocomposite has been effectively created among medication and ZLT joined with biodegradable polymers (e.g., chitosan, gelatin, and alginate) and delayed the medication discharges. Strikingly, Diclofenac sodium has been stacked into common ZLT and the medication discharged for 8 h.

HTC (anionic earth) is a two-dimensional layered twofold hydroxides material. The medication can be stacked into the interlayers because of its high fondness for carbonate particles in corrosive medium. HTC has been created as a medication bearer utilizing ketoprofen, sodium diclofenac, and chloramphenicol succinate tranquilize discharge up to 24 h. HTC was utilized as a controlled discharge detailing utilizing ineffectively water-solvent medication and discharged the medication in a controlled discharge way in intestinal condition.

4.4.3 NANOCLAY FOR ANTICANCER DRUGS

Paclitaxel (hostile to malignancy medicate) stacked poly(D,L-lactide-*co*-glycolide)/MMT nanoparticles were set up by the emulsion/dissolvable vanishing strategy and it was tried for in vitro medication discharge. The underlying explosion of 22% on the first day and in quite a while, about 36% of medication was discharged with a somewhat decreased starting burst and speed discharge. The adsorption and desorption of natural particles and surfactants on layered silicates demonstrates that these materials can be utilized for medicate conveyance. The arrival of buformin from buformin/MMT intricate and unadulterated buformin hydrochloride in counterfeit intestinal juice has indicated an arrival of 70% buformin with a lower rate when contrasted with an unadulterated compound in 3 h.

4.4.4 NANOCLAY FOR MEDICAL DEVICES

The medical devices contain polymer transporter (as network) and medication intercalated layered silicates (as fortification). Nanoclay gives controlled arrival of remedial operator in restorative gadgets, for example, fix, implantable, or insertable medicinal gadget. Intercalation of fenbufen in a layered twofold hydroxide covering with Eudragit S 100 gives a composite material that shows the controlled arrival of the medication under in vitro conditions.

4.4.5 NANOCLAY FOR GENE THERAPY

To treat hereditary insufficiencies and perilous infections, quality treatment has been utilized. A transporter framework, for example, intercalated layer is required for the effective presentation of outside DNA into cells. As of late, it has been effectively exhibited that novel layered twofold hydroxide could frame a nanohybrid by intercalating with bimolecular anion, for example, mononucleotides, DNA which shows that antisense oligonucleotide atoms bundled in the layered twofold hydroxide can enter cells, apparently through phagocytosis or endocytosis. The leukemia cells were utilized to investigate the layered twofold hydroxide's potential as a quality transporter.

4.4.6 NANOCLAY IMPROVES DRUG DISSOLUTION RATE

Adsorption expands the dissolvability of the medication. Finely partitioned strong particles extraordinarily increment the surface territory accessible to the disintegration medium. Smectites were seen as successfully upgraded the in vitro disintegration pace of nonionic and acidic insoluble medications. Phenytoin MMT adsorbates had the option to improve the bioavailability of the medication in people in examination with phenytoin sodium cases.

4.4.7 NANOCLAY IN COSMETIC USES

Clay is a significant therapeutic and corrective apparatus and has been utilized by the world for over a considerable length of time. Clay, as French green mud (illite) and rhassoul mud (smectite), are utilized remotely for topical purposes and for restorative purposes. In Mesopotamia, blends of clays were utilized as cleanser, another facial and skin-care treatment. In Egypt, "red ochre" (a color produced using normally tinted mud minerals, with generally a lot of hydrated Fe(III)) is utilized for lips and cheeks [22, 23].

4.4.8 FUTURE ASPECTS OF NANOCLAY

Nanocomposites dependent on muds and natural mixes are extending because of their favorable circumstances. Nanoclays get extraordinary enthusiasm for applications dependent on their ability for particular adsorption of atoms. In this way, they have been utilized for application in chromatography detachments, to expel natural contaminations from air and water, and to create improved plan for pesticides, as synthetic sensors and subatomic strainers, and so forth. Among other p significant advancement in the improvement of clay has been made in the course of recent years. The nanoclay medication conveyance framework holds a splendid future in different pharmaceutical and restorative applications. These are utilized as symptomatic operators and against bacterial specialists. Nanoclays have the incredible potential when contrasted with polymers and other nanotechnological sedate conveyance frameworks like dendrimers, carbon nanocylinder, liposomes and neosomes, and so on for tranquilize conveyance applications.

Greater part of the orally controlled medications that are ingested into the foundational dissemination are inadequately dissolvable, along these lines display low oral bioavailability. To conquer this serious issue, there are numerous procedures utilized for the upgrade of disintegration through different medication conveyance frameworks. Nanoclay, a novel methodology has been utilized for improving the disintegration of ineffectively dissolvable medication. At the point when medication discharges from the clay surface and advances or ascends by the powerless holding and because of hydrophilic properties of the earth, which thus improved in tranquilize solvency in the clay.

The objective of this study is to formulate and characterize aceclofenac (AC) using nanoclay for improved oral bioavailability using MMT and polyvinyl alcohol [24, 25].

The specific objectives of the present investigation are as follows:

1. To formulate AC-loaded nanoclay
2. To characterize physicochemical AC-loaded nanoclay
3. To evaluate AC-loaded nanoclay for
   a. In vitro cytotoxicity studies
   b. In vivo bioavailability studies

4.5 METHODOLOGY

4.5.1 PREFORMULATION STUDIES

Preformulation might be depicted as a period of the innovative work process where the definition researcher describes the physical, synthetic, and mechanical properties of another medication substance, so as to create steady, sheltered, and powerful dose structures. In a perfect world, the preplan stage starts from the get-go in the revelation procedure with the end goal that suitable physical, substance information is accessible to help in the determination of new compound elements that enter the advancement procedure. During this assessment, conceivable cooperation with different dormant fixings proposed for use in conclusive dose structure is additionally considered.

4.5.2 SOLUBILITY STUDIES

The solubility of AC was determined in various solvents by adding an excess amount of drug to 10 mL of solvents in conical flasks. The flasks were kept at 25 ± 0.5 °C in isothermal shaker for 24 h to reach equilibrium. The equilibrated samples were removed from the shaker and centrifuged at 4000 rpm for 15 min. The supernatant was taken and filtered through 0.45-μm membrane filter. The concentration of AC was determined in the supernatant after suitable dilution by using UV–visible spectrometer at 276 nm.

4.5.3 COMPATIBILITY STUDIES

4.5.3.1 FOURIER TRANSFORM INFRARED SPECTROSCOPY (FTIR)

IR spectroscopy can be utilized to examine and foresee any physiochemical association between various parts in a definition and thusly it very well may be applied to the determination of reasonable substance good excipients while choosing the fixings, we would pick, those which are steady, perfect, cosmetically, and remedially worthy. Infrared spectra coordinating methodology was utilized for location of any conceivable synthetic collaboration between the medication, lipid, and surfactants. A physical blend of medication, lipid, and surfactants was arranged and blended in with appropriate amount of potassium bromide. This blend was packed to shape a straightforward pellet utilizing a water-driven press at 15 tons pressure. It was checked from 4000 to 400 cm^{-1} of every FTIR spectrophotometer (FTIR 8400 S, Shimadzu). The IR range of the physical blend was contrasted and those of unadulterated medication, lipid, surfactants, and pinnacle coordinating were done to identify any appearance or vanishing of peaks.

4.5.3.2 DIFFERENTIAL SCANNING CALORIMETRY

Differential scanning calorimetric examination was utilized to describe the warm conduct of the medication, lipids, and their physical blends. Test was pleated in standard aluminum skillet and warmed from 20 to 400 °C at a warming pace of 10 °C/min under consistent cleansing of dry nitrogen at 30 mL/min. A vacant skillet, fixed similarly as the example, was utilized as a source of perspective. DSC thermograms were acquired utilizing a programmed warm analyzer framework. Temperature adjustment was performed utilizing indium alignment reference standard.

4.5.3.3 DEVELOPMENT OF CALIBRATION CURVE

Standard stock solution was set up by dissolving 50 mg of medication in 50 mL of ethanol to get centralization of 1 mg/mL. 10 mL of stock arrangements were additionally weakened to 100 mL with phosphate cushion (pH 6.8, prepared according to I.P.) to get a working arrangement of focus 100 μg/mL of AC, individually. The arrangements were additionally weakened with phosphate cradle and examined in the wavelength extend 276 nm. The medications comply with Beer's law inside the fixation scope of 10–50 μg/mL for AC. Standard bend among focus and absorbance was plotted.

4.5.3.4 PREPARATION OF AC/ MMT NANOCLAY

In the preparation of the AC/MMT, 1 g of MMT was dispersed in 30 mL of distilled water with vigorous stirring for 0.5 h at room temperature. Various amounts of AC were dissolved in 150 mL of distilled water (pH value adjusted to 8.5 with H_3PO_4). Then, the two solutions were mixed together and kept at 80 °C with vigorous stirring for 3 h. After centrifugation, the sediment was washed with deionized water, and then further dried in a vacuum oven at 80 °C for 24 h.

4.5.3.5 PREPARATION OF DRUG-LOADED BATCHES

Drug-loaded nanoclay was prepared by MMT as polymers, AC as drug, and F-127 was used as hydrophilic surfactant.

4.6 EVALUATION OF NANOCLAY

4.6.1 PARTICLE SIZE AND ZETA POTENTIAL

Particle size and zeta potential of the strong nanoclay were estimated by photon connection spectroscopy utilizing a Malvern Zetasizer Nano ZS90 (Malvern Instruments, Worcestershire, UK), which takes a shot at the Mie hypothesis. All size and zeta potential estimations were done at 25 °C utilizing dispensable polystyrene cells and expendable plain collapsed slender zeta cells, individually, after fitting weakening with unique scattering arrangement medium.

4.6.2 POLYDISPERSITY INDEX

Polydispersity was determined according to the equation,

$$\text{Polydispersity} = D\,(0.9) - D\,(0.1)/D\,(0.5)$$

where,

D (0.9) corresponds to particle size immediately above 90% of the sample.

D (0.5) corresponds to particle size immediately above 50% of the sample.

D (0.1) corresponds to particle size immediately above 10% of the sample [26].

4.6.3 INTERCALATION EFFICIENCY AND DRUG CONTENT

Accurately weighed amount of drug was added to accurately weighed nanoclay in a suitable solvent .the solution was stirred for 30 min. The solution was then centrifuged. The supernatant layer was taken and measured in UV spectrophotometer. The absorbance obtained was extrapolated in the calibration graph to get the concentration. Moreover, the concentration is kept in the formula to get loading and entrapment efficiency.

%IE = [(drug added – unintercalated drug)/drug added] × 100

%LC = [intercalated drug/nanoclay weight] × 100

4.7 MORPHOLOGICAL STUDY (SCANNING ELECTRON MICROSCOPY)

External morphology of nanoparticles was resolved utilizing scanning electron microscopy (SEM). Tests were weakened with ultrapurified water to get an appropriate fixation. At that point, the examples were spread on an example holder and dried utilizing vacuum. They were in this manner covered with gold (JFC 1200 fine coater, Japan) and inspected by an SEM.

4.7.1 IN VITRO RELEASE STUDIES

The release of AC from the nanoclay was considered under sink conditions. AC stacked nanoclay identical to 40 mg were suspended in 1 mL of disintegration media (phosphate buffer 6.8) and put in dialysis sacks (MWCO 12000, HiMedia). The dialysis sacks were set in 100 mL of disintegration medium and mixed under attractive mixing at 37 C. About 5 mL aliquots of the disintegration medium were pulled back at each time interim and a similar volume of new disintegration medium was added to keep up a consistent volume. Tests pulled back from pH 6.8 phosphate support were examined for AC content spectrophotometrically against solvent blank.

4.7.2 RELEASE KINETICS

In vitro dissolution has been perceived as a significant component in sedate advancement. Under specific conditions, it tends to be utilized as a surrogate for the appraisal of bioequivalence. A few hypotheses/motor models portray medicate disintegration from prompt and changed discharge measurement structures. There are a few models to represent the medication disintegration profiles where f_t is the capacity of t (time) identified with the measure of medication broke down from the pharmaceutical dose framework. To look at disintegration profiles between two medication items model ward (bend fitting), measurement investigation, and model-free strategies can be utilized [27]. In a request to explain the mode and instrument of medication discharge, the in vitro information was changed and deciphered at graphical interface built utilizing different motor models. The zero request discharge, Equation (4.1) depicts the medication disintegration of a few kinds of adjusted discharge pharmaceutical measurement structures, as on account of transdermal frameworks, grid tablets with low dissolvable medications, covered structures, osmotic frameworks, and so forth, where the medication discharge is free of fixation.

$$Q_t = Q_o + K_o t \quad (4.1)$$

where Q_t is the amount of drug released in time t, Q_o is the initial amount of the drug in the solution, and K_o is the zero-order release constant

The first-order equation (4.2) describes the release from the system where release is concentration-dependent, for example, pharmaceutical dosage forms containing water-soluble drugs in porous matrices.

$$\log Q_t = \log Q_o + K_1 t/2.303 \quad (4.2)$$

where Q_t is the amount of drug released in time t, Q is the initial amount of drug in the solution, and K is the first s order release constant.

Higuchi described the release of drug from insoluble matrix as a square root of time

$$Q_t = K_H \sqrt{t} \quad (4.3)$$

where Q_t is the amount of drug released in time t and K_H is Higuchi's dissolution constant.

4.7.3 HEMOCOMPATABILITY STUDIES

Blood samples of healthy human volunteers were obtained from blood bank of government hospital, Ooty in evacuated siliconized glass tube containing sodium citrate. Red blood cells were separated by centrifugation at 1500 rpm for 10 min and then washed 3 times with saline. Stock solution of erythrocytes in saline water was prepared such that the cell count was 1 × 108 cells/mL. Equal volumes of RBC suspension and nanoparticles dispersion were suspended in a microcentrifuge tube such that the final concentrations of nanoparticle dispersion and nanoparticles were 150–1000 μg/mL and incubated separately at 37 °C for 1 h. 1% triton X and saline water were used as positive and negative controls respectively. After 1 h, the tubes were centrifuged at 1500 rpm for 10 min and the hemoglobin released in the supernatant was detected by UV absorbance at 276 nm. All measurements were performed in triplicate (n = 3) and the SD was calculated. The percent hemolysis was calculated by the formula:

$$\% \text{ hemolysis} = \frac{\text{ABS sample} - \text{ABS } 0\%}{\text{ABS } 100\% - \text{ABS } \%}$$

where Abs sample is the absorbance of supernatant of erythrocyte and nanoparticles suspension.

Abs0% is the absorbance of supernatant of erythrocyte and PBS suspension.

Abs100% is the absorbance of supernatant of erythrocyte and triton X.

4.8 IN VITRO AND IN VIVO STUDIES

4.8.1 DETERMINATION OF MITOCHONDRIAL SYNTHESIS BY MTT ASSAY

1. The monolayer cell culture was trypsinized and the cell check was changed in accordance with 1.0 × 105 cells/mL utilizing Dulbecco's modified medium (DMEM) containing 10% f*etal bovine serum (*FBS).
2. To each well of a 96 well microtitre plate, 100 μL of the weakened cell suspension (roughly 10,000 cells/well) was included.
3. Following 24 h, when a halfway monolayer was framed, the supernatant was flicked off, the monolayer was washed once with medium and 100 μL of various test focuses arranged in upkeep media were added per well to the fractional monolayer in microtitre plates. The plates were then brooded at 37 C for 48 h in 5% CO_2 environment, and minuscule assessment was done and perceptions recorded like clockwork.
4. Following 48 h, the example arrangements in the wells were disposed of and 20 mL of MTT (2 mg/mL) in MEM-PR (MEM without phenol red) was added to each well.
5. The plates were delicately shaken and brooded for 3 h at 37 °C in 5% CO_2 air.
6. The supernatant was evacuated and 50 mL of isopropanol was included and the plates were delicately shaken to solubilize the framed formazan.
7. The absorbance was estimated utilizing a microplate per user at a wavelength of 540 nm.

The rate development restraint was determined utilizing the accompanying equation and convergence of medication or tests expected to hinder cell development by half qualities were created from the portion reaction bends for every cell line.

% cell viability = mean OD of individual test group ×100

mean OD of control group [28].

4.8.2 HIGH-PERFORMANCE LIQUID CHROMATOGRAPHY (HPLC) BIOANALYTICAL METHOD

Reverse phase HPLC method is the most popular mode for analytical and preparative separations of the compounds in chemical, biological, pharmaceutical, and food samples. In reversed-phase mode, the stationary phase is nonpolar and the mobile phase is polar. The polar compounds get eluted first in this mode and nonpolar compounds are retained for longer time. In the present study, methods for the estimation of AC present in the blood plasma samples were developed and validated. For the estimation of AC in blood plasma, the chromatographic variables, namely pH, solvent strength, solvent ratio, flow rate, addition of peak modifiers in mobile phase, nature of the stationary phase, detection wavelength, and internal standard were studied and optimized for the separation and retention of the drug. The following are the optimized chromatographic conditions, preparation of standard and sample solutions, and the methods used for the estimation of AC in plasma [29].

Chromatographic conditions: Shimadzu gradient HPLC system was used with following configurations:

- LC-20 AD solvent delivery system (pump)
- Manual injector 25 μL (rheodyne)
- SPD-20A UV detector
- Isocratic data station

Stationary phase: Hypersil BDS C18 (250 cm × 4.6 mm i.d., 5 μm)

Mobile phase: Acetonitrile: 25 mM potassium dihydrogen orthophosphate (pH 3.25)

Mobile phase ratio: 60:40

Flow rate: 1 mL/min

Sample volume: 20 μL

Detection: 276 nm

Data station: isocratic solutions

The mobile phase was filtered through 0.22 μm nylon membrane and degassed using ultrasonicator. All the experiments were carried out at room temperature.

Preparation of AC Standard Stock Solution: About 10 mg of AC was transferred into a 10 mL volumetric flask and the volume was made upto the mark with mobile phase to give 1 mg/mL (1000 μg/mL) solution. From this stock solution, 10 mL of 100 μg/mL solution was prepared and again from this solution 10 mL of 10 μg/mL was prepared.

Preparation of Analytical Calibration Curve Solutions: From the standard stock solution 0.25–2 μg/mL standard solutions were prepared and stored below 8 °C until further analysis.

Preparation of Blank Plasma: Blank plasma (0.2 mL) was

transferred into 2.0 mL centrifuge tube and 0.2 mL of precipitating agent (10% perchloric acid) were added. Finally made upto 2 mL with the mobile phase. The resulting solution was vortexed for 5 min and centrifuged at 4000 rpm for 10 min. The supernatant layer was separated and analyzed.

Preparation of Bioanalytical Calibration Curve Samples: A total of 0.2 mL of AC solutions were transferred to 2.0 mL centrifuge tube respectively, to this 0.2 mL of plasma and 0.2 mL of precipitating agent were added. The resulting solution was vortexed for 5 min and centrifuged at 4000 rpm for 10 min. The supernatant layer was separated and analyzed.

Preparation of Plasma Samples: Plasma samples (0.2 mL) obtained from study subjects were transferred into 2.0 mL centrifuge tube and 0.2 mL of precipitating agent was added. The resulting solution was vortexed for 5 min and centrifuged at 4000 rpm for 10 min. The supernatant layer was separated and analyzed.

4.9 METHOD OF ANALYSIS

The plasma concentration–time data was analyzed by one compartmental model, using Kinetica software (5.0, Thermo Scientific). Pharmacokinetic parameters like total area under the curve (AUC) 0–8 h, terminal phase half-life ($t_{1/2}$), peak plasma concentration(C_{max}), and time to reach the maximum plasma concentration (T_{max}) were determined. The relative bioavailability of AC-loaded nanoclay after oral administration was calculated in comparison to aceclofenac suspension (AC-SUSP).

4.9.1 *IN VIVO* ORAL BIOAVAILABILITY STUDIES

Male Wistar rats weighing 200–250 g were used for oral bioavailability studies. All animal experiments were approved by the Institutional Animal Ethical Committee, J.S.S. College of Pharmacy, Ooty (Proposal no. JSSCP/IAEC/M.PHARM/PH.CEUTICS/06/2016-17). All the rats fasted for 12 h before the experiments but had free access to water.

GROUPINGS

Group 1: control group
Group 2: pure drug of AC
Group 3: F1 nanoclay formulation treated group

TABLE 4.4 Groupings

Groups	No. of Animals	Formulation
I	2 males	Control
II	2 males	Pure drug of AC
III	2 males	F1

Prior to oral drug administration, four group of rats (total number = 6) fasted overnight (>12 h). For comparison of pharmacokinetic parameters, pure drug of AC, optimized formulation of nanoclay F1 was given to the rats via oral gavage (10 mg/kg). Blood samples (0.5 mL) were collected from the retro-orbital plexus at time points 0.5, 1, 1.5, 2, 4, 6, and 8 h after dosing. Immediately plasma was harvested by centrifugation at 3500 rpm for 5 min and transferred to a fresh Eppendorf tubes containing 30 µL of heparin and frozen to −20 °C prior to analysis [30, 31].

4.10 RESULTS AND DISCUSSION

4.10.1 PREFORMULATION STUDIES

4.10.1.1 SOLUBILITY STUDIES

The saturation solubility of AC was determined in different solvents viz. phosphate buffer pH 1.4, pH 6.8, and distilled water. The results are given in Table 4.5. AC was found to be highly soluble in phosphate buffer pH 6.8. The solubility in phosphate buffer pH 1.4 and distilled water

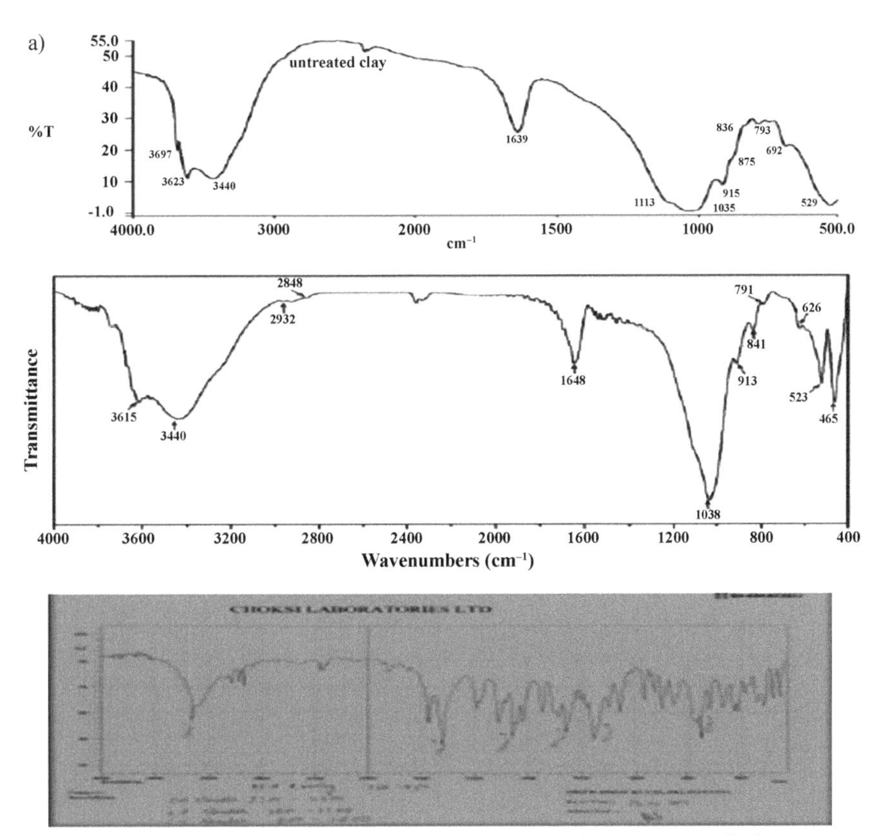

FIGURE 4.9 (a) IR spectra of AC, (b) IR spectra of MMT, (c) IR spectra of mixture.

was found to be 24.35 μg/mL and 12.85 μg/mL, respectively.

TABLE 4.5 Solubility Profile of AC in Different Media

Solvent	Solubility
Phosphate buffer pH 1.4	24.35 μg/mL
Phosphate buffer pH 6.8	55.46 μg/mL
Distilled water	12.85 μg/mL

4.10.2 COMPATIBILITY STUDIES

4.10.2.1 FTIR STUDIES

After interpretation of the above spectra it was confirmed that there was no major shifting, loss, or appearance of functional peaks between the spectra of drug, polymer, physical mixture of drug, and polymer (3615 cm^{-1}, 3440 cm^{-1}, 2440 cm^{-1}, 1648 cm^{-1}, and 1038 cm^{-1}). From the spectra, it was concluded that the drug was entrapped into the polymer matrix without any chemical interaction. From the IR study it was concluded that, the selected polymer MMT and polyvinyl alcohol (PVA) were found to be compatible in entrapping the selected drug AC.

Figure 4.9 represents (a) IR spectra of AC, (b) IR spectra of MMT, (c) IR spectra of mixture.

4.10.2.2 DSC STUDIES

It was concluded that the drug was entrapped into the polymer matrix without any chemical interaction. From the DSC (Figure 4.10) study, it was concluded that the polymers were found to be compatible in entrapping the selected drug AC. The melting of the mixture proved that the crystalline form of the drug is converted to amorphous form and hence more solubility.

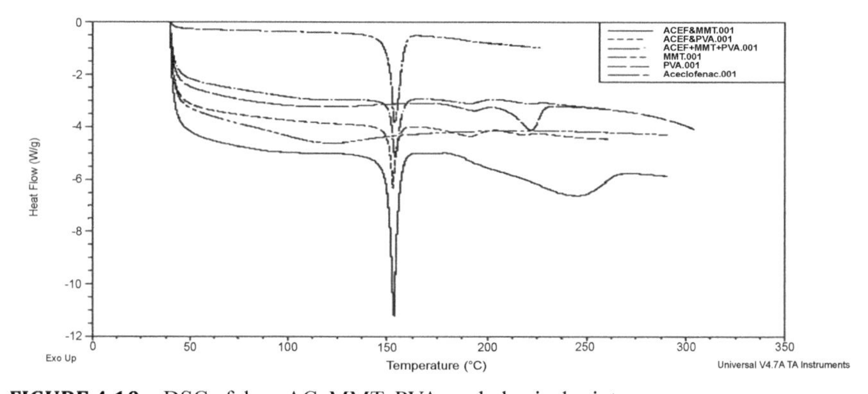

FIGURE 4.10 DSC of drug AC, MMT, PVA, and physical mixture.

4.10.2.3 DEVELOPMENT OF CALIBRATION CURVE

Calibration curve of the medication was created to establish out the linearity between convergence of medication in the arrangement and its optical thickness. It was presumed that the ideal linearity between the focus and absorbance was seen when the fixation extend was from 10 to 50 μg/mL. Table 4.6 and Figure 4.11 show the alignment of AC utilizing phosphate buffer pH 6.8.

TABLE 4.6 Calibration Curve of AC in Phosphate Buffer pH 6.8 (λ_{max} = 276 nm)

Sr. No.	Concentration (μg/mL)	Absorbance
1	10	0.115
2	20	0.172
3	30	0.314
4	40	0.443
5	50	0.587

4.10.3 EVALUATION OF MMT NANOCLAY

4.10.3.1 ZETA POTENTIAL, SIZE REPORT, AND POLYDISPERSITY

Zeta potential is the potential contrast between the scattering medium and the stationary layer of liquid connected to the scattered molecule. The criticalness of zeta potential is that its worth can be identified with the security of colloidal scatterings. The zeta potential demonstrates the level of aversion between adjoining, correspondingly charged particles in scattering. For atoms and particles that are sufficiently little, a high zeta potential will present solidness, for example, the arrangement or scattering will oppose accumulation. At the point when the potential is low, fascination surpasses aversion and the scattering will break and flocculate.

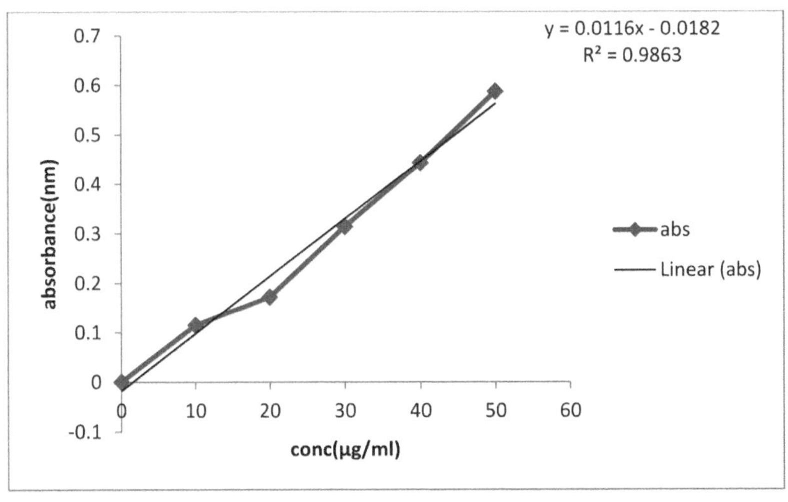

FIGURE 4.11 Calibration curve of AC in phosphate buffer pH 6.8 (λ_{max} = 276 nm).

4.10.3.2 POLYDISPERSITY

The polydispersity index (PI) is the measure of size distribution of the nanoparticle formulation. PI was measured using Malvern zetasizer. PI values range from 0.000 to 1.000, that is, monodisperse to very broad particle size distribution. PI values of all the formulations indicate that particle size distribution was unimodal. The optimized F1 batch having least particle size (286.9 nm) had a PI of 0.09. Figure 4.12 and Figure 4.13 represent zeta potential report of F1 and size report of F1, respectively.

Results

		Mean (mV)	Area (%)	St Dev (mV)
Zeta Potential (mV): -14.1	Peak 1:	2.32	13.3	9.90
Zeta Deviation (mV): 138	Peak 2:	-26.2	10.7	8.18
Conductivity (mS/cm): 0.0389	Peak 3:	-93.2	9.8	7.88

Result quality: See result quality report

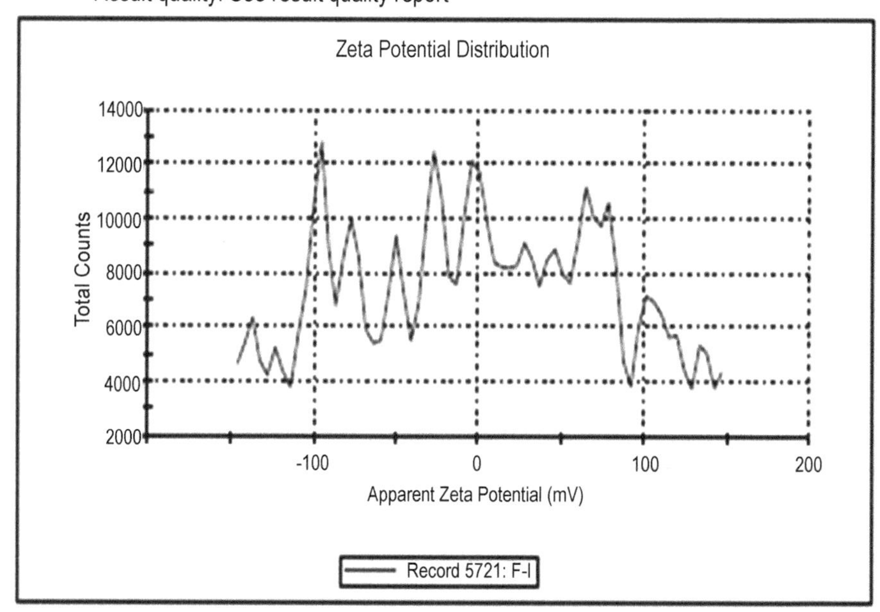

FIGURE 4.12 Zeta potential report of F1.

Results

			Size (d.nm):	% Intensity:	St Dev (d.n...
Z-Average (d.nm):	281.7	Peak 1:	286.9	100.0	60.71
Pdt:	0.099	Peak 2:	0.000	0.0	0.000
Intercept:	0.708	Peak 3:	0.000	0.0	0.000
Result quality:	Good				

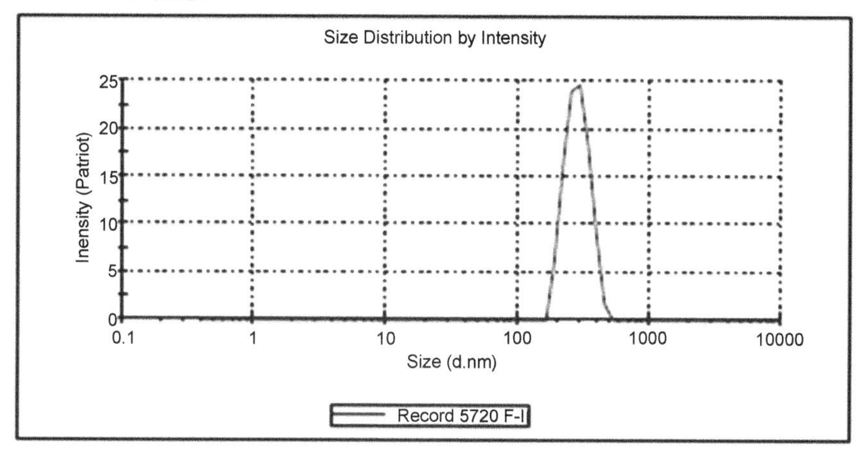

FIGURE 4.13 Size report of F1.

4.10.3.3 INTERCALATION EFFICIENCY AND DRUG LOADING

TABLE 4.7 Drug Loading and Intercalation Efficiency of Drug-Loaded Batches Prepared with Different Concentration of Drugs in Different Concentration

Formulation	Intercalation Efficiency (%)	Drug Loading (%)
F1	78	11.94
F2	12	1.7
F3	22	3.6
F4	54	15.1
F5	53	7.5
F6	45	9
F7	10	0.9
F8	20	1.5

4.10.3.4 SCANNING ELECTRON MICROSCOPY

The external morphological studies (SEM) uncovered that greatest nanoparticles were about round fit as a fiddle (Figure 4.14). The nanoparticle size saw by SEM corresponded well with the molecule size estimated by zeta sizer (Malvern instrument).

FIGURE 4.14 Scanning electron micrograph (SEM) of F1 nanoclay.

4.10.3.5 DSC OF FORMULATION (FIGURE 4.15)

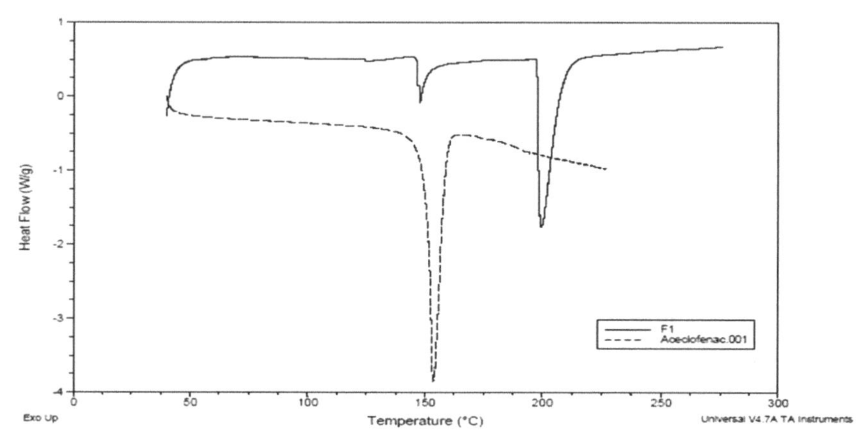

FIGURE 4.15 DSC of formulation.

4.10.3.6 IN VITRO RELEASE STUDIES

In vitro dissolution studies were carried out in phosphate buffer pH 6.8. The release profiles indicate that nanoclay formulations showed a retarded release of the drug from the MMT polymer when compared with plain aceclofenac solution (AC-SOL). The in vitro release data and graph of nanoclay formulations and AC-SOL in phosphate buffer pH 6.8 is shown in Table 4.8 and Figure 4.16. It was observed that AC-SOL showed 95.21 release in 4 h compared to 46.48 release for formulation F1 at the end of same time ($p < 0.05$). This is due to fact that there is no barrier for diffusion at dialysis membrane interface for AC molecules. Hence, higher release was observed in case of AC-SOL. But, formulation F1 released the drug in a sustained manner.

TABLE 4.8 Comparative In Vitro Release Data for MMT Nanoclay Formulations Compared to Pure Drug in Phosphate Buffer pH 6.8

Time (h)	Cumulative Percent Release (%)	
	F1	**Pure Drug**
0	0	0
0.5	8.28	33.56
1	16.45	65.42
2	33.25	85.46
4	46.48	95.21
6	57.17	
8	63.85	
12	76.42	
24	89.83	

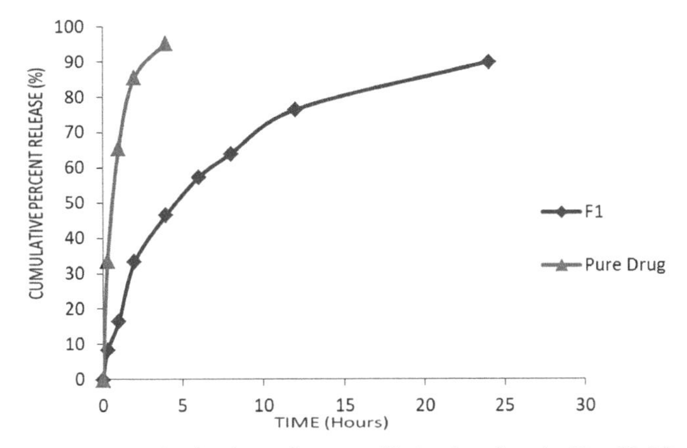

FIGURE 4.16 Comparative in vitro release profile in phosphate buffer pH 6.8, time (h).

4.10.3.7 RELEASE KINETICS

In order to elucidate mode and mechanism of drug release, the in vitro data was transformed and interpreted at graphical interface constructed using various kinetic models. The in vitro release data obtained for F1 formulation, in phosphate buffer pH 6.8, was fitted into various kinetic models. The results are shown in Table 4.9. Figures 4.17–4.21 represent various kinetics models of kinetics of F1 batch.

TABLE 4.9 R^2 Values and Slope Values for Applied Values

Sr. No.	Models	R^2 Values	Slope Values
1	Zero order	0.778	3.581
2	First order	0.971	−0.010
3	Higuchi	0.959	20.183
4	Hixson Crowell	0.920	−0.102
5	Korsemayer–Peppas	0.611	−0.689

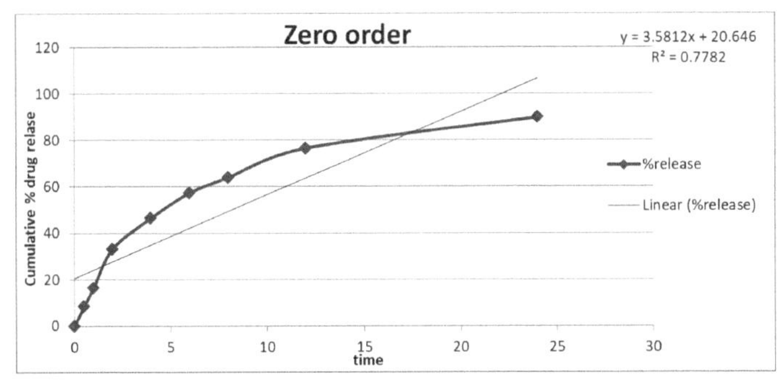

FIGURE 4.17 Zero-order kinetics of F1 batch.

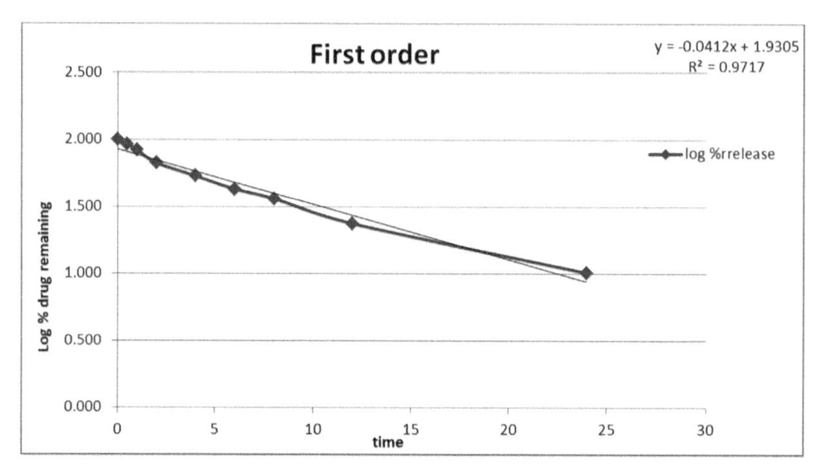

FIGURE 4.18 First-order kinetics of F1 batch.

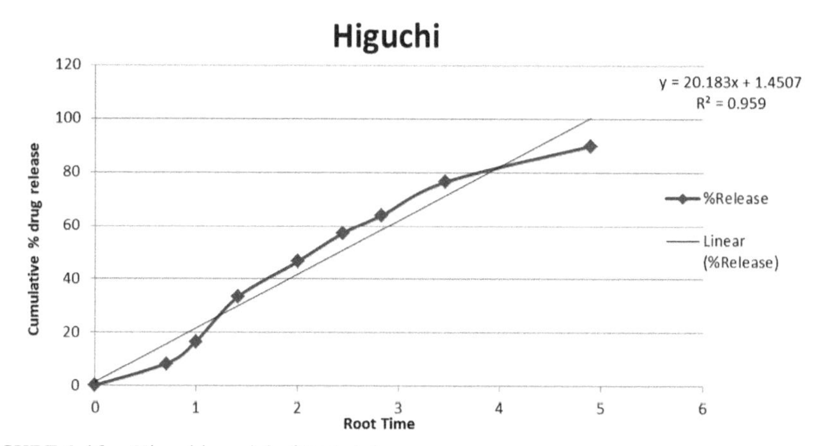

FIGURE 4.19 Higuchi model of F1 batch.

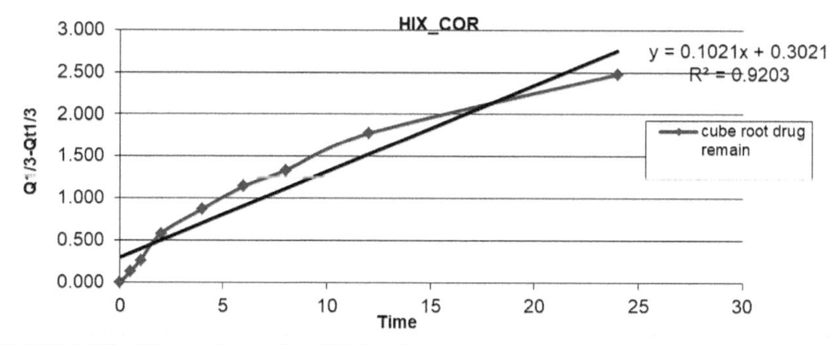

FIGURE 4.20 Hixson Crowells of F1 batch.

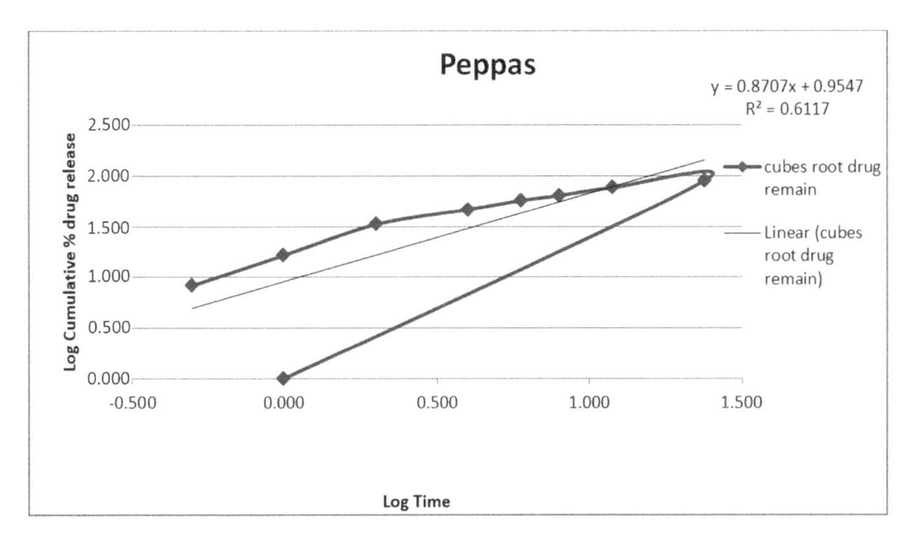

FIGURE 4.21 Korsemayer–Peppas model of F1 batch.

The dissolution kinetics of optimized batch was applied to various dissolution models such as zero order, first order, Higuchi, Korsemayer–Peppas, and Hixson Crowell.

The best-fitted model gives the highest R^2 value and least slope value. Thus, first-order model fits best for the dissolution data of the optimized batch as it showed the highest value for R^2.

4.10.3.8 HEMOCOMPATABILITY STUDIES

Nanoclay was subjected to rigorous blood biocompatibility tests. Erythrocyte-induced hemolysis in vitro can be considered to be a simple and reliable measure for estimating the membrane damage caused in vivo. Percent hemolysis was determined spectrophotometrically, detecting plasma free hemoglobin derivatives after incubating the particles with blood and then separating the undamaged cells by centrifugation. Typically less than 5% hemolysis is considered acceptable for blood biocompatibility. The concentration of 125–1000 μg/mL was subjected to determined percentage hemolysis. In this 1% triton X 100 was used as positive control and saline water as a negative control. Positive control showed percentage hemolysis of 3.714 ± 0.65 whereas negative control of 0.023 ± 0.001. On increasing the concentration from 150 μg/mL to 1000 μg/mL, there was no significant increase in % hemolysis. (Table 4.10). The results indicated that the AC nanoclay were hemocompatible and did not produce any toxic effects.

TABLE 4.10 Result of Hemolysis

Batches	% Hemolysis		
	125 μg/mL	500 μg/mL	1000 μg/mL
F1	0.011±0.001	0.015±0.001	0.022±0.001

4.10.3.9 MTT ASSAY

The percentage growth inhibition was calculated using the following formula and concentration of drug or test samples needed to inhibit cell growth by 50% values were generated from the dose-response curves for each cell line.

$$\% \text{ Cell viability} = 100 - \frac{\text{Mean OD of individual test group}}{\text{Mean OD of control group}} \times 100$$

The IC 50 value of the formulation F1 is about 245 mg/mL. Figure 4.22. represents percentage cell inhibition for F1.

4.10.4 BIOANALYTICAL METHOD DEVELOPMENT AND ANALYSIS

Bioanalytical calibration curve of AC was prepared in mobile phase and peak area was calculated. The results are given in Table 4.11 and Figure 4.23.

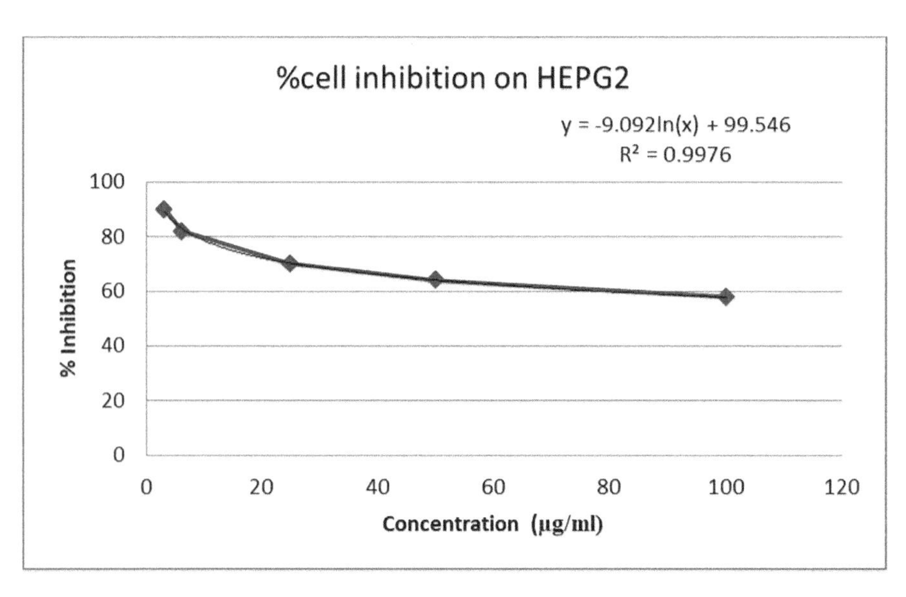

FIGURE 4.22 Percentage cell inhibition for F1.

TABLE 4.11 Bioanalytical Calibration Curve of AC

Concentration (μg/mL)	Peak Area
0.3	4,267,340
0.6	7,010,633
1.2	9,179,178
1.8	11,238,678
2.4	13,196,251
3	16,595,313
6	33,090,627

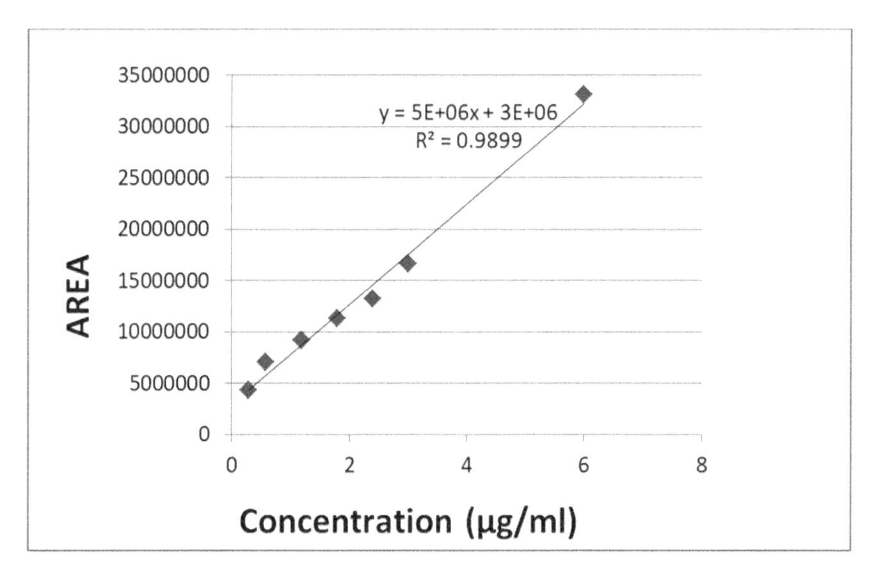

FIGURE 4.23 Bioanalytical calibration curve of AC.

The chromatograms of AC showed a stable baseline. The regression equation in the range of 0.5–2 μg/mL was as follows: $R^2 = 0.989$.

The concentration of AC was determined in plasma samples separated at different time intervals by HPLC analysis. The concentration of plasma samples was determined from the area of the chromatographic peak using the calibration graph. Chromatograms of blank plasma, pure drug, and F1 shown in Figure 4.24, 4.25 and 2.46, respectively.

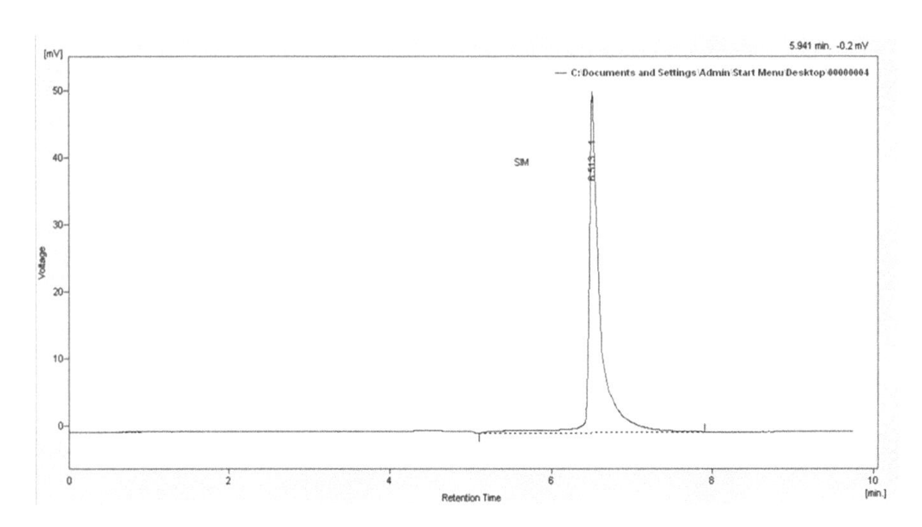

FIGURE 4.24 Chromatogram of pure drug.

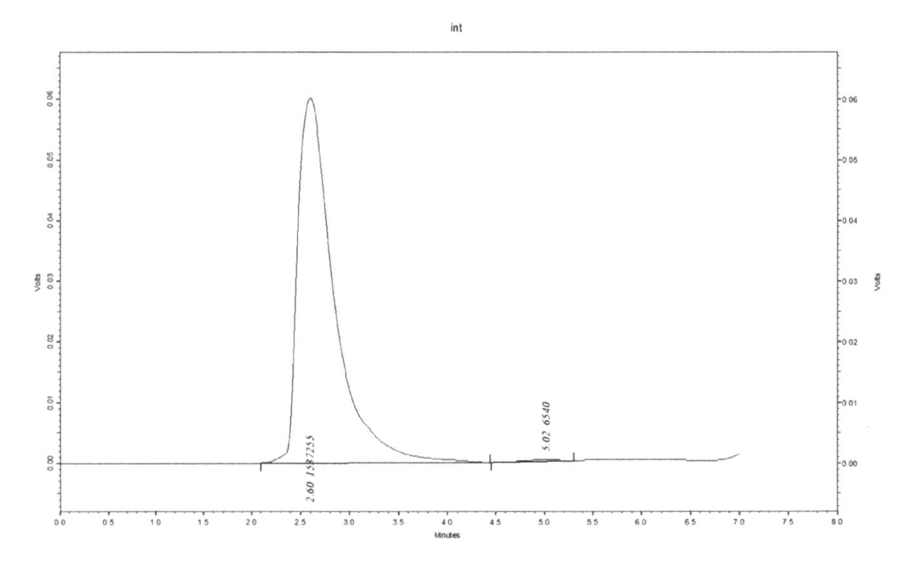

FIGURE 4.25 Chromatogram of plasma.

4.10.4.1 DATA ANALYSIS

Pharmacokinetic parameters of AC after oral organization appear in Table 4.12. Pinnacle focus (C_{max}) and time of pinnacle fixation (T_{max}) were acquired straightforwardly from the individual plasma-fixation time profiles. The zone under the focus time bend from time zero to time *t*

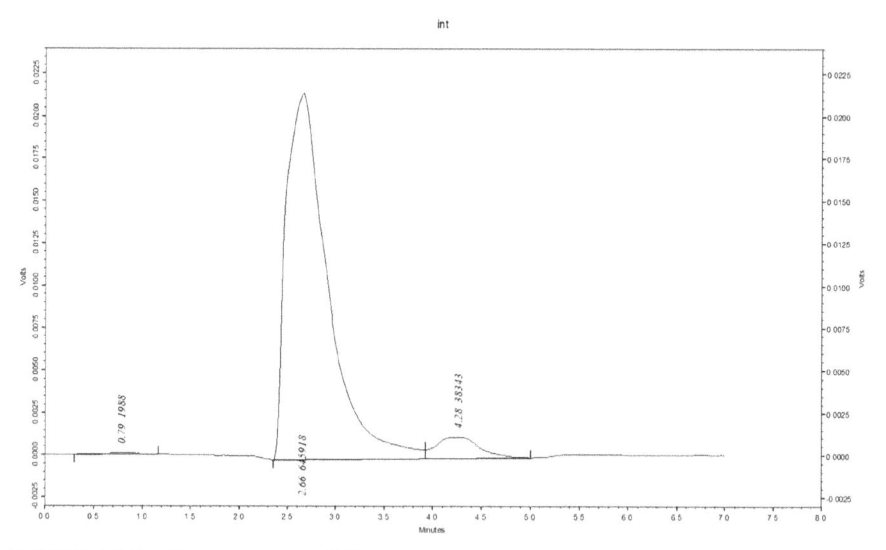

FIGURE 4.26 Chromatogram of F1.

($AUC0{\rightarrow}t$) was determined utilizing the trapezoidal strategy. The region under the bend (AUC) decides the bioavailability of the medication for the given similar portion in the detailing. The zone under the absolute plasma focus time bend from time zero to limitlessness was determined by

$$AUC_{0\rightarrow\infty} = AUC_{0\rightarrow t} + C_t/K_e$$

where C_t is the AC focus seen finally time, and K_e is the evident end rate steady acquired from the terminal slant of the individual plasma fixation time bends after logarithmic change of the plasma focus esteem and use of direct relapse. The relative bioavailability (F_r) at a similar portion was determined as F_r = $AUC_{SLN',\ 0\rightarrow t}AUC_{SOL',\ 0\rightarrow t}$

The mean residence time (MRT) was estimated from MRT = $AUMC_{0\rightarrow\infty/}AUC_{0\rightarrow\infty}$

TABLE 4.12 Pharmacokinetic Parameters of AC After Oral Administration (Mean ± S.D.) ($^*p < 0.05$)

Parameter	Aceclofenac	F1
C_{max} (μg/mL)	2.19 ± 0.73	3.12 ± 0.98
T_{max} (h)	1.583 ± 0.083	4.833 ± 0.166
$t_{1/2}$ (h)	2.14	3.45
$AUC_{0\rightarrow t}$ (μg h/mL)	10.90 ± 0.374	56.93 ± 0.905
$AUC_{0\rightarrow\infty}$ (μg h/mL)	9.11 ± 7.54	15.03 ± 7.57

Figure 4.27 shows the concentration–time curve of F1 and AC-SUSP. It is evident that there was increased absorption of AC. The C_{max} value

of AC in nanoclay (2.19 ± 0.73) was significant ($p < 0.05$) than that observed with F1 (3.12 ± 0.98). Twenty-four hours after oral administration the AC plasma concentration was still more than 0.5 μg/mL. The $AUC_{0\rightarrow t}$ values of AC after oral administration of F1 were 56.93-fold higher than those obtained with AC-SUSP.

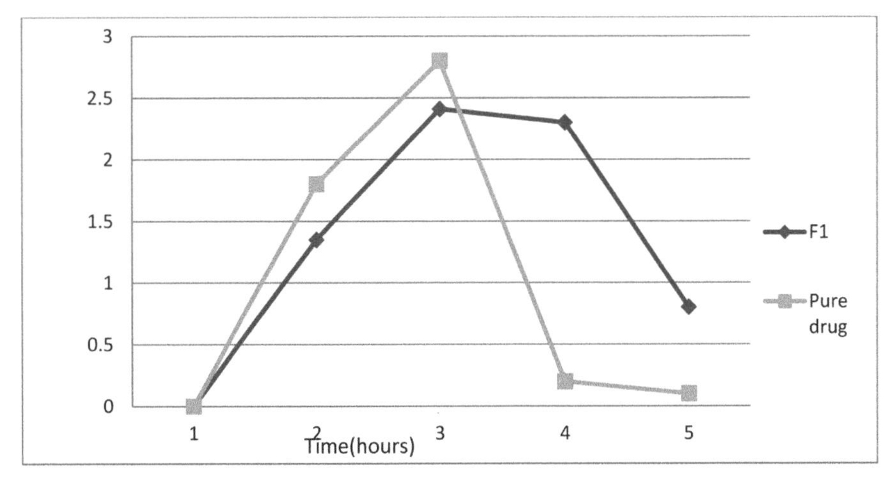

FIGURE 4.27 Concentration time profile after oral administration of F1 and AC-SUSP.

4.11 SUMMARY

1. This thesis deals with the investigation carried out on the formulation and evaluation of nanoclay containing AC so as to improve its oral bioavailability, to decrease dosing frequency, maintain prolonged therapeutic levels of the drug following dosing, and to reduce the dose to achieve same pharmacological effect.
2. Based on the literature studies various excipients, namely, polymers (MMT and PVA) and surfactants (F-127) were selected for the preparation of MMT nanoclay.
3. In preformulation studies, saturation solubility studies were carried out in different solvents and the solubility was found to be the highest in phosphate buffer pH 6.8.
4. Compatibility of the selected polymers and AC were carried by DSC and FTIR peak matching method.
5. In formulation development, MMT nanoclay was prepared and the effect of certain process on particle size was studied.
6. MMT nanoclay was evaluated for zeta potential, entrapment efficiency, drug loading, SEM studies, PI, and in vitro release studies.

7. Oral bioavailability studies were carried out in albino Wistar rats where by formulations, namely, F1 and AC-SUSP were administered orally and pharmacokinetic parameters were calculated.

All the above investigations brought out many facts which lead to following conclusions:

1. Solubility of phosphate buffer pH 6.8 was found to be 55.46 μg/mL.
2. Selected clay was found to be compatible with AC based on FTIR peak matching method and DSC.
3. Batch prepared with 50 mg drug concentration, 4 h stirring time, 2000 rpm stirring speed, 0.5 mg MMT clay, and 1% w/v surfactant concentration showed minimum particle size and was identified as an ideal batch.
4. MMT nanoclay formulations released the drug in sustained manner over a period of 24 h.
5. Zeta potential is −14.1, intercalation efficiency and drug content of F1 was found to be 78% and 11.94, SEM was found to be 5 μg/mL, and polydispersity index (PI) to be −0.099.
6. Pharmacokinetic parameters of AC after oral administration $C_{max,}$ T_{max} $AUC_{0\rightarrow t,}$ $AUC_{0\rightarrow\infty}$ of AC and F1 was found to be C_{max} value of AC in nanoclay (2.19 ± 0.73) was significant ($p < 0.05$) than that observed with F1 (3.12 ± 0.98), T_{max} value of AC in nanoclay (1.583 ± 0.083) than that observed with F1 (4.833 ± 0.166), The $AUC_{0\rightarrow t}$ values of AC after oral administration of F1 were 56.93-fold higher than those obtained with AC-SUSP.

4.12 CONCLUSION

To conclude, the preparation of nanoclay with MMT was suitable for producing MMT nanoclay. Lipophilic drugs like AC can be successfully incorporated into the clay. The formulated MMT nanoclay showed a significant increase in oral bioavailability compared to pure drug suspension. When drug releases from the clay surface and promoted or raises by the weak bonding and due to hydrophilic properties of the clay may increase the dissolution leads to enhanced bioavailability. The dose of the AC nanoclay needs to be corrected in accordance with increased bioavailability, to minimize its dose-related adverse effects. Nanoclay provided sustained release of the drugs, and these systems are the preferred drug carriers for lipophilic drugs to overcome the oral bioavailability problem of drug.

KEYWORDS

- **nanoclay**
- **aceclofenac**
- **oral bioavailability**
- **in-vivo studies**

REFERENCES

1. Singh, S.; Chakraborty, S; Shukla, S; Mishra, B. Lipid An Emerging Platform for Oral Delivery of Drugs with Poor Bioavailability. *Eur. J. Pharm. Biopharm.* **2009**, *73*, 1–15.
2. Suresh, R.; Borkar, S. N.; Sawant, V. A.; Shende, V. S.; Dimble, S. K. Nano Clay Drug Delivery System. *Int. J. Pharm. Sci. Nanotechnol.* **2010**, *3(2)*, 901–905.
3. Garrido, E. G.; Theng, B.; Mora, M. L. Clays and Oxide Minerals as Catalysts and Nano- Catalysts in Fenton-like Reactions—A Review. *Appl. Clay Sci.* **2010**, *47(3–4)*, 182–192.
4. Gomes, C. D.; Silva, J. B. Minerals and Clay Minerals in Medical Geology. *Appl. Clay Sci.* **2007**, *36(1–3)*, 4–21.
5. Carretero, M. I.; Pozo, M. Clay and Non-Clay Minerals in the Pharmaceutical and Cosmetic Industries Part II. Active Ingredients. *Appl. Clay Sci.* **2010**, *47(3–4)*, 171–181.
6. Viseras, C.; Lopez-Galindo, A. Pharmaceutical Applications of Some Spanish Clays (Sepiolite, Palygorskite, Bentonite): Some Preformulation Studies. *Appl. Clay Sci.* **1999**, *14*, 69–82.
7. Patel, H. A.; Rajesh, S. S.; Hari, C. B.; Raksh, V. J. Nucleation of Polymer Synthesis and Characterization of Silicates and Catalysis Division. *Polym. Bull.* **1999**, *42*, 118–122.
8. Krikorian R., inventor; Synthesis Studios Inc, assignee. Mobile Proximity-Based Notifications. United States patent application US 11/956,018. 2008 Jul 31.
9. Joshi, G. V.; Kevadiya, B. D.; Patel, H. A.; Bajaj, H. C.; Jasra, R.V. Montmorillonite as a Drug Delivery System: Intercalation and In Vitro Release of Timololmaleate. *Int. J. Pharm.* **2009**, *374*, 53–57.
10. Choy Y., Fyer A. J., Lipsitz J. D. Treatment of specific phobia in adults. Clinical psychology review. 2007 Apr 1;27(3):266-86.
11. Mittal V., Kamakura W.A. Satisfaction, Repurchase Intent, and Repurchase Behavior: Investigating the Moderating Effect of Customer Characteristics. *J. Marketing Res.* **2001**, *38*(1), 131-42.
12. De Paiva, L. B.; Morales, A. R.; Valenzuela Diaz, F. R. Organoclays: Properties, Preparation and Applications. *Appl. Clay Sci.* **2008**, *42(1–2)*, 8–24.
13. Aguzzi, C.; Cerezo, P.; Viseras, C.; Caramella, C. Use of Clays as Drug Delivery System: Possibilities and Limitation. *Appl. Clay Sci.* **2007**, *36(1–3)*, 22–36.
14. Wang, X. Y.; Du, Y. M.; Luo, J. W. Biopolymer/Montmorillonite Nanocomposite: Preparation, Drug-Controlled Release Property and Cytotoxicity. *Nanotechnology*. **2008**, *19*, 1–7.
15. Aguzzi, C.; Cerezo, P.; Viseras, C.; Caramella, C. Use of Clays as Drug Delivery Systems: Possibilities and Limitations. *Appl. Clay Sci.* **2007**, *36*, 22–36
16. Ainurofiq, A.; Choiri, S. Application of Montmorillonite, Zeolite, and Hydrotalcite Nanocomposite Clay-Drug as Drug Carrier of Sustained Release Tablet Dosage Form. *Indonesian J. Pharm.* **2014**, *25(3)*, 125–131.

17. Carretero, M. I.; Pozo, M. Clay and Non-Clay Minerals in the Pharmaceutical Industry: Part Excipients and Medical Applications. *Appl. Clay Sci.* **2009**, *46*, 73–80.
18. Khalil, H.; Mahajan, D.; Rafailovich, M. Polymer–Montmorillonite Clay Nanocomposites. Part 1. Complexation of Montmorillonite Clay with a Vinyl Monomer. *Polym. Int.* **2005**, *54*, 423–427.
19. Gao, J. M.; Gu, Z.; Song, G. J.; Li, P. Y.; Liu, W. D. Preparation and Properties of Organo- Montmorillonite/Fluoro Elastomer Nanocomposites. *Appl. Clay Sci.* **2008**, *42*, 272–275.
20. Aguzzi, C.; Cerezo, P.; Viseras, C.; Caramella, C. Use of Clays as Drug Delivery Systems: Possibilities and Limitations. *Appl. Clay Sci.* **2007**, *36*, 22–36.
21. Lin, F. H.; Lee, Y. H.; Jian, C. H.; Wong, J. M.; Shieh, M. J.; Wang, C. Y. A Study of Purified Montmorillonite Intercalated with 5-Fluorouracil as Drug Carrier. *Biomaterials*. **2002**, 23, 1981.
22. Abdel-Mohsen, M.; Mohamed, H.; Wadood, H. Study of the Effect of Montmorillonite and Florite on the Dissociation Constant, Release and Local Anaesthetic Activity of Lidocaine. *STP Pharm. Sci.* **2001**, *11*, 295–300.
23. Ambrogi, V.; Nocchetti, M.; Latterini, L. Promethazine–Montmorillonite Inclusion Complex to Enhance Drug Photostability. *Langmuir*. **2014**, *30*, 14612–14620.
24. Francis, D; Rita, L. Rapid Colorimetric Assay for Cell Growth and Survival: Modifications to the Tetrazolium Dye Procedure Giving Improved Sensitivity and Reliability. *J. Immunol. Meth.* **1986**, *89*, 271–277.
25. Maus S.; Yin F.; Lühr H.; Manoj C.; Rother M.; Rauberg J.; Michaelis I.; Stolle C.; Müller R.D. Resolution of direction of oceanic magnetic lineations by the sixth-generation lithospheric magnetic field model from CHAMP satellite magnetic measurements. Geochemistry, Geophysics, Geosystems. 2008 Jul;9(7).
26. Jescheniak J. D.; Meyer A.S.; Levelt W. J. Specific-word frequency is not all that counts in speech production: Comments on Caramazza, Costa, et al. (2001) and new experimental data.
27. Yin, L.; Wang, F.; Fu, J. Effects of Post-Synthesis Treatments on the Pore Structure and Stability of MCM-41 Mesoporous Silica. *Mater. Lett.* **2007**, *61*, 3119–3123.
28. Shaikh, K.; Devkhile, A. Simultaneous Determination of Aceclofenac, Paracetamol, and Chlorzoxazone by RP-HPLC in Pharmaceutical Dosage Form. *J. Chromatogr. Sci.* **2008**, *46*, 649–652.
29. Mutalik, S.; Anju, P.; Manoj, K.; Usha, A. N. Enhancement of Dissolution Rate and Bioavailability of Aceclofenac: A Chitosan-Based Solvent Change Approach. *Int. J. Pharm. Sci.* **2008**, *350*, 279–290.
30. Musmade, P.; Subramanian, G.; Srinivasan, K. High-Performance Liquid Chromatography and Pharmacokinetics of Aceclofenac in Rats. *Anal. Chim. Acta.* **2007**, *585*, 103–109.
31. Saba N., Jawaid M., Asim M. (2016) Recent Advances in Nanoclay/ Natural Fibers Hybrid Composites. In: Jawaid M., Qaiss A., Bouhfid R. (eds) Nanoclay Reinforced Polymer Composites. Engineering Materials. Springer, Singapore. https://doi.org/10.1007/978-981-10-0950-1_1
32. de Azeredo, HMC, Mattoso, LHC, and McHugh, TH (March 22nd 2011). Nanocomposites in Food Packaging – A Review, Advances in Diverse Industrial Applications of Nanocomposites,

Boreddy Reddy, IntechOpen, DOI: 10.5772/14437. Available from: https://www.intechopen.com/chapters/14374

33. Batra M., Gotam S, Dadarwal P., Nainwani R., Sharma M. Nano-clay as polymer porosity reducer: a review. J. Pharm. Sci. Technol. 2011;3(10):709-16.

34. Maina E., Wanyika H., Gacanja A. Natural pyrethrum extracts photo-stabilized with organo clays. Journal of Scientific Research and Reports. 2016:1-20. Source:https://www.researchgate.net/publication/287464989_Natural_Pyrethrum_Extracts_Photo-stabilized_with_Organo_Clays

CHAPTER 5

Study on Internet of Nanotechnology (IoNT) in a Healthcare Monitoring System

V. J. CHAKRAVARTHY,[1] P. HAKKIM DEVAN MYDEEN,[1] and M. SEENIVASAN[2]

[1]*P G Department of Computer Science, The New College, Chennai, India*

[2]*Department of Mathematics, Annamalai University, Tamil Nadu, India*

[*]*Corresponding author. E-mail:*

ABSTRACT

Nowadays, nanotechnology has provided more efficient and enhanced solutions for various applications, namely industry, agriculture, healthcare, and military applications. The development of nanomachines has a small component containing the set of molecules which can be performing the predetermined tasks of nanotechnology. The next generation of nanosensors and nanodevices has led to the development of standard IoT technology in terms of enhanced nanotechnology with the Internet called Internet of nanotechnology (IoNT). This chapter proposes monitoring healthcare system based on how IoNT will construct diagnostics and treatment, e-health monitoring, additional modified, suitable, and well-timed. This development will be followed by radically reduced costs and rising the accessibility and quality of medical care. Therefore this approach is highly essential for future enhancement of medical care.

5.1 INTRODUCTION

In the healthcare sector, nanotechnology has various applications including the growth of imaging application, diagnostic and treatment, more powerful medicine, implants, mechanism of drug delivery and devices [1]. Analysis of component-based computing was studied by Kumaraguru and et al. [2]. Therefore the analysis of nanomedicine using nanoscale-structured technique has a painful injury, treating, diagnosing, preventing diseases, and simple nanodevice can be contrived today. In the biological system, the interaction of materials as the core-based nanodevices and structured nanoscale. Nowadays, the longer period may be 10–20 years, the type of nanomachines may give the nanorobots and molecular machine system as a type might join the medical armamentarium and finally the maximum powerful tools possible to ill-health, ageing, and human disease by giving physicians [3].

In future, cognitive enhancement of nanorobots is nanomedicine instrumentation. Various classes of e-health-based nanorobots are designed now namely vasculoids, clottocytes, reciprocates, and micro bivores. It can perform various functions in the body of augmentation and maintaining the system [4]. Nanoscience provided an electronic nanodevice and interface between electronic nanodevice and biological phenomenon on existing biological nanomachines, namely nanoactuators and nanosensors. Nanosensors can be used to compute, identify, and detect the body fluids of biological substances in the diagnostics area. These are leading to detect the ecological contaminant in the body as well as the ability to detect early disease and prior treatments [5, 6]. Nanomachines can be remotely controlled over the Internet by an external user which can be arranged inside the human body that creates the new networking model of IoNT.

5.2 IONT ARCHITECTURE FOR HEALTHCARE APPLICATIONS

Internet-based healthcare is carried out in various forms which can be enabled to track the health information for some patients. This will present more valued data and produces for opening the smarter devices by reducing the interaction request for a direct patient in healthcare professional in Figure 5.1 [7]. This IoNT is faster and has provided the best medical services at reduced costs to more people. Benefits of IoNT health namely better insights, improving the patient care by the providers, efficiencies of the supply chain, hospital management, and chronic disease controlling [8].

The IoNT applications are connected and interact with each other and are quickly changing the development of healthcare by focusing on the way of people to suggest the favorable solutions for development on privacy and security models of the healthcare industry that makes an additional model for healthcare [9]. With the help of some e-health development by fast information and communication technologies advancements and the growing number of smart devices namely portable devices and sensors. E-health solution can be used to make the actionable decision that provides a great wealth of information. Internet powered e-health system generates a lot of chances to increase

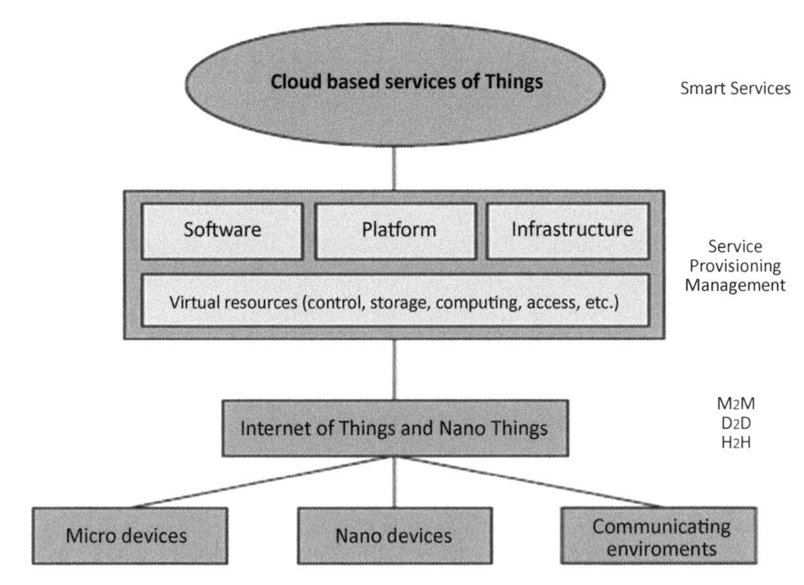

FIGURE 5.1 The Internet of NanoThings architecture.

efficiency, develop results by linking information, people, devices, processes, context, and makes the healthcare system reasonable. Tools or sensors used to track and record the personal data are readily obtainable to the general public. The main blocks of IoNT model of the network design in healthcare application are shown in Figure 5.2.

- Nanonodes have inadequate memory to perform a simple computation and smallest nanomachines that can communicate short distances. Nanonodes are based on the design of bio-inspired and nanomaterials [10].
- In general, classification of nanosensors are of three types: (1) Physical nanosensors belong

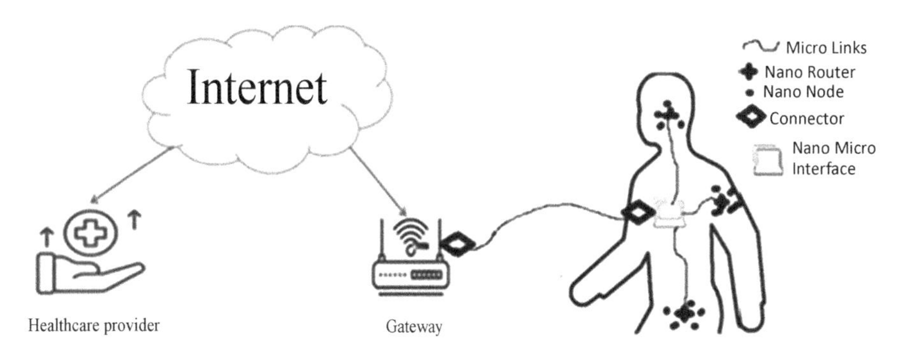

FIGURE 5.2 Architecture of IoNT in healthcare application [5].

to thermal, optical, acoustical, mechanical, magnetic, and radiation; (2) Chemical nanosensors are molecular energies and atomic energies; (3) Biological nanosensors belong to DNA interaction, enzymatic interaction, and antigen interaction.

- Nano-routers—It can be used to control the performance of nanonodes and nanodevices coming from limited nanomachines which collect information by exchange of exact simple mechanism instructions namely sleep, read-value on and off, and so forth.
- Nano-micro interface devices—These are used to convey it to microscale in which devices are collecting the information from nano-routers and vice versa. This device can be used for nano-communication techniques which is capable of transmitting in the nanoscale and model of traditional communiqué in traditional network communication.
- Gateway—The entire system of this device allows remote control over the Internet.

5.3 IMPLEMENTATION OF THE IONT IN MEDICAL APPLICATIONS

The concept of IoT is the implementation of IoNT. It is attained in different objects by integrating nanosensors using nano-networks. The working of the medical application model is shown in Figure 5.3, this provides to sense or by use of definite instruments to access the data in situ places are previously isolated that were difficult to their past bulky sensor size. It will collect the environmental data and new medical data, new findings, possibly leading to the improvement of existing knowledge and enhanced medical diagnostics [11]. The technology of nanoatennas is based on graphene that has been operating at the frequency of terahertz [12]. It describes the great level of operating attenuation at these frequencies and also networking at nanolevel [12]. In IoNT, the function of each task namely sensing actuation can be done by "nano-machine," the described dimensions are in the range from 1 to 100 nm [13].

In the world, it can be seen that IoT will not only be expanded but also the scales are not visible to the naked eye of human by the use of IoNT and Internet of Bio-NanoTechnology (IoBNT). These will be used for medical field at cellular level and industrial fields at water purification for filtration work or dialysis. In order to overcome the main obstacle of IoBNT from the unified merger will follow of IoNT-based IoT systems with the existing healthcare as well as network [14]. IoBNT is being stemmed from synthetic biology as well as the arrangement

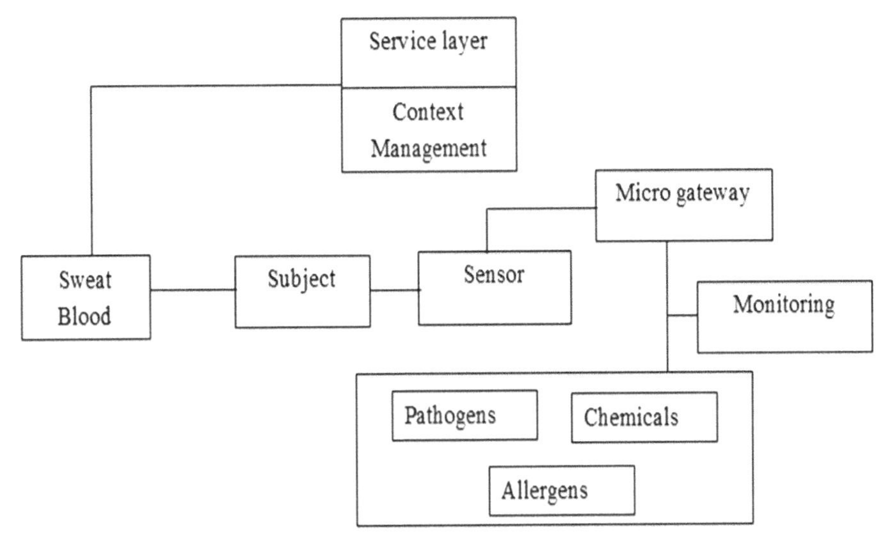

FIGURE 5.3 IoNT in medical applications [11].

of nanotechnology tools and techniques to permits the engineering of implanted-based biological devices [15]. This will decrease the risk of undesired effects on healthcare environment.

5.4 CONCLUSION

This chapter concludes nanotechnology focused on the scientific discipline of design characterization, synthesis and application of nanoscale materials and devices. These possibilities are to develop the healthcare through production, design, description, intelligent materials and their application within medicine. The future goal of healthcare is the detection and diagnosis of the early disease, operative therapy personalized to the patient, precise and attended with cost reduction is enabled.

KEYWORDS

- **Internet of nanotechnology (IoNT)**
- **healthcare**
- **monitoring**
- **nanotechnology**
- **nanomachines**

REFERENCES

1. Miller G., Kearnes M. Nanotechnology, Ubiquitous Computing and The Internet of Things: Challenges to Rights to Privacy and Data Protection, Draft Report to the Council of Europe, 2012.
2. Kumaraguru P.V., Chakravarthy P.J., and Seenivasn M. Analysis of Component based Computing.

International Journal of Engineering & Technology, **2018**, 7(4), 133–136.

3. Freitas R.A., Current Status of Nanomedicine and Medical Nanorobotics, Journal of Computational and Theoretical Nanosciences, **2005**, 2, 1–25.
4. Swan M. Nanomedical Cognitive Enhancement, 2014. Online]: http://ieet.org/index.php/IEET/more/swan20140711
5. Akyildiz I. F., Jornet J.M. Internet of Nano Things, IEEE Wireless Communications, **2010**, 5, 58–63.
6. Wagner V., Hüsing B., Gaisser S., and Bock A.K. Nanomedicine: Drivers for Development and Possible Impacts, European Commission—Joint Research Centre, Institute for Prospective Technological Studies, 2008.
7. Balasingham I. IoT in E-health: Challenges—Opportunities, 2014, [Online]: http://www.internet-ofthings.no/Presentasjoner_02_april_2014/I_Balasingham_OUH_Conference_IoT_Oslo_02_April_2014_Final.pdf
8. Transforming Patient Care with IoT, [Online]: http://www. microsoft.com/en-us/server-cloud/solutions/internet-of-things-health.aspx
9. Will the Internet of Things Analytics Revolutionize the Healthcare Industry? A Saviance Technologies Whitepaper 2014.
10. Enisa Omanović-Mikličanin, Mirjana Maksimović and Vladimir Vujović, The Future of Healthcare: Nanomedicine and Internet of Nano Things, Folia Med. Fac. Med. Univ. Saraeviensis **2015**, 50(1), 23–28.
11. Bala Subramaniam S. and Kangasharju, J. Realizing the Internet of Nano Things: Challenges, Solutions, and Applications. Computer **2013**, 46, 62–68.
12. Chaudhry J., Qidwai U., Miraz, M.H., Ibrahim, A., and Valli, C. Data Security among I ISO/IEEE 11073 Compliant Personal Healthcare Devices through Statistical Fingerprinting. In Proceedings of the 9th IEEE-GCC, Conference and Exhibition 2017, Manama, Bahrain, 9–11 May 2017, pp. 319–324.
13. Ali N.A., Abu-Elkheir M. Internet of Nano-Things Healthcare Applications: Requirements, Opportunities, and Challenges. Proceedings of 11th IEEE International Conference on Wireless and Mobile Computing, Networking and Communications (WiMob' 2015), Abu Dhabi, UAE, 19–21 October 2015.
14. Akyildiz I.F., Pierobon M., Balasubramaniam S., and Koucheryavy Y. The Internet of Bio-Nano Things. IEEE Commun. Mag. **2015**, 53, 32–40.
15. Miraz M.H. Maaruf Ali, Excell P.S., and Picking R. Internet of Nano-Things, Things and Everything: Future Growth Trends Future Internet **2018**, 10, 68.

CHAPTER 6

Safety Assessment and Nanotoxicology of Nanomaterials in a Drug Delivery System

SHARIDA FAKURAZI[1,2,*], SULEIMAN ALHAJI MUHAMMAD[2,5], MAS JAFRI MASARUDIN[3], and MOHD ZOBIR HUSSEIN[4]

[1]*Department of Human Anatomy, Faculty of Medicine and Health Sciences, Universiti Putra Malaysia, 43400 UPM, Serdang, Selangor Darul Ehsan, Malaysia*

[2]*Laboratory of Vaccine and Immunotherapeutics, Institute of Bioscience, Universiti Putra Malaysia, 43400 UPM, Serdang, Selangor Darul Ehsan, Malaysia*

[3]*Department of Cell and Molecular Biology, Faculty of Biotechnology and Biomolecular Sciences, Universiti Putra Malaysia, 43400 UPM, Serdang, Selangor Darul Ehsan, Malaysia*

[4]*Material Synthesis and Characterisation Laboratory, Institute of Advanced Technology, Universiti Putra Malaysia, 43400 UPM, Serdang, Selangor Darul Ehsan, Malaysia*

[5]*Department of Biochemistry, Usmanu Danfodiyo University, Sokoto, Nigeria*

[*]*Corresponding author. E-mail: E-mail: sharida@upm.edu.my*

ABSTRACT

Advancement of interdisciplinary aspect of nanotechnology has boosted the expansion of development of nanomedicine which is able to increase therapeutic efficacy. Hypothetically, when healthy tissue is unharmed while delivery of administered nanomedicine is specific to disease site, nanotechnology has been accepted as a prospective strategy to reduce toxicity associated with particular drugs. However, toxicological assessment of nanomaterials has not been conducted in parallel to the speed to its development when there has been lack of reports on their safety evaluations. Looking at the potential of utilizing nanotechnology in the development of nanomedicine, more toxicological data of nanomedicine is required before any therapeutic substance is available for human use. The advantage of nanomedicine over its conventional counterparts lies on its ultra-small size, with high surface area to volume ratio. The molecules may consist of hydrophilic or hydrophobic compounds with some atypical physicochemical properties; which is able to alter physical and chemical properties as well as the pharmacokinetics. Similarly, nanoparticles have the potential to cross blood-brain barriers due to their nanosize. Alteration in physicochemical properties of nanomedicine influences bioavailability, transport fate of the substance, cellular uptake, biotransformation and elimination process of the materials involved with increasing complexity of in vivo environment. A comprehensive evaluation of safety profile would allow a wide range of understanding of toxicological significance of the nanomedicine. The evaluation of these profiles is definitely useful and required for a comprehensive evaluation of therapeutic benefits for safer development of nanodrug delivery system.

6.1 INTRODUCTION

The use of nanoscale size materials or drugs in the field of medicine, particularly in the area of drug delivery, has revolutionalized the diagnostic, therapeutic, and preventive approaches to manage especially chronic diseases. Besides, the usage is also reported to be advantageous in the field of surgery, prostheses, and implantable medical devices, such as stents and catheters. The extremely small composition or of "nano-size" with innovative strategies manipulated on the delivery systems produces materials of special physical, chemical, and biological properties, which permit access to a different part of the body. The nanoscale materials learned to integrate well into biomedical devices because most biological organelles are also nanosized. At a nano scale size, the delivery of a therapeutic substance to areas like the brain is made permissible even in the presence of an intact blood-brain barrier (BBB) [1]. The blood-brain barrier has a low permeability to drugs which limit the transportation of drug substance across the system. Therefore, brain diseases such as glioblastoma, brain cancer, Alzheimer and Parkinson disease, strokes, and multiple sclerosis at present are poorly treated. This is due to the impediments of BBB to foreign substances and drug materials. Recent development of nanotechnology allows the fabrication of drug materials within a size range of 1–100 nm, [2] may be available to overcome the limitation of drug delivery in brain disease [2].

Conventional cancer therapy encounters a similar setback. Established chemotherapeutic regime concedes side effects faced by patients when they are exposed to treatment [4]. Extensive exposure of systemic side effects precludes the effectiveness of such treatment, besides patients avoid undergoing such treatment.

The creation and synthesis of nanomedicine which is a therapeutic agent of less than 100 nm size, offers exceptional pharmacological potential for disease management strategies. The technology allows surface and matrix modification and innovations have extensively attracted the attention of researchers and clinicians all over the world. The technology has become especially popular since this technology allows improvement in disease cells targeting leaving the healthy cells intact [5]. In addition, it also has the ability to maintain the steady state therapeutic levels over an extended time period [6, 7]. Furthermore, the unwanted side effects are also being restricted when the healthy tissues are left unharmed, and diseased tissues are specifically targeted. On contrary, conventional drug delivery allows unspecific systemic distribution of drugs which has limited control of release rate, with fluctuation in concentration of drug in the blood and subsequently in cells and tissues [8]. This permits inconsistent distribution of conventional drugs to various parts of the body, at the same time, allowing metabolic degradation of drugs and premature drug clearance [9]. Therefore, nanomedicine offers the potential in surmounting the limitations that the present conventional system might possess at the same time offering numerous opportunities and advances in therapeutic strategies and make significant contributions to health industries [10].

The progress and expansion of nanotechnology and the availability of nanomedicine have unfolded potential developments in chemotherapy and cancer management or even in the diagnosis of cancer. The growth of the technology offers valuable means of target therapy to selectively target the neoplasms or cancer tissues. Utilizing the technology leads to obliterating the cancer tissue with minimal damage to healthy tissues. The invention also allows the detection and elimination of cancer cells before the formation of malignant tumors [7, 8].

Despite extensive development on nanotechnology in biomedical applications and drug delivery, toxicological assessment and safety aspects of the synthesized nanomaterials have not been elucidated in parallel to the speed of its development. The ultrasmall size of the materials contributed to high surface area to volume ratio changes the kinetics of biodistribution, perhaps allowing a wider distribution after the administration of the materials compared to its counterpart. Besides producing a potential therapeutic benefit, the nanoscale materials may confer a potential threat to human and the environment [11].

There is a lot of report on drug delivery system or host in fulfilling the purpose of specific targeting of the disease. There are delivery systems that have been approved by the US Food and Drug Administration (FDA) [12]. There are various other systems and drugs that are queuing and in the clinical trial waiting for approval.

6.2 PHYSICOCHEMICAL CHARACTERIZATION OF NANOMEDICINE

The main feature of a nanoparticle is predominantly their nanosize structure where they incline to have a greater tendency to form agglomeration. The process of agglomeration allows the sticking of particles to one another or to solid surfaces, occurs due to attractive van der Waals forces between particles. The nanoparticles are particles between 1 and 100 nanometres (nm) in size with a surrounding interfacial layer, acquire high surface area to volume ratio which provides a very high surface energy. To minimize its surface energy, the nanoparticle creates aggregation and agglomeration.

In addition, the nanoparticle holds a surface charge which is known as zeta potential, which is a measure of the effective electrical charge of the nanoparticle surface. It provides information on the stability imposed by the system. A zeta potential between −10 mV and +10 mV is considered neutral while nanoparticles greater than +30 mV or less than −30 mV are considered as strongly cationic and strongly anionic, respectively. The value of zeta potential attained from nanoparticles is particularly dependent on pH and the conductivity of the dispersing medium. Zeta potential also holds essential strength and characteristic on the measures and means of the interaction of nanoparticles with cell surface membrane and the duration that it stays in the blood microcirculation.

It is essential that the physicochemical properties of nanomedicine are evaluated and interpreted correctly particularly when the vitro assays are not directly translated to in vivo micro and microenvironment. This may be the case when most of in vitro data is not translated to preclinical in vivo and human clinical trial. Most of the drug delivery materials are synthesized from biocompatible metal [13–15], nonmetal [16], organic [17], or inorganic materials [18, 19] with pH-dependent solubility. This is determined by the pK_a value which governed the degree of ionization of the materials in the tissue environment. The pK_a value or a dissociation constant is an index to express the acidity of weak acids, which measure the strength of a solution. When a drug is more likely to be in unionized form, the more likely it is to be lipid soluble and transferred via passive diffusion through the membrane [20]. A low percentage of hydrogen bonding groups within molecules renders them hydrophobic. Small size molecule is able to traverse pores in the membrane which then accounts for the high permeability of the cell membrane to the nanoparticles.

Extracellular pH of tumor tissue is in acidic microenvironment which is around an average of pH 6.0–7.0 compared to the pH of extracellular pH of normal and healthy tissue which is around pH 7.4 [21]. The declining of extracellular pH is contributed by the high rate of glycolysis in cancer cells [22], even though in hypoxic condition. The difference in pH gradient allows distinct pH sections between intracellular and extracellular tumor cells, and also between tumor mass and host tissues. The environment favors the pH-dependent controlled released character of nanoparticles when drugs are only released at favorable pH. Many studies have reported the inclination of drugs to be released based on acidity or alkalinity of the cellular microenvironment, in a controlled released manner [23].

In tumorigenesis and cancer network microenvironment, the morphology of tumor vascular and pH differs from the morphology of normal environment. The nanosize drug delivery system (20–200 nm) extravasate and accumulate in the extrastitial fluid [24]. The nanomaterial is retained in the tumor tissues due to inefficient lymphatic system drainage [25] which impairs efficient removal from the tumor microenvironment. The phenomenon known as enhanced permeability retention effect promotes passive delivery of the nanomaterials. The enhanced permeability retention is the foundation of tumor-associated localization of many therapeutics and imaging contrast agents, which has lacked benefit of surface targeting molecule. Sufficiently small size of nanoparticle inclines to passively extravasate and accumulate in tumor parenchyma [26].

Ideal physicochemical structure of nanocarrier should be able to avoid immune surveillance and hide from reticuloendothelial system to avoid opsonization and phagocytosis [27]. Besides, the clearance process removes the nanoparticle carrier system from systemic circulation which is performed by the liver or the kidneys. Therefore, a nanocarrier that is larger than 10 nm would avoid filtration by the kidneys [28] while to avoid elimination by the liver, requires the nanocarrier to be smaller than 100 nm [29]. Meanwhile, for efficient extravasation from the leaky vascular membrane, the nanocarrier should be less than 400 nm. For efficient clearance of nanocarrier and to avoid deposition in the systemic circulation, the charge of the nanoparticle should be neutral or anionic [30].

In tumor microenvironment, drainage of the lymphatic system is inefficient [31]. The interstitial fluid pressure is high in the tumor environment together with an increase in osmotic pressure. This has also posed a barrier for efficient delivery of anticancer drugs especially with bigger molecular weight [32]. Elevated interstitial fluid pressure is higher in the centre, which diminishes as approaching the periphery, reduces the transcapillary transport in the tumor which hinders efficient delivery of the anticancer agent to the cancer site. The situation is in favor to the nanosize material which is less affected by the barrier, in fact this facilitates the extravasation of the nanocarrier system which could otherwise be precluded from the cancer microenvironment.

Besides, there are also some limitation factors of transportation via passive diffusion, such as the degree of tumor vascularization and extend of angiogenesis is different. Thus, the efficiency of delivery and extravasation differ, which is dependent on tumor types, anatomical sites, and the stage of cancer that the patient is going through. The presence of interstitial fluid hypertension in solid tumors which disallows homogenous distribution of drugs in the tumor sites at the same time may also limit the successful uptake.

Despite potentially efficient and promising potential of nanodelivery system in unfolding the limitation of conventional drugs, the possible implication on toxicological aspect of the system must be extensively studied before it is widely available to the general public. The fabricated system must be treated as a new drug entity which ought to undergo preclinical and clinical testing stages. Sufficient evaluation on the potential toxicological implications on the system which has been engineered in such a way may have on human biological structure and functions. The subsequent sections are deliberating on toxicological evaluations which are potentially available during preclinical testing, in vitro and in vivo.

6.3 IN VITRO TOXICOLOGY EVALUATION

Data obtained from in vitro experiments may not reflect or be directly translated into in vivo model. Thorough characterization of the synthesized nanoparticles is required to link the potential of toxicity likely to be exerted in cells in vitro which may be associated with in vivo. Therefore, confirmation and verification of the data in vivo experiments is essential to substantiate on how nanoparticle interact with the biological system.

In vitro toxicological MTT assay depicted the percentage of viable cells following incubation of a range of concentrations of nanoparticles with the cell under study compared to similar concentrations of a complementary pristine drug. Although there might be a similar percentage of cell viability or cell toxicity observed in the assay conducted, the concentration of nanoparticle is not directly the same as the concentration of the pristine drug in the media [33, 34]. Cytotoxicity reports have indicated that some nanoparticles, for example, superparamagnetic iron oxide upon cell internalization results in disruption of the organization of cell cytoskeleton [35]. Depending on the coated polymers together with the concentration of iron oxide, in biological activity is reported due to the difference in the release profile [36]. It was reported that iron oxide nanocomposites coated with gallic acid and polyethylene glycol polymer demonstrated higher anticancer activity on the breast cancer cell lines (MCF7) compared to the one coated with polyvinyl alcohol. The internal organelles are reported to be affected by iron oxide administration [37].

The type of nanocomposite used determines the concentration of active drug following incubation of nanoparticles in the mixture. The concentration will also be influenced by the percentage of drug loading, the pH of media, controlled release property of the nanoparticle which determines the percentage drug release over time [38]. Change in the parameter adopted in designing the nanocomposite changes the physicochemical characters of the nanocomposite [34]. Therefore, in some in vitro assays conducted, pristine bioactive compounds demonstrate higher biological activity or cytotoxicity against cancer cells [23, 33]. This is due to the pristine bioactive compounds reacting with the cells, either by passive diffusion through the cell membrane or via carrier-mediated cellular uptake, immediately reacting with the cellular organelles.

In this sense, a comprehensive characterization of the synthesized nanodelivery material is necessary in order to allow valid comparison with a pristine drug. Besides, mounting expenditure, efforts, and suffering plus ethical concerns over animal usage in vitro pharmacological studies of the nanodelivery, with careful consideration and planning may be able to represent the in vivo system. This can operate as a simple system to examine the effect of such materials without endangering animals especially at screening phase.

Initial screening of in vitro assays conducted to assess cell viability and metabolic activity included MTT assay. The assay is a colorimetric assay which is based on the ability of nicotinamide adenine dinucleotide phosphate-dependent cellular oxido-reductase enzymes to reduce tetrazolium dye MTT (yellow) to insoluble formazan, which is purple in color. This reduction action may only occur in living cells, not occurring in dead cells. The absorbance of the color solution is then measured at a certain wavelength (between 500 and 600 nm). Although this assay is largely being used, there are times when the assay which is largely dependent on the activity of mitochondria reductase could indicate large errors. The redox active surface of SPIONs alters mitochondria function as well as decreased cell viability.

Similar purpose is also seen with Cell Counting Kit-8 (CCK-8) and neural red assay. The CCK-8 is also a colorimetric assay which utilizes highly water-soluble tetrazolium salt, WST-8 is reduced by dehydrogenase activities in cells to give a yellow-color formazan dye which is soluble in the tissue culture media. The amount of formazan dye is directly proportional to the number of living cells. The detection sensitivity of CCK-8 is higher than the MTT.

Neutral red assay is another assay which provides quantitative estimation of the number of viable cells. The principle of the assay is based on the detection of viable cells via the uptake of neutral red dye in lysosomes. Absorbance of the dye is detected using a spectrophotometer. This procedure may be cheaper and more sensitive than other cytotoxicity tests.

Another way of identifying cell viability assay is to do tryphan blue assay. The assay utilizes tryphan blue to quantify live cells by labeling dead cells. The dye penetrates the porous cell membrane of the dead cells, which then enters the cytoplasm. The dead cells appear blue under a light microscope, which can then be quantified, and the number of live cells can be obtained.

Lactate dehydrogenase (LDH) enzymes leakage may also be used as an indicator of cytotoxicity and cell damage. High level of LDH in a single cell culture assay means that there is some potential rupture of plasma membrane, or cell necrosis has occurred. It is a glycolytic enzyme which is found widely in almost every tissue, especially skeletal muscle, heart, liver, kidneys, brain, lungs, and red blood cells. Since LDH is found in many types of tissues in the body, high level indicates some forms of tissue damage. Further test must be done to determine the location of tissue damage.

Preliminary cytotoxicity assessment may also be carried out using brine shrimp *(Artemia salina L.)* lethality assay [39, 40] used as a screening test for potential toxicity. The assay has been used to screen various materials inclusive of nanoparticles, marine natural products, metal ions [41], dental materials [42], plant extracts [43–45]. The technique utilizes small number of subject materials, low number of test model and importantly, it is economical [46]. Some have claimed that the assay has demonstrated a good correlation between the results obtained for lethal concentration that kills 50% of the exposed population in BSLA and the results obtained in Acute Oral Toxicity Assay in mice [47]. Besides, it is a simple assay which requires no special training for the assay and equipment. Data is rapidly obtained after 24 h of exposure and monitoring could be extended to a maximum of 60 h. At the start of every assay, the *Artemia nauplii* must be at the same age, and the cysts obtained before hatching must also come from the same geographical region. Besides, standardization of the experimental condition is essential in order to obtain repeatable results, especially, salinity, aeration, light, and pH condition of the water. Variations in pH condition of the water, particularly a decrease in pH contributes to the lethality of *Arterima larvae* [48]. In order to maintain conformity with the standard procedure, positive and negative controls are the essential part of the assay and must be included in the experiment.

Another alternative method in toxicity testing is to utilize Hen's egg test on the chorioallantoic membrane; using chicken egg to assess irritability of the chorioallantoic membrane in the presence of a large number of blood vessels. The Hen's egg test on the chorioallantoic membrane is a well-known basic toxicity test for embryotoxicity and also relevant on systemic toxicity and immunopathology [49, 50]. The test provides a good correlation with the reported data from Draize tests, including observation on electron microscopy. The test has been approved, by regulatory agencies in France, Germany, the Netherlands, and the UK, along with appropriate histological examination. Besides, the test may be a useful and suitable alternative model for acute toxicity screening before embarking of comprehensive toxicological investigations in rodents [50].

Some nanoparticles may affect the DNA constituents leading to DNA damage and mutation. A number of techniques for detecting DNA damage (e.g., micronuclei, mutations, structural chromosomal aberrations) have been used to identify biomaterials or nanocomposites with genotoxicity. Comet assay is an assay to measure the presence of DNA damage. DNA is a double helix supercoiled structure developed by Cook and co-workers in the 1970s [51]. In the event of DNA damage, there is a DNA strand break, and when an electrophoretic field is imposed, the DNA loop is able to extend toward the anode. The percentage of incidents of DNA strand breaks induced by xenobiotic compounds or anticancer agents are dose-dependent. When more breaks are present, more DNA loops are relaxed, and more DNA appears in the comet tail. The relative intensity of tail DNA fluorescence is measured as an index of DNA break frequency after staining with a suitable dye. A more formal name of this assay is single-cell gel electrophoresis when the damaged cells form a comet-shaped pattern after electrophoresis.

The assay was conducted and first reported in 1984 [52], which employed cell lysis and electrophoresis at pH 9.5. The pH of the buffer system was modified to alkaline pH, to >pH 13.0 (with 0.3 M NaOH) which is then essentially the assay that is widely utilized these days. This assay is versatile and appealing to researchers especially when the assay is also simple and sensitive. The comet assay can be applied to cells in cultures, blood samples from animals or humans, sperm, disaggregated animal tissues, or even from plant tissues.

The assay is mainly used in genotoxicity testing to screen new drugs, chemicals for potential carcinogenic properties. In vitro test can be done on suitable cultured cell lines [43, 53, 54], and analysis can also be done on various tissues from experimental animal models [54, 55] or collected human specimen [56, 57].

The other frequently applied genotoxicity testing can be conducted to detect gene mutations using Ames test (or Salmonella typhimurium reverse mutation assay). The assay utilizes mutant bacteria, which is unable to synthesize histidine amino acid. Therefore, the bacteria are unable to grow and proliferate in a medium lacking histidine. When tested chemicals are employed to the bacteria, the condition is reversed, and the bacteria are able to grow in medium lacking histidine. The observation suggests that the employed chemical has a mutagenic potential which is able to cause a reversal of mutation in the bacterial cells [58].

Micronucleus test is also a test used to screen for potential genotoxicity. This assay is widely used, the most successful, and reliable assay and accepted by OECD guidelines for genotoxicity in chemicals.

Nanoparticles are becoming more popular each day especially in the expansion of the use of nanoparticles as a drug delivery system in nanomedicine. The expansion of bigger size drugs from micro to nanosized materials have observed that size matters in delivering the bioactive materials to the disease site of interest, leaving the healthy tissues intact. Most delivery systems allow controlled and prolonged release of drugs, allow sufficient drugs to be delivered at the mandatory disease tissue. The efficacy of the delivery system is influenced by its structural features, importantly the biocompatibility of the materials used to synthesize the system. Simultaneously, the circumstances applied to the toxicological significance of the system apply to cells, or during the administration of live animals or human.

Many nanoparticles consist of metal inorganic materials we found to cause chromosomal aberrations, oxidative DNA damage, strand breaks as well as mutations[59]. Due to factors such as inconsistencies in the synthesis methods or physicochemical characters, there have been some discrepancies in the literature reported. Silver nanoparticles which have been widely reported to have potent antibacterial properties, offer an appealing alternative to antibiotics including treatment of otitis media, which is an inflammatory condition of the middle ear [60]. Conversely, the auditory cells are reported to be more sensitive and inversely affected by silver nanoparticles [61] compared to keratinocytes cells. The toxicity is due to the production of excessive reactive oxygen species, found to be size- and dose-dependent, when a high level of DNA lesions was observed in cochlear cells [61]. Genotoxicity of nickel nanoparticles has also been reported in the Drosophila model, when a significant increase of small single spots and total mutant spots was observed, suggesting potential toxicity of the nanoparticles [62].

Although there are advantages and disadvantages of in vitro, where the in vitro assays provide robust and rapid screening, the absence of physiological biokinetics by the in vitro system may contribute to some misinterpretation of data. Most of the in vitro data, especially on anticancer work is performed on secondary cancer cells, which may significantly function abnormally compared to the cancer cells which are currently available in vivo. Eventually, in vivo animal model is essential and required prior to any human clinical trial.

6.4 IN VIVO TOXICOLOGY EVALUATION

In addition to being developed as a drug delivery system, some of the nanomaterials are also highly utilized in industries. The condition allows potential environmental, occupational, and possible customer exposure. Systemic toxicity exerted by nanoparticles could be unintentional or premeditated. It ranges from inhaled nanoparticles following occupational and environmental exposure to direct intended exposure following administration of a theranostic substance to image disease tissue for diagnosis purposes, or application of nanomedicine intended for therapeutic purposes.

The interaction of the nanoparticles with live animals or humans is dependent on the route of exposure and administration. The size of particles is directly linked to their prospective potential for causing health problems. Small size particles especially of less than 10 μm in diameter create the utmost hazard since they are simply inhaled and can get deep into the lungs, and enable them to get into the bloodstream, and to be distributed systemically. Besides, these particles would be able to be carried long distances by wind and then settle on the ground or water which may then contribute to environmental damage. The effects of the settling may include, affecting the diversity of the ecosystem, contributes to acid rain effects, making the lakes and streams more acidic, or changing the balance of nutrient balance in coastal water.

Following exposure through inhalation, leads to size-dependent pulmonary deposition. A large fraction of nanoparticles is removed by mucociliary escalators from upper respiratory airways. A small fraction of the nanoparticle is swallowed which then results in secondary excretion via oral route excretion. Some fractions are deposited in the liver and other secondary organs.

Data related to animal in vivo toxicological evaluation has started to accumulate recently. Appropriate study must be designed for better understanding on how the nanoparticles interact in a comprehensive biological system. A wide range of understanding of cellular interaction is much required for pronounced perception and comprehension on the action exerted by the nanoparticle drug delivery system in terms of efficacy and toxicity.

Consequently, drug application to be utilized in the respiratory system must be of appropriate and appropriate size that is able to be delivered to the demanded tissues at the respiratory, without being eliminated by body immune system.

Additionally, the site of nanoparticle action also plays some roles in the intensity of consequences of administration and exposure.

Biocompatibility is an imperative aspect which must be taken into consideration. A biocompatible nanoparticle refers to the ability of a biomaterial in performing a desired function with the most appropriate beneficial cellular or tissue response with respect to therapeutic application, and devoid of provoking any undesirable local or systemic effects in the host. (Poly)acrylonitrile-butadiene-styrene modified silver nanoparticle was fabricated into polymeric implants, a medical device, to be applied as middle ear prostheses. The substance was developed into prototype, administered, and evaluated in vivo on rats between 4 and 48 weeks [60]. Microscopic evaluations of tissue morphology, histochemical staining of enzymes or protein biomarkers can be conducted to indicate biocompatibility.

Gold nanoparticle (AuNP) has been reported to be a useful biosensor, a good candidate used to deliver drugs and genes [63]. The nanoparticle has also been used as a localized photothermal agent mediating tumor cell necrosis hyperthermia [64]. Injection of 13 nm AuNP to the animal was shown to induce acute liver inflammation with neutrophil influx [65]. Besides inducing inflammation, AuNP was shown to induce apoptosis in the liver and the nanoparticle was found to accumulate in the liver and spleen after 7 days. There was the presence of cytoplasmic vesicles and lysosomes in the Kupffer cells and spleen. The plasma half-life was also shown to be prolonged. In a particular case, hepatotoxicity may be observed when the translocated biomaterials are deposited in the liver, ability to trigger excessive production of reactive oxygen species consequently permit the occurrence of DNA damage [66].

Toxicity implications in vivo depicted potential harm of the nanoparticle to the whole animal and human. The potential toxicity of the nanoparticle can be studied in vivo, either material induced acute, subacute, or chronic toxicity. Prior to conducting any testing on animals, the study must be approved by relevant Institute Animals Ethics Committee and the proposed protocols should adhere and satisfy the guidelines of the local governing body. Animals are then to be acclimatized to laboratory conditions (temperature preferably at 22 °C ± 2 °C). Animals are maintained with adequate oxygen, appropriate animal cage, sufficient change of animal beddings, adequate supply of water, and food (unless the study requires the removal of food and water).

Acute toxicity is a rapid onset of adverse effects results after oral or dermal route either from a single exposure or from multiple exposures within 24 h, or inhalation exposure within 4 hours duration.

To be classified as acute toxicity, the adverse effects are observed within 14 days of the nanoparticle administration. The test is conducted in rodents which use mortality as the main observational end point. During the experimental period, physical evaluation of the animal, morphological, biochemical, pathological, and histological changes of the tissues are evaluated. During the observation episode, animals should be observed for the presence of tremors, convulsions, salivation, sleep, lethargy, diarrhea, and coma.

If there is any mortality during the study period, necropsy must be done on the animal to investigate on the cause of death. Generally, it is recommended that acute toxicity testing is conducted with two different animal species, one rodent and one non-rodent. Acute toxicity testing also permits determination of lethal dose 50% (LD50). The LD50 is a derived dose at which 50% of the rested animal model will be expected to experience mortality.

The most likely route of administration is via oral route. Nevertheless, the particular route decided during the test should represent the route which the nanoparticle is developed for. Testing of substance for application directly to the lungs should be considered for inhalation or nasal instillation to the rodents. Sometimes, topical application of test substance is required where eye irritation test or skin irritation test can be conducted using Draize test. This test utilizes rabbits or guinea pigs, used to evaluate the harmfulness of chemicals for topical applications. A volume of test substance is applied to the animal's eyes or to the surface of the skin. Evaluations on the physical appearance such as redness, swelling, discharge, ulceration, hemorrhage, or blindness could be assessed on the eyes. While the application of a substance to the skin, may result in erythema or edema. Assessment and observation on the animals are conducted for a 14 days period.

An alternative to Draize test on the skin, murine local lymph node assay can be conducted. This is a widely accepted method where the test substance is applied on the surface of the ears of a mouse for three consecutive days, and the proliferation of lymphocytes in the draining lymph node is then measured.

Meanwhile, subacute toxicity is defined as adverse effects observed after multiple or continuous exposure between 24 h and 28 days. The tests are intended to evaluate the toxicity of chemicals after repeated administration at the same time help to establish doses for longer-term toxicity analysis such as subchronic or chronic toxicity evaluations. The test substance must be administered regularly at a consistent time each day, to avoid nocturnal variation in the animals. Usually, rodents of any gender are preferable for this test, age 5–6 weeks. Prior to administration of test substance, the baseline of behavioral and biochemical parameters of the animals should be recorded, for the purpose of analysis in percentage changes. Following consecutive administration of test substance, most body organs should be removed, and tissues are to be analyzed for any histopathological changes. Similar parameters mentioned earlier can be repeated. In this study, changes in immune system parameters can be investigated. Parameters such as macrophage function, delayed-type hypersensitivity, primary antibody response to T-cell dependent antigen, mitogen- or antigen-stimulated lymphocyte proliferative responses are to be assessed in immunotoxicological studies.

In addition, mutagenicity testing can be conducted in repeated dosing studies, to assess potential minuscular changes to base sequence of DNA, aberration of chromosome, or structural aberration to DNA inclusive of duplications, inversions, insertions, and translocations. Alteration in DNA structure, sequences, or changes in base pairs, results in the development of carcinogenesis.

The subchronic systemic toxicity is defined as adverse effects occurring after repeated or continuous administration of a test sample or medical devices for up to 90 days or not exceeding 10% of the animal's lifespan. The decision on the toxicity test must be based on the overall testing strategy as well as the purpose, the nature of exposure, and clinical duration of use of nanomedicine or clinical device. Subsequently, chronic toxicity is defined as adverse reactions occurring after repeated administration of the test sample for a major part of the life span.

Body weight changes, biochemical and physiological variations, physical, and behavioral evaluations on the animals are observed. Gross pathological changes and histological evaluations are then subjected to histopathological analysis upon sacrifices of the animals.

In chronic toxicity study, the duration of the study is usually considered to be 6 months, study design and end points evaluated are usually similar to subchronic toxicity. The objective of the study is to determine the effects of substance following prolonged and repeated exposure, where the animal model is continuously exposed to test materials and progressive accumulation of damage in one or more critical target organ is expected.

The most common species utilized for in vivo toxicity model is the mice or rat model. Prior to inauguration to clinical practice, pharmaceutical substances must undergo rigorous toxicity testing to ensure and establish their safety. This is achievable via in vivo animal models with different studies mentioned earlier, for the sake of different purposes. Results obtained from these studies should at least establish a safe starting dose for clinical trials, and provide some understanding into certain biomarkers for clinical monitoring. These studies should at least assess potential organ toxicity and in some ways to indicate its reversibility. Importantly, these studies should provide information on a drug treatment regimen which would produce the least toxicity.

It is critical to assume that the strength and weakness of any particular animal model is essential to understand the relevance of specific findings to human. The selection of an appropriate model to be employed is the key to accurately predict the outcome seen in humans. Nevertheless, some drugs showing tolerated safety and decent efficacy in preclinical animal model may show variation and different pharmacological properties when they are administered to humans.

Acute toxicity study of silver nanoparticles in rats indicated that the materials accumulated in the organs involved in immune system, with mild irritation in thymus and spleen. The liver and kidneys were observed to be the most affected organs when the tested silver nanoparticles were administered via intravenous administration. Significant level of chromosomal breakage and polyploidy cells were also detected which implied potential carcinogenic effects of silver nanoparticles [67]. The toxicity effect of silver nanoparticles was influenced by size, where those measuring 10 nm in diameter or smaller had higher acute toxicity effect in rodents [65].

A report on layered double hydroxide nanoparticle, which is an anionic clay, indicates that the nanoparticles have a promising potential as a drug delivery carrier, when layered double hydroxide was reported to not exhibiting acute oral toxicity in mice up to 2000 mg/kg for 14 days. There were no significant changes on behavior of the animals, weight gain, survival rate, or organosomatic index. No increment in serum biochemical parameters, and there was no accumulation of layered double hydroxide in any specific organ [68].

6.5 THE SUITABILITY AND ADEQUACY OF IN VITRO AND IN VIVO DISEASE MODEL

The conundrum that faces researchers is when the results seen in in vitro are not being translated into in vivo animal model. Similarly, the results observed in the animal model are not translated by results seen when the study is conducted with human subjects. One of possible vindication is when the in vitro or in vivo modeling do not mimic the human disease condition as close as possible.

Some animal models may seem inappropriate when there is a lack of appropriate drug targets in the preclinical animal model tested. Testing of therapeutic antibodies, the relevant species used should express target antigen and tissue cross-reactive profile similar to that in human.

Furthermore, when testing for sex-specific drugs, the model used should have appropriate sex for the intended use. It is inappropriate to use the drug for prostate cancer in the female animal model or female fertility drugs which act specifically in regulating female reproductive hormone is to be tested in male rats.

The metabolic fate of the model is different from that of humans. Carcinogenic activity of tamoxifen is higher in rats than in humans when the drug produces liver cancer in rats and has not been shown to produce DNA adducts or liver tumors in human patients. The enzymatic pathway responsible for the production of tamoxifen reactive metabolites which form DNA adducts is higher in rats compared to humans. Besides, the detoxification enzymes pathways are lower in rats than in humans. Consequently, the carcinogenic effect of tamoxifen has limited relevance to the assessment of the safety of tamoxifen in humans.

The model may also have different susceptibility to infection due to specific pathogens. The bottom line is that suitable and accepted animal model(s) must be technically picked. Choosing an inappropriate animal model is a wasteful of resources, particularly of time, money, and effort. It is unquestionably unethical when performing such act to animals from an animal welfare point of view.

Studies must be conducted to establish the relationship of nanomaterial risk with their physical chemistry. The modification of physical chemistry of the engineered nanoparticles particularly in terms of size and variation of materials manipulated in drug delivery system and nanomedicine imposed intrinsic toxicity, following bioaccumulation of nanoparticle. The most common toxicological implication of nanoparticle is the generation of oxygen species, membrane disruption, the release of toxic components and immunological responses.

Some nanoparticles are recognized as being as nontoxic [69]. On the other hand, there have been a number of reports on their toxicological indications [70], which may have been contributed to the modification of physicochemical properties.

Nanoparticles may well be identified as foreign by the immune cells, causing the cells to react against either surface or core components to launch and intensify an inflammatory response, which involves secretion of signaling cytokines, triggering a cascade of the event to attract more cells to destroy the foreign substances [71]. However, many of these molecular events are still under extensive study.

Basic quality for nanomaterial considered for biological application, biosafety in cells, and live biosystems in terms of some physiological parameters such as adsorption, distribution, metabolism, excretion, and toxicity becomes essential. The in vivo toxicity studies have shown that biocompatibility of graphene and its derivatives greatly depends on its lateral size, dosage, functionalization, charge, and reactive oxygen species [23]. Graphene oxide (GO) accumulation can mainly occur in the liver, lungs, spleen, and kidneys can be cleared quickly from the bloodstream and accumulation in the liver can be eliminated by liver secretion into the bile tract system [72]. GO purification via several washing steps, polymeric modifications, conjugation strategies through surface modifications, also enhances the biocompatibility and circulation times in vivo. Graphene and its derivatives considered as "biosafe" material, by its size distribution, amount of oxygen-containing groups which are directly related to the method of production, optimum dosage, its aggregation formation tendency, and surface coating.

KEYWORDS

- **safety assessment**
- **nanotoxicology**
- **drug delivery system**
- **pharmacokinetics**

REFERENCES

1. Furtado, D., Björnmalm, M., Ayton, S., Bush, A. I., Kempe, K., and Caruso, F. (2018). Overcoming the blood–brain barrier: The role of nanomaterials in treating neurological diseases. *Advanced Materials*. https://doi.org/10.1002/adma.201801362
2. Schaeublin, N. M., Braydich-Stolle, L. K., Schrand, A. M., Miller, J. M., Hutchison, J., Schlager, J. J., and Hussain, S. M. (2011). Surface charge of gold nanoparticles mediates mechanism of toxicity. *Nanoscale*. https://doi.org/10.1039/c0nr00478b
3. Teleanu, D. M., Chircov, C., Grumezescu, A. M., Volceanov, A., and Teleanu, R. I. (2018). Blood–brain delivery methods using nanotechnology. *Pharmaceutics*. https://doi.org/10.3390/pharmaceutics10040269
4. Nurgali, K., Jagoe, R. T., and Abalo, R. (2018). Editorial: Adverse effects of cancer chemotherapy: Anything new to improve tolerance and reduce sequelae? *Frontiers in Pharmacology*. https://doi.org/10.3389/fphar.2018.00245
5. Djearamane, S., Lim, Y. M., Wong, L. S., and Lee, P. F. (2018). Cytotoxic

effects of zinc oxide nanoparticles on cyanobacterium Spirulina (Arthrospira) platensis. *PeerJ*. https://doi.org/10.7717/peerj.4682

6. Tan, J. M., Arulselvan, P., Fakurazi, S., Ithnin, H., and Hussein, M. Z. (2014). A review on characterizations and biocompatibility of functionalized carbon nanotubes in drug delivery design. *Journal of Nanomaterials*. https://doi.org/10.1155/2014/917024
7. Usman, M. S., Hussein, M. Z., Fakurazi, S., and Ahmad Saad, F. F. (2017). Gadolinium-based layered double hydroxide and graphene oxide nano-carriers for magnetic resonance imaging and drug delivery. *Chemistry Central Journal*. https://doi.org/10.1186/s13065-017-0275-3
8. Usman, M. S., Hussein, M. Z., Kura, A. U., Fakurazi, S., Masarudin, M. J., and Saad, F. F. A. (2018). Synthesis and characterization of protocatechuic acid-loaded gadolinium-layered double hydroxide and gold nanocomposite for theranostic application. *Applied Nanoscience (Switzerland)*. https://doi.org/10.1007/s13204-018-0752-6
9. Harsha, S., Al-Dhubiab, B. E., Nair, A. B., Attimarad, M., Venugopala, K. N., and Sa, K. (2017). Pharmacokinetics and tissue distribution of microspheres prepared by spray drying technique: Targeted drug delivery. *Biomedical Research (India)*.
10. Senapati, S., Shukla, R., Tripathi, Y. B., Mahanta, A. K., Rana, D., and Maiti, P. (2018). Engineered cellular uptake and controlled drug delivery using two dimensional nanoparticle and polymer for cancer treatment. *Molecular Pharmaceutics*. https://doi.org/10.1021/acs.molpharmaceut.7b01119
11. Keller, A. A., Adeleye, A. S., Conway, J. R., Garner, K. L., Zhao, L., Cherr, G. N., ... Zuverza-Mena, N. (2017). Comparative environmental fate and toxicity of copper nanomaterials. *NanoImpact*. https://doi.org/10.1016/j.impact.2017.05.003
12. Zhong, H., Chan, G., Hu, Y., Hu, H., Ouyang, D. A (2018) Comprehensive map of FDA-approved pharmaceutical products. *Pharmaceutics*. 10(4):263. Published 2018 Dec 6. doi:10.3390/pharmaceutics10040263
13. Nanda, S. S., Yi, D. K., and Kim, K. (2016). Study of antibacterial mechanism of graphene oxide using Raman spectroscopy. *Scientific Reports*. https://doi.org/10.1038/srep28443
14. Yong, Y. K., Moheimani, S. O. R., and Petersen, I. R. (2010). High-speed cycloid-scan atomic force microscopy. *Nanotechnology*. https://doi.org/10.1088/0957-4484/21/36/365503
15. Acosta-Torres, L. S., Lpez-Marín, L. M., Núñez-Anita, R. E., Hernández-Padrón, G., and Castaño, V. M. (2011). Biocompatible metal-oxide nanoparticles: Nanotechnology improvement of conventional prosthetic acrylic resins. *Journal of Nanomaterials*. https://doi.org/10.1155/2011/941561
16. Chen, S., Yuan, R., Chai, Y., and Hu, F. (2013). Electrochemical sensing of hydrogen peroxide using metal nanoparticles: A review. *Microchimica Acta*. https://doi.org/10.1007/s00604-012-0904-4
17. Van Schooneveld, M. M., Gloter, A., Stephan, O., Zagonel, L. F., Koole, R., Meijerink, A., ... De Groot, F. M. F. (2010). Imaging and quantifying the morphology of an organic–inorganic nanoparticle at the sub-nanometre level. *Nature Nanotechnology*. https://doi.org/10.1038/nnano.2010.105
18. Giner-Casares, J. J., Henriksen-Lacey, M., Coronado-Puchau, M., and Liz-Marzán, L. M. (2016). Inorganic

nanoparticles for biomedicine: Where materials scientists meet medical research. *Materials Today*. https://doi.org/10.1016/j.mattod.2015.07.004
19. Chen, Y., Chen, H., and Shi, J. (2014). Inorganic nanoparticle-based drug codelivery nanosystems to overcome the multidrug resistance of cancer cells. *Molecular Pharmaceutics*. https://doi.org/10.1021/mp400596v
20. Smith, D., Artursson, P., Avdeef, A., Di, L., Ecker, G. F., Faller, B., … Testa, B. (2014). Passive lipoidal diffusion and carrier-mediated cell uptake are both important mechanisms of membrane permeation in drug disposition. *Molecular Pharmaceutics*. https://doi.org/10.1021/mp400713v
21. Kato, T., Totsuka, Y., Ishino, K., Matsumoto, Y., Tada, Y., Nakae, D., … Wakabayashi, K. (2013). Genotoxicity of multi-walled carbon nanotubes in both in vitro and in vivo assay systems. *Nanotoxicology*. https://doi.org/10.3109/17435390.2012.674571
22. Helmlinger, G., Sckell, A., Dellian, M., Forbes, N. S., and Jain, R. K. (2002). Acid production in glycolysis-impaired tumors provides new insights into tumor metabolism. *Clinical Cancer Research*.
23. Barahuie, F., Hussein, M. Z., Gani, S. A., Fakurazi, S., and Zainal, Z. (2015). Synthesis of protocatechuic acid-zinc/aluminium-layered double hydroxide nanocomposite as an anticancer nanodelivery system. *Journal of Solid State Chemistry*. https://doi.org/10.1016/j.jssc.2014.09.001
24. Bertz, A., Wöhl-Bruhn, S., Miethe, S., Tiersch, B., Koetz, J., Hust, M., … Menzel, H. (2013). Encapsulation of proteins in hydrogel carrier systems for controlled drug delivery: Influence of network structure and drug size on release rate. *Journal of Biotechnology*. https://doi.org/10.1016/j.jbiotec.2012.06.036
25. Tee, J. K., Yip, L. X., Tan, E. S., Santitewagun, S., Prasath, A., Ke, P. C., … Leong, D. T. (2019). Nanoparticles' interactions with vasculature in diseases. *Chemical Society Reviews*. https://doi.org/10.1039/C9CS00309F
26. Miao, L., and Huang, L. (2015). Exploring the tumor microenvironment with nanoparticles. *Cancer Treatment and Research*. https://doi.org/10.1007/978-3-319-16555-4_9
27. Moghimi, S. M., and Szebeni, J. (2003). Stealth liposomes and long circulating nanoparticles: Critical issues in pharmacokinetics, opsonization and protein-binding properties. *Progress in Lipid Research*. https://doi.org/10.1016/S0163-7827(03)00033-X
28. Brannon-Peppas, L., and Blanchette, J. O. (2012). Nanoparticle and targeted systems for cancer therapy. *Advanced Drug Delivery Reviews*. https://doi.org/10.1016/j.addr.2012.09.033
29. Steichen, S. D., Caldorera-Moore, M., and Peppas, N. A. (2013). A review of current nanoparticle and targeting moieties for the delivery of cancer therapeutics. *European Journal of Pharmaceutical Sciences*. https://doi.org/10.1016/j.ejps.2012.12.006
30. Yuan, Y. Y., Mao, C. Q., Du, X. J., Du, J. Z., Wang, F., and Wang, J. (2012). Surface charge switchable nanoparticles based on zwitterionic polymer for enhanced drug delivery to tumor. *Advanced Materials*. https://doi.org/10.1002/adma.201202296
31. Brannon-Peppas, L. and Blanchette, J. O. (2012). Nanoparticle and targeted systems for cancer therapy. Advanced Drug Delivery Reviews. https://doi.org/10.1016/j.addr.2012.09.033.
32. Jain, A. K., Singh, D., Dubey, K., Maurya, R., Mittal, S., and Pandey, A. K. (2017). Models and methods for in

vitro toxicity. In *In Vitro Toxicology* (pp. 45–65). https://doi.org/10.1016/B978-0-12-804667-8.00003-1
33. Dorniani, D., Hussein, M. Z. Bin, Kura, A. U., Fakurazi, S., Shaari, A. H., and Ahmad, Z. (2012). Preparation of Fe3O4 magnetic nanoparticles coated with gallic acid for drug delivery. *International Journal of Nanomedicine*. https://doi.org/10.2147/IJN.S35746
34. Rosman, R., Saifullah, B., Maniam, S., Dorniani, D., Hussein, M., and Fakurazi, S. (2018). Improved anticancer effect of magnetite nanocomposite formulation of Gallic Acid (Fe_3O_4-PEG-GA) against lung, breast and colon cancer cells. *Nanomaterials*. https://doi.org/10.3390/nano8020083.
35. Soenen, S. J., Rivera-Gil, P., Montenegro, J. M., Parak, W. J., De Smedt, S. C., and Braeckmans, K. (2011). Cellular toxicity of inorganic nanoparticles: Common aspects and guidelines for improved nanotoxicity evaluation. *Nano Today*. https://doi.org/10.1016/j.nantod.2011.08.001
36. Dorniani, D., Kura, A. U., Hussein-Al-Ali, S. H., Hussein, M. Z. Bin, Fakurazi, S., Shaari, A. H., and Ahmad, Z. (2014). Release behavior and toxicity profiles towards leukemia (WEHI-3B) cell lines of 6-mercaptopurine-PEG-coated magnetite nanoparticles delivery system. *Scientific World Journal*. https://doi.org/10.1155/2014/972501
37. Patil, U. S., Qu, H., Caruntu, D., O'Connor, C. J., Sharma, A., Cai, Y., and Tarr, M. A. (2013). Labeling primary amine groups in peptides and proteins with N-hydroxysuccinimidyl ester modified Fe_3O_4@SiO_2 nanoparticles containing cleavable disulfide-bond linkers. *Bioconjugate Chemistry*. https://doi.org/10.1021/bc400165r
38. Kura, A. U., Cheah, P. S., Hussein, M. Z., Hassan, Z., Tengku Azmi, T. I., Hussein, N. F., and Fakurazi, S. (2014). Toxicity evaluation of zinc aluminium levodopa nanocomposite via oral route in repeated dose study. *Nanoscale Research Letters*. https://doi.org/10.1186/1556-276X-9-261
39. Baravalia, Y., Vaghasiya, Y., and Chanda, S. (2012). Brine shrimp cytotoxicity, anti-inflammatory and analgesic properties of Woodfordia fruticosa Kurz flowers. *Iranian Journal of Pharmaceutical Research*.
40. Mayorga, P., Pérez, K. R., Cruz, S. M., and Cáceres, A. (2010). Comparison of bioassays using the anostracan crustaceans Artemia salina and Thamnocephalus platyurus for plant extract toxicity screening. *Revista Brasileira de Farmacognosia*. https://doi.org/10.1590/s0102-695x2010005000029
41. Kokkali, V., Katramados, I., and Newman, J. D. (2011). Monitoring the effect of metal ions on the mobility of Artemia salina Nauplii. *Biosensors*. https://doi.org/10.3390/bios1020036
42. Pelka, M., Danzl, C., Distler, W., and Petschelt, A. (2000). A new screening test for toxicity testing of dental materials. *Journal of Dentistry*. https://doi.org/10.1016/S0300-5712(00)00007-5
43. Sharma, S., and Vig, A. P. (2013). Evaluation of in vitro antioxidant properties of methanol and aqueous extracts of Parkinsonia aculeata L. leaves. *The Scientific World Journal*. https://doi.org/10.1155/2013/604865
44. Reddi Mohan Naidu, K., Khalivulla, S. I., Rasheed, S., Fakurazi, S., Arulselvan, P., Lasekan, O., and Abas, F. (2013). Synthesis of bisindolylmethanes and their cytotoxicity properties. *International Journal of Molecular Sciences*. https://doi.org/10.3390/ijms14011843.

45. Gadir, S. A. (2012). Assessment of bioactivity of some Sudanese medicinal plants using brine shrimp (Artemia salina) lethality assay. *Journal of Chemical and Pharmaceutical Research.*
46. Pisutthanan, S., Plianbangchang, P., Pisutthanan, N., Ruanruay, S., and Muanrit, O. (2004). Brine shrimp lethality activity of thai medicinal plants in the family meliaceae. *Naresuan University Journal.*
47. Arslanyolu, M., and Erdemgil, F. Z. (2006). Evaluation of the antibacterial activity and toxicity of isolated arctiin from the seeds of Centaurea sclerolepis. *Ankara Universitesi Eczacilik Fakultesi Dergisi.*
48. Lagarto Parra, A., Silva Yhebra, R., Guerra Sardiñas, I., and Iglesias Buela, L. (2001). Comparative study of the assay of Artemia salina L. And the estimate of the medium lethal dose (LD50 value) in mice, to determine oral acute toxicity of plant extracts. *Phytomedicine.* https://doi.org/10.1078/0944-7113-00044
49. Luepke, N. P., and Kemper, F. H. (1986). The HET-CAM test: An alternative to the draize eye test. *Food and Chemical Toxicology.* https://doi.org/10.1016/0278-6915(86)90099-2
50. Kuen, C., Fakurazi, S., Othman, S., and Masarudin, M. (2017). Increased loading, efficacy and sustained release of silibinin, a poorly soluble drug using hydrophobically-modified chitosan nanoparticles for enhanced delivery of anticancer drug delivery systems. *Nanomaterials.* https://doi.org/10.3390/nano7110379
51. Cook, P. R., Brazell, I. A., and Jost, E. (1976). Characterization of nuclear structures containing superhelical DNA. *Journal of Cell Science.*
52. Ostling, O., and Johanson, K. J. (1984). Microelectrophoretic study of radiation-induced DNA damages in individual mammalian cells. *Biochemical and Biophysical Research Communications.* https://doi.org/10.1016/0006-291X(84)90411-X
53. Hackenberg, S., Scherzed, A., Kessler, M., Hummel, S., Technau, A., Froelich, K., … Kleinsasser, N. (2011). Silver nanoparticles: Evaluation of DNA damage, toxicity and functional impairment in human mesenchymal stem cells. *Toxicology Letters.* https://doi.org/10.1016/j.toxlet.2010.12.001
54. Kim, S. C., Chen, D. R., Qi, C., Gelein, R. M., Finkelstein, J. N., Elder, A., … Pui, D. Y. H. (2010). A nanoparticle dispersion method for in vitro and in vivo nanotoxicity study. *Nanotoxicology.* https://doi.org/10.3109/17435390903374019
55. Naya, M., Kobayashi, N., Mizuno, K., Matsumoto, K., Ema, M., and Nakanishi, J. (2011). Evaluation of the genotoxic potential of single-wall carbon nanotubes by using a battery of in vitro and in vivo genotoxicity assays. *Regulatory Toxicology and Pharmacology.* https://doi.org/10.1016/j.yrtph.2011.07.008
56. Senapati, S., Shukla, R., Tripathi, Y. B., Mahanta, A. K., Rana, D., and Maiti, P. (2018). Engineered cellular uptake and controlled drug delivery using two dimensional nanoparticle and polymer for cancer treatment. Molecular Pharmaceutics. https://doi.org/10.1021/acs.molpharmaceut.7b01119.
57. Su, J. K., Byeong, M. K., Young, J. L., and Hai, W. C. (2008). Titanium dioxide nanoparticles trigger p53-mediated damage response in peripheral blood lymphocytes. *Environmental and Molecular Mutagenesis.* https://doi.org/10.1002/em.20399
58. Jain R.K., Tong R.T., and Munn, L.L., (2007). Effect of vascular normalization by antiangiogenic therapy on interstitial

hypertension, peritumor edema, and lymphatic metastasis: Insights from a mathematical model. *Cancer Research,* 15 2007 (67) (6) 2729-2735; DOI: 10.1158/0008-5472. CAN-06-4102
59. Lebedová, J., Hedberg, Y. S., Odnevall Wallinder, I., and Karlsson, H. L. (2018). Size-dependent genotoxicity of silver, gold and platinum nanoparticles studied using the mini-gel comet assay and micronucleus scoring with flow cytometry. *Mutagenesis*. https://doi.org/10.1093/mutage/gex027
60. Ziąbka, M., Dziadek, M., Menaszek, E., Banasiuk, R., and Królicka, A. (2017). Middle Ear Prosthesis with Bactericidal Efficacy-In Vitro Investigation. *Molecules(Basel,Switzerland)*.https://doi.org/10.3390/molecules22101681
61. Perde-Schrepler, M., Florea, A., Brie, I., Virag, P., Fischer-Fodor, E., Vâlcan, A., … Maniu, A. (2019). Size-dependent cytotoxicity and genotoxicity of silver nanoparticles in cochlear cells in vitro. *Journal of Nanomaterials*. https://doi.org/10.1155/2019/6090259
62. Carmona, E. R., García-Rodríguez, A., and Marcos, R. (2018). Genotoxicity of copper and nickel nanoparticles in somatic cells of drosophila melanogaster. *Journal of Toxicology*. https://doi.org/10.1155/2018/7278036
63. Kim, E. Y., Schulz, R., Swantek, P., Kunstman, K., Malim, M. H., and Wolinsky, S. M. (2012). Gold nanoparticle-mediated gene delivery induces widespread changes in the expression of innate immunity genes. *Gene Therapy*. https://doi.org/10.1038/gt.2011.95
64. Zhang, Y., Zhan, X., Xiong, J., Peng, S., Huang, W., Joshi, R., … Min, W. (2018). Temperature-dependent cell death patterns induced by functionalized gold nanoparticle photothermal therapy in melanoma cells. *Scientific Reports*, *8*(1), 8720. https://doi.org/10.1038/s41598-018-26978-1
65. Cho, W. S., Cho, M., Jeong, J., Choi, M., Cho, H. Y., Han, B. S., … Jeong, J. (2009). Acute toxicity and pharmacokinetics of 13 nm-sized PEG-coated gold nanoparticles. *Toxicology and Applied Pharmacology*. https://doi.org/10.1016/j.taap.2008.12.023
66. Modrzynska, J., Berthing, T., Ravn-Haren, G., Jacobsen, N. R., Weydahl, I. K., Loeschner, K., … Vogel, U. (2018). Primary genotoxicity in the liver following pulmonary exposure to carbon black nanoparticles in mice. *Particle and Fibre Toxicology*. https://doi.org/10.1186/s12989-017-0238-9
67. Wen, H., Dan, M., Yang, Y., Lyu, J., Shao, A., Cheng, X., … Xu, L. (2017). Acute toxicity and genotoxicity of silver nanoparticle in rats. *PLoS ONE*. https://doi.org/10.1371/journal.pone.0185554
68. Yu, J., Chung, H. E., and Choi, S. J. (2013). Acute oral toxicity and kinetic behaviors of inorganic layered nanoparticles. *Journal of Nanomaterials*. https://doi.org/10.1155/2013/628381
69. Connor, E. E., Mwamuka, J., Gole, A., Murphy, C. J., and Wyatt, M. D. (2005). Gold nanoparticles are taken up by human cells but do not cause acute cytotoxicity. *Small*. https://doi.org/10.1002/smll.200400093
70. Malik, M. A., Wani, M. Y., Hashim, M. A., and Nabi, F. (2011). Nanotoxicity: Dimensional and morphological concerns. *Advances in Physical Chemistry*. https://doi.org/10.1155/2011/450912
71. Halamoda-Kenzaoui, B., and Bremer-Hoffmann, S. (2018). Main trends of immune effects triggered by nanomedicines in preclinical studies. *International Journal of*

Nanomedicine, *13*, 5419–5431. https://doi.org/10.2147/IJN.S168808
72. Huang, H., Guo, X., Li, D., Liu, M., Wu, J., and Ren, H. (2011). Identification of crucial yeast inhibitors in bio-ethanol and improvement of fermentation at high pH and high total solids. *Bioresource Technology*, *102*(16), 7486–7493. https://doi.org/10.1016/j.biortech.2011.05.008

CHAPTER 7

Sustainable Development of Polyhydroxyalkanoates and Their Application as an Effective Tool in Drug Delivery and Tissue Engineering†

ANUPAMA SHRIVASTAVA[1], SONAM DUBEY[2], ALKA MISHRA[1], SOURISH BHATTACHARYA[2*], and SANDHYA MISHRA[2]

[1]*Department of Microbiology, Parul Institute of Applied Sciences, Parul University, Vadodara, Gujarat, India*

[2]*Division of Biotechnology and Phycology, CSIR-Central Salt and Marine Chemicals Research Institute (CSIR-CSMCRI), Bhavnagar, Gujarat, India*

[*]*Corresponding author. E-mail: E-mail: sourishb@csmcri.res.in*

ABSTRACT

Biopolymers are great alternatives of synthetic polymers. A biopolymer has various forms, such as ε-polylysine, polyhydroxyalkanoate (PHA), poly-lactic acid (PLA), starch-based plastics, and so on, which are totally biodegradable in nature that makes them compatible for living organisms. Of all the bioplastics, PHA has lower acidity and bioactivity so it is more acceptable for biomedical purposes compared to other forms of bioplastics such as PLA and poly-glycolic acid. PHA is produced by various microorganisms under unfavorable conditions, such as higher carbon availability and limitation of mineral nutrients. Therefore, this chapter summarizes some of the most recent uses of PHA (bioplastics) in various biomedical applications, such as drug carriers (sodium dodecyl sulfate), biodegradable implants, tissue engineering, medical devices, antimicrobial application, and coating material for biochip and bioelectronics.

†Anupama Shrivastava and Sonam Dubey contributed equally to this work.

7.1 INTRODUCTION

Biological materials such as wool, leather, silk, and cellulose are traditional commodity polymers. Currently, modern biotechnological methods such as recombinant, bacterial fermentation and advancing genetic engineering, and nanotechnology provide new or specifically required properties, which are given to the natural polymers and create the usage potential for uncommon biopolymers with their derivatives and blends. Present research on medical applications focused on using biopolymers such as, particularly, ε-polylysine, poly-lactic acid (PLA), poly-glycolic acid (PGA), chitin, chitosan, hyaluronan, and their derivatives, blends, and so on.

Nowadays, we mostly use synthetic plastics and their consumption is increasing in our daily life because of technology advancement and population growth. But, these synthetic plastics, such as polyethylene, polybutylene, polystyrene, polyvinyl chloride, and so on, have harmful effects on the environment as they are nonbiodegradable in nature. Hence, there is a demand for greener plastics or biodegradable plastics that can act as alternative of petroleum-based plastic [1] Bioplastics are an example of biodegradable plastics that are ecofriendly and biocompatible in nature. Bioplastics can easily be degraded by environmental microorganisms by breaking down their chemical bonds present in the structure of biopolymer [2] They are totally biocompatible in nature when they are used in biomedical applications. Most of the biopolymers have esters, orthoesters, anhydride, and carbonate like hydrolyzable linkage in the backbones. Those aliphatic polyesters which have an ester bond have very high biocompatibility and they have various physical, chemical, and biological properties. Polyhydroxyalkanoate (PHA) is the most commonly used biopolymer in biomedical applications due to its high biocompatibility [3].

Biocompatibility term refers to the appropriate host response toward the use of material in a particular application [1]. It includes a detailed description of biological mechanisms like cytotoxicity, immunogenic, carcinogenic, and thrombogenic response of the host. These responses depend on material–tissue interactions that were closely interdependent. In biomedical applications, it is necessary to minimize and adjust the material–tissue interaction for biologically stable materials and permanently implanting the devices. These interactions should be suitable and stable for long-time performance and treatments. So, biocompatibility is the most important factor of bioplastic that should be examined before using biomedical purposes [4].

7.1.1 CATEGORIES OF PHA

PHA is a cytoplasmic storage material of microorganisms under unfavorable growth conditions. First, PHA was observed by Beijerinck in 1888 inside a bacterial cell as inclusion refractile bodies. It is most widely used as a great alternative to synthetic plastic in agriculture, marine, and biomedical applications. It is completely degraded into CO_2 and H_2O by microorganisms present in the environment, such as soil, water, and sewage [5]. Mainly, there are two categories of PHAs—one is short chain length (SCL) PHAs with 3 to 5 carbon atoms per repetition unit and the another one is medium chain length (MCL) with 6 to 14 carbon atoms per repetition unit (Figures 7.1 and 7.2). These two differ in their mechanical behavior. SCL-PHAs are brittle in nature and tend to have high crystallinity whereas MCL-PHAs are more flexible but exhibit low mechanical strength as compared to SCL-PHAs. We can overcome these shortcomings by physical blending, chemical functionalization, and copolymerization. PHA can be blended with different types of biodegradable polymers such as starch, cellulose derivatives, lignin, PLA, and PCL, and chemical modification includes controlled, living polymerization techniques [6–8].

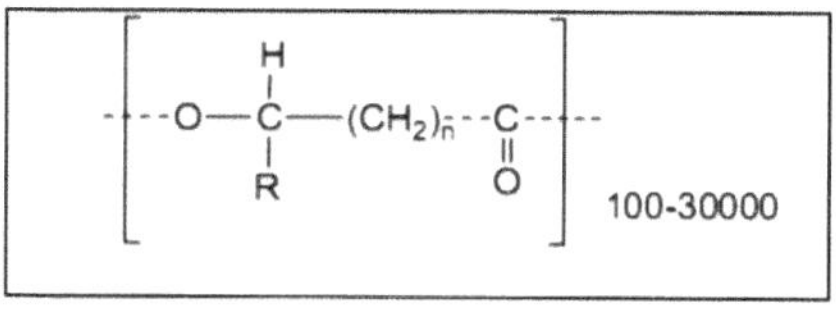

$$\left[-O-\underset{R}{\overset{H}{C}}-(CH_2)_n-\underset{O}{\overset{}{C}}- \right]_{100\text{-}30000}$$

FIGURE 7.1 General structure of PHA [5].

n=1 R=methyl:polymer = poly(3-hydroxybutyrate)

R=ethyl: polymer = poly(3-hydroxyvalerate)

n=2 R=hydrogen polymer = poly(4-hydroxybutyrate)

n=3 R= hydrogen = poly(5-hydroxybutyrate)

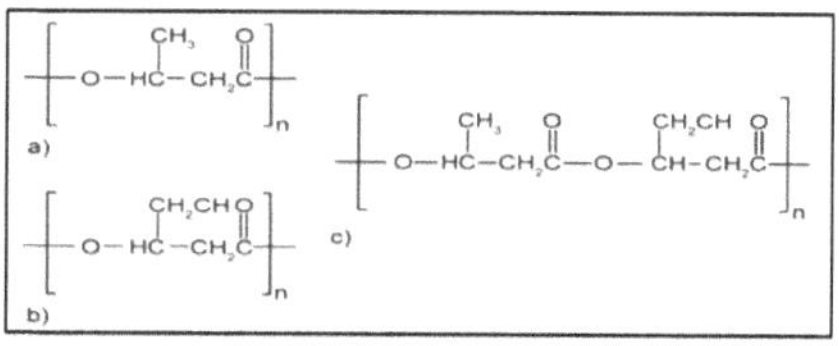

a) $\left[-O-HC(CH_3)-CH_2C(=O)- \right]_n$

b) $\left[-O-HC(CH_2CH_3)-CH_2C(=O)- \right]_n$

c) $\left[-O-HC(CH_3)-CH_2C(=O)-O-CH(CH_2CH_3)-CH_2C(=O)- \right]_n$

FIGURE 7.2 Chemical structure of some polyhydroxyalkanoates, (a) PH3B, (b) PHV, (c) PHBV (from [5]).

Poly-3-hydrxybutyrate (PHB) is the most common member of the PHA family. It is homo-polymer of 3-hydroxybutyrate or linear polyester of D(−)-3-hydrxybutyric acid, which was first discovered in bacteria by Lemoigne in 1925 [9]. It belongs to SCL PHA family. PHA can be produced by numerous bacteria; there are more than 300 different microorganisms known which can produce PHA [10]. *Ralstonia eutrophus* or *Bacillus megaterium* are common PHB-producing microorganisms. In the condition of physiological stress, PHB can be produced by either pure culture or mixed culture of bacteria [11]. PHB has a high degree of biodegradability due to its semicrystalline isotactic stereoregular polymer property with 100% R configuration [12]. Further, it has high melting temperature of 160–175 °C (Table 7.1).

General properties of polyhydroxyalkanoates:

1. PHA is soluble in chloroform and other chlorinated hydrocarbons whereas insoluble in water.
2. It is biocompatible in nature so it is applicable in biomedical purposes.
3. It is nontoxic in nature and its degradation is easy.
4. It has good resistance to ultraviolet (UV) rays, but it has poor acid–base resistance.

TABLE 7.1 Range of Typical Properties of PHAs

Property[a] (units)	Values
Tg (°C)	2
Tm (°C)	160–175
X_{cr} (%)	40–60
E (GPa)	1–2
σ (MPa)	15–40
ε (%)	1–15
WVTR (g mm/m^2 day)	2.36
OTR (cc mm/ m^2 day)	55.12

[a]T_g: Glass transition temperature; T_m: melting temperature; X_{cr}: crystallinity degree; E: Young's modulus; σ: tensile strength; WVTR: water vapor transmission rate; OTR: oxygen transmission rate.

In the biomedical field, PHA is widely applicable as it is compatible with mammalian blood and tissue. It can be reabsorbed by the human body so it is applicable in surgical implants as well as in surgery as seam thread for the blood vessels and healing of wounds. It can also be used as microcapsule and material for cell and tablet packaging in the pharmaceutical industry [13].

7.2 FACTORS AFFECTING PHA SYNTHESIS

7.2.1 NUTRIENT AVAILABILITY AND OPTIMIZATION

PHA production increases in nutrient-limiting media. Low concentration of nitrogen and phosphorus is suitable for production of PHA whereas higher concentration of nitrogen leads to increase in biomass growth with no PHA production [14]. It was reported that 45.1% and 54.2% of production of PHA occurred under limiting nitrogen and phosphorus concentration, respectively. After optimization of same culture condition, 14% increase in copolymer P3(HB-co-HV) synthesis was observed [15]. Excess carbon source inhibits the accumulation of PHA. We can say that aerobic, nitrogen excess or limited environment, phosphorus excess or limited environment, or carbon source affect the production of PHA [16].

Determining carbon/nitrogen (C/N) variable ratio is very important during PHA production process. Different (C/N) ratio is required for different cultures with different carbon and nitrogen source [17]. Like if sucrose is carbon source and ammonium sulfate is nitrogen source for *Alkaligenes latus* ATCC 29713, the C/N ratio is 28.3 that is required for increasing 1.8-folds production of PHA at 33 °C temperature, 6.5 pH [18]. For activated sludge, C/N ratio of 144 is optimum. Generally, optimum C/N ratio of 25 showed maximum PHA production [19].

With the nutrient media, the designing of cultural condition is also important because it allows us to use nutrients to increase PHA production. For example, for the maximum production of PHA (1.45 $g \cdot L^{-1}$) from *Bacillus megaterium,* Box–Behnken design is used with the ammonium sulfate as nitrogen source, glucose as carbon source, and Na_2HPO_4 and KH_2PO_4 as phosphorus source in a nutrient media [20]. Similarly, from activated sludge, a five-level three-factor central composite rotary design was used for PHA production. In *Halomonas boliviensis*, factorial design was employed to study the effect of ammonium, phosphate, and yeast extract on PHB production in fed-batch culture [21].

Different acids can also act as a suitable carbon source for the production of PHA. For example, citrate (0.5%) and acetate (0.5%) acted as carbon source and led to formation of 51% and 77% of PHB by 5 days of dark incubation in *Aulosira fertilissima* under phosphate-deficient condition [22].

Also, type of polymer synthesized depends on type of substrate used [23]. Like if higher propionate is used as carbon source, it leads to higher hydroxyvalerate (HV) content formation, whereas butyric acid forms only hydrxybutyrate (HB) units [24].

7.2.2 EFFECT OF PH

pH is a very important factor during PHA production. In acidogenic fermentation, it was observed that initial alkaline pH of 9 is good for production of PHA [25]. pH also affects the polymer composition of PHA especially on HV content. Usually, microorganisms maintain their pH of cytoplasm for optimization of cell functioning in response to change in external pH, and after changing in pH from 5.5 to 9.5, their HV content is changed from 10% to 30% mol. But optimum pH for production of PHA is neutral (pH = 7) [26].

7.2.3 FEEDING STRATEGY

Feeding strategy is also a very important factor in the production and composition of PHA. It mainly affects the monomer composition of PHA and it was studied by ^{13}C-NMR spectroscopy [27]. When acetate was used as carbon source in aerobic condition, it shows no effect on microbial composition. Substrate utilization rate is observed higher in pulse-feeding mode rather than continuous-feeding mode [28]. But in continuous-feeding mode, PHA accumulation up to 64.5% is observed in activated sludge collected from wastewater treatment plant and food waste [25]. In two-stage fed-batch system, the production of PHB was increased in case of *Pseudomonas putida* species when pregrown cells in glucose were transferred to medium containing octanoate under nitrogen- and oxygen-limiting environment [29].

7.3 UPSTREAM PROCESS OF PHA PRODUCTION

There are a lot of microorganisms that can produce PHA, but only a limited number of them are used for the production of PHA based on their high PHA accumulation ability. The strain we choose is normally capable of utilizing large range of sugar for carbon source consisting glucose, sucrose, fructose, fatty acids, and gluconate for PHA synthesis. *Bacillus cereus* can utilize this wide range of sugar [30]. Some of the microorganisms namely *Azotobacter vinelandii* are not good for industrial-scale production of PHA because they have low yield and capsule-forming characteristics. *Alkaligens latus* strain UWD of *Azotobacter vinelandii is* considered as good PHA-producing strain for industrial purposes because it can store 70% to 80% PHA and UWD strain is also a capsule-negative mutant [31].

Currently, PHA production at industrial scale is very low, and the production is done via bacterial fermentation process. So, bacterial fermentation plays a crucial step in the production of PHA. This process is divided into batch fermentation and fed-batch fermentation. Batch fermentation is further divided into two categories: one-stage cultivation and two-stage cultivation. Fed-batch cultivation requires fed continuously. It is also divided into two types—one is for growth-associated products and another one is for growth nonassociated products. *Cupriavidus necator* follows growth nonassociated pattern. In this, first growth is achieved then PHA accumulation starts here. Accumulated PHA can be used as an energy source by microorganisms in the presence of limited carbon substrate; therefore, there is a requirement of appropriate feeding strategies for the carbon substrate to obtain a fed-batch culture with high productivity [32, 33].

7.3.1 EXAMPLE OF SOME FEEDING STRATEGIES FOR PRODUCTION OF PHA

Alderete and co-workers [34] studied the production of copolymer poly(3HB-co-3HV) using fed-batch fermentation using mutant strain of *Alkaligenes eutrophus* which has the ability to utilize alcohols. They applied feeding strategies of addition of 1-propanol to the medium containing ethanol which resulted in the total production of 15.1 mol% 3HV units under phosphate-limiting condition. They found improved copolymer yield after switching between alcohol mixture also.

Kim and co-workers [34] reported the production of copolymer (3HB-co-3HV) by using propionic acid and glucose in *Ralstonia eutropha.* At 20% air saturation, increase of propionic acid to glucose ratio from 0.17 to 0.15 mol/mol led to an increase in the production of 3HV polymer from 4.3 to 14.3 mol%. They also observed decrease in production rate and yield from 2.55 to 1.64 g/L/h and 0.33 to 0.28 mol/mol, respectively.

Ghosh et al. [35] and Shrivastav et al. [36] report the utilization of biodiesel waste residue as raw substrate for PHA production in *Halomonas hydrothermalis.* The same biodiesel waste stream was effectively utilized by *Bacillus licheniformis* for the simultaneous production of both 0.2 g/L ε-polylysine and 64.6% (w/w) polyhydroxyalkanoate [37]. Bera and co-workers[38] found that adding 0.35% crude levulinic acid as co-substrate with 2% biodiesel waste residue and 10% hydrolysate profoundly added 81% valerate units with 125 °C mp and 464 kDa molecular weight (Figure 7.3). Also, when pure levulinic acid was added in the production media, it drastically decreased the biomass and PHA productivity in *Halomonas hydrothermalis* MTCC 5445.

In another report by Dhangdhariya et al. [39], dry sea mix was used as a natural sea substrate for the production of PHA in *Bacillus megaterium.* The statistical optimization showed 56.77% of PHA accumulation within 24 h of incubation. This indicates the possibility of exploring marine bioresources as a potential substrate for bacterial growth [40].

Shang and co-workers [41] reported PHB production under phosphate-limiting condition. They observed higher PHB accumulation in cell-growth stage and low-glucose concentration-favored PHB synthesis in *Ralstonia eutropha.* Therefore, limitation of carbon substrate can also cause PHA accumulation.

In *Ralstonia eutropha* REZ, a pH-stat fed batch is used for co-expressing the phbC and zwf genes. After applying the feeding of

10 g/L of fructose, the pH level 7 was maintained. Then, 60.1% production of poly(3HB-co-3HV) was observed with cell mass of 14 g/L [42].

Optimization of nutrient feed and concentration and addition time was studied by Khanna and Srivastava [43] for synthesis of PHA. They developed a mathematical model to describe the batch kinetics. This model was then extrapolated for the fed-batch fermentation and these strategies resulted in the production of 32 g/L biomass and 14 g/L PHB with a productivity of 0.28 g/L/h.

Optimization is the process at which we obtained the maximum benefits by improving the performance of process, system, or product and this process has been used from very long time in fermentation industry to increase the productivity [44]. It allows us to interpret the critical value for each factor that indulges the response. Optimization is used widely in analytical applications, industry, and bioprocessing [45].

There are many techniques used for optimization of fermentation industry. Traditionally, we used one-variable-at-a-time optimization. Here, we maintained one factor at a time on an exponential response. But this technique does not include the interactive effects among the variable studies and it is very time consuming process because of lots of experiments [46].

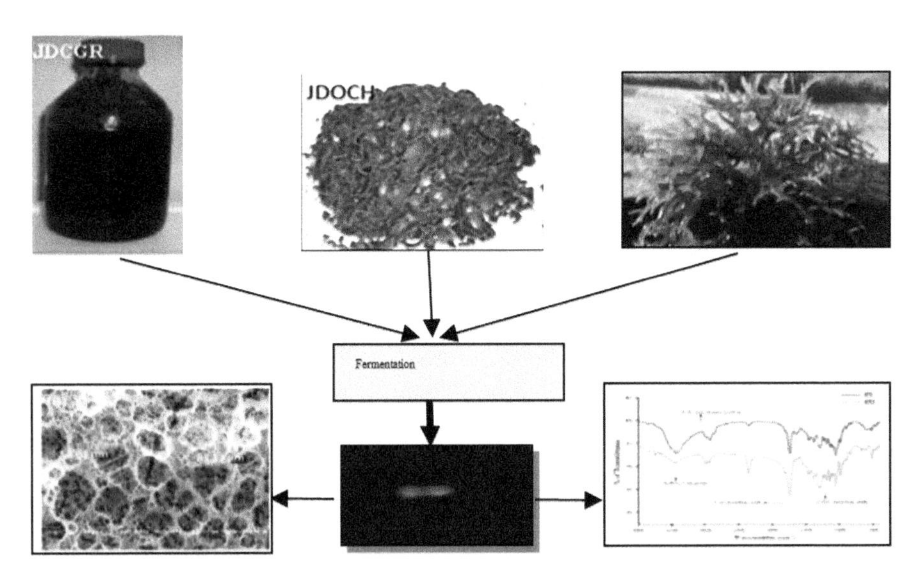

FIGURE 7.3 Graphical representation indicating utilization of raw substrates for PHA production.

Therefore, widely we used response surface methodology (RSM) for optimization. RSM provides statistical techniques for designing experiments, building models, and finding a critical value for each factor that satisfies the response. In this process, we can evaluate interactive effect among the variables and can generate contour plots and three-dimensional (3D) surface from the linear and quadratic effects of the variables. An equation is generated by using software which is to be further applied for calculation of optimal response of the system [47, 48].

From economic point of view, the cost of carbon substrate majorly contributes in the production cost of PHA. So, nowadays many researchers are trying to find out cheap carbon substrate and for this industrial waste products, residual compounds from biotechnological processes, and agro-industrial waste might serve as potential carbon sources for production of PHA.

Presence of PHA in bacterial cell can be identified by various techniques. The most common one is gas chromatography (GC). Others are Sudan black B staining, Nile red fluorescence staining, and carbon Fuchsin staining. These all methods are used for determining PHB producing organisms.

7.4 DOWNSTREAM PROCESS OF PHA PRODUCTION

PHA containing cell must be separated from broth after fermentation process, and there are several techniques for doing so like centrifugation, flocculation, and filtration which are applied for separation process. Then cell is disrupted to recover the PHA. There are many methods for extraction of PHA. The most common one is extraction of the polymer from the biomass with the solvents. The solvents used in this procedure are chloroform, dichloroethane, methylene chloride, and so on [52]. But, this method has limited use because it requires huge amount of solvents and solvents create hazards for the operators and the environments. Other most effective method is extraction with sodium hypochlorite. Sodium hypochlorite digests the non-PHA cellular biomass which causes degradation of PHA and reduction in the molecular weight [49, 50].

Nonsolvent recovery processes can also be used for extraction of PHA. This procedure was developed by ZENECA for the commercial production of poly(3HB) and poly(3HB-co-3HV) from *A. eutrophus* [51]. This method involved two steps: first one was the exposure of cells to a temperature of 80 °C and the second step was treatment with a cocktail of various hydrolytic enzyme such as lysozyme, phospholipase, proteinase, lecithinase, and so on. Most of the cellular biomass was degraded or hydrolyzed by using these enzymes, whereas the PHA remained intact. Then PHA was recovered by washing, flocculation, and drying in form of white powder and this white powder was melted, extruded, and changed into the form in which the PHA was supplied to the manufacturers.

Ionic liquids are the potential extractants for PHAs from bacterial cells. A methodology was developed to utilize 1-ethyl-3-methylimidazolium diethylphosphate as green solvent for the downstream processing of the bacterial biomass. Herein, the recovery of 60% ± 2% PHA was obtained in *Halomonas hydrothermalis* MTCC 5445 indicating similar thermal and mechanical properties compared with standard PHB (Table 7.2). Figure 7.4 indicates the digital image of the film prepared from recovered PHB using ionic liquids [52].

TABLE 7.2 Various Polymer Recovery Methods that have Been Studied

Extraction Method	Comments	Strain	Results	References
Solvent extraction	Chloroform	*Cupriavidus necator* DSM 545	Purity: 95% Yield: 96%	Fiorese et al., 2009 [53]
	Chloroform	*Bacillus cereus* SPV	Purity: 92% Yield: 31%	Valappil et al., 2007 [54]
	Acetone water		Yield: 80%–85%	Narasimhan et al., 2008 [55]
	Methyl tert-butyl ether	*Pseudomonas putida*	Yield: 15%–17.5%	Wampfler et al., 2010 [56]
	Methylene chloride	*C. nectar*	Purity: 98%	Zinn et al., 2003 [57]
Surfactants	SDS	Recombinant *E. coli*	Yield: 89% Purity: 99%	Choi and Lee, 1999 [58]
Sodium hypochlorite	Sodium hypochlorite	*C. necator*	Purity: 86%	Hahn et al., 1995 [59]

TABLE 7.2 *(Continued)*

Extraction Method	Comments	Strain	Results	References
Dispersion of sodium hypochlorite and chloroform	Sodium hypochlorite and chloroform	*B. cereus* SPV	Yield: 30% Purity: 95%	Valapalli et al., 2007 [54]
Enzymatic digestion	SDS -EDTA+ enzyme	*P. putida*	Purity: 93%	Kathiraser et al., 2007 [60]
	Bromolain; pancreatin	*C. necator*	Purity: 90%	Kapritchkoff et al., 2006 [61]
	Sonication	*B. flexus*	Purity: 92%	Divyashree et al., 2009 [62]
Gamma irradiation	Radiation with chloroform	*B. flexus*	Yield: 45%–54%	Divyashree et al., 2009 [62]
Ionic liquid	1-Ethyl-3-methylimidazolium diethylphosphate	*H. hydrothermalis*	Yield: 60% ± 2%	Dubey et al., 2017 [52]

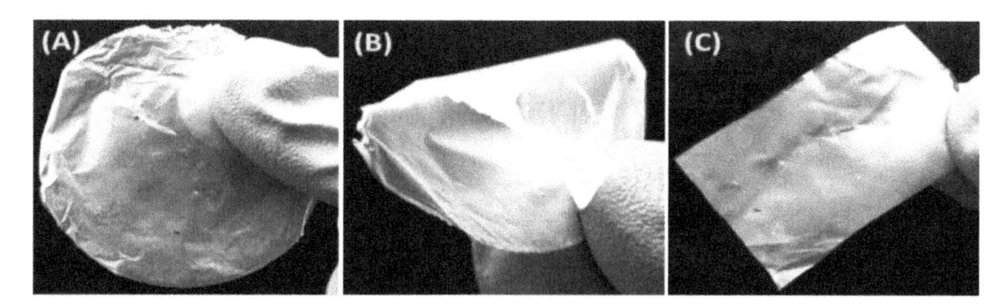

FIGURE 7.4 PHA film prepared using green extraction route (ILs).

Source: Reprinted with permission from Ref. [52]. © 2017 American Chemical Society.

For analysis of PHA granules, we used several techniques and methods, namely Fourier transform infrared (FTIR) analysis (the field emission scanning electron microscopy) is used to determine PHA granules in cell suspension. Other techniques include GC-MS analysis, HPLC, and NMR.

7.5 ANTIMICROBIAL APPLICATION OF PHA

Antimicrobial agents are those substances that kill or inhibit the growth of microorganisms. These microorganisms include bacteria, fungi, and protozoa. Due to the antibiotic resistance property, many of the microorganisms get adapted to the new synthetic drugs or other antibiotics, it also evolves into superantigen so there is a need for developing new type of drug0 or infection-fighting strategies to control microbial infections.

Now, the most-widely spread antibiotic-resistant bacteria is methicillin resistant *Staphylococcus aureus* and *Salmonella* spp [63–65]. Allen and coworkers[66] reported that, MCL (R)-3-hydroxycarboxylic acid produced from *J. curcas* oil by *Sterptomyces* spp. have bactericidal activity toward E. coli and baceriostatic for *S. typhimurium* and *L. monocytogenes* and other several antibiotic resistant pathogens. (R)-3-hydroxycarbooxylic acids [R-3HAs] is chiral enantiomers of PHA with 3-, 4- 5-, and 6-hydroxyalanoic acids in the (R)-configuration.

Hema and coworkers [67] studied that PHA copolymer poly(3-hydroxybutyrate-co70%4-hydroxybutyrate) obtained from *Cupriavidus* sp. also shows antibacterial properties against pathogens such as *S. aureus*.

Fernandes and co-workers[68]. reported that 99% to 100% reduction occured in *E. coli* and *S. aureus* culture by using an antifungal agent chorohexidine with PHB/PEO fibres of 1 wt% concentration.

P3HB and P4HB can also be used for healing wound and treating skin infections [69]. It was also reported that D-peptide associated with 3HAs had anticancer property [70]. In intestine of giant tiger prown, it was found that PHA acts as antimicrobial when it was supplemented as a food product.

Therefore, now we can use directly or in modified form (chemically, biologically, cross-linking, or as nanomaterials hybrid) of PHA as an effective anti-microbial agent (Figure 7.5).

The MCL3HA has bactericidal effect on pathogen as it effects cytoplasmic membrane and membrane proteins, including electron transport chain and kills the microorganisms. By using PHA as antibiofilm, we can also control vibrio in aquaculture field because vibrio acts as common dreadful pathogenic organism in the aquaculture field by forming biofilm [71]. Eugenol is a compound that is incorporated with PHB and extracted from *B. mycoides* and this was synthesized by Narayanan et al. [72]. This compound has antimicrobial properties and it is effective against food spoilage bacterial and fungal pathogen such as *B. cereus, Salmonella typhimurium, Aspergillus niger, A. flavus,* and so on.

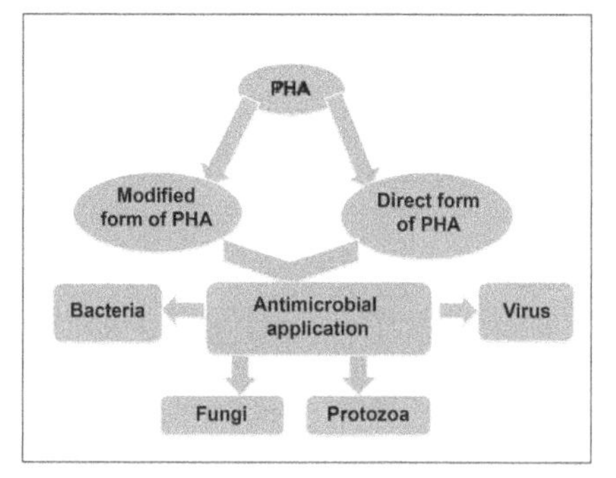

FIGURE 7.5 Schematic presentation of antimicrobial application of PHAs.

Composites of PHA also have valuable effect in controlling many dreadful pathogens. These composites are synthesized by depolymerization of PHA. Hema and co-workers [73] discovered copolymer P(3HB-co-4HB) produced from *Cupravidus* sp. that had antimicrobial activity against *S. aureus* or Gram-positive bacteria. Cerqueira and co-workers [74] synthesized poly(3-hydroxybutyrate-*co*-3-hydroxyvalerate (PHBV) copolymer which was modified by using corn zein protein and cinnamaldehyde through electrospinning process and it exhibited high antimicrobial activity against *Listeria monocytogenes*.

Composite material with nanoparticle has very useful properties like high-surface area, porous nature, and toxic nature. Therefore, it has been used in many applications such as antimicrobial agents and biocatalyst [75]. When PHA is mixed with nanoparticles (NPs), such as silver, zinc, copper, and titanium, it worked as antimicrobial and antiviral agents [76]. ZnO, cellulose, and P(3HB-co-4HB) composite showed wide-spectrum antibacterial activity. Similarly as against feline calicivirus and murine norovirus, Castro-Mayorga and co-workers [77] used as an antiviral drug that was made up of an Ag/PHBV3/PHBV18 composite.

7.6 PHA AS DRUG CARRIER

For the high efficiency of drug on target system, it should be delivered in controlled and targeted manner. PHAs can act as that material which has desirable properties such as high biocompatibility and hydrophobicity. So, they can be converted into porous matrices, films, microcapsules, and microsphere NPs. When the drugs such as anticancer agents, antibiotics, anti-inflammatory agents, hormones and so on, are carried through PHA, its efficiency is increased [78].

Rifamycin with PHB microsphere is applied as hemoembolizing agent. Due to microsphere, it was very effective in releasing or delivery of drug. For antibiotic, implanted rods made up of with PHB, PHBV, and their copolymer are used as carriers for more effectiveness [79]. It was found that MCL-PHA is more efficient carrier for drug delivery system due to its lower crystallinity and melting point. So, these have been used for transdermal drug delivery. PHA beads are also applied in diagnosis of bovine tuberculosis as tuberculin skin test reagents [80, 81].

Gangrade and Price [82] reported very efficient incorporation between steroid (progesterone) and carrier microsphere PHB or P(3HB-3HV). A study showed that PHB microsphere with rubomycin (an antitumor drug) inhibits the proliferative property of Ehrlich's carcinoma in mice [83].

PHB microsphere was used as carrier material with Tramadol in epidural administration for pain relief by Salman and co-workers [84]. Efficiency of tetracycline was increased when it was incorporated with PHA copolymer microsphere against microorganisms associated with periodontal disease [85]. Microsphere P(3HB-3HV) conjugated Paclitaxel proved to be toxic towards primary ovarian cancer cell and endometrial [86]. Also, anticancer activity of peptides and anticancer drug was proved to be increased when it was conjugated with PHA copolymer 3HD. It was also proved that 3HB monomer enhanced calcium level within the cell by protecting the mitochondria and it could be used as novel drug for bone disease and nervous system [87].

Shishatskaya and team [88] reported that 3HB/4HB monomer was effective in skin repair treatment for treating inflammation by forming polymer film and electrospun membranes, and it can also increase angiogenic characteristic of skin and enhances wound healing process. Biopolymer can be also used as scaffolds for tarsal repair in eyelid reconstruction in rat.

Application of biopolymer in anticancer drug delivery system is also very important. By using PHA, it could be very efficient and targeted, for example, docetaxel (DTXL)-loaded P(3HB) produced from *B. cereus* sp. plays very effective role in delivery of DTXL (anticancer drug) drug [89]. PHB loaded with NPs also acts as efficient carrier for delivery of photosensitizer drug in photodynamic therapy (Figure 7.6).

7.7 APPLICATION OF PHA IN TISSUE ENGINEERING

Tissue engineering is an emerging technology. It combines biology, material sciences, and surgical reconstruction to help in enhancement and maintenance of tissue function via repairing and surgical method. It is applied in bone replacement,

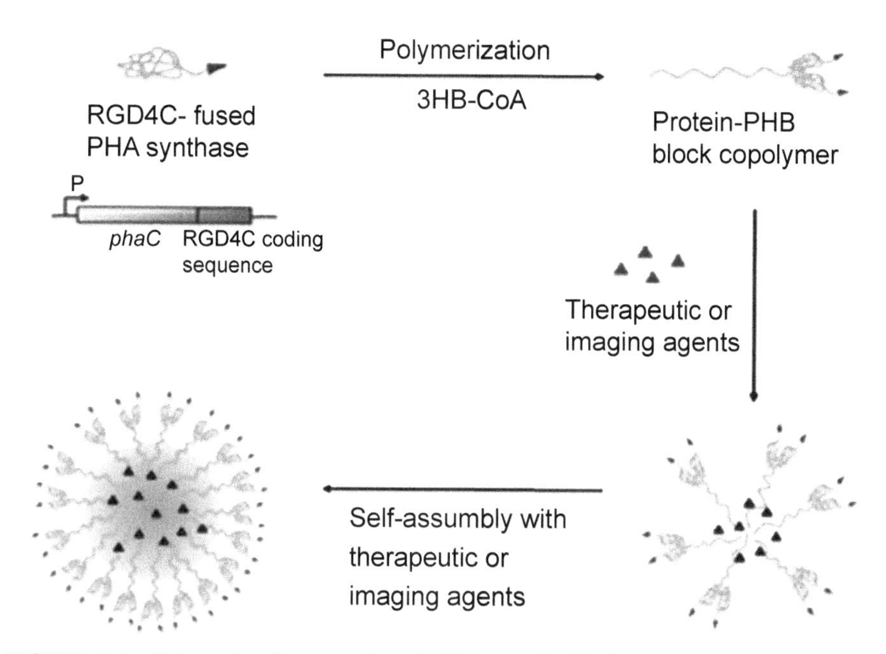

FIGURE 7.6 Schematic of preparation of PHA–protein block copolymers, their further modification by therapeutic and/or imaging agents and self-assembly to micelles [90].

fracture nonunion, and pathological loss of bones [91]. Biopolymer could be useful in this engineering due to its biocompatibility, mechanical strength, and degradation rate. Actually tissue engineering system copied the function of extracellular network (ECMs) by putting the cells along with growth factors in synthetic scaffold, which act as temporary ECMs. After the new bone formation, the temporary scaffold gets degraded and gets absorbed by the body [92] [93].

Roles of PHA and PHB are significant in tissue engineering field. They help in scaffold formation, subcutaneous tissue engineering, nerve tissue engineering, cartilage tissue-tendon and ligament tissue engineering, skin tissue engineering, and so on. PHA is a suitable candidate for tissue engineering because it has good mechanical properties, biocompatibility, and sufficient microchemistry of surface.

Mainly tissue engineering is done in three steps: first one is cell substitute, second one is used to induce tissue substance, and the last one is application of scaffold for implantation of cells.

Scaffold mainly helps in repair and regeneration of defective tissues. It also helps in formation of ECMs. It is made up of bioactive molecule like biodegradable polymer [94]. UV–Vis and attenuated total reflection (ATR)/FTIR spectrophotometry are used to determine PHBV scaffolds. P(3HB-co-3HHx) scaffold is used in cartilage tissue engineering as a support material [95] and in fibroblast growth and capsulation.

It can also be brought in use for tarsal repair and when this polymer is blended with hydroxyapatite, it is used in chondrocytes proliferation, osteoblast growth, migration, and cartilage repairing [96]. There is a layer of bonelike appetite formed within very less time on HA/PHB composite after its immersion in an acellular simulated body fluid at 37 °C. The mechanical strength of PHB/HA and PHBV/HA composites is 62 mPa, which is same as human bones, so it is a suitable material for fracture fixation. Scaffold also helps in liver tissue engineering, tissue remodeling, and so on. PHA composites also play effective role in neural tissue engineering, such as P(3HB) which helped in axonal regeneration with low level of inflammatory infiltration [97]. P(3H-co-3HHx) is used in neuronal regeneration. Lu et al. demonstrated improvement during differentiation of neural stem cells and neural progenitor cell into neurons after applying PHA-coated films.

PHA scaffolds were found to be helpful in repairing injury of the central nervous system [98]. In bone marrow cell, P(3HB-*co*-3HV) and P(3HB-*co*-3HHx) were found to be best for osteoblast attachment, proliferation, and differentiation.

PHA also plays novel role in cartilage tissue engineering such as after implantation, it helps in neo-cartilage formation and regeneration of hyaline cartilage at the defective site.

It was proved that PHBV film acts as a temporary substrate for growing retinal pigment epithelial cells by forming organized monolayer before their subretinal transplantation. PHBV film became hydrophilic by oxygen plasma treatment that allows attachment of D407 cell to the film surface in efficient manner [99]. Doyle et al. [100] observed that no inflammatory response was seen up to 12 months after implantation of PHB material in bone tissue. It was also observed that bone is rapidly formed in highly organized manner, with up to 80% of the deposition occurring directly on the new bone [100, 101] (Figure 7.7).

7.8 APPLICATION OF PHA IN FORMATION OF MEDICAL DEVICES

Medical devices include simple devices, test equipment, and implants, such as rivets, orthopedic pins, stents, staples, mesh, suture fasteners, and so on. For developing any medical devices, we have to

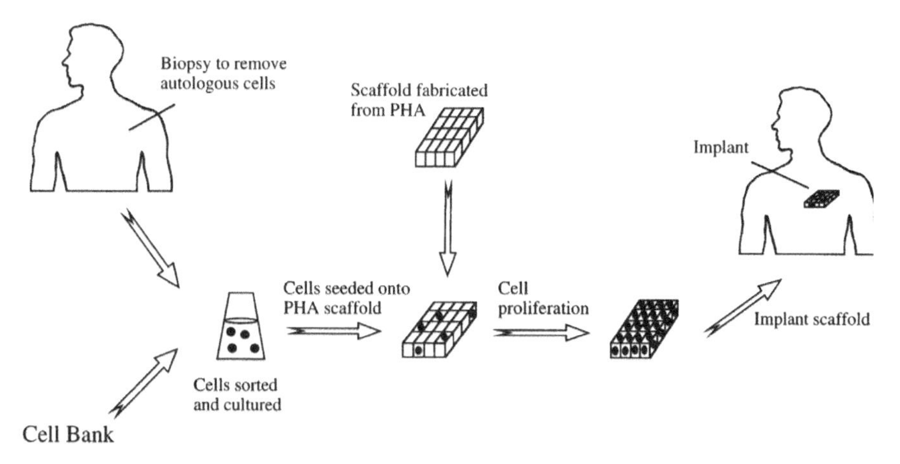

FIGURE 7.7 Role of PHAs in tissue engineering.
Source: Reprinted with permission from Ref. [101]. © 1999 Elsevier.

consider many properties of manufacturing material such as physical and mechanical properties, thermal properties, electrical properties, chemical resistance, long-term durability, and the biopolymer should be of medical-grade polymer. After degradation of PHB polymer in animal host, 3-hydroxybutyrate is formed that is a natural metabolite of animal body and it helps in formation of ketone body. Therefore, sutures, medical implants, and so on, made from PHA do not produce immune response in our body. Sterilization method does not affect mechanical strength and tensile strength of PHA. So, these properties make PHA a suitable material for manufacturing medical devices [102]. PHB and P(HB-co-HV) sutures are used in muscle-facial wound. PHA staples is used for the fixation of soft tissues, biodegradable adhesion barrier is used for preventing undesirable adhesion in general surgery, PHA stents are used in vessels to prevent reocclusion of a vessel, bulking agents are used for filling the defects in plastic surgery, orthopedic pins of PHA are used in bone, and soft tissue fixation as a bone-filling augmentation material [103]. Several medical devices of PHA are made by Tepha, Inc. in Cambridge, MA, USA. The first TephaFLEX® suture was manufactured by using P4HB polymer and it was first approved by FDA. This company also manufactures surgical mesh and films fabricated from PHA (Figure 7.8). All the products have suitable mechanical properties for use in surgical applications (www.tepha.com).

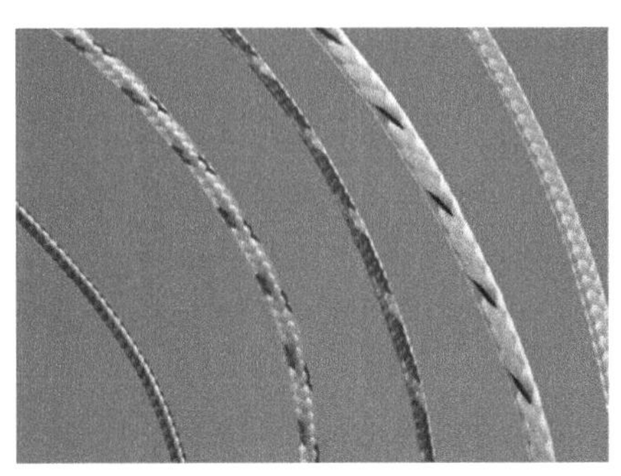

FIGURE 7.8 Picture of TephaFLEX® suture.

Source: https://www.meddeviceonline.com/doc/teleflex-medical-granted-european-fiber-suture-0001

7.9 USE OF PHA IN BIODEGRADABLE IMPLANTS

Implants are those materials which can disintegrate into our body. They are used in many fields such as orthopedic surgery, general surgery, cardiology, urology, and so on. The materials used as implants should be absorbable, resorbable, and degradable. Biodegradable implantation is an emerging field nowadays. It offers biodegradable implants over metal implants.

Metal implants have many side effects and disadvantages. They impart in toxicity of cell by secreting metal ions in surrounding of tissues. Removal and surgical implantation procedure are also complicated in case of metal implants. Whereas biodegradable implants provide proper fixation strength and degrade after formation of host tissue [101, 102]. Examples of biodegradable implants are sutures, staples, tacks, anchors, and so on. Now, there are 40 different biopolymers known that are used as biodegradable implants. PCL, polyorthoester, PLLA, PGA, PLLA-co-PHA, and PHA are some of the biopolymers that are used as implants [103].

There are five phases found during biodegradable implants and tissue reaction [104]. These phases are given below.

(1) Healing phase
(2) Latency phase
(3) Protected resorptive
(4) Progressive resorptive phase
(5) Recovery phase.

PHAs are widely applicable in biodegradable implantation due to their biodegradability, biocompatibility, tissue strength, and they do not have any carcinogenic effects on tissue or cells.

PHBHHX NP composites were used by Peng et al. 2012 [105] for insulin delivery and they observed that only 20% of insulin was delivered within 31 days and high amount (5.42%) was released during first 8 hours. So, this method served as long-lasting insulin delivery formation.

Chitosan–poly(hydroxybutyrate-*co*-valerate) hydrogel with chondroitin sulfate NPs which could withstand varying stress activities was effectively used in nucleus pulposus tissue engineering by Nair and co-workers [106].

Due to the flexible nature of PHA, it is widely used in artificial blood vessel and heart valve formation. For example, P3HB4HB is regarded as suitable material for forming artificial blood vessel. PHBHHx mimic cellular microenvironment for cell culture and it provides good tool to study relationship between microstructure and cell function [107].

7.10 CHALLENGES IN INDUSTRIAL SCALE PRODUCTION OF PHA

PHAs have extremely wonderful and unique characteristics but their real-life application is still limited because of its low-level production at industrial scale. There are several factors that are responsible for low-level production.

Major one is cost of carbon substrate. The carbon substrate we used in PHA fermentation has food-based origin, so cost of it automatically increases. So, the main challenge is to find out low-cost substrate for production of PHA.

There are more than 300 PHA-producing species that are known, but only few of them have real potential to produce PHA in good amount, and this property is totally dependent on gene. So, another challenge is how to select a host organism to express genes which are responsible for PHA synthesis.

Another one challenge is physical and chemical properties of PHA which include attention to lowering of melting point, glass transition temperature, tensile strength, load carrying capacity, and so on, which can be overcome by forming copolymer of high-molecular weight [108].

Another challenge includes how to manipulate feed composition for better production of PHA to develop techniques for improving physicochemical property and develop method to modify the product generated from metabolism of PHAs [109].

7.11 CONCLUSION

PHA promises a good biodegradable and biocompatible alternate of synthetic plastics and it has wide application in different fields. However, the major demand in medical field is that PHA can be conjugated with other biodegradable polymers for improving its strength. At present, PHA is being used in various biomedical applications such as drug delivery system (sodium dodecyl sulfate), antibacterial agents, tissue engineering, and so on. Still many researches are currently going on improving PHA properties to enable it for treating diseases and reducing loaded use of synthetic plastics. In future context, the use of PHA can be extended into wide range such as cancer therapy, treatment of neurodegenerative, metabolic disorder, and so on.

KEYWORDS

- **polyhydroxyalkanoate**
- **biocompatibility**
- **biodegradability**
- **drug carrier**
- **biomedical application**

REFERENCES

1. Williams, D. F. On the mechanisms of biocompatibility. *Biomaterials*. 2008, 29, 2941–2953.
2. Babu, R. P.; O'connor, K.; Seeram, R. Current progress on bio-based polymers and their future trends. *Prog Biomaterials*. 2013, 2, 8.
3. Tokiwa, Y.; Calabia, B. P. Review degradation of microbial polyesters. *Biotechnol Lett*. 2004, 26, 1181–1189.
4. Vert, M. Polymeric biomaterials—Strategies of the past vs strategies of the future. *Prog Polym Sci*. 2007, 32, 755–761.
5. Shrivastav, A.; Mishra, S. K.; Mishra, S. *Utilization of cheap carbon source for microbial production of polyhydroxyalkanoates*. Nova Science Publishers, Inc., 2011, Hauppauge, New York, United States, ISBN: 978-1-61324-672-6.
6. Chanprateep, S. Current trends in biodegradable polyhydroxyalkanoates. *J Biosci Bioeng*. 2010, 110, 621–632
7. Chanprateep, S.; Kulpreecha S. Production and characterization of biodegradable terpolymer poly(3-hydroxybutyrate-co-3-hydroxyvalerate-co-4-hydroxybutyrate) by *Alcaligenes* sp. A-04. *J Biosci Bioeng*. 2006, 101, 51–56.
8. De Koning G. J. M.; Van Bilsen H. M. M.; Lemstra, P. J.; Hazenberg, W.; Witholt, B.; Preusting, H.; Van der Galiën, J. G.; Schirmer, A.; Jendrossek, D. A biodegradable rubber by crosslinking poly(hydroxyalkanoate) from *Pseudomonas oleovorans*. *Polymer*. 1994, 35, 2090–2097.
9. Lemoigne, M. *Bull Soc Chimique Belgique*. 1926, 8, 770.
10. Steinbüchel, A.; Füchtenbusch, B. Bacterial and other biological systems for polyester production. *Trends Biotechnol*. 1998, 16, 419–427.
11. Laycock, B.; Halley, P.; Pratt, S.; Werker, A.; Lant, P. Thechemomechanical properties of microbial polyhydroxyalkanoates. *Prog Polym Sci*. 2013, 38, 536–583.
12. Noda, I.; Marchessault, R. H.; Terada, M. *Polymer data handbook*. Oxford University Press, London, 1999.
13. Misra, S. K.; Ohashi, F.; Valappil, S. P.; Knowles, J. C.; Roy, I; Silva, S. R. P.; Salih, V.; Boccaccini, A. R. Characterization of carbon nanotube (MWCNT) containing P(3HB)/ bioactive glass composites for tissue engineering applications. *Acta Biomaterialia*. 2010, 6, 735–742.
14. Albuquerque, M. G. E.; Torres, C. A. V.; Reis, M. A. M. Polyhydroxyalkanoate (PHA) production by a mixed microbial culture using sugar molasses: effect of the influent substrate concentration on culture selection. *Water Res*. 2010, 3419–3433.
15. Reddy, M. V.; Mohan, S. V. Influence of aerobic and anoxic microenvironments

on polyhydroxyalkanoates (PHA) production from food waste and acidogenic effluents using aerobic consortia. *Bioresource Technol.* 2012, 103, 313–321.
16. Shi, H. P.; Lee, C. M.; Ma, W. H. Influence of electron acceptor, carbon, nitrogen, and phosphorus on polyhydroxyalkanoate (PHA) production by *Brachymonas sp.* P12. *World J Microbiol Biotechnol.* 2007, 23, 625–632.
17. Mokhtarani, N.; Ganjidoust, H.; Farahani, E. V. Effect of process variables on the production of Polyhydroxyalkanoates by activated sludge. *Iranian J Environ Health Sci Eng.* 2012, 9, 6.
18. Grothe, E.; Moo-Young, M.; Chisti, Y. Fermentation optimization for the production of poly (β-hydroxybutyric acid) microbial thermoplastic. *Enzyme Microb Technol.* 1999, 25, 132–141.
19. Kumar, M. S.; Mudliar, S. N.; Reddy, K. M. K.; Chakrabarti, T. Production of biodegradable plastics from activated sludge generated from a food processing industrial wastewater treatment plant. *Bioresour Technol.* 2004, 95, 327–330.
20. Berekaa, M. M.; Al Thawadi, A. M. Biosynthesis of polyhydroxybutyrate (PHB) biopolymer by *Bacillus megaterium* SW1-2: Application of Box-Behnken design for optimization of process parameters. *Afr J Microbiol Res.* 2012, 6, 2101–2108.
21. Hong, C.; Hao, H.; Haiyun, W. Process optimization for PHA production by activated sludge using response surface methodology. *Biomass Bioenerg.* 2009, 33,721–727.
22. Samantaray, S.; Nayak, J. K.; Mallick, N. Wastewater utilization for poly-β-hydroxybutyrate production by the cyanobacterium *Aulosira fertilissima* in a recirculatory aquaculture system. *Appl Environ Microbiol.* 2011, 77, 8735–8743.
23. Johnson, K.; Jiang, Y.; Kleerebezem, R.; Muyzer, G.; van Loosdrecht, M. C. Enrichment of a mixed bacterial culture with a high polyhydroxyalkanoate storage capacity. *Biomacromolecules.* 2009, 10, 670–676.
24. Lemos, P. C.; Viana, C.; Salgueiro, E. N.; Ramos, A. M.; Crespo, J. P. S. G.; Reiszcorr, M. A. M. Effect of carbon source on the formation of polyhydroxyalkanoates (PHA) by a phosphate-accumulating mixed culture. *Enzyme Microb Technol.* 1998, 22, 662–671.
25. Chen, H.; Meng, H.; Nie, Z.; Zhang, M. Polyhydroxyalkanoate production from fermented volatile fatty acids: effect of pH and feeding regimes. *Bioresour Technol.* 2013, 128, 533–538.
26. Dionisi, D.; Beccari, M.; Di Gregorio, S.; Majone, M.; Papini, M. P.; Vallini, G. Storage of biodegradable polymers by an enriched microbial community in a sequencing batch reactor operated at high organic load rate. *J Chem Technol Biotechnol.* 2005, 80, 1306–1318.
27. Ivanova, G.; Serafim, L. S.; Lemos, P. C.; Ramos, A. M.; Reis, M. A. M.; Cabrita, E. J. Influence of feeding strategies of mixed microbial cultures on the chemical composition and microstructure of copolyesters P (3HB-co-3HV) analyzed by NMR and statistical analysis. *Magn Reson Chem.* 2009, 47, 497–504.
28. Ciggin, A. S.; Rossetti, S.; Majone, M.; Orhon, D. Effect of feeding regime and the sludge age on the fate of acetate and the microbial composition in sequencing batch reactor. *J Environ Sci Health A.* 2012, 47, 192–203.
29. Kim, G. J.; Lee, I. Y.; Yoon, S. C.; Shin, Y. C.; Park, Y. H. Enhanced yield and a high production of medium-chain-length poly (3-hydroxyalkanoates)

in a two-step fed-batch cultivation of *Pseudomonas putida* by combined use of glucose and octanoate. *Enzyme Microb Technol.* 1997, 20, 500–505.
30. Valappil, S. P.; Peiris, D.; Langley, G. J.; Herniman, J. M.; Boccaccini, A. R.; Bucke, C.; Roy, I. J. Polyhydroxyalkanoate (PHA) biosynthesis from structurally unrelated carbon sources by a newly characterized *Bacillus spp. Biotechnol.* 2007, 127, 475.
31. Page, W. J. Production of polyhydroxyalkanoates by *Azotobacter vinelandii* UWD in beet molasses culture. *FEMS Microbiol Rev.* 1992, 9, 149–157.
32. Chen, G.; Zhang, G.; Park, S.; Lee, S. Industrial scale production of poly (3-hydroxybutyrate-co-3-hydroxyhexanoate). *Appl Microbiol Biotechnol.* 2001, 57, 50–55.
33. McNeil, B.; Harvey, L. M. (Eds.). *Fermentation: a practical approach.* New York: IRL Press, 1990; pp. 136–140.
34. Alderete, J. E.; Karl, D. W.; Park, C. H. Production of poly (hydroxybutyrate) homopolymer and copolymer from ethanol and propanol in a fed-batch culture. *Biotechnol Prog.* 1993, 9, 520–525.
35. Ghosh P. K.; Mishra, S.; Gandhi, M. R.; Upadhyay, S. C.; Paul, P.; Anand, P. S.; Popat, K. M.; Shrivastav, A.; Mishra, S. K.; Ondhiya, N.; Maru, R. D.; Gangadharan, D.; Brahmbhatt, H.; Boricha, V. P.; Chaudhary, D. R.; Rewari, B. Integrated process for the production of *Jatropha* methyl ester and by-products. *Eur Patent.* 2014, 2, 475, 754.
36. Shrivastav, A.; Mishra, S. K.; Shethia, B.; Pancha, I.; Jain, D.; Mishra, S. Isolation of promising bacterial strains from soil and marine environment for polyhydroxyalkanoates (PHAs) production utilizing *Jatropha* biodiesel byproduct. *Int J Biol Macromolec.* 2010, 47, 283–287.
37. Bhattacharya, S.; Dubey, S.; Singh, P.; Shrivastava, A.; Mishra, S. Biodegradable polymeric substances produced by a marine bacterium from a surplus stream of the biodiesel industry. *Bioengineering.* 2016, 3, 34.
38. Bera, A.; Dubey, S.; Bhayani, K.; Mondal, D.; Mishra, S.; Ghosh, P. K. Microbial synthesis of polyhydroxyalkanoate using seaweed-derived crude levulinic acid as co-nutrient. *Int J Biol Macromol.* 2015, 72, 487–494.
39. Dhangdhariya, J. H.; Dubey, S.; Trivedi, H. B.; Pancha, I.; Bhatt, J. K.; Dave, B. P.; Mishra, S. Polyhydroxyalkanoate from marine *Bacillus megaterium* using CSMCRI's Dry Sea Mix as a novel growth medium. *Int J Biol Macromol.* 2015, 76, 254–261.
40. Ghosh P. K.; Upadhyay, S. C.; Mishra, S. C.; Mohandas, V. P.; Srivastava, D. N.; Shahi, V. K.; Sanghavi, R. J.; Thampy, S.; Makwana, B. S.; Pancha, I.; Pal, R.; Sen, R. A process for the preparation of natural salt formulations for seawater 374 substitution, mineral fortification, WO/2013/098857, 2013.
41. Shang, L.; Jiang, M.; Chang, H. N. Poly (3-hydroxybutyrate) synthesis in fed-batch culture of *Ralstonia eutropha* with phosphate limitation under different glucose concentrations. *Biotechnol Lett.* 2003, 25, 1415–1419.
42. Choi, J. C.; Shin, H. D.; Lee, Y. H. Modulation of 3-hydroxyvalerate molar fraction in poly (3-hydroxybutyrate-3-hydroxyvalerate) using *Ralstonia eutropha* transformant co-amplifying phbC and NADPH generation-related zwf genes. *Enzyme Microb Technol.* 2003, 32, 178–185.
43. Khanna, S.; Srivastava, A. K. Optimization of nutrient feed

concentration and addition time for production of poly (β-hydroxybutyrate). *Enzyme Microb Technol.* 2006, 39, 1145–1151.
44. Bezerra, M. A.; Santelli, R. E.; Oliveira, E. P.; Villar, L. S.; Escaleira, L. A. Response surface methodology (RSM) as a tool for optimization in analytical chemistry. *Talanta.* 2008, 76, 965–977.
45. Candioti, L. V.; De Zan, M. M.; Camara, M. S.; Goicoechea, H. C. Experimental design and multiple response optimization. Using the desirability function in analytical methods development. *Talanta.* 2014, 124, 123–138.
46. Mallick, N.; Gupta, S.; Panda, B.; Sen, R. Process optimization for poly (3-hydroxybutyrate-co-3-hydroxyvalerate) co-polymer production by *Nostoc muscorum. Biochem Eng J.* 2007, 37, 125–130.
47. Barham, P. J.; Selwood, A. *U.S. Patent No. 4,391,766.* Washington, DC: U.S. Patent and Trademark Office, 1983.
48. Williamson, D. H.; Wilkinson, J. F. The isolation and estimation of the poly-β-hydroxy-butyrate inclusions of *Bacillus* species. *Microbiology,* 1958, 19, 198–209.
49. Berger, E.; Ramsay, B. A.; Ramsay, J. A.; Chavarie, C.; Braunegg, G. PHB recovery by hypochlorite digestion of non-PHB biomass. *Biotechnol Tech.* 1989, 3, 227–232.
50. Byrom, D. Polymer synthesis by microorganisms: technology and economics. *Trends Biotechnol.* 1987, 5, 246–250.
51. Holmes, P.A.; Lim, G.B. European Patent Application, 1985.
52. Dubey, S.; Bharmoria, P.; Gehlot, P. S.; Agrawal, V.; Kumar, A.; Mishra, S. 1-Ethyl-3-methylimidazolium diethylphosphate based extraction of bioplastic "Polyhydroxyalkanoates" from bacteria: green and sustainable approach. *ACS Sustain Chem & Eng.* 2017, 6, 766–773.
53. Fiorese, M. L.; Freitas, F.; Pais, J.; Ramos, A. M.; de Aragão, G. M.; Reis, M. A. Recovery of polyhydroxybutyrate (PHB) from *Cupriavidus necator* biomass by solvent extraction with 1,2-propylene carbonate. *Eng Life Sci.* 2009, 9, 454–461.
54. Valappil, S. P; Misra, S. K; Boccaccini, A. R; Keshavarz, T; Bucke, C; Roy, I. Large-scale production and efficient recovery of PHB with desirable material properties, from the newly characterized *Bacillus cereus* SPV. *J Biotechnol.* 2007, 132, 251–258.
55. Narasimhan, K; Cearley, A. C; Gibson, M. S; Welling, S. J. Process for the solvent-based extraction of polyhydroxyalkanoates from biomass. *U.S. Patent 7378266,* USA, 2008.
56. Wampfler, B.; Ramsauer, T.; Rezzonico, S.; Hischier, R.; Köhling, R.; ThönyMeyer, L.; Zinn, M. Isolation and purification of medium chain length poly(3-hydroxyalkanoates) (MCL-PHA) for medical applications using nonchlorinated solvents. *Biomacromole.* 2010, 11, 2716–2723.
57. Zinn, M.; Weilenmann, H. U.; Hany, R.; Schmid, M.; Egli, T. Tailored synthesis of poly([R]-3-hydroxybutyrateco-3-hydroxyvalerate) (PHB/HV) in Ralstonia eutropha DSM 428. *Acta Biotechnol.* 2003, 23,309–316.
58. Choi, J. I.; Lee, S. Y. Efficient and economical recovery of poly(3-hydroxybutyrate) from recombinant *Escherichia coli* by simple digestion with chemicals. *Biotechnol Bioeng.* 1999, 62, 546–553.
59. Hahn, S. K.; Chang, Y. K.; Lee, S. Y. Recovery and characterization of poly(3-hydroxybutyric acid) synthesized in *Alcaligenes eutrophus* and recombinant *Escherichia coli. Appl Env. Microbiol.* 1995, 61, 34–39.

60. Kathiraser, Y.; Aroua, M. K.; Ramachandran, K. B.; Tan, I. K. P. Chemical characterization of medium chain length polyhydroxyalkanoates (PHAs) recovered by enzymatic treatment and ultrafiltration. *J Chem Technol Biotechnol.* 2007, 82, 847–855.
61. Kapritchkoff, F. M.; Viotti, A. P.; Alli, R. C. P.; Zuccolo, M.; Pradella, J. G. C.; Maiorano, A. E.; Miranda, E. A.; Bonomi, A. Enzymatic recovery and purification of polyhydroxybutyrate produced by *Ralstonia eutropha. J Biotechnol.* 2006, 122, 453–462.
62. Divyashree, M. S.; Shamala, T. R.; Rastogi, N. K. Isolation of polyhydroxyalkanoate from hydrolyzedcells of *Bacillus flexus* using aqueous two-phase system containing polyethylene glycol and phosphate. *Biotechnol Bioprocess Eng.* 2009, 14, 482–489.
63. Deurenberg, R. H.; Vink, C.; Kalenic, S.; Friedrich, A. W.; Bruggeman, C. A.; Stobberingh, E. E. The molecular evolution of methicillin resistant *Staphylococcus aureus. Clin Microbiol Infect.* 2007, 13, 222–235.
64. He, X.; Ahn, J. Survival and virulence properties of multiple antibiotic-resistant *Salmonella typhimurium* under simulated gastrointestinal conditions. *Int J Food Sci Tech.* 2011, 46, 2164–2172.
65. Levy, S. B. Factors impacting on the problem of antibiotic resistance. *J Antimicrob Chemother.* 2002, 49, 25–30.
66. Allen, A. D.; Daley, P.; Ayorinde, F. O.; Gugssa, A.; Anderson, W. A.; Eribo, B. E. Characterization of medium chain length (R)-3-hydroxycarboxylic acids produced by *Streptomyces* sp. JM3 and the evaluation of their antimicrobial properties. *World J Microbiol Biotechnol.* 2012, 28, 2791–2800.
67. Hema, R.; Ng, P. N.; Amirul, A. A. Green nanobiocomposite: reinforcement effect of montmorillonite clays on physical and biological advancement of various polyhydroxyalkanoates. *Polym Bull.* 2013, 70, 755–771.
68. Fernandes, J. G.; Correia, D. M.; Botelho, G.; Padrão, J.; Dourado, F.; Ribeiro, C.; Lanceros-Méndez, S.; Sencadas, V. PHB-PEO electrospun fiber membranes containing chlorhexidine for drug delivery applications. *Polym Test.* 2014, 34, 64–71.
69. Shishatskaya, E. I.; Nikolaeva, E. D.; Vinogradova, O. N.; Volova, T. G. Experimental wound dressings of degradable PHA for skin defect repair. *J Mater Sci Mater Med.* 2016, 27, 165.
70. O'Connor, S.; Szwej, E.; Nikodinovic-Runic, J.; O'Connor, A.; Byrne, A. T.; Devocelle, M.; O'Donovan, N.; Gallagher, W. M.; Babu, R.; Kenny, S. T.; Zinn, M. The anti-cancer activity of a cationic anti-microbial peptide derived from monomers of polyhydroxyalkanoate. *Biomaterials.* 2013, 34(11), 2710–2718.
71. Kiran, G. S.; Jackson, S. A.; Priyadharsini, S.; Dobson, A. D.; Selvin, J. Synthesis of Nm-PHB (nanomelanin-polyhydroxy butyrate) nanocomposite film and its protective effect against biofilm-forming multi drug resistant *Staphylococcus aureus. Sci Rep.* 2017, 7(1), 9167.
72. Narayanan, A.; Ramana, K. V. Synergized antimicrobial activity of eugenol incorporated polyhydroxybutyrate films against food spoilage microorganisms in conjunction with pediocin. *Appl Biochem Biotechnol.* 2013, 170, 1379–1388.
73. Hema, R.; Ng, P. N.; Amirul, A. A. Green nanobiocomposite: reinforcement effect of montmorillonite clays on

physical and biological advancement of various polyhydroxyalkanoates. *Polym Bull.* 2013, 70, 755–771.
74. Cerqueira, M. A.; Fabra, M. J.; Castro-Mayorga, J. L.; Bourbon, A. I.; Pastrana, L. M.; Vicente, A. A.; Lagaron, J. M. Use of electrospinning to develop antimicrobial biodegradable multilayer systems: encapsulation of cinnamaldehyde and their physicochemical characterization. *Food Bioprocess Technol.* 2016, 9, 1874–1884.
75. Patel, S. K. S.; Kalia, V. C.; Choi, J. H.; Haw, J. R.; Kim, I. W.; Lee, J. K. Immobilization of laccase on SiO2 nanocarriers improves its stability and reusability. *J Microbiol Biotechnol.* 2014, 24, 639–647.
76. Rodriguez-Contreras, A.; Garcia, Y.; Manero, J. M.; Ruperez, E. Antibacterial PHAs coating for titanium implants. *Eur Polym J.* 2017, 90, 66–78.
77. Castro-Mayorga, J. L.; Freitas, F.; Reis, M. A. M.; Prieto, M. A.; Lagaron, J. M. Biosynthesis of silvernanoparticles and polyhydroxybutyrate nanocomposites of interest in antimicrobial applications. *Int J Biol Macromol.* 2018, 108, 426–435.
78. Nobes, G. A. R.; Maysinger, D.; Marchessault, R. H. Polyhydroxyalkanoates: materials for delivery systems. *Drug Deliv.* 1998, 5, 167–177.
79. Türesin, F.; Gürsel, I.; Hasirci, V. Biodegradable polyhydroxyalkanoate implants for osteomyelitis therapy: in vitro antibiotic release. *J Biomater Sci Polym Ed*, 2001, 12, 195–207.
80. Chen, S.; Parlane, N. A.; Lee, J.; Wedlock, D. N.; Buddle, B. M.; Rehm, B. H. New skin test for detection of bovine tuberculosis on the basis of antigen-displaying polyester inclusions produced by recombinant *Escherichia coli. Appl. Environ Microbiol.* 2014, 80, 2526–2535.
81. Parlane, N. A.; Chen, S.; Jones, G. J.; Vordermeier, H. M.; Wedlock, D. N.; Rehm, B. H.; Buddle, B. M. Display of antigens on polyester inclusions lowers the antigen concentration required for a bovine tuberculosis skin test. *Clin. Vaccine Immunol.* 2016, 23, 19–26.
82. Gangrade, N.; Price, J. C. Poly (hydroxybutyrate-hydroxyvalerate) microspheres containing progesterone: preparation, morphology and release properties. *J Microencapsul.* 1991, 8, 185–202.
83. Shishatskaya, E. I.; Goreva, A. V.; Voinova, O. N.; Inzhevatkin, E. V.; Khlebopros, R. G.; Volova, T. G. Evaluation of antitumor activity of rubomycin deposited in absorbable polymeric microparticles. *Bull Exp Biol Med.* 2008, 145, 358–361.
84. Salman, M. A.; Sahin, A.; Onur, M. A.; Oge, K.; Kassab, A.; Aypar, U. Tramadol encapsulated intopolyhydroxybutyrate microspheres: in vitro release and epidural analgetic effect in rats. *Acta Anaesthesiol Scand.* 2003, 47, 1006–1012.
85. Sendil, D.; Gürsel, I.; Wise, D. L.; Hasirci, V. Antibiotic release from biodegradable PHBV microparticles. *J Control Release.* 1999, 59, 207–217.
86. Vilos, C.; Morales, F. A.; Solar, P. A.; Herrera, N. S.; Gonzalez-Nilo, F. D.; Aguayo, D. A.; Mendza, H. L.; Comer, J.; Bravo, M. L.; Gonzalez, P. A.; Kato, S. P. Paclitaxel-PHBV nanoparticles and their toxicity to endometrial and primary ovarian cancer cells. *Biomaterials.* 2013, 34, 4098–4108.
87. Shrivastav, A.; Kim, H. Y.; Kim, Y. R. Advances in the applications of polyhydroxyalkanoate nanoparticles for novel drug delivery system. *BioMed Res Int.* 2013, 2013, 581684.

88. Shishatskaya, E. I.; Nikolaeva, E. D.; Vinogradova, O. N.; Volova, T. G. Experimental wound dressings of degradable PHA for skin defect repair. *J Mater Sci Mater Med.* 2016, 27, 165.
89. Mascolo, D. D.; Basnett, P.; Palange, A. L.; Francardi, M.; Roy, I.; Decuzzi, P. Tuning core hydrophobicity of spherical polymeric nanoconstructs for docetaxel delivery. *Polym Int.* 2016, 65, 741–746.
90. Michalak, M.; Kurcoka, P., Hakkarainenb, M. Polyhydroxyalkanoate-based drug deliverysystems. Polyhydroxyalkanoate-based drug delivery systems. *Polym Int.* 2017, 66: 617–622.
91. Lin, C. Y.; Schek, R. M.; Mistry, A. S.; Shi, X.; Mikos, A. G.; Krebsbach, P. H.; Hollister, S. J. Functional bone engineering using ex vivo gene therapy and topology-optimized, biodegradable polymer composite scaffolds. *Tissue Eng.* 2005, 11, 1589–1598.
92. Kalfas, I. H. Principles of bone healing. *Neurosurg Focus.* 2001, 10 (4), 1–4.
93. Dimar, J.; Glassman, S.; Burkus, K.; Carreon, L. Clinical outcomes and fusion success at 2 years of single-level instrumented posterolateral fusions with recombinant human bone morphogenetic protein-2/compression resistant matrix versus iliac crest bone graft. *Spine.* 2006, 31, 2534–2539.
94. Jagur-Grodzinski, J. Polymers for tissue engineering, medical devices, and regenerative medicine. Concise general review of recent studies. *Polym Adv Technol.* 2006, 17, 395–418.
95. Ye, C.; Hu, P.; Ma, M. X.; Xiang, Y.; Liu, R. G.; Shang, X. W. PHB/PHBHHx scaffolds and humanadipose-derivedstem cells for cartilage tissue engineering. *Biomaterials.* 2009, 30, 4401–4406.
96. Wang, Y. W.; Wu, Q.; Chen, J.; Chen, G. Q. Evaluation of three-dimensional scaffolds made of blends of hydroxyapatite and poly (3-hydroxybutyrate-co-3-hydroxyhexanoate) for bone reconstruction. *Biomaterials.* 2005, 26, 899–904.
97. Wang, Y. W; Yang, F.; Wu, Q.; Cheng, Y. C.; Yu, P. H. F.; Chen, J.; Chen, G. Q. Effect of composition of poly(3-hydroxybutyrate-co-3-hydroxyhexanoate) on growth of fibroblast and osteoblast. *Biomaterials.* 2005, 26, 755–761.
98. Xu, X. Y.; Li, X. T.; Peng, S. W.; Xiao, J. F.; Liu, C.; Fang, G.; Chen, G. Q. The behaviour of neural stem cells on polyhydroxyalkanoate nanofiber scaffolds. *Biomaterials.* 2010, 31, 3967–3975.
99. Tezcaner, A.; Bugra, K.; Hasırcı, V. Retinal pigment epithelium cell culture on surface modified poly (hydroxybutyrate-co-hydroxyvalerate) thin films. *Biomaterials.* 2003, 24, 4573–4583.
100. Doyle, C.; Tanner, E. T.; Bonfield, W. In vitro and in vivo evaluation of polyhydroxybutyrate and of polyhydroxybutyrate reinforced with hydroxyapatite. *Biomaterials* 1991, 12, 841–847.
101. Williams, S. F.; Martin, D. P.; Horowitz, D. M.; Peoples, O. P. PHA applications: addressing the price performance issue I. Tissue engineering. *Int J Biol Macromol.* 1999, 25, 111–121.
102. Modjarrad, K.; Ebnesajjad, S. (eds). *Handbook of polymer applications in medicine and medical devices.* Elsevier, Amsterdam, Netherlands, eBook, ISBN 9780323221696, 2013.
103. Williams, S. F.; Martin, D. P. Applications of PHAs in medicine and pharmacy. *Biopolymers.* 2002, 4, 91–127.
104. Peng, Q.; Zhang, Z. R.; Gong, T.; Chen, G. Q.; Sun, X. A rapid-acting, long-acting insulin formulation based

on a phospholipid complex loaded PHBHHx nanoparticles. *Biomaterials*. 2012, 33, 1583–1588.

105. Nair, M. B.; Baranwal, G.; Vijayan, P.; Keyan, K. S.; Jayakumar, R. Composite hydrogel of chitosan–poly (hydroxybutyrate-co-valerate) with chondroitin sulfate nanoparticles for nucleus pulposus tissue engineering. *Colloids Surf. B*. 2015, 136, 84–92.
106. Chen, G. Q.; Zhang, J. Microbial polyhydroxyalkanoates as medical implant biomaterials. *Artif Cell Nanomed B*. 2018, 46, 1–18.
107. Godovsky, D. Device applications of polymer-nanocomposites. *Adv Polym Sci*. 2000, 163–205.
108. Ray, S.; Kalia, V. C. Microbial cometabolism and polyhydroxy-alkanoate co-polymers. *Indian J Microbiol*. 2017, 57 (1), 39–47.
109. Singh, M.; Kumar, P.; Ray, S.; Kalia, V. C. Challenges and opportunities for customizing polyhydroxyalkanoates. *Indian J Microbiol*. 2015, 55, 235–249.

CHAPTER 8

Tailoring of Intrinsic Physicochemical Properties of Nanoparticles for Antimicrobial Therapy

SWAROOP CHAKRABORTY[1,*] and SUPERB K. MISRA[2]

[1]*Biological Engineering, Indian Institute of Technology Gandhinagar, Gandhinagar 382355, Gujarat, India*

[2]*Material Science & Engineering, Indian Institute of Technology Gandhinagar, Gandhinagar 382355, Gujarat, India*

[*]*Corresponding author. E-mail: E-mail. swaroop.ch@iitgn.ac.in*

ABSTRACT

The global issue of antibiotic resistance has raised various concerns, and there is an urge for the discovery and development of novel antibiotics to control the infection pandemics. The application of nanoparticles in terms of biomedicine has gained significant attention since past two decades. The development of engineered nanomaterial has given an avenue to target the pathogenic microorganism with an enhanced efficacy of bacterial killing. There are various schools of thoughts with regards to mechanism of nanoparticle's antibacterial activities such as production of reactive oxygen species, disruption of cell membrane, enhancing deoxyribonucleic acid damage, or the combination of all of the above. The physicochemical properties (size, shape, surface properties, solubility, etc.) of nanoparticles have a big impact on generating the biological responses from the bacterial cells. Tuning the physicochemical properties can significantly change mode of action of nanoparticles within a biological system. Therefore, it is important to understand the impact of physicochemical properties of nanoparticles to make an effective use of these nanoparticles for the antimicrobial actions. The scope of this chapter is to highlight the efficacy of nanoparticles as antibacterial agents as compare to the conventional antimicrobial therapy. This chapter will highlight various physicochemical properties, namely, size, shape, surface properties and solubility, and its role in tailoring antibacterial action.

8.1 INTRODUCTION

Infections caused by antibiotic-resistant pathogens have increased in recent times and are becoming a primary cause for global health concern. According to the reports of Center for Disease Control and Prevention, the total number of infections due to multidrug resistance *Staphylococcus aureus* (MRSA) has increased from 127,000 to 278,000 between the years 1999 and 2005 [50]. Furthermore, the cause of death annually due to MRSA strain has significantly increased from 11,000 to 17,000 during the same period. Due to the implementation of various preventive measures (viz., proper sanitization, disinfection of hospital wastes, isolation of MRSA-infected patients, etc.), a decline was reported in the number of cases of MRSA infection from the period of 2005 to 2008 [54]. Due to the improper and excessive use of the antibiotics, there is a significant rise in strains of multidrug-resistant (MDR) pathogens over the past few years. Microorganisms especially bacteria have a very rapid growth rate and the ability to develop numerous mechanism of antibiotic resistance when a particular subset of the population is exposed to antibiotics [16]. Several mechanisms have been reported through which the microorganisms develop antibiotic resistance. These mechanisms include inactivation or alteration of the antibiotics by microorganisms, target site alteration of antibiotics, and alteration in the metabolic pathway where the antibiotics are supposed to act on. Due to these reasons, the accumulation of antimicrobials is getting reduced inside the body and getting cleared through the excretory system. Owing to the failure of traditional antibiotics in treating drug-resistant bacterial strains, there is a need for an entirely novel approach in the development of antimicrobials to cope with the problem of antimicrobial resistance in microorganisms [91]. One such innovative approach utilizes the nanoparticles as carriers for antimicrobial applications. Due to several unique physicochemical properties of nanoparticles like size, shape, surface properties, etc., nanoparticles can be exploited for the antimicrobial therapies. Several reports suggest that nanoparticles (silver, gold, copper oxide, zinc oxide) themselves act as an antimicrobial agent and can also be an effective carrier (solid-lipid nanoparticles, liposome) to deliver antimicrobial drugs [1, 5, 105]. With an increasing concern toward MDR wherein the microbes are getting resistant to various antibiotics, nanoparticles are becoming potentially the problem solver for the issue of MDR pathogens [112]. The nanoparticles as drug carriers facilitate targeted drug delivery inside the body. Therefore, these unique properties of nanostructured materials are therefore proving to be a promising approach to overcome the problems of various microbial infections.

8.1.1 OVERVIEW OF THE MECHANISM OF ANTIBIOTIC RESISTANCE

Mutation is the most common and major cause for generating antibiotic resistance among various bacterial species. These mutations primarily occur due to inappropriate and random usage of bacteriostatic and bactericidal agents resulting in high selection pressure among bacterial species exposed to it. With continuous selection pressure, bacterial colonies mutate and generate resistance against these antibacterial agents. The mutated genes are then transferred to their next generation with the characteristics of antibiotic resistance [50]. The resistance capabilities of bacteria are mainly divided into four categories. Firstly, due to enzymatic inactivation of drugs, for example, beta-lactamase case [16]. Secondly, due to alteration of drug target sites leading to resistance as noticed in the case of penicillin-binding proteins in MRSA. In the third mechanism, bacterial cell walls are altered due to a mutation in several genetic mutations and therefore, the antimicrobials are not capable of binding to specific targets. The fourth mechanism is the lowering of drug uptake by the bacterial cell [91]. In this particular case, para-aminobenzoic acid acts as an important precursor for the synthesis of folic acid and bacterial nucleic acids. The sulphonamide-resistant bacterial strains do not require this precursor and it utilizes folic acids present in mammalian cells for nucleic acid synthesis. Due to this, there is a significant decrease in drug permeability and also an increase in the active efflux of drugs nearby the bacterial cell surface, resulting in low drug accumulation in the cellular compartment [68].

A bacterial cell has efflux pumps that may help in extruding antimicrobial agents near the cell surfaces so that the antimicrobial agents do not reach the target site. Mechanism of resistance may play a major role in reducing the efficiency of antimicrobials (Figure 8.1). Further, the antimicrobial resistance is enhanced by overproduction of efflux pumps and contributes to the development of MDR pathogens. Therefore, the suppression of most of these mechanisms of antimicrobial resistance has been targeted by the research community to find novel antimicrobials having multiple functions.

8.1.2 CHALLENGES FACED BY EXISTING THERAPIES

Conventional therapies include chemotherapeutic drugs and antibiotics, which are responsible for the killing of pathogens and interfering in their growth mechanism. Since 1928, when the first antibiotic penicillin was introduced commercially, there has been a plethora of antibiotics that are developed for treating various infections [98]. These antibiotics are used in various forms ranging from ointment (Neosporin) to injections. Due to the high efficiency of antibiotics in killing pathogens and treating several infections, these drugs were proven to be a landmark discovery in the 20th century. Several mechanisms of antimicrobial action are discussed later in this chapter. Although being effective in treating infectious diseases, there are some limitations, which include tolerance of antimicrobials in the body, narrow spectrum range of antibiotics, toxicity and allergens to humans and inefficient delivery mechanism to the site of the target. Due to inefficient drug delivery through conventional antimicrobial

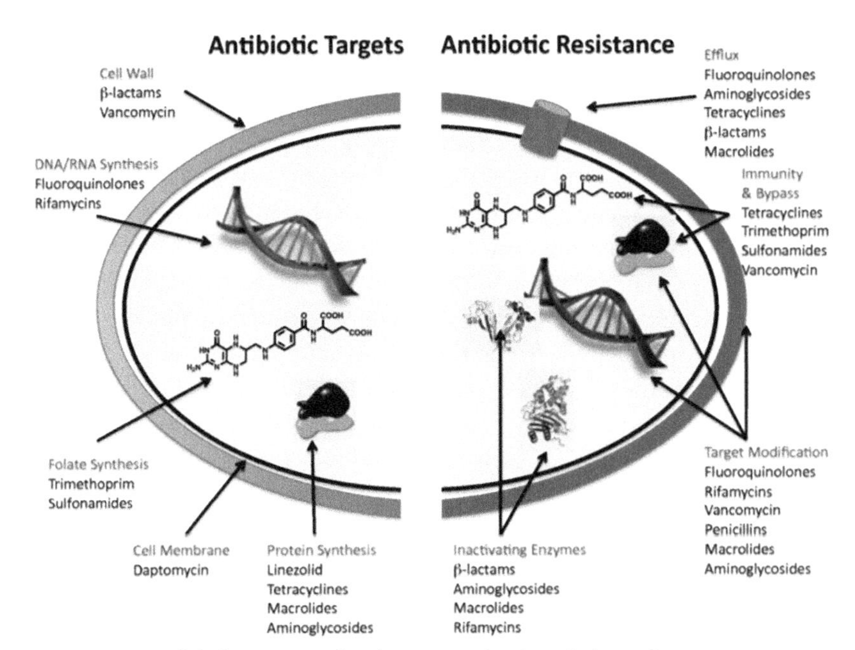

FIGURE 8.1 Antibiotic targets and resistance mechanisms in bacteria.

Source: Reprinted from Ref. [115]. (http://creativecommons.org/licenses/by/2.0)

therapy, most of the drugs, which are administered orally in the form of tablet and capsules, get distributed nonspecifically to other parts of the body and may give rise to several side effects. Often, the drugs are administered at higher doses and most of the drug molecules are not able to reach the targeted site. Further, delivery of hydrophilic antibiotics inside the cells becomes a complicated issue as these drug molecules are not able to pass through the lipid bilayer of the cell membrane. Another major drawback of the conventional antimicrobial therapies arose due to the development of bacterial resistance. Research suggests that almost 70% of the bacteria that causes severe infections have now become resistant to most of the conventional antibiotics used against its treatment. Some of the pathogens are so reluctant that only experimental treatment with the potential toxic drug can inhibit their growth in vitro [83]. However, other therapies including antibody therapy and bacteriophage therapies are highly efficient to treat infections, but due to their higher cost it becomes an expensive form of treating microbial infections in human beings. With the development of MDR strains of various life-threatening pathogens, serious clinical issues are arising and giving rise to the need for developing novel antimicrobial therapies against these infectious pathogens. The major problems in existing therapies are as follows: (1) unwanted biodistribution of drug molecules leading to administration of drugs at higher doses; (2) antibiotic resistance among major classes of pathogens; (3) nontargeted drug delivery affecting the systemic function of the body; and (4) delivery of hydrophilic antibiotic inside the cells.

8.1.3 NANOPARTICLE-RELATED ANTIMICROBIAL THERAPIES

Nanotechnology is known for modifying and manipulating the properties and structure of materials at its nanoscale. At the nanoscale, the discreteness in the electronic energy level gives rise to unique physical, chemical, and biological properties. Unique properties of nanoparticles (1–100 nm) combined with the growth inhibitory mechanism have led to its increased application in the field of medicine. Nanotechnology has revolutionized the field of medicine by contributing in the way we diagnose and treat microbial disease. One of the oldest uses of materials driven antimicrobial treatment is demonstrated through the use of silver for treating burns and copper for killing bacteria and germs to make water potable. Engineered silver nanoparticles (Ag NPs) are being extensively used in medical (as coatings to surgical instruments, antimicrobial agents) as well as consumer products (nano-silver embedded socks, water filters, washing machines, etc.) due to the proven bactericidal ability of silver. Antimicrobial activity of nanoparticles is influenced by the surface area of nanoparticles in contact with microorganisms. To this cause, the high surface area to volume ratio of nanoparticles enhances its interaction with the microbial cells. Another method through which nanoparticles show the high bactericidal effect is by generating reactive oxygen species (ROS), which locally kills the bacteria. ROS is produced from nanoparticles by the release of metal ions or through their interaction with UV light. For example, iron nanoparticles generate ROS by decomposing hydrogen peroxide to hydroxyl radical by Fenton reaction. Hydroxyl radicals effectively degrade the deoxyribonucleic acid (DNA) strands and inactivate enzymes. The production of ROS in metal oxides depends upon the number of defect sites present in the structure. Nanoparticle-based treatment has shown better results compared to the antimicrobial drug alone, for several drug-resistant microbes. Moreover, the problem associated with unspecific biodistribution and nontargeted delivery of drugs has been promisingly addressed by the use of nanoparticles as a drug carrier. Some of the commonly used antitherapeutics are shown in Figure 8.2. This chapter discusses how the physicochemical properties of nanoparticles can be tailored to affect its efficacy for antimicrobial treatment. The mechanism by which some of the common types of nanoparticles cause toxicity to the pathogenic microorganisms has also been highlighted.

8.2 MECHANISM OF BACTERICIDAL ACTION

As shown in Figure 8.1, there are primarily five pathways that have been targeted to develop novel antimicrobials. These pathways include disrupting protein synthesis, disrupting cell wall synthesis, disrupting ribonucleic acid (RNA) synthesis, disrupting DNA synthesis, and other intermediately metabolic pathways. The most successful clinical use of antimicrobials has been achieved by primarily targeting these major pathways of pathogenic microorganisms. Due to ever increasing cases of drug resistance of conventional antimicrobials, it is essential to have an understanding of each pathway in detail for getting a broader scope of developing novel antimicrobials to combat against new and resistant strains of pathogenic microorganisms [91]. Various reported mechanisms of bactericidal action are represented in Figure 8.4.

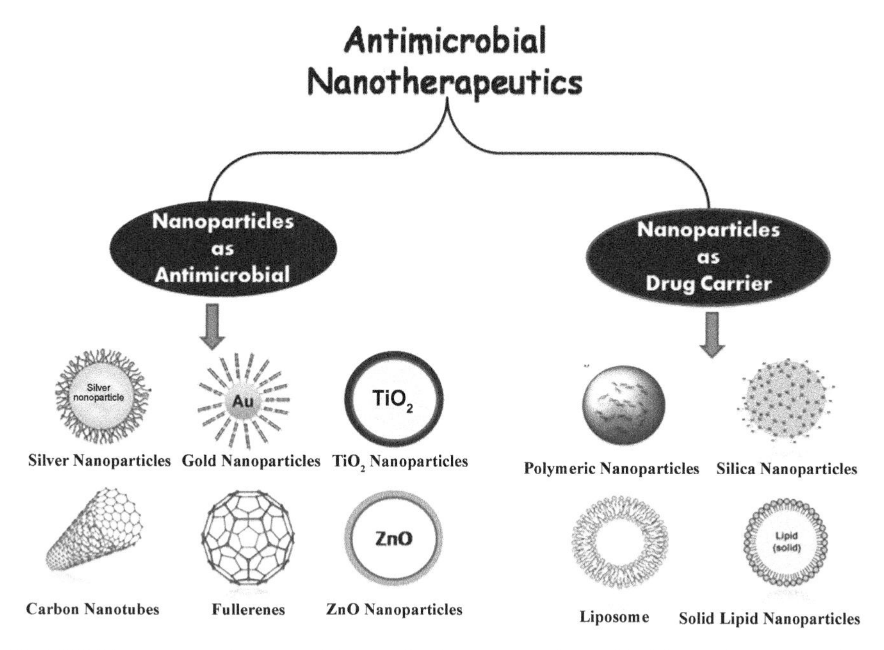

FIGURE 8.2 Types of nanoparticles used either as an antimicrobial themselves or as a drug carrier for antimicrobial therapy.

8.2.1 CELL WALL SYNTHESIS AND CELL MEMBRANE FUNCTION INHIBITORS

Bacterial cells are protected by cell walls, which provide a specific advantage for the survival of the microorganisms and are very critical for their life cycle. This unique feature works to our advantage, as antimicrobial drugs can be developed that target the cell walls of pathogenic microorganism explicitly and inhibit its growth. Conventional antimicrobials (cephalosporins, penicillins, vancomycin, and bacitracin) have particularly shown to hurt the cell walls synthesis in bacteria. The cell membranes act as an essential protection barrier for a cell by regulating and segregating the extra- and intracellular exchange of various substances. The disruption of these cellular structures may result in potential leakage important cellular contents vital for bacterial cell survival. These cellular structures are found in prokaryotic and eukaryotic cells and because of this fact the possible action of certain classes of antimicrobials is not specific and very poorly selective that may have a potential toxic effect on the host cells. The clinical use of such antimicrobials is therefore kept for only specific applications.

Examples of such antimicrobials include colistin and polymixin B [29, 50].

8.2.2 PROTEIN SYNTHESIS INHIBITORS

The cellular structure of microorganisms and enzymes are composed of proteins. Hence, the synthesis of proteins becomes a vital necessity for the survival and reproduction of an organism. There are several antibiotics that specifically target these protein synthesis pathways by binding itself to either of the ribosomal subunits (the 30S or 50S), during the process of translation. Due to this binding, the activity of the ribosome gets disrupted and inhibits the synthesis of essential intracellular proteins in bacteria. The inability of bacterial cells to multiple blocks and eventually leads to cell death. Examples of antimicrobials reported to target protein synthesis pathways are chloramphenicol, tetracycline, macrolides, and aminoglycosides [115].

8.2.3 NUCLEIC ACID SYNTHESIS INHIBITORS

Nucleic acid is the key factor for the replication process of microbial cells in almost all living forms survives on this planet, including bacteria. DNA and RNA are the principal sources of transferring genetic information as well as the signaling process in the cells. Some antimicrobials bind specifically to the major components in the DNA and RNA synthesis pathways, causing severe interference in their synthesis and blocking the normal cellular processes. Due to this, the ability of multiplication and survival of the microorganism gets compromised, which ultimately leads to cell death. Some of the antimicrobials that have been highly exploited by the researchers and as commercial medicines are rifampicin, metronidazole, and quinolones [29].

8.2.4 METABOLIC PROCESSES INHIBITION

Antimicrobials can also target highly essential metabolic processes for survival and growth of the pathogenic microbes. The example includes the use of trimethoprim and sulfonamides antibiotics that inhibit the formation of folic acid synthesis. The synthesis of folic acid in bacterial cells is an essential process and acts as a precursor of DNA synthesis pathway. The antibiotics composed of sulfonamides target an enzyme known as dihydropteroate synthase and bind to it causing competitive inhibition and disrupting the formation of folic acid. Similarly, the drug trimethoprim inhibits the action of dihydrofolate reductase, which is another critical enzyme of folic acid synthesis pathway [115].

8.2.5 BACTERIOSTATIC ACTION

Bacteriostatic antibiotics are the one that interfere with protein production, cellular metabolism processes, and DNA replication [69] of bacterial cells and limit their ability to grow. The group of antibiotics that has bacteriostatic action includes sulfonamides, tetracyclines, trimethoprim, spectinomycin, macrolides, chloramphenicol, and lincosamides. These antibiotics work in close proximity with the immune system of host and help in removing the microbes from the body. There has not been a significant distinction made between the bacteriostatic and bactericidal antibiotics. In various cases, at high concentrations bacteriostatic agents become bactericidal [66].

8.3 COMMONLY USED SOLUTIONS

8.3.1 PROBIOTICS

Probiotics are beneficial microorganisms that are similar to the microflora present inside the human body. These classes of microorganisms have been reported to have several functions that take place in the human gut environment and are having potential antimicrobial activities. There are three mechanisms that help in achieving the antimicrobial activity inside the human bodies: the modulation of microflora inside the gut microenvironment, potential ability to replace the pathogens responsible for causing diseases, and production of secondary metabolites that itself act as an antimicrobial agent and also act as an immune system booster [4]. Probiotics have shown the highest efficacy as an antimicrobial agent when consumed through the oral route. There are several probiotic supplements available. Capsules, drinks and food, and these supplements are also reported to have beneficial strains of bacteria such as *Lactobacillus bulgaricus*, *L. bifidus,* and *Streptococcus thermophiles* [99]. These strains of beneficial microorganisms have been extensively used for the treatment of diarrhea, acute gastroenteritis, and deficiency of nutrients. Apart from the digestive system, these probiotic therapies have also been reported to be used for certain respiratory disorders such as cystic fibrosis, pneumonia, etc. based on the immunomodulation capabilities of the probiotics strains. The route of administration is the most important parameter to get comprehensive benefits of the probiotics [27]. In the recent scenario, due to the unique physicochemical properties and payload carrier applications of nanoparticles, efforts are being made to use these systems as the carrier of probiotics. Although not much work has been done in this context by looking at the importance of antimicrobial potential of probiotics, there is a possibility of using nano-technology for the application of probiotic delivery.

8.3.2 ANTIMICROBIAL PEPTIDES

Antimicrobial peptides (AMPs) can be naturally derived molecules or synthesized in the laboratory according to the applications. Some of the naturally derived AMPs from mammals, insects, and amphibians include magainin, cathelicidin, brevinin, and cecropin peptides. Synthetic AMPs are produced by the combination of various natural AMP sequences. As the mechanism suggests, these peptides attack the cell membrane of the microorganism directly and tend to damage it by forming pores on membrane surfaces. The formation of pores leads to osmotic imbalance and leaking of important molecules out from the cell and eventually causing cell death. Research has shown these AMPs to also cause cell death by interacting with the pathways responsible for the synthesis of important protein and nucleic acid, which are the building blocks of any organism [79]. Effective delivery of AMPs is a major challenge. Water et al. [114] designed polymeric nanoparticle-based AMP delivery system for combating against severe infections. They impregnated cationic peptides known as plectasin poly (lactic-co-glycolic acid) nanoparticles with 71% drug encapsulation and achieved a sustained release of peptides for up to 24 hours. The PLGA-plectasin nanoparticles when tested against *S. aureus* showed improved antimicrobial activity as compared to free plectasin peptides [114].

8.3.3 BACTERIOPHAGES

In the 20th century, Frederick discovered bacteriophages, which are the viruses that infect bacteria. Introduction of antibiotics led to discontinuation of the usage of bacteriophage due to its high cost and immunogenicity. With the increase in MDR bacteria, bacteriophages are being reintroduced for developing possible therapeutics [17]. The major advantage of this treatment is that it is relatively simpler to develop a new phage in case of phage resistance than to develop new antibiotics against antibiotic-resistant bacteria. Further, due to its high penetration capability in the area of infection, bacteriophages are preferred more for therapeutic applications. There are certain disadvantages of bacteriophages, as it is not concentration dependent and not eliminated from the body through metabolic pathways. Due to the production of antiphage antibodies by the immune system against the bacteriophage, it may result in severe allergic side effects [96]. The bacteria develop resistance against phage but due to its prompt replacement with effective new bacteriophage, the problems of resistance become completely irrelevant. There are many successful cases observed for effective bacteriophage treatment. In the year 1990, a dog with otitis infection and a human with burn infection have shown a positive outcome after phage therapy [94]. Yacoby et al. [118] demonstrated the use of bacteriophage like nanoparticles for antibody-targeted nanomedicines. The bacteriophage nanoparticles were loaded with antibiotics such as chloramphenicol, which was further connected to the neomycin aminoglycoside for targeted antimicrobial therapy.

8.3.4 ANTIBODY THERAPY

Antibody therapy utilizes serum-derived immunoglobulins which generally consist of polyclonal antibodies, out of which a certain fraction attacks the intended pathogen responsible for causing an infection. These polyclonal antibodies consist of multiple variations, which attack a group of antigen and not meant for targeting a particular antigen. Serum-based antibody therapy was the first ever antibody therapy used against infectious diseases in which the antibodies are used against pathogens due to their antimicrobial properties. In contrast to conventional antimicrobial therapies, monoclonal antibody (mAB) therapy has brought a big revolution in the field of antimicrobial therapeutics. mAB is produced by the process of hybridoma technology. These mAB are still not under full-fledged utilization commercially due to reasons such as high cost of production and increasing antigenic variation in a resistant microorganism. Since no therapy is available against the MDR bacteria, there is an emergence of mAB therapy by forming various cocktail formulations of antibodies [88]. Antibodies conjugated to nanoparticles can be useful in generating combined properties as these antibodies can be used to a specific selection of the target and the nanoparticles with a small size can be used for antimicrobial drug delivery purposes. This hybridization of antibody conjugated with nanoparticles can show a limited specificity for the antimicrobial application. Therefore, antibody conjugated with nanoparticles offers new hope for designing hybrid nanocarriers for antimicrobial therapies (Figure 8.3).

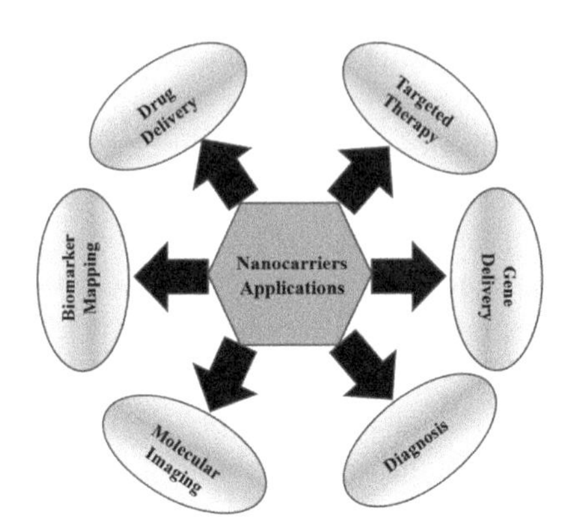

FIGURE 8.3 Application of nanocarriers for antibacterial therapy.

8.4 NANOMATERIAL-BASED ANTIMICROBIAL THERAPY

Material (metals, ceramics, metal oxides, etc.)-based antimicrobial therapy has been successfully demonstrated, wherein certain metal-based ions have shown to be antimicrobial to a range of micro-organisms. Some of these metals are silver, copper, gallium [49], and certain classes of nanobioceramics include phosphate-based glasses and bioactive glasses [87]. The growth of manufactured nanoparticles enables us to increase the antimicrobial efficacy of these materials through controlled manipulation of physicochemical parameters in their nanoscale form. The reasoning behind this expectation stems from the fact that nanoparticles are highly reactive, compared to their bulk counterpart. Nanomaterials are being used for antimicrobial treatment either as a carrier of antimicrobial agents or due to the inherent antimicrobial potency of the nanomaterials themselves [77]. Controlled release of drugs and drug delivery at a targeted site are two significant properties to make a system ideal for the drug delivery application. In that context, nanoparticles provide a promising alternative in the development of the capable and potential therapeutic system. There is an optimal concentration of drug which is required to act on the targeted site for the effective treatment of an infection. The size, shape, and surface properties of nanoparticles are the major factors for achieving a targeted and controlled release of therapeutic molecules according to the optimum regime of drugs [123]. There are several ways through nanocarriers can be used for antibacterial therapy (Table 8.3). Some of the key advantages of nanoparticle-based antimicrobial therapeutics are as follows (Figure 8.4):

- Sustained and controlled release of therapeutic molecules at a targeted site and thereby increasing the efficacy of the drug.
- The drug can be bound to the surface of nanocarriers for selective attachment of drugs and its tailored release.

- Provision of delivery of various hydrophilic, hydrophobic, and lipophilic drug molecules by altering the physicochemical properties of nanoparticles.
- Due to the ultra-small size of the nanoparticle system, it can easily penetrate the cell walls of microorganism.
- By tuning the physicochemical properties of nanoparticles, release and degradation profile of loaded drugs can be monitored to get zero or first-order kinetics.
- Nanoparticles can carry multiple drugs which can be delivered to the same target at a single time point, e.g. core–shell nanoparticles.

8.5 PHYSICOCHEMCIAL PROPERTIES OF NANOPARTICLES

The properties of a material significantly change from its bulk state when the size of the material approaches to nanometer range. With the lowering of size up to nanometer range, the percentage of atom on the surface increases compared to the bulk material. It leads to a drastic change in its surface reactivity. At the nanoscale, the surface to volume ratio increases that gives rise to additional physicochemical properties of the materials (such as quantum confinement, surface plasmon resonance, and magnetic properties). Nanoparticles have shown significant activity against many pathogens because of

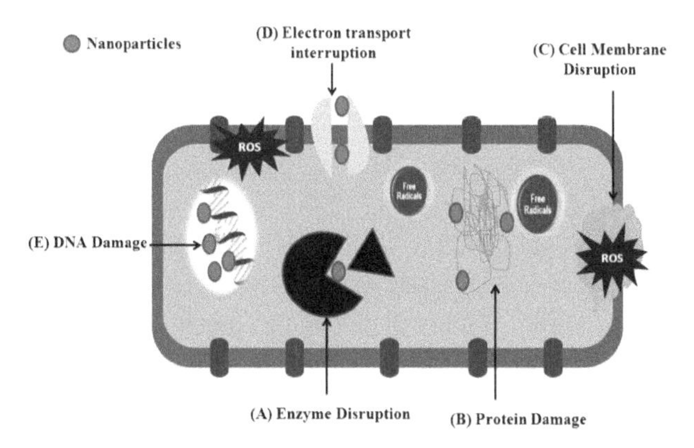

FIGURE 8.4 Mechanism of antimicrobial activity of nanoparticles; important pathways targeted by various nanoparticles for their antimicrobial action. (A) Nanoparticles interfering with the structure of the enzyme and inhibit its activity. (B) Protein damage through ROS generation and binding to tRNA. Ribosome complex inhibits the translation process. (C) Nanoparticles generate ROS which damages the integrity of the cell membrane and also destroys the cell walls. (D) Nanoparticles also hinder the electron transport chain by interfering with the electron transport channel. (E) Formation of free radicals and ROS resulting in DNA damage.

their smaller size and higher surface to volume ratio. The physicochemical properties (size, shape, crystallinity, crystal phase, surface chemistry, etc.) of nanoparticles have a significant influence on the particles dynamics and application potential. The properties including shape, size, and surface properties of the nanoparticles play a significant role in influencing physiological transport behavior of nanoparticles in blood flow, its interaction with various macromolecules, its clearance from the body, and the biodistribution of the drug near intended target site for therapy. The physicochemical properties of nanoparticles can also affect the endocytotic pathways through which nanoparticles gets internalized in the cell and its clearance from it. Therefore, a detailed understanding of the physicochemical properties of nanoparticles will help to design nanocarriers with different configurations, geometries, and functionalities for antimicrobial applications. Some of the particle properties that affect the antimicrobial activities of nanoparticles are size, shape, surface properties, and solubility. Figure 8.5 demonstrates a general mechanism of toxicity of metal/metal oxide nanoparticles in a cell.

8.5.1 PARTICLE SIZE

Size control of nanoparticles is by far the most common approach used for getting better antimicrobial properties. The close resemblance of small nanoparticles to biomolecules has become an exciting proposition for various in vivo research purposes. Nanoparticles themselves have been highly exploited as an antimicrobial against various pathogenic microorganisms and give an alternative to combat against MDR pathogens. The size of the nanoparticles and nanocarriers plays a significant role in its in vivo distribution, the response generated against it by the body, and its elimination from the system. The uptake of the nanoparticles is through the attachment or strong adhesion with the plasma membrane of cells followed by processing of these particles through endocytic pathways. Figure 8.6 shows the various pathways of nanoparticles internalization.

Copper oxide nanoparticles of 20 and 29 nm were tested against a range of Gram-positive (*Bacillus subtilis* and *S. aureus*) and Gram-negative bacteria (*Escherichia coli* and *Pseudomonas aeruginosa*). The zone of inhibitions study showed that 20 nm CuO nanoparticles were comparable to the results obtained from tetracycline (antimicrobial agent) alone, for all tested bacteria [5]. The study also demonstrated that by increasing the size from 20 to 29 nm, the toxicity toward the bacterial cell reduced. Often the nanoparticles that are tested for antimicrobial studies possess a wide size distribution, which makes it difficult to ascertain size-dependent

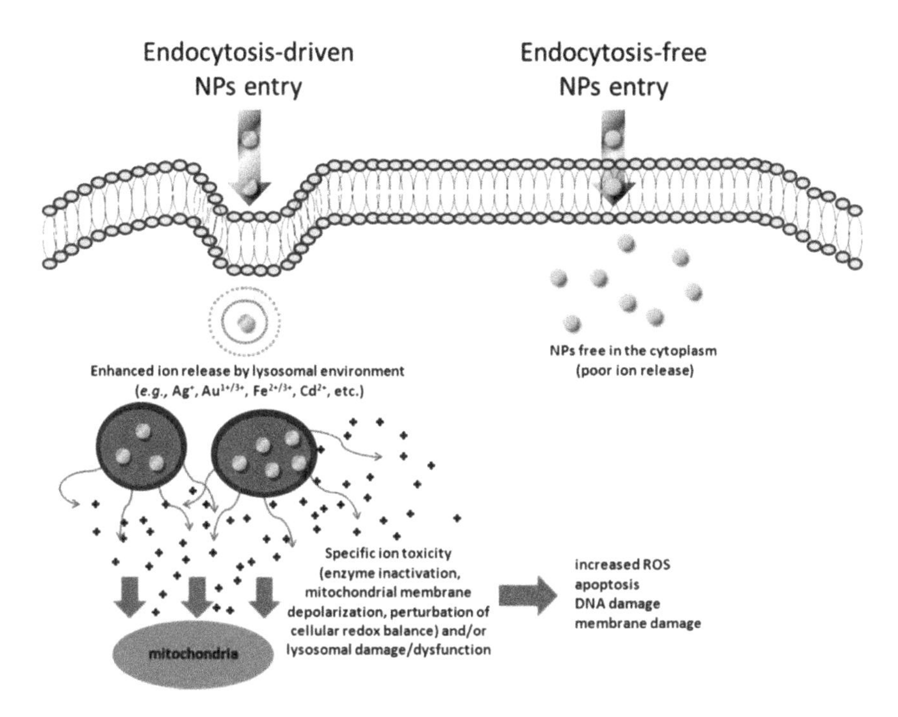

FIGURE 8.5 Scheme showing general mechanism of toxicity of metal and metal oxide nanoparticles.

Source: Reprinted from Ref. [86]. (http://creativecommons.org/licenses/by/2.0)

microbial toxicity. To address this issue, Adams et al. [3] synthesized palladium nanoparticles with narrow size distribution 2 ± 0.2 nm, 2.5± 0.1 nm, and 3.1 ± 0.1 nm. Twenty-four hours colony forming unit study demonstrated that 2.5 nm particles are more toxic to *S. aureus* compared to 2 and 3.1 nm particles. However, for *E. coli*, 2 nm particles were proved to be more toxic. Another study elucidated that the surface area of bacteria available for nanoparticle adsorption was critical for bacterial toxicity of these sub 10 nm nanoparticles. Concentration-based study on dextrose-encapsulated gold nanoparticles (GNPs) (25 ± 5, 60 ± 5, and 120 ± 5 nm) showed dose-dependent toxicity (both bacteriostatic and bactericidal) toward *S. epidermidis* and *E. coli* [6]. In this case, the 120 nm particles showed higher toxicity toward the bacterial cell growth and the mode of toxicity was through disruption of the cell membrane. Size-dependent studies have also been conducted for zinc oxide nanoparticles, wherein the antibacterial activity was inversely proportional to the size of the nanoparticles in *S. aureus* (including MRSA) [82]. Kaweeteerawat et al. [42] tested the mode of toxicity for a range of copper-based compounds

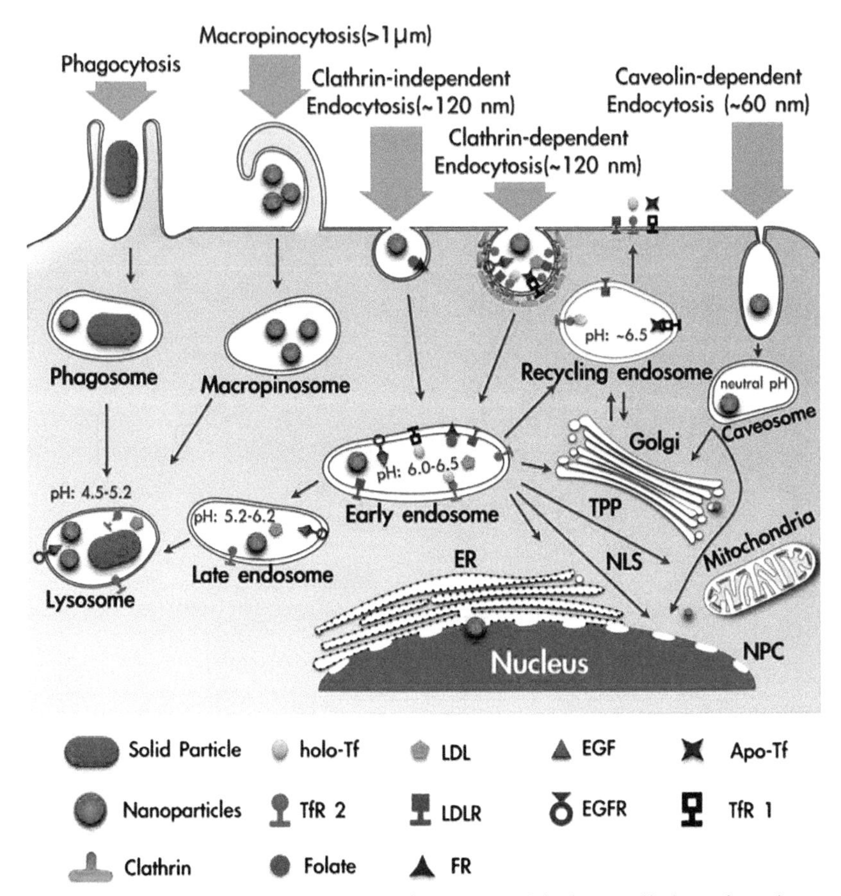

FIGURE 8.6 Various entry mechanisms for nanoparticle in a cell through endocytotic pathways.

Source: Reprinted with permission from [127]. © 2014 Elsevier.

(nano- and micro-copper; nano and micro copper oxide; ionic copper and copper hydroxide) and the results demonstrated that (1) copper and copper oxide nanoparticles are more toxic than micron-sized particles; and (2) the mode of toxicity for nanoparticles was also different from that induced by micron-sized and ionic copper.

8.5.2 PARTICLE SHAPE

The number of research papers investigating size-dependent antimicrobial properties is far more than the papers looking into the effect shape can have on the antimicrobial properties of nanoparticles. It is primarily because controlling shape of nanoparticles is a more complicated process at the synthesis stage. Reported results suggest that shape of the nanoparticles can become an important parameter for effective drug delivery as well as for interaction of nanoparticles with the biological system [80]. Although nanoparticles of different shapes have been synthesized (flowers, tripods, antennas etc.), the variants used for antimicrobial studies are a handful (i.e., rods, spheres, plates, and sheets). Shape-dependent toxicity of these materials is highly influenced by the process of phagocytosis and endocytosis in vivo. Spherical particles are more easily facilitated by the endocytic pathways as compared to other nanostructured materials like rods or fibers [12]. For example, gold nanorods are used for hyperthermia therapy, but the similar application is less used in the case of spherical and the rod-shaped nanoparticles. It is because rod-shaped nanoparticles have higher infrared absorption as compared to the spheres. Further, it has been reported that irrespective of the nature of particles (homogeneous or heterogeneous), the spherical particles are comparatively lesser cytotoxic than the other shaped materials. It is because most of the nonspherical nanoparticles can cause several consequences while being disposed of via blood capillaries. The rod-shaped materials have the ability to block potassium ion channels with higher efficiency as compared to spherical nanoparticles.

Pal et al. [74] tested rod, spherical, and triangular-shaped nanoparticles and found a considerable difference in antimicrobial action. Triangular-shaped Ag NPs had the highest antimicrobial activity due to its higher atomic density on its surfaces. In a separate study, rod-shaped and truncated Ag NPs were tested for antimicrobial properties on *E. coli* and truncated triangular silver particles showed higher antibacterial efficacy [84]. Titania nanoparticles (nanotubes, nanorods, nanosheets, and nanoparticles) were used to study the effect of shape on the viability of *E. coli* and *Aeromonas hydrophila*, Nanotubes and nanosheets were shown to be less phototoxic compared to the rod and spherical titania nanoparticles [102]. The effect of shape can be better understood by examining the antimicrobial properties of carbon-based nanomaterials (i.e., rod-shaped carbon nanotubes, spherical fullerenes and sheet-like structures of graphene oxides). Dizaj et al. [15] have summarized the antimicrobial properties of all these carbon-based nanomaterials in their recent review, and a summary is shown in Table 8.1. A recent novel study demonstrated a hybrid antimicrobial nanocarrier system, wherein antimicrobial Ag NPs were grown on graphene sheets [70]. The hybrid carrier exploited the cell entrapment mechanism of graphene oxide and the bactericidal ionic pathway of Ag NPs.

TABLE 8.1 A Summary of the Mechanism through Which Carbon-Based Nanomaterials Show Their Antimicrobial Properties

Type of Nanomaterial	Mechanism of Antimicrobial Action	Factors That Affect Antimicrobial Activity
Fullerene	Inhibit growth by impairing the respiratory chain; inhibition of energy metabolism	Size, surface charge, surface chemistry (functional groups)
Nanotubes	Physical interaction with the cell membrane, cell membrane disruption	Nanotube diameter, aspect ratio, length, agglomeration, surface charge, the functional group
Graphene oxide	Cell entrapment mechanism, oxidative stress	Size (area), functional group

8.5.3 SURFACE PROPERTIES

Antimicrobial activity of nanoparticles depends upon the contact area with microorganisms, and the high surface to volume ratio enhances this interaction. It implies that the interaction of nanoparticles with the biological system will be highly influenced by its surface chemistry. There are various factors such as protein corona, colloidal behavior of nanoparticles (agglomeration), adsorption of chemical/biological entities on particles, and capping agents/surfactant attached to the particle surface that affect transmembrane permeability of the nanocarriers and its antimicrobial efficacy [45]. The nanoparticles having positive surface charge are better able to interact with the negatively charged cell membrane and also have better interaction with plasma protein as compared to neutral or negatively charged nanoparticles. Surface properties are also important for dispersion, aggregation/agglomeration and solubility of nanoparticles, when released into certain exposure medium [34]. Zinc oxide nanoparticles capped with oxalic acid showed higher antibacterial activity compared to noncapped ZnO nanoparticles [48]. To illustrate the pivotal role surface chemistry plays, researchers capped Ag NPs with a variety of surfactants (chitosan, lactate, polyvinylpyrrolidone, polyethylene glycol, gelatin, sodium dodecylbenzene sulfonate, citrate, dexpanthenol, and carbonate) to study the effect of surface chemistry on photosynthesis of freshwater algae (*Chlamydomonas reinhardtii*) [65]. The surfactants played a mediatory role in controlling the available silver ions (which is known to be antimicrobial). These findings indicate that if we tailor the surface chemistry of soluble particles, such that rate of dissolution of nanoparticles is altered, then we can also control its antimicrobial behavior (assuming ion-mediated antimicrobial properties). Porosity is another surface property that has gained much attention to antimicrobial drug delivery applications. Recently, mesoporous nanomaterials (silica or silicon-based) have got much attention to circumvent the problems associated with the current therapeutic nanocarriers used for antimicrobial drug delivery and imaging applications. Without compromising the health of the patient, it has been reported that the mesoporous nanoparticles have the properties of controlled and sustained drug release at a targeted site. The pore size of these particles is about 2–50 nm, which can be tuned according to the application and quantity of therapeutic molecules to prevent the premature release of the drug [107]. An example of a composite particle system has been illustrated by Tian et al. [100], whereby 2–10 nm Ag NPs were dispersed in the porous framework of mesoporous silica nanoparticles. These composite nanocarriers showed higher antimicrobial properties because of the sustained release of silver ions from the embedded particles for a period of up to 1 month. Agglomeration of nanoparticles is another very important property, which is highly influenced by the size and surface properties of nanoparticles. The toxicity of nanoparticles inside a microbial cell is greatly affected by the amount of agglomeration taking place inside the intracellular compartment. It has been reported that the aggregated forms of carbon nanotubes show the harsher effect on cells as compared to the well dispersed form of carbon nanotubes [119].

8.5.4 THE SOLUBILITY OF NANOPARTICLES

Solubility being a derived property has a significant role in the antimicrobial activity, as demonstrated by dissolution of Ag NPs into Ag^+ ions to generate a highly toxic response inside the bacterial cells. The solubility of nanoparticles is much more complex than any of the other properties that have been mentioned, yet ion-mediated antimicrobial property is common to several of the nanoparticles (silver, zinc oxide, copper oxide). Nanoparticle solubility is influenced by various other physicochemical properties such as size, shape and surface chemistry, as shown in Figure 8.7. If the nanoparticles solubilize in certain media, then there will be a generation of ions resulting in a change in the mechanism of its uptake inside the cells. Ag NPs have been widely reported to have antibacterial activity due to several proposed mechanism. One of the proposed mechanisms involves the dissolution of Ag NPs into Ag^+ ions. These Ag^+ ions interfere with the functioning of respiratory enzymes and lead to the death of the pathogenic bacteria. A 2010 paper by Jaiswal et al. [39] demonstrated another mechanism wherein the higher antimicrobial activity for β-cyclodextrin capped Ag NPs was due to a Trojan-horse mechanism. Under this mechanism, Ag NPs release silver ions inside the bacterial cell. Figure 8.8 demonstrates a general overview of the antimicrobial behavior of metal nanoparticles.

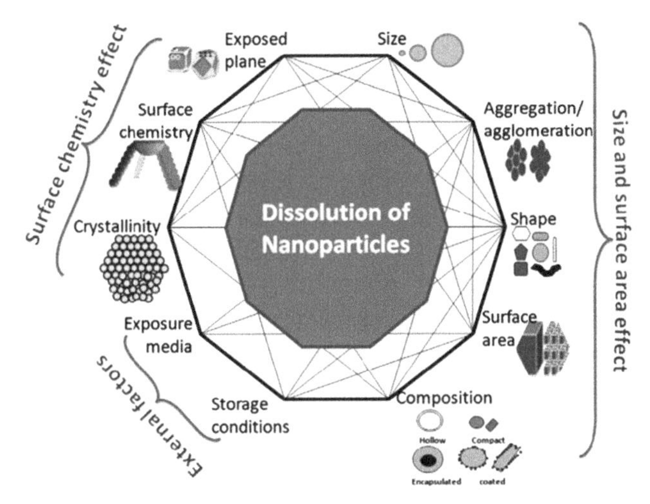

FIGURE 8.7 The complex role physicochemical properties of nanoparticles play in affecting dissolution of nanoparticles.
Source: Reprinted with permission from Ref. [128]. © Elsevier 2012.

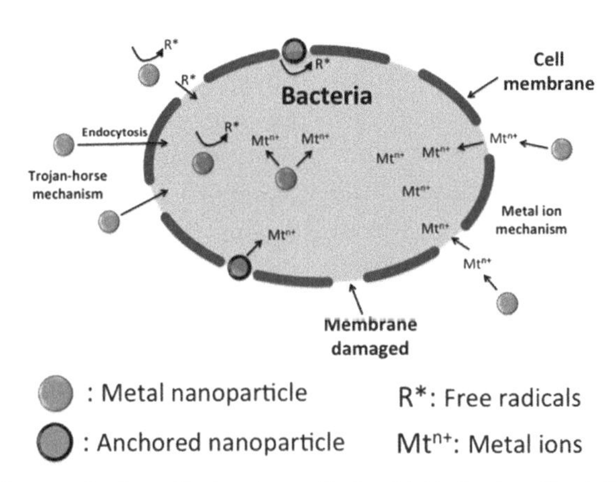

FIGURE 8.8 A general schematic showing antimicrobial behavior of metallic nanoparticles.
Source: Reprinted from Ref. [75]. https://creativecommons.org/licenses/by/4.0/

8.6 CASE STUDIES

Metal-based particles in the form of silver have been conventionally used in its bulk as well as an ionic form for antimicrobial therapies. However, it has been demonstrated that upon its size reduction from 10 μm to 10 nm, there is an enhancement in its antimicrobial activities [74]. The metal nanoparticles are stable at any temperature due to its hard structure and this enables its application in many antimicrobials therapies. The commonly used nanoparticles for antimicrobial therapies are Ag NPs and GNPs. The intrinsic antimicrobial property of metal oxide nanoparticles has gained much attention in recent years for the application of antimicrobial therapeutics. The most widely studied metal oxide nanoparticles exhibiting antimicrobial properties include ZnO, TiO_2, SiO_2 and CuO nanoparticles. Due to their size, crystal structure, morphology, and composition, these particles have also been widely used for the therapeutic purposes as nanomedicine and cosmetic industries (Table 8.2).

TABLE 8.2 Nanoparticle is Exploiting Physicochemical Properties for Antimicrobial Actions

Sr. No.	Nanoparticles	Against Pathogen	Mechanism of Antimicrobial Action
1.	Silver nanoparticles	*L. monocytogenes, E. coli, S. typhimurium, V. parahaemolyticus, M. grisea* and *Bipolaris sorokiniana, S. aureus*	Formation of free radicals Generation of ROS Reactivity of silver ions with sulfur and phosphate group of DNA bases
2.	Gold nanoparticles	*E. coli, S. typhimurium, Vibri, P. aeruginosa*	Interaction with the cell membrane and disrupts its function GNPs inhibit a subunit of the ribosome that binds to tRNA for the process of translation, Photothermal killing of microbial cells
3.	Zinc oxide nanoparticles	*P. aeruginosa, E. coli, S. typhimurium*	Production of ROS by photocatalytic activity Interaction of zinc ions with enzyme functions
4.	Titanium oxide nanoparticles	*P. aeruginosa, E. coli, Salmonella typhimurium, C. albicans*	Develop oxidative stress by generating ROS leads to disruption of the cell membrane and causes leakage of intracellular components
5.	Iron oxide nanoparticles	*P. aeruginosa, K. pneumonia*	Production of ROS after entering into the cell
6.	Carbon nanotubes and fullerenes	*E. coli, Cupriavidus metallidurans*	Severe cell membrane damage and resulted in the inactivation of the bacterial cell Photodynamic activity by fullerenes

TABLE 8.3 Summary of the Selected Nanocarrier System

S. No.	Nanocarrier System	Physicochemical Properties	Carrier Type	Drug-Loaded	Against Pathogen	Observation	Ref
1.	Liposome	Size: 20–500 nm Shape: spherical (unilayered, multilayered) Surface properties (hydrophobic and hydrophilic)	Hydrogenated soy phosphatidylcholine, cholesterol, and distearoylphosphatidyl glycerol	Amphotericin B	*A. fumigatus*	Supports several routes of administration Possibility to load lipophilic drugs	[97]
			1,2-Dipalmitoyl-sn-glycerol-3-phosphocholine (DPPC) and cholesterol	Polymyxin B	*P. aeruginosa*	Less toxicity and no immunogenic response Surface modification possible	[71]
			Egg phosphatidylcholine, diacetyl phosphate, and cholesterol	Vancomycin or teicoplanin	*Methicillin-resistant S. aureus*	Biodegradable and biocompatible	[72]
2.	Solid lipid nanoparticles	Size: 100–1000 nm Surface properties (hydrophobic and hydrophilic) Water retention capacity Surface modification	Stearic acid	Rifampicin, isoniazid, pyrazinamide	*Mycobacterium tuberculosis*	Controlled and sustained release of the hydrophilic drug	[76]
			Glyceryl tripalmitate and tyloxapol	Clotrimazole	*Aspergillus* sp.	Biocompatible and biodegradable	[95]
			Stearic acid, soya phosphatidylcholine, and sodium taurocholate	Ciprofloxacin hydrochloride	*E. coli, S. aureus*	Ability to pass lipid bilayer Targeted drug delivery enhancing the pharmacokinetics of the drugs	[38]
			Stearic acid, soya phosphatidylcholine, and sodium taurocholate	Tobramycin	*P. aeruginosa*	Both lipophilic and hydrophilic drugs can be impregnated	[11]

TABLE 8.3 *(Continued)*

S. No.	Nanocarrier System	Physicochemical Properties	Carrier Type	Drug-Loaded	Against Pathogen	Observation	Ref
3.	Polymer nanoparticles	Size: 100–1000 nm Shape: spherical, ellipsoidal	PLGA	Antimicrobial peptides (Plectasin)	*S. aureus*	Stable structure Can sustain harsh biological conditions	[114]
		Surface properties: tunable	Alginate nanoparticle	Rifampicin, isoniazid, pyrazinamide, ethambutol.	*M. tuberculosis*	The possibility of size and surface properties tuning Targeted and sustained drug delivery	(Ahmad et al. 2006)
			Glycosylated polyacrylate nanoparticle	Ciprofloxacin	*S. aureus*		[1]
			poly(ε-caprolactone) PCL	Amphotericin B	*L. donovanihas*		[23]
			Chitosan	Ciprofloxacin	*E. coli, S. aureus, B, thuringiensis*		[56]
4.	Silica nanoparticles	Size: 60–350 nm Shape: spherical	Silica- Xerogel-PEG	Gentamycin	*Salmonella enterica*	Enhanced adsorption of drug	[89]
		Surface: porosity, mesoporous, surface functionalization	Metal-silica nanoparticles	Silver	*E. coli, S. aureus, P. aeruginosa*	Higher drug loading capacity due to the presence of mesopore	[52, 113]
			Bioglasses and bioceramics	Rifampicin and isoniazid	*M. tuberculosis*	Slow and sustained release of drugs	[125]
			Mesoporous silica-N halamine	N-Halamine and quaternary ammonium group	*E. coli, S. aureus*	Biocompatible	[114]

8.6.1 SILVER NANOPARTICLES

Antimicrobial properties of silver have been exploited in various sectors ranging from textile industries to many coating-based biomedical applications. Ag NPs are a promising solution for antimicrobial applications because of its low toxicity to human cells, useful broad-spectrum antibacterial activity, high thermal stability, and multiple bactericidal mechanisms. It has shown promising activity against antibiotic-resistant bacteria due to several mechanisms like the generation of ROS and free radicals. Ag NPs are quite potent in its antibacterial effect, as it is effective to inhibit the growth of both Gram-negative and Gram-positive bacteria. Zarei et al. [122] demonstrated the use of Ag NPs as an antimicrobial therapeutics, by testing on foodborne pathogens such as *Listeria monocytogenes*, *E. coli*, *Salmonella typhimurium*, and *Vibrio parahaemolyticus*. The efficiency of Ag NPs has also been tested by Jo et al. [41], against various plant funguses, such as *Magnaporthe grisea* and *Bipolaris sorokiniana*. The results suggested that the particle size of 20–30 nm had good penetration capacity and gets itself internalized inside the fungal cells. Therefore, apart from the human pathogens, it has a high ability to control the spore formation in plant pathogens and comparatively less toxic than various fungicides. Size and shape of Ag NPs have a great emphasis on its antimicrobial activity. Smaller Ag NPs have greater penetration through the cell surface which significantly enhances the antimicrobial activity and makes Ag NPs an excellent tool for the antimicrobial therapeutics. Ivask et al. [37] demonstrated size-dependent bioavailability of Ag NPs by taking *E. coli* as the model organism. According to their analysis, Ag NPs of size 10 nm were shown to be more bioavailable to the *E. coli* cells than silver nitrate salt and generate more toxicity due to the release of silver ions inside the cells. Surface coating on Ag NPs has reported to enhance the substrate adhesion. Brobbey et al [9] demonstrated that plasma coating on the surface of Ag NPs leads to adhesion of nanoparticles on the substrate. It also limits the environmental exposure of Ag NPs. It was further reported that the plasma-coated Ag NPs were highly antibacterial in case of Gram-negative strain (*E. coli*) even at a plasma coating thickness of 195 nm. However, the same results were not observed in case of Gram-positive strain (*S. aureus*) [9].

Several mechanisms are proposed to be the probable mechanisms of antimicrobial action of Ag NPs. Balakumaran et al. [8] treated Ag NPs against eight different pathogens and found that the accumulation of Ag NPs on the surface of the cell membrane is responsible for the disruption of cell walls of the pathogen. Due to the small size of Ag NPs, it can penetrate the bacterial cell wall and cause structural changes resulting in an increase of cell membrane permeability and ultimately death of the bacterial cell. The accumulation of Ag NPs on the cell surface forms "pits." Treated *E. coli* cells were significantly damaged due to the formation of pits on the cell walls. The formation of these irregular pits increases the unwanted membrane permeability and causes the release of membrane proteins and lipo polysaccharides (LPS) molecules [93]. Formation of free radicals is another mechanism by which Ag NPs can trigger antimicrobial action. Based on electron spin resonance spectroscopy studies, free radicals are generated when Ag NPs come in contact with the bacterial cells [15, 43]. Due to these free radicals, there is pore formation on the cell membrane which leads to the death of the cell. Another mode, by which Ag NPs can work, is its dissolution in the extracellular matrix of bacterial cell walls resulting in the release of silver ions which are known to inactivate many vital enzymes by interacting with the thiol group of enzymes. The inactivation of enzymes leads to inhibition of major functions required for the growth of bacteria and leads to the death of the organism. Further, there is a generation of ROS due to the inhibition of respiratory enzymes by the dissolution of Ag NPs into its ionic form. ROS is generally toxic to the bacterial cells and it leads to the death of the microorganism. The high reactivity of silver with the phosphorus and sulfur present in DNA of a bacterial cell is responsible for the inhibition of DNA replication which causes death of the microorganism.

8.6.2 GOLD NANOPARTICLES

GNPs are another class of metal nanoparticles that are extensively used for antimicrobial therapies. Several physicochemical properties like small size (10–50 nm), shape (spherical, rod, star), surface charge, etc., make GNPs as one of the best choices as an antimicrobial agent carrier. GNPs are reported to be nontoxic to human cells at a lower concentration, have high functionalization abilities, and its photothermal property makes it easy to detect. In contrast to Ag NPs, the antimicrobial activity of GNPs does not involve the generation of ROS. The actions of GNPs for antimicrobial therapies are mostly due to its interaction with the cell membrane of the microbes, which result in a modification in the level of ATP inside the microbial cells. It also involves in the inhibition of tRNA binding to ribosome resulting in breaking of the translational pathway. GNPs have shown greater efficacy as an antimicrobial against both bacteria and fungus [14, 101]. The microorganism against which it has shown promising activity includes *Micrococcus luteus, P. aeruginosa, S. aureus, E. coli, Aspergillus fumigatus and A. niger*. There are several examples where GNPs are tagged with surface markers for targeted therapeutics. One such example includes surface modification of GNPs with 5-fluorouracil, which turned out to increase the efficiency of antimicrobial action of GNPs [101]. Surface modification has always held the key property for GNPs, as GNPs can be surface modified with positively charged capping agent (poly-allylamine hydrochloride) as well as with negatively charged citrate. The positively surface charged poly-allylamine hydrochloride@ GNPs tailored a strong interaction with the negatively charged cell membrane of the bacteria resulting in higher internalization [26]. The presence of strong capping agent also initiates the self-assembly process of the GNPs which increases the cellular toxicity in bacteria [47]. The negatively surface charged citrate-capped GNPs result in agglomeration, which is conversely an undesirable property for antimicrobial action due to the increase in the size of nanoparticles. Therefore, size, shape, and surface properties of GNPs play a significant role in designing an efficient antimicrobial therapeutics. GNPs are also reported to be bactericidal at very high concentration and possible reason for antibacterial activity may include the presence of coexisting chemicals attached to the surface of GNPs that are not entirely removed from its surface. These chemicals may include gold ions, a surfactant or stabilizing agents as well as the precursors involved in the synthesis of GNPs [124].

There are two main mechanisms through which GNPs show their antimicrobial activity [14]. Firstly, GNPs interact with the cell membrane and disrupt the integrity of the membrane potential. A sudden disruption of the cell membrane causes leaking of metal ions such as K^+ and Na^+ from the cells. Further, due to the leaking of ATP from the intracellular compartment, there is a modification in the level of ATP inside the cellular compartment which leads to inhibition of cellular growth and ultimately death of bacterial cells. This was shown at the proteomic level and a downregulation in F-ATP synthase was observed. Secondly, GNPs inhibit a subunit of the ribosome that binds to tRNA for the process of translation. It disrupts the normal functioning of the bacterial cells and leads to the cell death. In addition due to the inherent property of infrared absorption of GNPs, it can also induce photothermal killing of pathogens. With certain modification on the surface of GNPs, antimicrobial agents are also attached to these nanoparticles and therefore used as effective drug carriers. A recent study by Manju et al. [55] showed that clavam stabilized GNPs when used as drug platform showed better inhibition of *E. coli* compared to use of clavam alone.

8.6.3 ZINC OXIDE NANOPARTICLES

Zinc oxide (ZnO) nanoparticles exhibiting various morphologies have a significant amount of antibacterial activities against a wide range of microorganism. Both microscale and nanoscale formulations of ZnO nanoparticles are extensively used as an antimicrobial agent. Interestingly, ZnO nanoparticles have shown negligible toxicity to human cell lines [13] and therefore have opened many acceptable utilities of these nanoparticles as therapeutics against pathogens [73]. ZnO particles have already seen a huge application in the cosmetic industry (sunscreens) and for designing antimicrobial therapeutic medicines. Further, ZnO nanoparticles are categorized under GRAS (Generally Regarded as Safe) by Food and Drugs Administration, and approved for various cosmetic therapies. A combination of two or more mechanism is enhancing the efficacy of antimicrobial activity of ZnO nanoparticles. Premanathan et al. [81] demonstrated that ZnO nanoparticles show more antimicrobial activity against Gram-positive bacteria compared to Gram-negative bacteria. ZnO nanoparticles were also reported to be nontoxic against mammalian peripheral blood mononuclear cells. Mostly, the toxicity involved in microbial cell death from ZnO is due to the generation of ROS. ZnO has the property of photocatalytic activity, which has been reported to be the major reason for generation of ROS inside the microbial cell [48]. In an aqueous environment, ZnO nanoparticles prone to UV radiation have phototoxic effects that tailor the production of superoxides (O^{2-}) and hydrogen peroxide (H_2O_2). Generations of these reactive species are very important to create a toxic environment inside the cell. These ROS molecules can penetrate the cell wall and kill the bacterial cells. To demonstrate the antibacterial mechanism of ZnO nanoparticles, Dwivedi et al. [22] used dichlorohydrofluorescein (DCFH-DA) dye which entered passively inside the bacterial cells because of the production of ROS by the ZnO nanoparticles and gets hydrolyzed by the enzyme esterases inside the cells. The DCFH eventually got oxidized to form DCF (dichlorofluorescein), a highly fluorescent molecule in the presence of ROS. The potential amount of ROS generated was measured by flow cytometry. Sevinç and Hanley

[90] have reported the release of Zn^{2+} ions from ZnO nanoparticles as another mechanism that has the potential to kill microbial cells. Zinc ions have certain negative impact on the active transport inhibition and enzyme functioning disruption. According to one article published by Wu et al. [116], a significant amount of toxicity is developed in the presence of ZnO nanoparticles in growth media due to the dissolution of particles into its ionic form.

8.6.4 TIO_2 NANOPARTICLES

Like ZnO nanoparticles, TiO_2 nanoparticles have also shown antimicrobial properties. However, in the case of titania nanoparticles, the bactericidal property depends on the physicochemical (size, shape, and crystal structures) property. Among the three crystalline phases of titania (anatase, rutile, and brookite), the anatase form has the highest antibacterial application as reported by Roy et al. [85]. TiO_2 nanoparticles have not been extensively studied for drug delivery application inside the body; rather it has been exploited for topical applications such as cosmetics for antimicrobial and anti-UV therapy. Under the action of UV, TiO_2 nanoparticles produce ROS. Production of ROS seems to be the main responsible reason behind the DNA damage, ultimately leading to microbial cell death. TiO_2 nanoparticles have shown promising activity against MRSA and due to the presence of TiO_2, the resistance against several beta-lactam antibiotics decreased [85]. Further, TiO_2 nanoparticles are very effective against fungal biofilms formed by *Candida albicans*, making it an attractive proposition for treating fungal infections [32]. TiO_2 can be incorporated into various medical devices required during surgery such as catheters. Titania nanoparticles develop toxicological effect by developing oxidative stress, which causes depolarization in the bacterial cell membrane leading to loss of cell membrane integrity. The loss of cell surface integrity leads to an increase in cell permeability and bacterial cell death. This mechanism was proved by Sohm et al. [92] by following both proteomic and transcriptomic approaches. It was further reported in their findings that due to the phenomenon of oxidative stress, the response toward osmotic control and cell metabolism was greatly affected. The nanotoxicity of TiO_2 nanoparticles causes a massive cell leakage of potassium and magnesium ions from the cell membrane of microorganism leads to increase in sodium ion uptake and also depletion of cellular ATP and causing the death of the pathogens [92].

8.6.5 SILICA NANOPARTICLES

Silica nanoparticles have been exploited for drug delivery applications due to its biocompatibility and have proven to become an efficient drug carrier for treating various biofilm-related infections [120]. The porosity that can be introduced in silica nanoparticles plays a key role in its use as a carrier for drug delivery application. Mesoporous silica nanoparticles are reported to be biocompatible toward mammalian cells. These characteristics are of course highly dependent on the concentration and the route of administration [110]. Due to the presence of large pore volumes and higher surface area of porous silica nanoparticles, the possibility of absorbing drugs is much larger in these particle types. This property helps in facilitating local drug delivery application. The highly reactive surface properties of silica nanoparticles are due to the presence of a silanol group on its surface. The process of silanization allows various modifications on the surface of silica nanoparticles which facilitate entrapment of a large amount of drug and enhance the control release mechanism [33]. There are few reports on the antibacterial activity of the silica nanoparticles. It includes the work done by Hun and Zhang (2007) to determine the *Staphylococcal* enterotoxin by using fluoroimmuno assay with the help of silica nanoparticles. The properties and the application of silica nanoparticles suggest that, at lower concentration and porous structure of these particles, it can be a biocompatible and an active carrier for drug delivery application. Liu et al. [52] developed highly disperse silver nanocluster-loaded mesoporous silica nanoparticles that facilitate an efficient and long-term release of silver ions. The nanoparticles show a promising antibacterial activity against *S. aureus* and *P. aeruginosa* having 17- to 27-folds higher potency as compared to bare Ag NPs.

8.6.6 IRON OXIDE NANOPARTICLES

Magnetic nanoparticles have gained advantageous properties that can be utilized for drug delivery applications. As its movement can be easily tracked in vivo, therefore it allows the clinicians to track its efficiency in targeting and delivering the drug [109]. Due to its magnetic properties, it has opened several doors for biomedical applications such as MRI, magnetically stimulated therapy, and hyperthermia therapies [59]. Research suggests that there is negligible toxicity of iron oxide nanoparticles toward eukaryotic cells but the toxicity level is quite high for the prokaryotes [67]. The physicochemical properties of magnetic nanoparticles like size, drug entrapment capacity, and magnetic strength play a vital role in designing iron oxide-based nanocarrier for drug delivery application [10]. The magnetite nanoparticles have been tested against urinary tract infection causing pathogen namely *S. saprophyticus* by Gu et al. [28], by surface attachment of vancomycin on the nanoparticles. Among most of the oxides tested against *E. coli, S. viridans, S. pyogenes, P. aeruginosa,* and *Klebsiella pneumonia*, magnetite nanoparticles have shown the highest antibacterial activity against *P. aeruginosa* with the possible generation of ROS as the mechanism of antimicrobial action. The mechanism of action of iron oxide nanoparticles is still a topic of debate as most of the report suggests that the killing of microorganism takes place by the generation of ROS in the microorganism and due to oxidative stress, there is a disruption in the cell membrane of the microorganisms [105].

8.6.7 LIPOSOMES

Liposomes are lipid-based ultrasmall spherical bilayered membrane vesicles which are surrounded by various molecules that are hydrophilic and hydrophobic. Depending on the requirement of the therapeutic system, liposomes can be formed as single layered (small unilamellar liposomes) and multiple layered (large unilamellared liposomes). The drug delivery system designed from liposomes can be made of either synthetic or natural lipid molecules. Phosphatidylcholine is the most commonly used lipid for synthesizing liposomes, which is a neutral phospholipid comprising of fatty acyl chain of various length and saturation. The preparation of liposome-based drug delivery system requires significant attention to its physicochemical properties such as size, polydispersity, surface charge and potential, shelf life, and its reproducibility. These nanosized liposomes are widely utilized for the antimicrobial drug and DNA delivery systems [103]. Due to the presence of lipid bilayer, this drug delivery system easily gets fused with the lipid layer present on the cell surface of the pathogens [123]. However, due to the smaller shelf life of liposomal drug delivery system, it affects the overall stability of the drug molecule inside the cell. Most of the vesicles get degraded inside the cell due to various ongoing enzymatic and physical processes [21]. Liposomes have many advantages over conventional therapy used for treating various infectious diseases. It helps in increasing the pharmacodynamics and pharmacokinetics of the drug molecules, targeted drug delivery, and shows an increase in efficacy against intracellular and extracellular pathogens [1]. Size and surface properties are vital in liposomal-based drug distribution [24]. The toxicity or the adverse effect of the drug can be minimized by knowing the therapeutic index and the concentration of drug required to act on the site of infection. There has been a significant contribution made by Deol and Khuller [18] by synthesizing lung-specific liposomal vesicles for the targeted drug delivery of anti-tuberculosis drugs. Liposomes are also widely used for the nucleic acid delivery system for treating various viral infections in human beings.

Surface properties and size play a major role in the drug delivery mechanism of the liposomal vesicles. It does not allow the dissolved hydrophilic drug molecules to leak out in the extracellular microenvironment. To deliver entrapped drug molecules, the outer hydrophobic lipid layer fuses with the lipid bilayer of the cell membrane of bacteria and thus gets inside the cellular compartments. With the digestion of these vesicles by intracellular enzymes, the drug oozes out in the cellular microenvironment and binds to the targeted sites. There are certain liposomal vehicles that are degraded by the process of endocytosis and cleared from the body [18]. Further, when the fusion of liposomes loaded drug happens with the cell membrane of the pathogen, there is a sudden burst out in the liposomal compartment helps in oozing out a large amount of drugs. High release of drug overwhelms the efflux pump of bacteria and hence suppresses the resistance mechanism. Another mechanism includes the binding of liposomal particles with the cell wall of pathogens and therefore acting as a drug reservoir that keeps on releasing drug molecules which diffuse inside the cellular compartments.

8.6.8 SOLID LIPID NANOPARTICLES

The development of solid lipid nanoparticles (SLNs) has made it become an alternative source to polymeric nanoparticles and liposomal vesicles for controlled drug delivery and therapeutic applications [2]. The size of the colloidal SLN particles ranges from 50 to 1000 nm, and it is mainly made up of lipid matrix dispersed in aqueous solution with hot/cool surfactant [61]. The main property of SLN nanocarriers is the capability to withhold a large amount of various hydrophilic drugs in its lipid matrix, making them an advanced nanocarrier system for delivering hydrophilic drugs [30]. Several other important properties of SLN for use in antimicrobial therapeutics are as follows [108]:

- Controlled and sustained release of hydrophilic drug molecules.

- Ability to pass lipid bilayer barrier in the plasma membrane.
- Nontoxic to the mammalian cells
- Both lipophilic and hydrophilic drugs can be impregnated in the SLN
- Relatively more comfortable production and lower cost.
- Biocompatible and biodegradable.
- Targeted drug delivery enhancing pharmacokinetics of the drugs.

Several pieces of evidence are reported in favor of SLN to have improved antimicrobial activities. One such evidence talks about the delivery of rifampicin-loaded SLNs for anti-mycobacterial with the encapsulation efficiency of about 82%. Sustained drug release was observed until 72 hours and there was a significant decrease in minimum inhibitory concentration (eight times) than the free rifampicin drug molecules. Nanostructured lipid carriers and SLN are extensively used for application in antifungal drug delivery (e.g. sustained release of clotirmazole) [95]. One very important application of SLN is to deliver azole drugs for antifungal therapy. The water-insoluble azole drugs include clotrimazole and econazole, which are impregnated in SLN for effective antifungal therapy. The occlusion property of SLN along with its small size makes it an ideal drug carrier with an extended residence time of drugs on the epidermis and highly efficient in penetrating the skin barrier. Reports suggest that there is a rapid penetration of antimicrobial drugs through SLN when subjected to in vivo evaluation of drug release. Due to the compatible interaction of SLN with keratinocytes and other metabolic activities, it offers a potential application for skin therapy. There are several antimicrobials drugs that have been incorporated within SLN (isoniazid, rifampicin, ciprofloxacin, gentamycin) and a good and sustained release of the drugs from SLN at the targeted site of infections has been obtained. SLN-based drug delivery has been highly studied for the anti-tuberculosis drug delivery application against *Mycobacterium.* Various anti-tuberculosis drugs have been impregnated on the SLN nanocarriers such as isoniazid and rifampimetc for both pulmonary and lymphatic systems [76]. González-Peredes et al. [25] developed novel solid lipid nanoparticles for the delivery of antimicrobial oligonucleotides. Based on their findings, there was a formation of "Transcription Factor Decoy" complex with SLN, and this complex offers a preferential bioaccumulation of TFDs in bacterial cells. The SLN was reported to be a promising class of antibacterial oligonucleotide nanocarrier and can help in fighting the complex issue of antimicrobial resistance (AMR) crisis [25]. However, there are few disadvantages of SLN like unwanted gelation, lower drug incorporation due to its crystalline structures and polymorphic transitions, which makes it a little unstable system compared to liposomes and polymeric nanoparticles.

8.6.9 POLYMERIC NANOPARTICLES

There are several applications of biologically and synthetically originated nanoparticles in their nature as well as modified form. Polymeric nanoparticles have been extensively exploited for antimicrobial therapies due to its biocompatibility. The synthetic and biological polymers are reported to tailor intrinsic bactericidal and bacteriostatic activities, particularly the polysaccharides are known to give a promising outcome regarding dressing and wound healing applications. Polymeric nanoparticles itself act as an antimicrobial agent (chitosan nanoparticles) and also as an antimicrobial drug carrier (poly(lactic-*co*-glycolic acid) [PLGA], poly caprolactone [PCL], etc). Biologically originated natural polysaccharides like chitosan have very important properties such as biodegradability, nontoxic, hemostatic activities, macrophage affecting functionalities, and natural antibacterial activities [7, 62]. Zahran and Marei [121] developed an innovative polymer metal nanocomposite for antimicrobial applications. The composite was formed with the incorporation of metal polymer nanoparticles through photoreduction, chemical reduction, or the thermal decomposition of metal salts within polymer matrix. In these methods, natural polymers such as starch, chitosan, dextran, gelatin, etc. were used. These polymers possess exceptional physicochemical properties such as high tensile strength, crystallinity, high surface area, and elasticity. Such properties also helped in enhancing the antibacterial properties against *E. coli* and *S. aureus* [121]. Antibacterial properties of quaternized chitosan nanoparticles reported by Ignatova et al. [36] have demonstrated a very high broad-spectrum bactericidal activity with a greater killing efficiency as compared to pure chitosan nanoparticles [40, 44]. A synthetic polymer such as polyvinyl alcohol possesses hydrophilicity and cross-linking ability which makes it more compatible for wound treatment. There can be a certain change in physicochemical properties made toward polymeric nanoparticles due to its flexible nature. Polymeric nanoparticles can be made into various forms (capsules, spheres) based on various methods of its synthesis. Some of the properties of polymeric nanoparticles that makes it suitable for antimicrobial therapy include [46] the following:

- Can be synthesized in various size and shape forms.
- Tailoring of surface properties.
- Surface modification is possible with polymeric nanoparticles with desirable manipulation of a surface functional group for targeted drug delivery applications.
- Core–shell particles can be prepared to allow hydrophobic cores being impregnated with drugs and the hydrophilic part provides shielding effect and prevents opsonization well as degradation of particles.
- Degradation of the polymer can be controlled.

Above-mentioned properties have been used in targeted drug delivery for treating various microbial infections. For instance, lectin conjugated with glycine polymeric nanoparticles can be targeted to carbohydrate toll-like receptors on *Helicobacter pyroli* cell surfaces [123]. Commonly used synthetic polymers for drug delivery include polycaprolactone, polylactic acid, chitosan, gelatin, poly-D, L-lactide co-glycolide, and polyvinyl alcohol. Polymeric nanocarrier systems have been used for targeted drug delivery against MRSA and *Mycobacterium tuberculosis.* Poly(ε-caprolactone) nanoparticles loaded with antibiotic amphoterin B have shown good efficiency in the killing of *Leishmania donovani,* when compared to the action caused by its stand-alone drugs counterparts [23]. More examples are highlighted in Table 8.3. For a controlled and sustained drug release from the polymeric nanocarriers, uniform distribution of the drug is an essential factor. In most of the cases, the controlled release of drugs occurs by diffusion or disruption of the polymeric matrix (acting as a sink). If the process of diffusion is faster than that of matrix disruption, then the whole mechanism of controlled release is governed by the process of diffusion throughout. There is a possibility of sudden release of the drug due to burst in the polymeric matrix or it is attributed by absorbed drug molecules on the surface of polymers. The methodology implemented for drug loading on polymeric nanocarrier greatly affects the overall profile of the drug release. The incorporation methodology is favorable in terms of better and sustained release of the drug molecules with a very lower burst effect. The diffusion process is further enhanced if the polymeric core is coated with some other polymer. Therefore, it is evident that solubility of polymeric nanoparticles and diffusion of the drug play a key role in the controlled release of the drug inside the body.

8.7 CONCLUSION

Nanoparticle-based antimicrobial therapy has opened multiple pathways of fighting against antimicrobial resistance, which has become a cause for global health concern. Most of the low water-soluble antimicrobial drugs with rapid clearance and low biodistribution inside the body are now showing greater efficacy when delivered through a nanocarrier system. Nanocarrier-based antimicrobial therapy is providing a platform of slow and sustained drug release, improved pharmacokinetics of drugs, and uniform distribution at the targeted site, as well as providing optimal time for clearance from the body. Several studies mentioned in the previous sections demonstrated the efficacy of nanoparticles itself as antimicrobial and as an antimicrobial drug carrier in overcoming the resistance and nontargeted drug delivery issues. These drug delivery systems are now subjected to clinical trials and many of them are already approved for the clinical uses. In the future of the nanoparticle-based antimicrobial system, undoubtedly a significant amount of research is still required to improve the treatment of life-threatening diseases like MDR bacterium such as *Mycobacterium tuberculosis*. The physicochemical properties of nanoparticles can be manipulated for engineered nanoparticles. Much emphasis is given in the field of designing multifunctional nanoparticles which will perform multiple functions such as targeting, diagnostics, imaging, and drug delivery at the same time. Imaging nanoparticles based on its properties (e.g. MRI for magnetic nanoparticles, quantum dots for cellular imaging, photothermal imaging using gold nanoparticles) are gaining huge importance for early detection and real-time monitoring of infectious and noninfectious diseases. Some of the current therapeutic measures such as probiotics and synthetic AMPs have shown auspicious activities against MDRI strains but again there is a significant problem of delivering those therapeutics at a targeted site that requires nanoparticles which are well known for their potential application as payload carriers. Already, there has been a significant shift in combating against deadly infections, from using conventional antibiotic therapies to more advanced nanocarrier-based antimicrobial therapy.

KEYWORDS

- **bactericidal action**
- **gold nanoparticles**
- **iron oxide nanoparticles**
- **liposomes**
- **nanomaterial-based antimicrobial therapy**
- **polymeric nanoparticles**

- **silica nanoparticles**
- **Silver Nanoparticles**
- **solid lipid nanoparticles**
- **TiO_2 nanoparticles**
- **Zinc oxide**

REFERENCES

1. Abeylath, S. C.; Turos, E.; Dickey, S.; Lim, D. V. Glyconanobiotics: Novel Carbohydrated Nanoparticle Antibiotics for MRSA and Bacillus Anthracis. *Bioorganic Med. Chem.* **2008**, *16* (5), 2412–2418. https://doi.org/10.1016/j.bmc.2007.11.052.
2. Abhilash, M. Potential Applications of Nanoparticles. *Int. J. Pharm. Biol. Sci.* **2010**, *1* (1).
3. Adams, C. P.; Walker, K. A.; Obare, S. O.; Docherty, K. M. Size-Dependent Antimicrobial Effects of Novel Palladium Nanoparticles. *PLoS One* **2014**, *9* (1). https://doi.org/10.1371/journal.pone.0085981.
4. Alexandre, Y.; Le Blay, G.; Boisramé-Gastrin, S.; Le Gall, F.; Héry-Arnaud, G.; Gouriou, S.; Vallet, S.; Le Berre, R. Probiotics: A New Way to Fight Bacterial Pulmonary Infections? *Medecineet Maladies Infectieuses.* **2014**, pp. 9–17. https://doi.org/10.1016/j.medmal.2013.05.001.
5. Azam, A.; Ahmed, A. S.; Oves, M.; Khan, M. S.; Memic, A. Size-Dependent Antimicrobial Properties of CuO Nanoparticles against Gram-Positive and -Negative Bacterial Strains. *Int. J. Nanomed.* **2012**, *7*, 3527–3535. https://doi.org/10.2147/IJN.S29020.
6. Badwaik, V. D.; Vangala, L. M.; Pender, D. S.; Willis, C. B.; Aguilar, Z. P.; Gonzalez, M. S.; Paripelly, R.; Dakshinamurthy, R. Size-Dependent Antimicrobial Properties of Sugar Encapsulated Gold Nanoparticles Synthesized by a Green Method. *Nanoscale Res. Lett.* **2012**, *7*. https://doi.org/10.1186/1556-276X-7-623.
7. Balakrishnan, B.; Mohanty, M.; Umashankar, P. R.; Jayakrishnan, A. Evaluation of an in Situ Forming Hydrogel Wound Dressing Based on Oxidized Alginate and Gelatin. *Biomaterials* **2005**, *26* (32), 6335–6342. https://doi.org/10.1016/j.biomaterials.2005.04.012.
8. Balakumaran, M. D.; Ramachandran, R.; Balashanmugam, P.; Mukeshkumar, D. J.; Kalaichelvan, P. T. Mycosynthesis of Silver and Gold Nanoparticles: Optimization, Characterization and Antimicrobial Activity against Human Pathogens. *Microbiol. Res.* **2016**. https://doi.org/10.1016/j.micres.2015.09.009.
9. Brobbey, K. J.; Haapanen, J.; Mäkelä, J. M.; Gunell, M.; Eerola, E.; Rosqvist, E.; Peltonen, J.; Saarinen, J. J.; Tuominen, M.; Toivakka, M. Effect of Plasma Coating on Antibacterial Activity of Silver Nanoparticles. *Thin Solid Films* 2019. https://doi.org/10.1016/j.tsf.2018.12.049.
10. Cao, Q.; Han, X.; Li, L. Enhancement of the Efficiency of Magnetic Targeting for Drug Delivery: Development and Evaluation of Magnet System. *J. Magn. Magn. Mater.* **2011**. https://doi.org/10.1016/j.jmmm.2010.11.058.
11. Cavalli, R.; Gasco, M. R.; Chetoni, P.; Burgalassi, S.; Saettone, M. F. Solid Lipid Nanoparticles (SLN) as Ocular Delivery System for Tobramycin. *Int. J. Pharm.* **2002**. https://doi.org/10.1016/S0378-5173(02)00080-7.
12. Champion, J. A.; Mitragotri, S. Role of Target Geometry in Phagocytosis. *Proc. Natl. Acad. Sci.* **2006**. https://doi.org/10.1073/pnas.0600997103.

13. Colon, G.; Ward, B. C.; Webster, T. J. Increased Osteoblast and Decreased Staphylococcus Epidermidis Functions on Nanophase ZnO and TiO_2. *J. Biomed. Mater. Res.—Part A* **2006**. https://doi.org/10.1002/jbm.a.30789.
14. Cui, Y.; Zhao, Y.; Tian, Y.; Zhang, W.; Lü, X.; Jiang, X. The Molecular Mechanism of Action of Bactericidal Gold Nanoparticles on *Escherichia coli*. *Biomaterials* **2012**. https://doi.org/10.1016/j.biomaterials.2011.11.057.
15. Danilczuk, M.; Lund, A.; Sadlo, J.; Yamada, H.; Michalik, J. Conduction Electron Spin Resonance of Small Silver Particles. *Spectrochim. Acta—Part A Mol. Biomol. Spectrosc.* **2006**. https://doi.org/10.1016/j.saa.2005.05.002.
16. Davies, J. Inactivation of Antibiotics and the Dissemination of Resistance Genes. *Science 264 (80).* **1994**. https://doi.org/10.1126/science.8153624.
17. Debarbieux, L.; Saussereau, E.; Maura, D. [Phagotherapy: A Nightmare for Bacteria, a Dream for Physicians?]. *BiolAujourdhui* **2013**. https://doi.org/10.1051/jbio/2013017. Epub 2013 Dec 13. http://dx.doi.org/10.1051/jbio/2013017.
18. Deol, P.; Khuller, G. K. Lung Specific Stealth Liposomes: Stability, Biodistribution and Toxicity of Liposomal Antitubercular Drugs in Mice. *Biochim. Biophys. Acta—Gen. Subj.* **1997**. https://doi.org/10.1016/S0304-4165(96)00088-8.
19. Dizaj, S. M.; Lotfipour, F.; Barzegar-Jalali, M.; Zarrintan, M. H.; Adibkia, K. Antimicrobial Activity of the Metals and Metal Oxide Nanoparticles. *Mater. Sci. Eng. C.* **20**14. https://doi.org/10.1016/j.msec.2014.08.031.
20. Dizaj, S. M.; Mennati, A.; Jafari, S.; Khezri, K.; Adibkia, K. Antimicrobial Activity of Carbon-Based Nanoparticles. *Adv. Pharm. Bull.* **2015**. https://doi.org/10.5681/apb.2015.003.
21. Drulis-Kawa, Z.; Dorotkiewicz-Jach, A. Liposomes as Delivery Systems for Antibiotics. *Int. J. Pharm.* **20**10. https://doi.org/10.1016/j.ijpharm.2009.11.033.
22. Dwivedi, S.; Wahab, R.; Khan, F.; Mishra, Y. K.; Musarrat, J.; Al-Khedhairy, A. A. Reactive Oxygen Species Mediated Bacterial Biofilm Inhibition via Zinc Oxide Nanoparticles and Their Statistical Determination. *PLoS One* **2014**. https://doi.org/10.1371/journal.pone.0111289.
23. Espuelas, M. S.; Legrand, P.; Loiseau, P. M.; Bories, C.; Barratt, G.; Irache, J. M. In Vitro Antileishmanial Activity of Amphotericin B Loaded in Poly(ε-Caprolactone) Nanospheres. *J. Drug Target.* **20**02. https://doi.org/10.1080/1061186021000060738.
24. Fielding, R. M. Liposomal Drug Delivery: Advantages and Limitations from a Clinical Pharmacokinetic and Therapeutic Perspective. *Clin. Pharmacokinet.* **1991**. https://doi.org/10.2165/00003088-199121030-00001.
25. González-Paredes, A.; Sitia, L.; Ruyra, A.; Morris, C. J.; Wheeler, G. N.; McArthur, M.; Gasco, P. Solid Lipid Nanoparticles for the Delivery of Anti-Microbial Oligonucleotides. *Eur. J. Pharm. Biopharm.* 20**19**. https://doi.org/10.1016/j.ejpb.2018.11.017.
26. Goodman, C. M.; McCusker, C. D.; Yilmaz, T.; Rotello, V. M. Toxicity of Gold Nanoparticles Functionalized with Cationic and Anionic Side Chains. *Bioconjug. Chem.* 20**04**. https://doi.org/10.1021/bc049951i.
27. Grehan, M. J.; Borody, T. J.; Leis, S. M.; Campbell, J.; Mitchell, H.; Wettstein, A. Durable Alteration of the Colonic Microbiota by the Administration of Donor Fecal

Flora. *J. Clin. Gastroenterol.* **2010**. https://doi.org/10.1097/MCG.0b013e3181e5d06b.

28. Gu, H.; Xu, K.; Xu, C.; Xu, B. Biofunctional Magnetic Nanoparticles for Protein Separation and Pathogen Detection. *Chem. Commun.* **2006**. https://doi.org/10.1039/b514130c.
29. Gualerzi, C. O.; Brandi, L.; Fabbretti, A.; Pon, C. L. *Antibiotics: Targets, Mechanisms and Resistance*; 2013. https://doi.org/10.1002/9783527659685.
30. Gupta, U.; Jain, N. K. Non-Polymeric Nano-Carriers in HIV/AIDS Drug Delivery and Targeting. *Adv. Drug Deliv. Rev*. **2010**. https://doi.org/10.1016/j.addr.2009.11.018.
31. Gurr, J. R.; Wang, A. S. S.; Chen, C. H.; Jan, K. Y. Ultrafine Titanium Dioxide Particles in the Absence of Photoactivation Can Induce Oxidative Damage to Human Bronchial Epithelial Cells. *Toxicology* **2005**. https://doi.org/10.1016/j.tox.2005.05.007.
32. Haghighi, F.; Mohammadi, S. R.; Mohammadi, P.; Hosseinkhani, S.; Shidpour, R. Antifungal Activity of TiO2 Nanoparticles and EDTA on Candida Albicans Biofilms. *Infect. Epidemiol. Med.* **2013**. https://doi.org/10.4489/MYCO.2010.38.4.328.
33. He, Q.; Shi, J. Mesoporous Silica Nanoparticle Based Nano Drug Delivery Systems: Synthesis, Controlled Drug Release and Delivery, Pharmacokinetics and Biocompatibility. *J. Mater. Chem.* **2011**. https://doi.org/10.1039/c0jm03851b.
34. Hoshino, A.; Fujioka, K.; Oku, T.; Suga, M.; Sasaki, Y. F.; Ohta, T.; Yasuhara, M.; Suzuki, K.; Yamamoto, K. Physicochemical Properties and Cellular Toxicity of Nanocrystal Quantum Dots Depend on Their Surface Modification. *Nano Lett.* **2004**. https://doi.org/10.1021/nl048715d.
35. Hun, X.; Zhang, Z. A Novel Sensitive Staphylococcal Enterotoxin C1 Fluoroimmunoassay Based on Functionalized Fluorescent Core-Shell Nanoparticle Labels. *Food Chem.* **2007**. https://doi.org/10.1016/j.foodchem.2007.03.068.
36. Ignatova, M.; Starbova, K.; Markova, N.; Manolova, N.; Rashkov, I. Electrospun Nano-Fibre Mats with Antibacterial Properties from Quaternised Chitosan and Poly(Vinyl Alcohol). *Carbohydr. Res.* **2006**. https://doi.org/10.1016/j.carres.2006.05.006.
37. Ivask, A.; Kurvet, I.; Kasemets, K.; Blinova, I.; Aruoja, V.; Suppi, S.; Vija, H.; Kak̈inen, A.; Titma, T.; Heinlaan, M.; et al. Size-Dependent Toxicity of Silver Nanoparticles to Bacteria, Yeast, Algae, Crustaceans and Mammalian Cells in Vitro. *PLoS One* **2014**. https://doi.org/10.1371/journal.pone.0102108.
38. Jain, D.; Banerjee, R. Comparison of Ciprofloxacin Hydrochloride-Loaded Protein, Lipid, and Chitosan Nanoparticles for Drug Delivery. *J. Biomed. Mater. Res. Part B Appl. Biomater*. **2008**. https://doi.org/10.1002/jbm.b.30994.
39. Jaiswal, S.; Duffy, B.; Jaiswal, A. K.; Stobie, N.; McHale, P. Enhancement of the Antibacterial Properties of Silver Nanoparticles Using β-Cyclodextrin as a Capping Agent. *Int. J. Antimicrob. Agents* **2010**. https://doi.org/10.1016/j.ijantimicag.2010.05.006.
40. Jia, Z.; Xu, W. Synthesis and Antibacterial Activities of Quaternary Ammonium Salt of Chitosan. *Carbohydr. Res*. **2001**.
41. Jo, Y.-K.; Kim, B. H.; Jung, G. Antifungal Activity of Silver Ions and Nanoparticles on Phytopathogenic Fungi. *Plant Dis*. **2009**. https://doi.org/10.1094/pdis-93-10-1037.
42. Kaweeteerawat, C.; Chang, C. H.; Roy, K. R.; Liu, R.; Li, R.; Toso, D.; Fischer,

H.; Ivask, A.; Ji, Z.; Zink, J. I.; et al. Cu Nanoparticles Have Different Impacts in *Escherichia coli* and Lactobacillus Brevis than Their Microsized and Ionic Analogues. *ACS Nano* **2015**. https://doi.org/10.1021/acsnano.5b02021.

43. Kim, J. S.; Kuk, E.; Yu, K. N.; Kim, J. H.; Park, S. J.; Lee, H. J.; Kim, S. H.; Park, Y. K.; Park, Y. H.; Hwang, C. Y.; et al. Antimicrobial Effects of Silver Nanoparticles. *Nanomed. Nanotechnol. Biol. Med.* **2007**. https://doi.org/10.1016/j.nano.2006.12.001.
44. Kim, J. Y.; Lee, J. K.; Lee, T. S.; Park, W. H. Synthesis of Chitooligosaccharide Derivative with Quaternary Ammonium Group and Its Antimicrobial Activity against Streptococcus Mutans. *Int. J. Biol. Macromol.* **2003**. https://doi.org/10.1016/S0141-8130(03)00021-7.
45. Kohli, A. K.; Alpar, H. O. Potential Use of Nanoparticles for Transcutaneous Vaccine Delivery: Effect of Particle Size and Charge. *Int. J. Pharm.* **2004**. https://doi.org/10.1016/j.ijpharm.2003.10.038.
46. Kumari, A.; Yadav, S. K.; Yadav, S. C. Biodegradable Polymeric Nanoparticles Based Drug Delivery Systems. *Colloids Surf. B Biointerfaces.* **2010**. https://doi.org/10.1016/j.colsurfb.2009.09.001.
47. Kundu, S.; Liang, H. Polyelectrolyte-Mediated Non-Micellar Synthesis of Monodispersed "aggregates" of Gold Nanoparticles Using a Microwave Approach. *Colloids Surfaces A Physicochem. Eng. Asp.* **2008**. https://doi.org/10.1016/j.colsurfa.2008.07.043.
48. Lakshmi Prasanna, V.; Vijayaraghavan, R. Insight into the Mechanism of Antibacterial Activity of ZnO: Surface Defects Mediated Reactive Oxygen Species Even in the Dark. *Langmuir* **2015**. https://doi.org/10.1021/acs.langmuir.5b02266.
49. Lemire, J. A.; Harrison, J. J.; Turner, R. J. Antimicrobial Activity of Metals: Mechanisms, Molecular Targets and Applications. *Nat. Rev. Microbiol.* **2013**. https://doi.org/10.1038/nrmicro3028.
50. Levy, S. B.; Bonnie, M. Antibacterial Resistance Worldwide: Causes, Challenges and Responses. *Nat. Med.* **2004**. https://doi.org/10.1038/nm1145.
51. Li, X.; Lu, Z.; Li, Q. Multilayered Films Incorporating CdTe Quantum Dots with Tunable Optical Properties for Antibacterial Application. *Thin Solid Films* **2013**. https://doi.org/10.1016/j.tsf.2013.09.088.
52. Liu, J.; Li, S.; Fang, Y.; Zhu, Z. Boosting Antibacterial Activity with Mesoporous Silica Nanoparticles Supported Silver Nanoclusters. *J. Colloid Interface Sci.* **2019**. https://doi.org/10.1016/j.jcis.2019.08.009.
53. Liu, M.; Gan, L.; Chen, L.; Zhu, D.; Xu, Z.; Hao, Z.; Chen, L. A Novel Liposome-Encapsulated Hemoglobin/Silica Nanoparticle as an Oxygen Carrier. *Int. J. Pharm.* **2012**. https://doi.org/10.1016/j.ijpharm.2012.02.019.
54. Syndor, M.; Perl, T. M. Hospital Epidemiology and Infection Control in Acute-Care Settings. *Clin. Microbiol. Rev.* **2011**. https://doi.org/10.1128/CMR.00027-10.
55. Manju, V.; Dhandapani, P.; GurusamyNeelavannan, M.; Maruthamuthu, S.; Berchmans, S.; Palaniappan, A. Tunable Release of Clavam from Clavam Stabilized Gold Nanoparticles Design, Characterization and Antimicrobial Study. *Mater. Sci. Eng. C* **2015**. https://doi.org/10.1016/j.msec.2015.01.047.
56. Marei, N.; Elwahy, A. H. M.; Salah, T. A.; El Sherif, Y.; El-Samie, E. A. Enhanced Antibacterial Activity of Egyptian Local Insects' Chitosan-Based Nanoparticles Loaded with

Ciprofloxacin-HCl. *Int. J. Biol. Macromol.* 20**19**. https://doi.org/10.1016/j.ijbiomac.2018.12.204.
57. Mayor, S.; Pagano, R. E. Pathways of Clathrin-Independent Endocytosis. *Nat. Rev. Mol. Cell Biol.* **20**07. https://doi.org/10.1038/nrm2216.
58. Misra, S. K.; Dybowska, A.; Berhanu, D.; Luoma, S. N.; Valsami-Jones, E. The Complexity of Nanoparticle Dissolution and Its Importance in Nanotoxicological Studies. *Sci. Total Environ.* **2012**. https://doi.org/10.1016/j.scitotenv.2012.08.066.
59. Mok, H.; Zhang, M. Superparamagnetic Iron Oxide Nanoparticle-Based Delivery Systems for Biotherapeutics. *Expert Opin. Drug Deliv*. **2012**. https://doi.org/10.1517/17425247.2013.747507.
60. Mroz, P.; Tegos, G. P.; Gali, H.; Wharton, T.; Sarna, T.; Hamblin, M. R. Photodynamic Therapy with Fullerenes. *Photochem. Photobiol. Sci.* **20**07. https://doi.org/10.1039/b711141j.
61. Mukherjee, S.; Ray, S.; Thakur, R. Solid Lipid Nanoparticles: A Modern Formulation Approach in Drug Delivery System. *Indian J. Pharm. Sci.* **2009**. https://doi.org/10.4103/0250-474x.57282.
62. Muzzarelli, R. A. A.; Guerrieri, M.; Goteri, G.; Muzzarelli, C.; Armeni, T.; Ghiselli, R.; Cornelissen, M. The Biocompatibility of Dibutyryl Chitin in the Context of Wound Dressings. *Biomaterials* **2005**. https://doi.org/10.1016/j.biomaterials.2005.03.006.
63. Nair, R.; Kumar, K. S. A.; Priya, K. V.; Sevukarajan, M. Recent Advances in Solid Lipid Nanoparticle Based Drug Delivery Systems. *J. Biomed. Sci. Res.* **2011**.
64. Nardello-Rataj, V.; Leclercq, L. Encapsulation of Biocides by Cyclodextrins: Toward Synergistic Effects Against Pathogens. *Beilstein J. Org. Chem.* **20**14. https://doi.org/10.3762/bjoc.10.273.
65. Navarro, E.; Wagner, B.; Odzak, N.; Sigg, L.; Behra, R. Effects of Differently Coated Silver Nanoparticles on the Photosynthesis of *Chlamydomonas reinhardtii*. *Environ. Sci. Technol.* **2015**. https://doi.org/10.1021/acs.est.5b01089.
66. Nemeth, J.; Oesch, G.; Kuster, S. P. Bacteriostatic versus Bactericidal Antibiotics for Patients with Serious Bacterial Infections: Systematic Review and Meta-Analysis. *J. Antimicrob. Chemother.* 20**15**. https://doi.org/10.1093/jac/dku379.
67. Neuberger, T.; Schöpf, B.; Hofmann, H.; Hofmann, M.; Von Rechenberg, B. Superparamagnetic Nanoparticles for Biomedical Applications: Possibilities and Limitations of a New Drug Delivery System. *J. Magn. Magn. Mater.* 2005. https://doi.org/10.1016/j.jmmm.2005.01.064.
68. Nikaido, H. Prevention of Drug Access to Bacterial Targets: Permeability Barriers and Active Efflux. *Science 264(5157).* **1994**. https://doi.org/10.1126/science.8153625.
69. Ocampo, P. S.; Lázár, V.; Papp, B.; Arnoldini, M.; Zur Wiesch, P. A.; Busa-Fekete, R.; Fekete, G.; Pál, C.; Ackermann, M.; Bonhoeffer, S. Antagonism between Bacteriostatic and Bactericidal Antibiotics Is Prevalent. *Antimicrob. Agents Chemother.* 20**14**. https://doi.org/10.1128/AAC.02463-14.
70. Ocsoy, I.; Paret, M. L.; Ocsoy, M. A.; Kunwar, S.; Chen, T.; You, M.; Tan, W. Nanotechnology in Plant Disease Management: DNA-Directed Silver Nanoparticles on Graphene Oxide as an Antibacterial against *Xanthomonas*

perforans. *ACS Nano* **2013**. https://doi.org/10.1021/nn4034794.

71. Omri, A.; Suntres, Z. E.; Shek, P. N. Enhanced Activity of Liposomal Polymyxin B against *Pseudomonas aeruginosa* in a Rat Model of Lung Infection. *Biochem. Pharmacol.* **2002**. https://doi.org/10.1016/S0006-2952(02)01346-1.
72. Onyeji, C. O.; Nightingale, C. H.; Marangos, M. N. Enhanced Killing of Methicillin-Resistant Staphylococcus Aureus in Human Macrophages by Liposome-Entrapped Vancomycin and Teicoplanin. *Infection***1994**. https://doi.org/10.1007/BF01715542.
73. Padmavathy, N.; Vijayaraghavan, R. Enhanced Bioactivity of ZnO Nanoparticles An Antimicrobial Study. *Sci. Technol. Adv. Mater.* **2008**. https://doi.org/10.1088/1468-6996/9/3/035004.
74. Pal, S.; Tak, Y. K.; Song, J. M. Does the Antibacterial Activity of Silver Nanoparticles Depend on the Shape of the Nanoparticle? A Study of the Gram-Negative Bacterium Escherichia Coli. *J. Biol. Chem.* **2015**. https://doi.org/10.1128/AEM.02218-06.
75. Palza, H. Antimicrobial Polymers with Metal Nanoparticles. *Int. J. Mol. Sci.* **2015**, pp. 2099–2116. https://doi.org/10.3390/ijms16012099.
76. Pandey, R.; Khuller, G. K. Solid Lipid Particle-Based Inhalable Sustained Drug Delivery System against Experimental Tuberculosis. *Tuberculosis* **2005**. https://doi.org/10.1016/j.tube.2004.11.003.
77. Park, E. J.; Yi, J.; Kim, Y.; Choi, K.; Park, K. Silver Nanoparticles Induce Cytotoxicity by a Trojan-Horse Type Mechanism. *Toxicol. Vitr.* **2010**. https://doi.org/10.1016/j.tiv.2009.12.001.
78. Patel, M. B.; Harikrishnan, U.; Valand, N. N.; Modi, N. R.; Menon, S. K. Novel Cationic Quinazolin-4(3H)-One Conjugated Fullerene Nanoparticles as Antimycobacterial and Antimicrobial Agents. *Arch. Pharm. (Weinheim).* **2013**. https://doi.org/10.1002/ardp.201200371.
79. Peters, B. M.; Shirtliff, M. E.; Jabra-Rizk, M. A. Antimicrobial Peptides: Primeval Molecules or Future Drugs? *PLoS Pathog.* **2010**. https://doi.org/10.1371/journal.ppat.1001067.
80. Petersen, E. J.; Nelson, B. C. Mechanisms and Measurements of Nanomaterial-Induced Oxidative Damage to DNA. *Anal. Bioanal. Chem.* **2010**. https://doi.org/10.1007/s00216-010-3881-7.
81. Premanathan, M.; Karthikeyan, K.; Jeyasubramanian, K.; Manivannan, G. Selective Toxicity of ZnO Nanoparticles toward Gram-Positive Bacteria and Cancer Cells by Apoptosis through Lipid Peroxidation. *Nanomed. Nanotechnol. Biol. Med.* **2011**. https://doi.org/10.1016/j.nano.2010.10.001.
82. Raghupathi, K. R.; Koodali, R. T.; Manna, A. C. Size-Dependent Bacterial Growth Inhibition and Mechanism of Antibacterial Activity of Zinc Oxide Nanoparticles. *Langmuir* **2011**. https://doi.org/10.1021/la104825u.
83. Ranghar, S.; Sirohi, P.; Verma, P.; Agarwal, V. Nanoparticle-Based Drug Delivery Systems: Promising Approaches against Infections. *Brazilian Arch. Biol. Technol.* **2014**. https://doi.org/10.1590/S1516-89132013005000011.
84. Rout, A.; Jena, P.K.; Sahoo, D.; and Bindhani, B.K.;. Green Synthesis of Silver Nanoparticles of different Shapes and its Antibacterial Activity against Escherichia coli.Int.J.Curr.Microbiol. App. Sci. 3, 2014; pp 374-383.
85. S. Roy, A.; Parveen, A.; R. Koppalkar, A.; Prasad, M. V. N. A. Effect of Nano Titanium Dioxide with Different Antibiotics against

Methicillin-Resistant Staphylococcus Aureus. *J. Biomater. Nanobiotechnol.* **2010**. https://doi.org/10.4236/jbnb.2010.11005.

86. Sabella, S.; Carney, R. P.; Brunetti, V.; Malvindi, M. A.; Al-Juffali, N.; Vecchio, G.; Janes, S. M.; Bakr, O. M.; Cingolani, R.; Stellacci, F.; et al. A General Mechanism for Intracellular Toxicity of Metal-Containing Nanoparticles. *Nanoscale* **2014**. https://doi.org/10.1039/c4nr01234h.
87. Sampath Kumar, T. S.; Madhumathi, K. Antibacterial Potential of Nanobioceramics Used as Drug Carriers. In *Handbook of Bioceramics and Biocomposites*; 2016; pp. 1333–1373. https://doi.org/10.1007/978-3-319-12460-5_58.
88. Saylor, C.; Dadachova, E.; Casadevall, A. Monoclonal Antibody-Based Therapies for Microbial Diseases. *Vaccine*. **20**09. https://doi.org/10.1016/j.vaccine.2009.09.105.
89. Seleem, M. N.; Munusamy, P.; Ranjan, A.; Alqublan, H.; Pickrell, G.; Sriranganathan, N. Silica-Antibiotic Hybrid Nanoparticles for Targeting Intracellular Pathogens. *Antimicrob. Agents Chemother*. **2009**. https://doi.org/10.1128/AAC.00815-09.
90. Sevinç, B. A.; Hanley, L. Antibacterial Activity of Dental Composites Containing Zinc Oxide Nanoparticles. *J. Biomed. Mater. Res. Part B Appl. Biomater*. **2010**, *94* (1), 22–31. https://doi.org/10.1002/jbm.b.31620.
91. Smith, A. "Bacterial resistance to antibiotics", In: Denyer S. P., Hodges N. A., Gorman S. P. (Eds.), Hugo and Russell's Pharmaceutical Microbiology. Blackwell Science, Malden, 2004; pp. 220–232.
92. Sohm, B.; Immel, F.; Bauda, P.; Pagnout, C. Insight into the Primary Mode of Action of TiO2 Nanoparticles on *Escherichia coli* in the Dark. *Proteomics* **2015**. https://doi.org/10.1002/pmic.201400101.
93. Sondi, I.; Salopek-Sondi, B. Silver Nanoparticles as Antimicrobial Agent: A Case Study on E. Coli as a Model for Gram-Negative Bacteria. *J. Colloid Interface Sci*. **2004**. https://doi.org/10.1016/j.jcis.2004.02.012.
94. Soothill, J. Use of Bacteriophages in the Treatment of Pseudomonas Aeruginosa Infections. *Expert Rev. Anti Infect. Ther*. **20**13. https://doi.org/10.1586/14787210.2013.826990.
95. Souto, E. B.; Wissing, S. A.; Barbosa, C. M.; Müller, R. H. Development of a Controlled Release Formulation Based on SLN and NLC for Topical Clotrimazole Delivery. *Int. J. Pharm*. **2004**. https://doi.org/10.1016/j.ijpharm.2004.02.032.
96. Summers, W. C. HERAPY William C. Summers. *Annu. Rev. Microbiol.* 20**01**, *55*, 437–451.Sydnor, E. R.
97. Takemoto, K.; Yamamoto, Y.; Ueda, Y.; Sumita, Y.; Yoshida, K.; Niki, Y. Comparative Studies on the Efficacy of AmBisome and Fungizone in a Mouse Model of Disseminated Aspergillosis. *J. Antimicrob. Chemother*. **2004**. https://doi.org/10.1093/jac/dkh055.
98. Taubes, G. The Bacteria Fight Back. *Science*. **20**08. https://doi.org/10.1126/science.321.5887.356.
99. Tejero-Sariñena, S.; Barlow, J.; Costabile, A.; Gibson, G. R.; Rowland, I. In Vitro Evaluation of the Antimicrobial Activity of a Range of Probiotics against Pathogens: Evidence for the Effects of Organic Acids. *Anaerobe* **2012**. https://doi.org/10.1016/j.anaerobe.2012.08.004.
100. Tian, Y.; Qi, J.; Zhang, W.; Cai, Q.; Jiang, X. Facile, One-Pot Synthesis, and Antibacterial Activity of Mesoporous Silica Nanoparticles Decorated with Well-Dispersed Silver Nanoparticles.

ACS Appl. Mater. Interfaces **2014**. https://doi.org/10.1021/am5026424.
101. Tiwari, P.; Vig, K.; Dennis, V.; Singh, S. Functionalized Gold Nanoparticles and Their Biomedical Applications. *Nanomaterials* **2011**. https://doi.org/10.3390/nano1010031.
102. Tong, T.; Shereef, A.; Wu, J.; Binh, C. T. T.; Kelly, J. J.; Gaillard, J. F.; Gray, K. A. Effects of Material Morphology on the Phototoxicity of Nano-TiO_2 to Bacteria. *Environ. Sci. Technol.* **2013**. https://doi.org/10.1021/es403079h.
103. Torchilin, V. Lipid-Core Micelles for Targeted Drug Delivery. *Curr. Drug Deliv.* **2005**. https://doi.org/10.2174/1567201054370221.
104. Torchilin, V. P. Recent Advances with Liposomes as Pharmaceutical Carriers. *Nature Reviews Drug Discovery*. **2005**. https://doi.org/10.1038/nrd1632.
105. Touati, D. Iron and Oxidative Stress in Bacteria. *Arch. Biochem. Biophys.* **2000**. https://doi.org/10.1006/abbi.1999.1518.
106. Tsao, N.; Luh, T. Y.; Chou, C. K.; Chang, T. Y.; Wu, J. J.; Liu, C. C.; Lei, H. Y. In Vitro Action of Carboxyfullerene. *J. Antimicrob. Chemother.* **2002**. https://doi.org/10.1093/jac/49.4.641.
107. Vallet-Regí, M. Nanostructured Mesoporous Silica Matrices in Nanomedicine. In *Journal of Internal Medicine*; 2010. https://doi.org/10.1111/j.1365-2796.2009.02190.x.
108. Varshosaz, J.; Ghaffari, S.; Khoshayand, M. R.; Atyabi, F.; Azarmi, S.; Kobarfard, F. Development and Optimization of Solid Lipid Nanoparticles of Amikacin by Central Composite Design. *J. Liposome Res.* **2010**. https://doi.org/10.3109/08982100903103904.
109. Veiseh, O.; Gunn, J. W.; Zhang, M. Design and Fabrication of Magnetic Nanoparticles for Targeted Drug Delivery and Imaging. *Advanced Drug Delivery Reviews*. **2010**. https://doi.org/10.1016/j.addr.2009.11.002.
110. Vivero-Escoto, J. L.; Slowing, I. I.; Lin, V. S. Y.; Trewyn, B. G. Mesoporous Silica Nanoparticles for Intracellular Controlled Drug Delivery. *Small*. **2010**. https://doi.org/10.1002/smll.200901789.
111. Wang, J. X.; Wen, L. X.; Wang, Z. H.; Chen, J. F. Immobilization of Silver on Hollow Silica Nanospheres and Nanotubes and Their Antibacterial Effects. *Mater. Chem. Phys.* **2006**. https://doi.org/10.1016/j.matchemphys.2005.06.045.
112. Wang, L.; Hu, C.; Shao, L. The Antimicrobial Activity of Nanoparticles: Present Situation and Prospects for the Future. *Int. J. Nanomed.* **2017**. https://doi.org/10.2147/IJN.S121956.
113. Wang, Y.; Yin, M.; Lin, X.; Li, L.; Li, Z.; Ren, X.; Sun, Y. Tailored Synthesis of Polymer-Brush-Grafted Mesoporous Silicas with N-Halamine and Quaternary Ammonium Groups for Antimicrobial Applications. *J. Colloid Interface Sci.* **2019**. https://doi.org/10.1016/j.jcis.2018.08.080.
114. Water, J. J.; Smart, S.; Franzyk, H.; Foged, C.; Nielsen, H. M. Nanoparticle-Mediated Delivery of the Antimicrobial Peptide Plectasin against Staphylococcus Aureus in Infected Epithelial Cells. *Eur. J. Pharm. Biopharm.* **2015**. https://doi.org/10.1016/j.ejpb.2015.02.009.
115. Wright, G. D. Q&A: Antibiotic Resistance: Where Does It Come from and What Can We Do about It? *BMC Biol.* **2010**. https://doi.org/10.1186/1741-7007-8-123.
116. Wu, B.; Wang, Y.; Lee, Y. H.; Horst, A.; Wang, Z.; Chen, D. R.; Sureshkumar, R.; Tang, Y. J. Comparative Eco-Toxicities of Nano-ZnO Particles under Aquatic and Aerosol Exposure

Modes. *Environ. Sci. Technol.* **2010**. https://doi.org/10.1021/es9030497.

117. Xie, S.; Tao, Y.; Pan, Y.; Qu, W.; Cheng, G.; Huang, L.; Chen, D.; Wang, X.; Liu, Z.; Yuan, Z. Biodegradable Nanoparticles for Intracellular Delivery of Antimicrobial Agents. *Journal of Controlled Release*. **2014**. https://doi.org/10.1016/j.jconrel.2014.05.034.
118. Yacoby, I.; Bar, H.; Benhar, I. Targeted Drug-Carrying Bacteriophages as Antibacterial Nanomedicines. *Antimicrob. Agents Chemother.* **2007**. https://doi.org/10.1128/AAC.00163-07.
119. Yang, S. T.; Wang, X.; Jia, G.; Gu, Y.; Wang, T.; Nie, H.; Ge, C.; Wang, H.; Liu, Y. Long-Term Accumulation and Low Toxicity of Single-Walled Carbon Nanotubes in Intravenously Exposed Mice. *Toxicol. Lett.* **2008**. https://doi.org/10.1016/j.toxlet.2008.07.020.
120. Yokoyama, R.; Suzuki, S.; Shirai, K.; Yamauchi, T.; Tsubokawa, N.; Tsuchimochi, M. Preparation and Properties of Biocompatible Polymer-Grafted Silica Nanoparticle. *Eur. Polym. J.* **2006**. https://doi.org/10.1016/j.eurpolymj.2006.08.015.
121. Zahran, M.; Marei, A. H. Innovative Natural Polymer Metal Nanocomposites and Their Antimicrobial Activity. *International Journal of Biological Macromolecules*. **2019**. https://doi.org/10.1016/j.ijbiomac.2019.06.114.
122. Zarei, M.; Jamnejad, A.; Khajehali, E. Antibacterial Effect of Silver Nanoparticles against Four Foodborne Pathogens. *Jundishapur J. Microbiol.* **2014**. https://doi.org/10.5812/jjm.8720.
123. Zhang, L.; Pornpattananangkul, D.; Hu, C.-M.; Huang, C.-M. Development of Nanoparticles for Antimicrobial Drug Delivery. *Curr. Med. Chem.* **2010**. https://doi.org/10.2174/092986710790416290.
124. Zhang, Y.; ShareenaDasari, T. P.; Deng, H.; Yu, H. Antimicrobial Activity of Gold Nanoparticles and Ionic Gold. *J. Environ. Sci. Heal. Part C Environ. Carcinog. Ecotoxicol. Rev.* **2015**. https://doi.org/10.1080/10590501.2015.1055161.
125. Zhu, M.; Wang, H.; Liu, J.; He, H.; Hua, X.; He, Q.; Zhang, L.; Ye, X.; Shi, J. A Mesoporous Silica Nanoparticulate/β-TCP/BG Composite Drug Delivery System for Osteoarticular Tuberculosis Therapy. *Biomaterials* **2011**. https://doi.org/10.1016/j.biomaterials.2010.11.025.
126. Hun, X.; Zhang, Z. A Novel Sensitive Staphylococcal Enterotoxin C1 Fluoroimmunoassay Based on Functionalized Fluorescent Core-Shell Nanoparticle Labels. *Food Chem.* **2007**, 105(4), 1623–1629.
127. Yameen, B., Choi, W. I., Vilos, C., Swami, A., Shi, J., Farokhzad, O. C. Insight into nanoparticle cellular uptake and intracellular targeting, Journal of Controlled Release, Volume 190, 2014, 485–499, https://doi.org/10.1016/j.jconrel.2014.06.038.
128. Misra, S. K., Dybowska, A., Berhanu, D., Luoma, S. N., Valsami-Jones, E. The complexity of nanoparticle dissolution and its importance in nanotoxicological studies, Science of The Total Environment, Volume 438, 2012, 225-232, https://doi.org/10.1016/j.scitotenv.2012.08.066.

CHAPTER 9

Bioinspired Nanofibers: Preparation and Characterization

POORNIMA BALAN*, JANANI INDRAKUMAR, ANBUTHIRUSELVAN SOLAIMUTHU, PADMAJA MURALI, and PURNA SAI KORRAPATI*

Biological Materials Laboratory, CSIR-Central Leather Research Institute, Adyar, Chennai, Tamil Nadu 600020, India

**Corresponding author.*
E-mail: purnasai@clri.res.in, purnasaik.clri@gmail.com

ABSTRACT

The contemporary research community on nanostructured materials and tissue engineering has evidenced its potential use and impact in a variety of fields, such as biological sciences, clinical diagnosis, and electronics and communication engineering. Among the variety of nanostructured materials, nanofibers pave the way for advancements, new level of versatility, and broader range of applications. Nanofibers can be fabricated from natural polymers or synthetic polymers or inorganic polymers. The secondary step of fabrication of nanofiber involves the functionalization and surface modification to exhibit the specific biochemical characteristics to the area of application. In this chapter, nanofiber fabrication through various different methods has been discussed in detail with special reference to electrospinning methods due to its myriad of tunable properties and applications. The factors entailed for the optimal synthesis of nanofibers are explained meticulously. Furthermore, we have described the various physicochemical characterizations required to understand the morphology, structure, strength, chemical composition, and interaction of functional groups of the nanofibers. This chapter summarizes in detail about the wide range of fabrication techniques of nanofibers and the spectrum of characterization methods such as physical, chemical, mechanical, and biological for their effective application in regenerative medicine.</Abs>

9.1 INTRODUCTION

Nanotechnology has revolutionized the global scientific approach in almost every field with special emphasis on medicine, energy applications, environmental science, information technology, and so on. The term nano typically refers to a unit which is one billionth a dimension (10^{-9}), although dimensions ranging from 100 to 500 nm are classified as nano in biology [1]. The large surface-area-to-volume ratio has led to the enormous transformation over the microforms. Further, the nano size simulates majority of the naturally occurring systems with special reference to biology such as cellular architecture and functional processes and hence, more became predominant in medicine and hence terms biomimetic and bioinspired have fascinated every individual in the present decade [2]. Biomimetic and bioinspired are the terminologies frequently used which essentially deal with either mimicking the natural occurring conditions or designing/modifying based on the structural and functional natural systems. Several of the nanotechnologies have been in vogue based on the understanding of the natural systems like structural colorations, stain and scratch resistant systems, self-cleaning, water filtration, catalysis, and so on, which led to wide range of nanomaterial applications ranging from health care to energy and aerospace manufacturing industry [3, 4].

Generally, nanomaterials are classified into four types namely carbon based, metal based, dendrimers, and composites. Carbon-based materials are mostly of carbon in the form of ellipsoids, hollow spheres (fullerenes), and tubes (nanotubes). Metal-based nanomaterials are usually quantum dots, metal oxides, nanosilver, nanogold, and so on. Dendrimers are nano-sized polymers built from branched units designed to perform specialized functions such as catalysis, drug delivery, and so on. Composites are a combination of one or more of metal based or polymeric based or carbon based and designed to utilize the multiple properties [5].

The most significant application in medicine would depend on the development of more predictive, preventive, and regenerative tools enabling early detection, multiplexed diagnostics, imaging, drug delivery, biomarker patterning, and discovery for the betterment of society. However, the major setback in the scientific advances of nanomedicine is due to the toxicological issues for health and environment. Hence, nanotechnology is on a tremendous stride to overcome this setback using a combinatorial approach of bio-informatics and biotechnology. The application of biomaterials and medical devices is becoming highly versatile resulting in the need for a symbiotic and interdisciplinary understanding of polymer chemistry, materials science, biomedical engineering, biophysics, and biochemistry leading to technological breakthroughs [6].

Material designing for a specific application is an ardent task since there is no universal material that can satisfy all the conditions required for different functionalities which depends on the nature of the material, structure, properties, and behavior in different microenvironments. Further, this is more predominant or conspicuous in the design of composite materials owing to the inter- and intra-molecular interactions between the blends. The choice of the materials as well as the methods of preparation highly influence the functional attributes. The speculation of functional attributes becomes more significant with reference to biological or biomedical applications owing to the complex and overlapping developmental phenomena.

This chapter, therefore, highlights the structural concepts of nanofibers that significantly influence the functional applications.

9.2 METHOD OF NANOFIBERS FABRICATION

Fabrication methods for nanofibers are distinct with the choice of material used like physical, chemical, electrostatic, and thermal fabrication approach. The corresponding methods for nanofibers preparation are briefly emphasized in the following sections (Figure 9.1).

9.2.1 BICOMPONENT EXTRUSION (ISLAND-IN-THE-SEA)

Bicomponent nanofibers are defined as squeezing two different polymer constituents through the same fiber spinneret. This form of nanofibers is also called as matrix fibers because in the cross-sectional view, they appear as one polymer injected into a matrix of a second polymer. In the island area, either uniform or nonuniform-sized fibers may be present. The core-sheath, islands-in-the-sea, eccentric, and segmented pie types of bicomponent nanofibers are shown in Figure 9.2.

Essentially, these fibers are spun from the amalgam of two polymeric solutions in the appropriate ratio where first solution is suspended as drop-by-drop in the second solution. Rapid cooling of the fiber under the spinneret holes improves the production of fibers. In some of the polymers, high concentration mixture

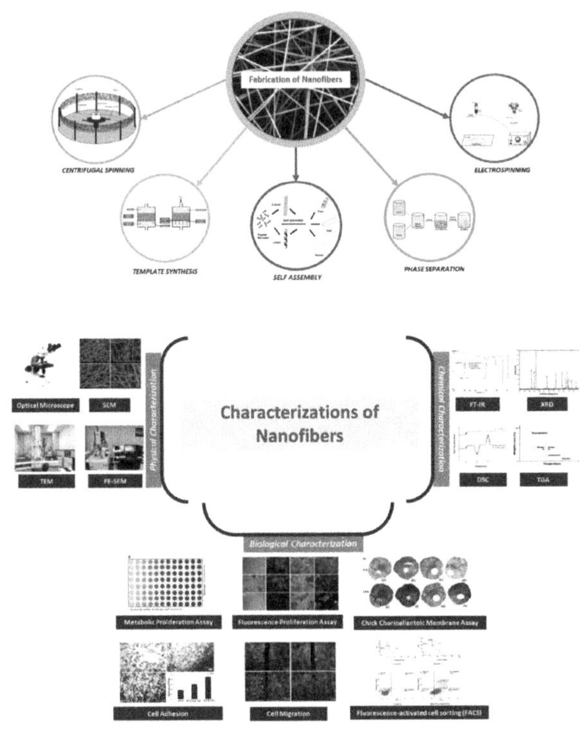

FIGURE 9.1 Methods for fabrication of nanofibers.

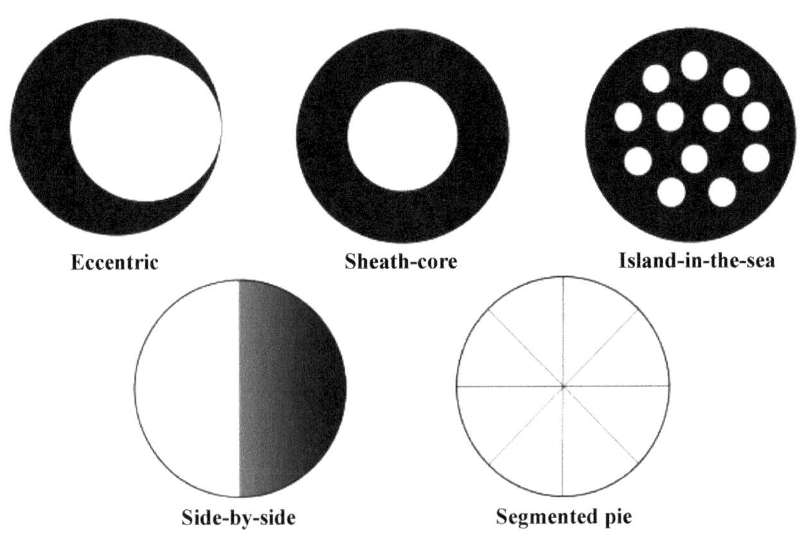

FIGURE 9.2 Cross-sections of bicomponent fibers (Island-in-the-sea). (Adapted from Ref. [7].)

makes it difficult for spinning because of heterogeneous blending. By using heat, solvents or a chemical or using any mechanical device, separation of individual fibers would be possible. The polymeric solutions are dispatched through a spinneret hole, split by a blade edge which arranges the two constituents by segments [7].

The pipe-in-pipe method is the most versatile technique for bicomponent nanofibers where the cascade of one constituent envelopes the other at the edge of the pipe. Nakata et al. [8] produced 39 diameter measurement of contentious polyethylene terephthalate nanofibers by sea–island type the flow-drawn fiber with further drawing and extrusion of the sea constituent.

9.2.2 PHASE SEPARATION (SOL–GEL PROCESS)

The phase separation process is a wet-chemical technique that a polymer solution is prepared with a solvent before initializing gelation process. The process consists of dissolution, gelation, solvent extraction, and freeze drying (Figure 9.3). The basic solid phase or a nanoscale porous foam is formed as a result. The crucial process involved in this method is the separation of phases corresponding to the physical inconsistency of the polymer solution. Later, the solvent phase is then extracted by omitting the residual phase (Figure 9.4). A detailed method for poly(L-lactic) acid (PLLA) nanofibers producing has been explained by Ma and Zhang [9].

FIGURE 9.3 Step-by-step process of phase separation technique.

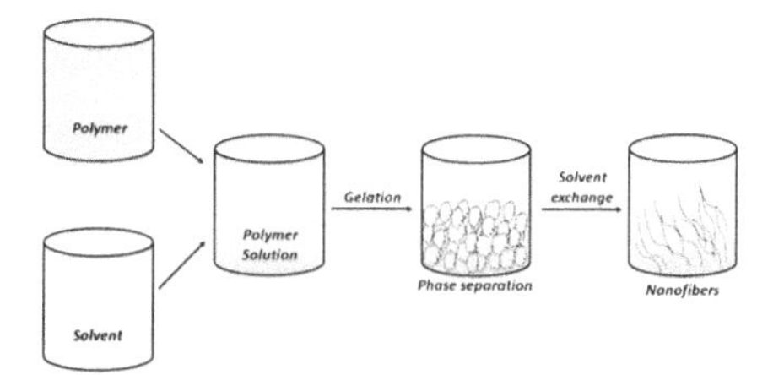

FIGURE 9.4 A schematic diagram of nanofiber formation by phase separation.

9.2.3 TEMPLATE SYNTHESIS

Template synthesis associates the use of a preconstruct to make a preferred structure or material (Figure 9.5). Thus, the casting technique and DNA replication can be believed as template-based synthesis. For, instance, the nanofiber fabrication by Feng et al. [10]. The template is mentioned to be metal oxide membrane with nanoscale diameter thickness pores. Freighting of the pressurized water along with a perforated membrane control develops extrusion of the polymer solution by contacting with solidifying agent and provides nanofibers whose diameters are respective to the pore size of the membrane. This is a commonly used method mostly to produce inorganic nanofibers, for example, carbon nanotubes and nanofibers or conductive polypyrrole (PPy) and polyaniline (PANI) [11].

9.2.4 SELF-ASSEMBLY

Self-assembly is a process in which individual and pre-existing components organize themselves into desired patterns and functions. A variety of peptides and proteins have been exhibited to make very stable nanofibers. These fibers are symmetrically ordered and acquire remarkable integrity and some helical periodicity. The diameter and the surface structure of the nanofibers can be altered by controlling the molecular structures [12]. The potential applications of these composite nanofibers include electronics, optics, sensing, and biomedical engineering. Figure 9.6 illustrates the self-assembly of various types of peptide materials.

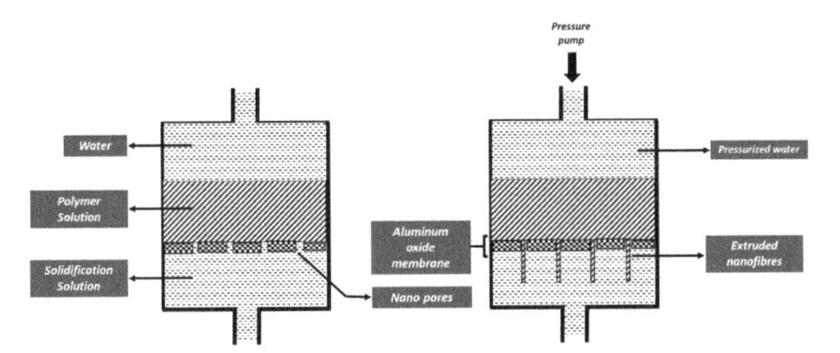

FIGURE 9.5 Schematic diagram showing the template preparation for nanofibers.

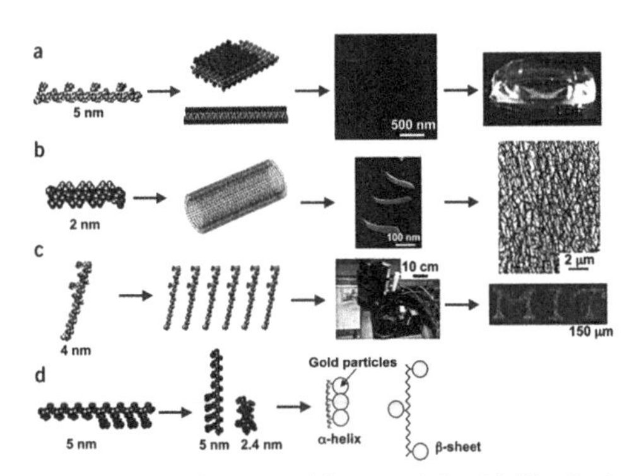

FIGURE 9.6 Self-assembly of various peptide materials. (a) The ionic self-complementary peptide, (b) a type of surfactant-like peptide, (c) surface nanocoating peptide, and (d) molecular switch peptide. (Reprinted with the permission from Zhang, S. Fabrication of novel materials through molecular self-assembly. Nat. Biotechnol. 21, 1171-1178, Nature Biotechnology November 2003 DOI: 10.1038/nbt874 Source: PubMed.)

9.2.5 DRAWING

Nanofibers production by virtue of drawing process can be defined as dry spinning at molecular level of polymer solution. The polymer material should be highly visco-elastic to attain high degree of fiber producibility but remain sufficiently solid to detain the developed stress during pulling. This method requires silicon dioxide (SiO_2) surface, a micropipette, and a micromanipulator to make nanofibers. By using micromanipulator, the micropipette was dipped into the droplet of polymer solution. The micropipette will be removed from contact with the solution droplet at the speed of approximately 1×10^{-4} ms^{-1} to stretch nanofibers. The drawn nanofiber was then deposited on the surface by using tip of micropipette. The drawing of nanofiber was frequently repeated on every droplet. The viscosity at the perimeter of the droplet of polymer solution is increased by evaporation of solvent. The viscoelastic behavior of the polymer material should be adequately cohesive with strong deformations to sustain the stress during fiber drawing [13].

9.2.6 CENTRIFUGAL SPINNING

Centrifugal spinning is a simple and is a nanofiber preparation method. It engraves considerable attention compared to traditional electrospinning essentially because of its higher production rate [14]. Other than electrostatic behavior of polymer solution, centrifugal force applied during the spinning process also plays a vital role in the higher production rate of nanofibers.This method also helps to fabricate nanofibers from the solution, in which the absence of the dielectric constant constraint and the involvement of high voltage electric field. Besides, carbon, ceramic, and metal fibers can also be fabricated using centrifugal spinning method [15]. The formation of nanofibers while spinning is proportional to the Laplace force (arising from surface curvature) and centrifugal force. Hooper et al. in 1924 [69] fabricated artificial silk fibers from viscose by implementing centrifugal enforcement to the highly viscous solution. The fiber formation process of centrifugal spinning relies upon the competition between centrifugal force and Laplace force (arising from surface curvature) [16].

These processes are divided into three crucial stages:

1. Turbulent injection of polymer solution into the orifice.
2. Spout elongation to intensify the surface area of the pressurized polymer stream.
3. Solvent evaporation to harden and shrink the polymer jet (Figure 9.7).

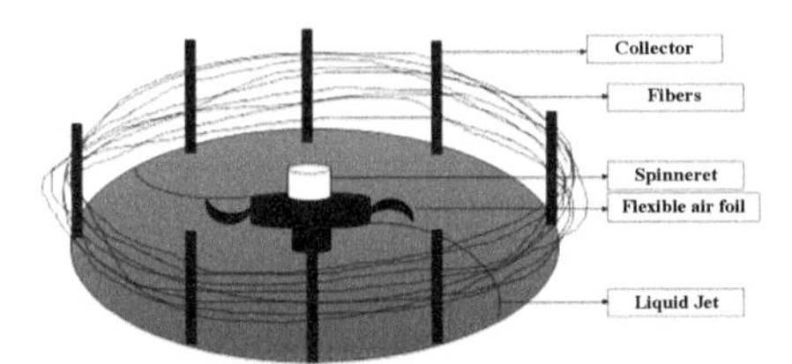

FIGURE 9.7 Schematic drawing of centrifugal spinning system.

In the first phase, consolidation of hydraulic pressure and centrifugal force at the orifice surpasses the resistance flow that charges the polymer solution through the nozzle as a jet [17]. The independent radial centrifugal force expands toward the collector wall, but the jet moves in a warped curve owing to rotation-dependent inertia. Elongation of the polymer jet is involved in the contraction of the length from the nozzle orifices to the collector plate. Concurrently, the solvents in the polymer solution evaporate, solidify, and contract the jet length. The rate of solvent evaporation is significant to its stability at the working environment. In the case of highly volatile solvents, it forms thick fiber diameter as the faster evaporation rate tends to rapid solidification which hinders the jet extension [18]. The parameters that influence the spinning process and the structural property of nanofibers involve angular velocity of spinneret, ambit of orifice, viscoelasticity, surface tension, rate of solvent evaporation, working temperature, and the tip-to-target distance.

9.2.7 *EXTRACTION*

Extraction method for nanofiber fabrication comprises the various chemical and mechanical treatments of natural extracts. With these treatments, the cellulose fibrils can be disintegrated from plant cell walls. For example, the cellulose nanofibers (CNFs) were extracted from wheat straw and soy hull with diameters of 10–120 nm and their chemical and mechanical treatments were explained by Beachley and Wen [19], the nanofiber extraction can also be performed from invertebrates.

9.2.8 *VAPOR-PHASE POLYMERIZATION*

Polymer-based micro/nanofibers have been fabricated from vapor-phase polymerization. Plasma-induced polymerization of vapor-phase vinyl trichlorosilane used to make organosiloxane fibers with diameters of 25 nm and typical lengths of 400–600 nm and cyanoacrylate fibers with diameters from 100 to 400 nm and lengths of hundreds of microns (Figure 9.8) [20].

9.2.9 *MELT BLOWN TECHNOLOGY*

Nanofiber fabrication through melt blown technology involves a single step that the molten polymer solution extruding into an orifice die and

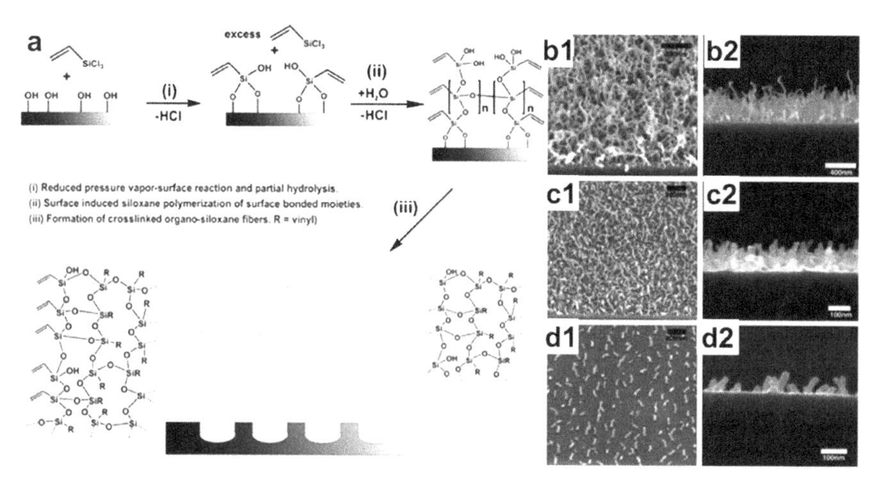

FIGURE 9.8 (a) Pictorial representation outlining a proposed mechanism for nanofiber formation. (b) Arial (1) and side views (2) of polymer nanofibers fabricated from vapor-phase polymerization at high (b), intermediate, (c) and low (d) packing densities. (Reprinted with the permission from [20].)

trailing the extrudate with hot air at the same temperature of molten polymer. The hot air prompts a drag force to attenuate the melt polymer extrudate into nanofibers, which are then accumulated in the form of nonwoven nanofiber mat (Figure 9.9). This technique provides the utilization of thermoplastic polymers in a fairly economic spinning process [21].

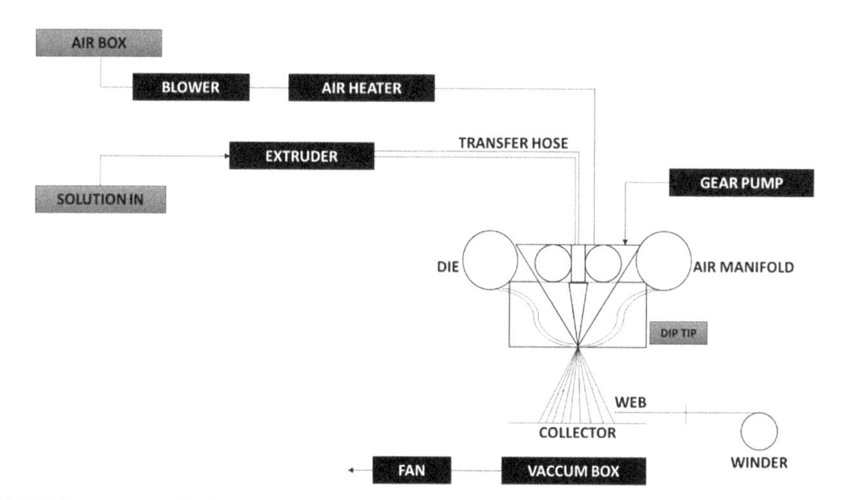

FIGURE 9.9 Detailed schematic diagram of melt blowing process.

9.3 ELECTROSPINNING

Electrospinning is a simple, useful, and versatile technique for the preparation of nanofibers. Its instrumental setup consists of three main components, such as syringe pump, high voltage source, and a ground collector. A basic setup of electrospinning is shown in Figure 9.10. The principle comprises administration of polymer solution held at its own surface tension at the end of the needle capillary attached to high-voltage electric field. When the potential of the high voltage is increased, the globular surface of the polymer solution at the edge of the capillary is drawn out and forms a conical-shaped Taylor cone. At the moment, the high voltage reaches an appropriate value where the gross electric force disables the surface tension. At the potential value, a charged jet of the polymer solution is expelled out from the apex of the Taylor cone. Therefore, an electric potential variation around 15– 40 kV should be applied for nanofiber formation. The solvents used for the dissolution of polymer evaporate at the extrudate moving toward the collector plate. The charged polymer fiber is randomly deposited on a collector [22, 23].

The jet of a solution comprises of four regions; the base, the jet, the spray, and the collection. The shape of the base depends on the surface tension of liquid. While increasing high voltage to the needle, charging the suspended drop of the polymer liquid takes on a conical shape. Additionally, it begins to evaporate and diameter of the jet decreases. The schematic step-by-step procedure of electrostatic effects for nanofiber formation has been explained in Figure 9.11. The jet is issued downs from the tip of the suspended drop of polymer solution and is attracted to the sharp frame of a collector disk revolving around a horizontal axis [24]. The collector plate is placed at an average distance of 15–20 cm from tip of the syringe needle to create a stronger converging electrostatic field.

Dimensions of the fibers can be affected by various parameters; flow rate, applied voltage, tip-to-collector ratio, molecular weight, viscosity, surface tension, conductivity of solution, and ambient parameters.

Nanofibers have opened a wide gateway for various application areas from health-care products such as wound dressings, drug delivery, adsorbent cloths, energy storage applications, and filtration. The key characteristics of nanofibers are large surface-area-to-volume ratio, higher mechanical strength, excellent flexibility, and porosity. To increase the mechanical stability of the nanofiber, mats have been crosslinked by various methods like irradiation, chemical, and dehydro thermal crosslinking.

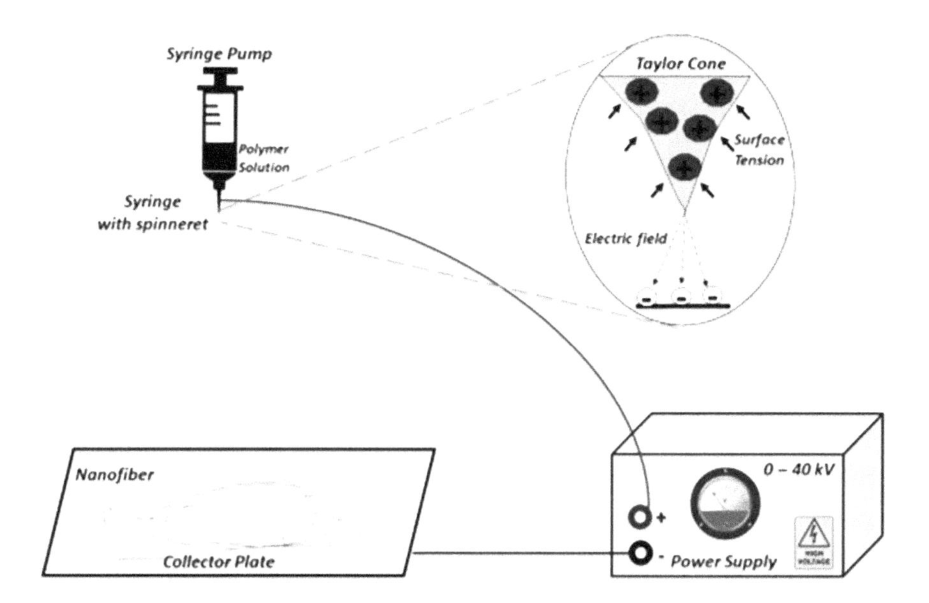

FIGURE 9.10 Schematic diagram of electrospinning setup.

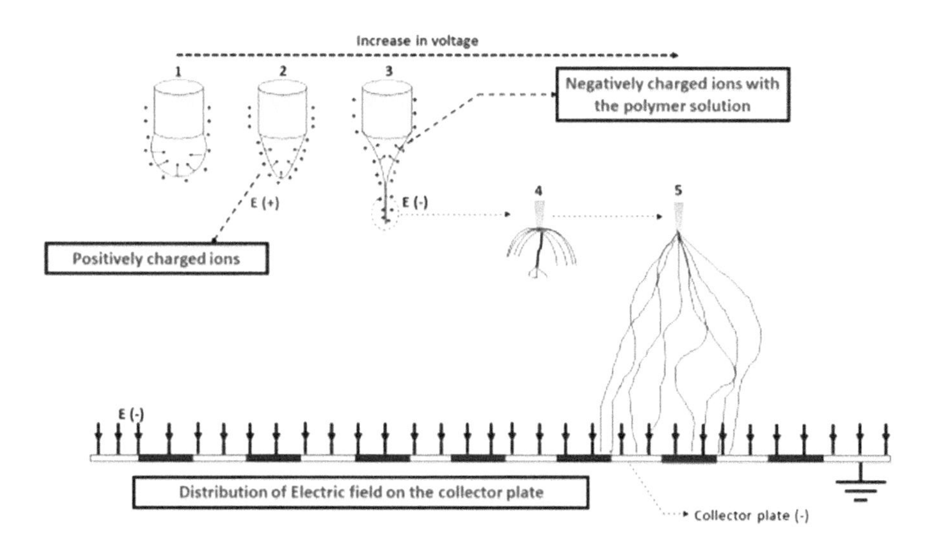

FIGURE 9.11 Step-by-step schematic procedure of nanofiber formation with reference to increase in voltage.

9.3.1 FACTORS AFFECTING ELECTROSPINNING PROCESS

Various parameters can alter the ability of nanofiber formations during electrospinning process. Those parameters can be differentiated in three different groups given below.

1. **Processing parameters:** Factors influencing in case of applied voltage, tip-to-target distance, and hydrostatic pressure applied to the spinning needle.
2. **Ambient parameters:** Surrounding condition, for instance, solution air flow, temperature, and relative humidity in the electrospinning chamber.
3. **Solution parameters:** The spinning solution properties such as molecular weight, concentration of the polymer's viscosity, conductivity, and surface tension.

In the following sections, the entire processing parameters involved in the electrospinning process are discussed in detail.

9.3.1.1 PROCESSING PARAMETERS

9.3.1.1.1 Tip-to-Target Distance

The structural morphology and fiber diameter of nanofibers are significant to the tip-to-target distance. When the distance from tip-to-target is too short, the polymer solution will solidify before reaching the collector plate, whereas if it is having long distance, the bead fiber can be formed. This is one of the crucial physical parameters of the electrospinning that needs to be optimized to achieve desired quality of nanofibers.

9.3.1.1.2 Applied Voltage

For the initialization of fiber formation, the applied voltage is the decisive factor. If the applied voltage is higher than the threshold voltage, the Taylor Cone formation ejected from charged jets can be occurred.

9.3.1.1.3 Feed Rate

Another important process parameter is the polymer solution feed flow rate within the syringe. In general, to allow required time for the polymer solution to polarize low-flow rate is preferable. If the feed rate is very high, lead fibers with thick diameter will form rather than the smooth fiber with thin diameter owing to the short drying time prior to reaching the collector and low-stretching forces.

9.3.1.1.4 Material of Choice as Collector

During the process of electrospinning, collectors act as the conductive matrix to obtain the charged fibers. Aluminum foil is used as a common collector, but it is complex to transfer the collected charged nanofibers to other collector upon different applications. With the need of fibers transferring, diverse collectors have been developed, including wire mesh, pin, grids, gridded bar, rotating rods or wheel, liquid bath, and so on.

9.3.1.2 SOLUTION PROPERTIES

9.3.1.2.1 Molecular Weight

Molecular weight of the polymers is also having potential impact on morphologies of electrospun fiber. Entanglement of the polymeric chain of the solution causes solution viscosity because of the molecular weight. A fixed concentration and low molecular weight of the polymers tend to form beads instead of forming smooth fiber. On the other hand, increasing molecular weight of the polymers results in the formation of smooth fiber. It should be noticeable that the formation of microribbons is obtained due to the massive molecular weight even at the low concentration of polymers.

9.3.1.2.2 Concentration of the Polymers

During the electrospinning process, the concentration of the polymer solution plays a vital role in the fiber formation. There are four critical concentrations from low to high conditions that are considerably noticeable.

1. At low concentration, polymeric micro(nano)-particles are formed. At this stage, electro spraying is conducted instead of electrospinning due to the low viscosity and high surface tensions of the solution.
2. At slightly increased concentration, the blend of beads and fibers will be obtained.
3. At the appropriate concentration, smooth nanofibers will be obtained.
4. At largely increased concentration, helix-shaped microribbon and nonnanoscaled fibers will be obtained.

Normally, the increase in the concentration of the solution results in the increase in fiber diameter when the electrospinning solution concentration is appropriate. Moreover, the solution's viscosity can also be altered by adjusting the solution concentration.

9.3.1.2.3 Viscosity of Solution

Viscosity of the solution is the significant key factor to determine the morphology of the fiber. It has been proven that at very low viscosity, continuous and smooth fibers cannot be formed. Increased viscosity results in the vigorous ejection of jets from the solution. So, it is important to maintain appropriate viscosity for electrospinning. Basically, viscosity of the solution is adjusted by varying the polymer concentration of the solution. The viscosity range of a different polymer or oligomer solution at which electrospinning is done is different.

Viscosity, polymer concentration, and polymeric molecular weight are related to each other which have to be optimized. For low viscosity solutions, surface tension is the noticeable factor that forms beads or beaded fibers. If the solution is having suitable viscosity, the continuous fiber will be formed.

9.3.1.2.4 Conductivity

Conductivity of the solution is solely determined by the types of the polymer, solvent sort, and the salt. Usually, natural polymers are polyelectrolytic in nature that increases the charge-carrying ability of the polymer jet by increasing the ions which have higher tension at electric field, resulting in the decreased amount of fiber formation.

9.3.1.2.5 Surface Tension

Surface tension is a significant factor due to function of the solvent compositions in electrospinning. It has proven that different concentration of the solvents provides different surface tension. At fixed concentration, reducing the surface tension of the solutions results in the conversion of beaded fibers into smooth fibers.

9.3.1.3 AMBIENT PARAMETERS

9.3.1.3.1 Temperature/ Relative Humidity

Low humidity may make the solvent totally dry and increase the velocity of the solvent evaporation. In contrast, high humidity is ended with thick fiber diameter because the charges on the jet get neutralized and the stretching forces become small.

9.4 SURFACE MODIFICATION OF NANOFIBERS

Nanofibers with distinct surface properties are of importance in various applications as the surface topographies affect functions like wettability, adsorption, electrical conductivity, and biocompatibility.

Surface modification of nanofibers has attracted a great deal of attention in recent decades as it improves the specific properties of the nanofibers. Various methodologies have been employed to modify nanofiber surfaces, including physical, chemical, and biological technologies.

9.4.1 PHYSICAL METHODS FOR SURFACE MODIFICATION

9.4.1.1 PLASMA TREATMENT

Plasma treatment methodology uses an ionized gas at either low pressure or atmospheric pressure for modification of nanofiber surfaces. It delivers a spotless and environmental-friendly approach to modify the surface of nanofibers without affecting the bulk properties of the treated nanofibers [25]. Liu et al. [26]. had fabricated pPLLA–Ag NP hybrid composite nanofibers using a simple and resourceful air plasma treatment by introducing hydrophilic groups onto hydrophobic PLLA electrospun nanofibers without varying the morphology of the nanofibers.

9.4.1.2 PHYSICAL VAPOR DEPOSITION

Physical vapor deposition (PVD) is a generally used coating technique that includes evaporation and sputtering these processes carried out in a vacuum chamber at working pressure (typically 10^{-2}–10^{-4} mbar). These processes consist of excitation of the element to be coated that enhances higher surface density to attain higher coating quality. Reactive gases such as nitrogen, acetylene, or oxygen are used during the coating process to generate multicompound or multilayer coating.

9.4.1.3 ION-BEAM IMPLEMENTATION

Ion-beam implementation refers to the process of treating the surfaces of nanofibers with a high beam of broad-energy spectrum. It increases the mechanical properties of the electrospun fibers. Wong et al. [27] experimented with the electrospun nanofibers of poly(vinyl alcohol) with 8.0 × 1015 ions/cm^2 of nitrogen ion and determined the change in elastic modulus using atomic force microscope (AFM) multipoint mechanical bending tests on individual fibers before and after ion beam implantation. The results confirmed that the fiber's elastic modulus increased by 30% It was determined that the surface modification by nitrogen ion treatment added the functional groups of amine and amide on the surface of the nanofibers.

9.4.2 CHEMICAL METHODS FOR SURFACE MODIFICATION

Surface chemistry plays a significant role in various applications of nanofibers. The surface modification of electrospun nanofibers by chemical methodologies can be subjectively divided into different categories: surface grafting, cross-linking, chelation, and surface chemical deposition.

9.4.2.1 Surface Grafting, Cross-linking, and Chelation

Başaran and Oral [28] reported the surface modification grafting of poly(ε-caprolactone) on electrospun gelatin nanofiber (GNF) by surface-established perforated polymerization using potentially cytotoxic cross-linkers or by applying plasma aminolysis reactions. After the fabrication of nanofiber, the perforated polymerization of ε-caprolactone has been initiated on the surface of GNF using tin(IV) isopropoxide and gelatin (primary amine functional groups) as an initiator and a coinitiator. The growing of polymer chains on GNF had been confirmed by both scanning electron microscope and energy dispersive X-ray spectroscopy (Figures 9.12 and 9.13). The experimental results also revealed that gelatin-grafted aligned polycaprolactone (PCL) nanofiber (NF) readily oriented endothelial cells along the fibers.

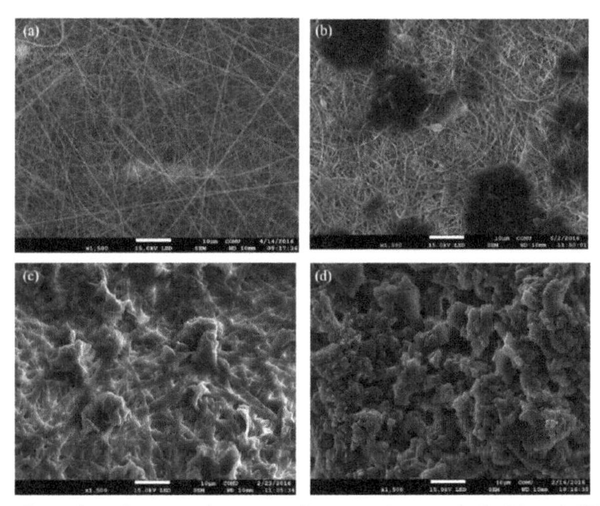

FIGURE 9.12 Scanning electron microscopy images of neat gelatin (a) and GNF-PCL-1h (b), GNF-PCL-4h (c), and GNF-PCL-24h (d). (Reprinted with the permission from İhsan Başaran and Ayhan Oral (2018): Grafting of poly(ε-caprolactone) on electrospun gelatin nanofiber through surface-initiated ring-opening polymerization, *International Journal of Polymeric Materials and Polymeric Biomaterials*, DOI: 10.1080/00914037.2017.1417287[28].)

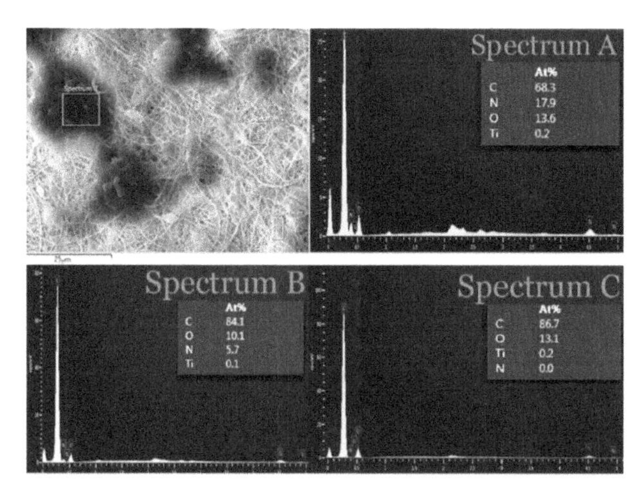

FIGURE 9.13 SEM-EDX spectra of GNF-PCL-1h. (Reprinted with the permission from İhsan Başaran and Ayhan Oral (2018): Grafting of poly(ε-caprolactone) on electrospun gelatin nanofiber through surface-initiated ring-opening polymerization, *International Journal of Polymeric Materials and Polymeric Biomaterials*, DOI: 10.1080/00914037.2017.1417287[28].)

9.4.2.2 Electroless Deposition

The surface modification of PA6 nanofibers through electroless deposition is done by Tao et al. [29] Oxygen low-temperature plasma treatment was used to add the conventional roughness by depositing concentrated sulfuric acid-potassium dichromate on the surface of nanofibers. The surface-modified PA6 nanofibers were characterized for structure and surface morphology by scanning electron microscopy (SEM) (Figure 9.14). The results show that there is no change in the porous structures and the diameters of the surface-modified nanofibers were increased after the Cu deposition. The Cu deposition has been deposited on the surface nanofibers and the Cu particles could clearly be recognized.

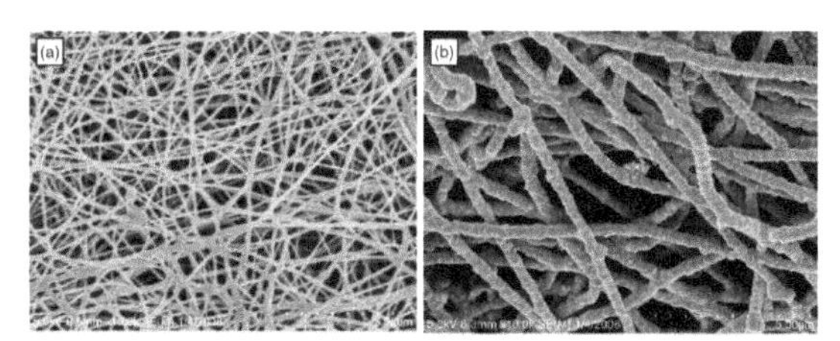

FIGURE 9.14 SEM images of the electrospun PA6 fibers before and after Cu coating: (a) as-spun and (b) after Cu coating. (Reprinted with the permission from [29].)

9.4.3 NANOTECHNOLOGICAL METHODS FOR SURFACE MODIFICATION

9.4.3.1 SOLUTION TO GELATION METHOD

Sol–gel methodology provides potentials to fabricate nanofibers with distinctive properties such as thermal, electrical, optic, chromogenic, photosensitive, and mechanical into nanofibers. Usually, metal alkoxides are used as a precursor for the preparation of 'Sol'. A series of hydrolysis and condensation reaction are applied to the precursor for the formation of inorganic domain. The specific cross-linking agent usage can help in prevention of the macroscopic phase separation. Silva et al. [30] fabricated the functionalized carbon nanofibers (CNFs) in H_2SO_4/HNO_3 followed by synthesis using $Fe(NO_3)3·9H_2O$. The results revealed that the iron content of the CNF/γ-Fe_2O_3 was increased twice from about 40 to about 87 mass %, correlation to the pristine CNF and oxidized CNF specimens, as verified by energy dispersive X-ray fluorescence and SEM images of CNF/γ-Fe_2O_3 sample surface, which has the highest iron mass percentage at cross-linked region. It was also proven by dispersive X-ray spectroscopy.

9.4.3.2 ATOMIC LAYER DEPOSITION

Atomic layer deposition (ALD) of nanofibers is related to the progressive reactions between the gas phase substrate and an exposed surface of nanofibers for thin layer deposition of precursor. This method empowers specific control of the surface chemistry. In addition, bonding of the atomic layer to the exposed surface is strong due to consecutive covalent bonding to the substrate.

For a wide range of applications, it is required to produce such nanofibers with predetermined surface properties. Polymer nanofibers with special characteristic features can alter the electrical conductivity, optical property, and biocompatibility.

9.4.3.3 MOLECULAR IMPRINTING

The imprinting of the molecular recognition constituents on the surface of the nanofibers is supposed as a modest method for the specific applications point-of-care strips, chromatographic methods, sensor fabrication, and catalysts. The specific properties of electrospun nanofibers can be enhanced by imprinting the derivatives of molecular recognition materials.

9.4.4 BIOLOGICAL METHODS FOR SURFACE MODIFICATION

Immobilizing bioactive molecules on the surface of electrospun nanofibers in recent years is in demand to achieve biologically functionalized materials for drug delivery, tissue engineering, and biosensors applications. Numerous biomolecules can be unified within tissue-engineered scaffolds to enhance their functional properties for biomedical applications. To make it more suitable for tissue engineered scaffold, Chen et al. [31] synthesized a slow-degrading silica nanofiber (SNF2) through an electrospun solution with an optimized tetraethyl ortho silicate to polyvinyl pyrrolidone (PVP) ratio. Laminin-modification of SNF2, namely SNF2-AP-S-L, was attained over a series of chemical reactions to attribute to the extracellular matrix (ECM) protein, laminin, to its surface. The SNF2-AP-S-L substrate was characterized by a combination of SEM, Fourier transform-infrared (FTIR) spectroscopy, nitrogen adsorption/desorption isotherms, and contact angle measurements.

Satish and Korrapati [32] fabricated a graft to support the peripheral neural regeneration by conjugating laminin-derived cell adhesive peptides to the aligned nanofibers made from polyvinyl cinnamate. The graft material helps to improve selective neural adhesion and regeneration and as a continual development-prepared triiodothyronine-encapsulated nanofiber to re-establish the functionality of the damaged nerve. The prepared nanofiber was characterized as shown in Figure 9.15 for its physicochemical, morphological, and topographical properties. The results show that the composite of nanofiber has a high latency to re-establish the injured nerve functions.

To achieve desired outcome for various applications like drug delivery, tissue engineering, and biosensors, the nanofibers where functionalized to its specific application. Bioactive peptide molecules functionalizing synthetic or natural polymeric materials have been used for the induction of biological regeneration therapies and wound healing.

9.5 CHARACTERIZATION OF NANOFIBERS

The availability of a wide range of bio-derived and synthetic biomaterials has broadened the scope for development of nanofibrous scaffolds, especially using the electrospinning technique. The characterization of the nanofiber is to understand the relationship between the structure and the features of nanofibers, and they are extensively applied in musculoskeletal tissue engineering (including bone, cartilage, ligament, and skeletal muscle [33]), skin tissue engineering,

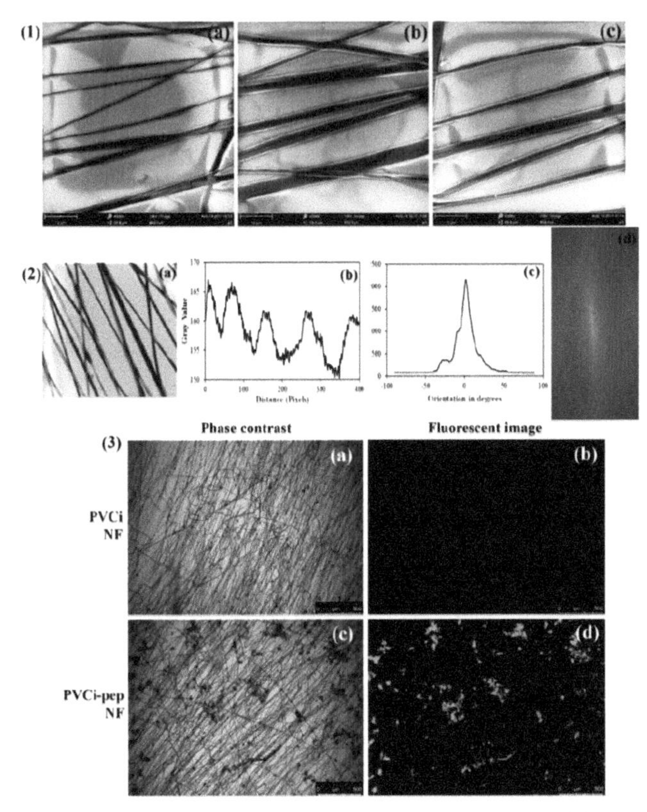

FIGURE 9.15 (1) SEM images of (a) PVCi nanofibers-P NF, (b) peptide conjugated PVCi nanofibers-PP NF, and (c) peptide conjugated, T3 incorporated PVCi nanofibers-PTP NF. (2) Analysis of alignment of the nanofibers, where (b) plot profile analysis is depicting alignment based on the peak height and placement, (c) orientation of aligned nanofiber determined using ImageJ (OrientationJ plugin), and (d) ImageJ FFT plots. (3) (c, d) Presence of surface peptide conjugation observed by FITC-tagged peptide in contrast to the (a, b) blank image obtained in control nanofibers viewed under fluorescence microscope. (Reprinted with the permission from [32].)

vascular tissue engineering, neural tissue engineering [34] and as carriers for the controlled delivery of drugs, proteins, and DNA [35] In order to characterize the nanofibers and also to control and optimize them, it is very important to measure the relevant properties of these biomaterials as well as their suspensions.

The most important attribute of scaffold is its morphology which includes specific surface area, and size of the pores, biocompatibility, controlled porosity, and permeability. The functional enhancement can be attained by characterizing the thermal properties, optical properties, and mechanical properties of nanofibers, because it is in these properties that they show particularly significant differences in the nano form compared to bulk form. The aim of this article is to focus on the characterization techniques that are particularly appropriate for characterizing nanofibers.

The structural and morphological properties of nanofibers can be characterized by SEM, transmission electron microscope (TEM), and AFM. The chemical characterization can be performed using FTIR, attenuated total reflectance-Fourier-transform infrared spectroscopy (ATR-FTIR) analysis, and Raman spectroscopy. By using the AFM, SEM, and TEM techniques, the bending test and Young's modulus can be obtained, which can be used to characterize the mechanical properties of nanofiber. Apart from the abovementioned methods, X-ray diffraction (XRD), FESEM, are TGA-DSC are some of the techniques that are also used to analyze the property of nanofibers [36] (Figure 9.16).

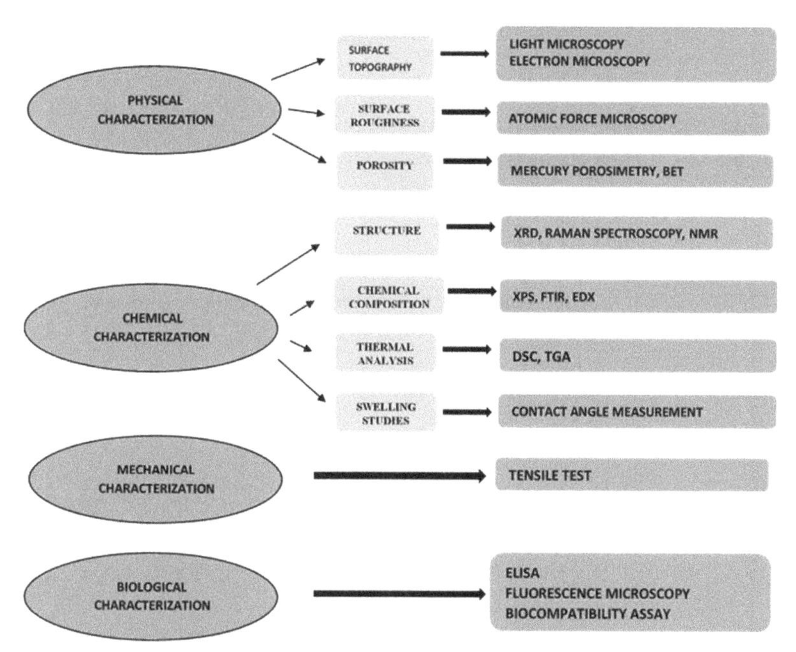

FIGURE 9.16 Characterization techniques for nanofibers (SEM: scanning electron microscopy, TEM: transmission electron microscopy, XRD: X-ray diffraction; DSC: differential scanning calorimetry; NMR: nuclear magnetic resonance; TGA: thermogravimetric analysis; FTIR: Fourier-transform infrared; XPS: X-ray photoelectron spectroscopy; EDX: energy-dispersive X-ray spectroscopy; ELISA: enzyme-linked immunosorbent assay; AFM: atomic force microscopy; BET: Brunauer–Emmett–Teller).

9.5.1 PHYSICAL CHARACTERIZATION

9.5.1.1 OPTICAL MICROSCOPE

In optical microscopy, light waves are used to view the samples and produce images by a system of "optical lenses." The resolving power of microscope is 0.2 μm. It is one of the simplest and lowest cost methods for analysis. Since the imaging takes place under atmospheric pressure, samples at swollen state can even be monitored and they appear same in in vitro and in vivo experiments. Information such as size, uniformity, and shape of the nanofiber can be obtained from the recorded images. The resolution of an object is limited in optical microscope because of its wavelength [35].

9.5.1.2 ELECTRON MICROSCOPE

Electron microscopes examine samples of very fine scale, as small as 0.001 μ (=10 Å). Thus, the resolution and magnification of an electron microscope is 200 times higher than the optical microscope. The types of the electron microscopes are SEM and TEM. In a SEM, the secondary electrons (SEs) generated by the sample are detected to capture an image that contains topological features of the specimen. The TEM, on the other hand, is generated by the electrons that transmit through a thin sample (Figure 9.17).

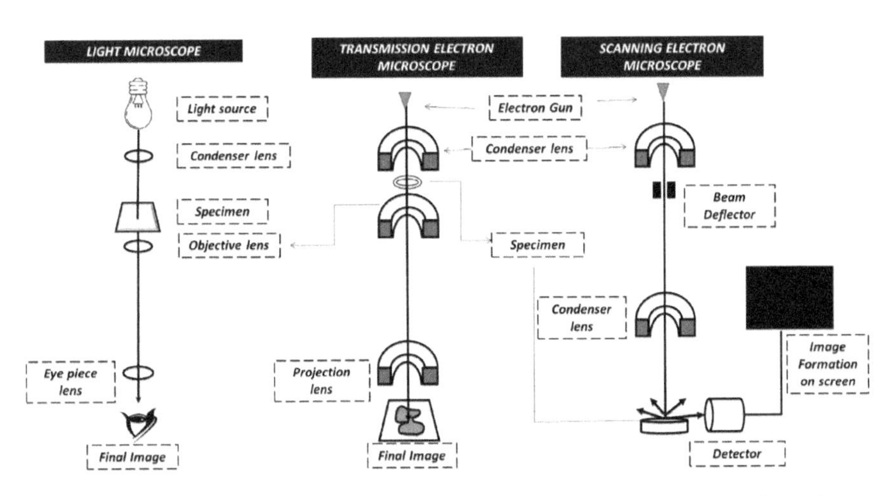

FIGURE 9.17 Principal features of an optical microscope, a transmission electron microscope, and a scanning electron microscope.

9.5.1.3 SEM—SURFACE TOPOGRAPHY

SEM provides extensive possibilities for the analysis of fiber morphologies. SEM micrographs of several electrospun fibers reveal a structure that consists of pores, beads, fibers, or a combination of pores, beads, and fibers. The beads may be spherical or elongated (spindle-like) and pores may be round, while the fibers may be round or flat. The combination of elongated beads, pores, and fibers, known as the bead-on-string morphology, is observed under many conditions. The fibers may exhibit a variety of coils, loops, and bends resulting in different size and shape fibers [37]

Shoba et al. [38] had encapsulated the papain enzyme in polyvinyl alcohol (PVA) nanofiber, and they had shown that the papain concentration more than 30 wt% to polymer produces beaded nanofibers (Figure 9.18c and d). The ideal composite was obtained at 10% PVA at a potential of 25 kV. Based on the quality of fiber obtained, the ratio of urea and papain was optimized as 2:1. The oriented continuous nanofibers with an average diameter of 200–400 nm was obtained at 10% PVA with 30 wt% papain and 15 wt% urea mixture (Figure 9.18a and b).

Qi et al. [39] designed a scaffold with PVA, metallocene polyethylene (mPE), and plectranthusamboinicus (PA) for bone tissue engineering application. It was shown that the PVA/mPE/PA scaffold had a reduced fiber diameter and better bone tissue growth compared to PVA scaffold.

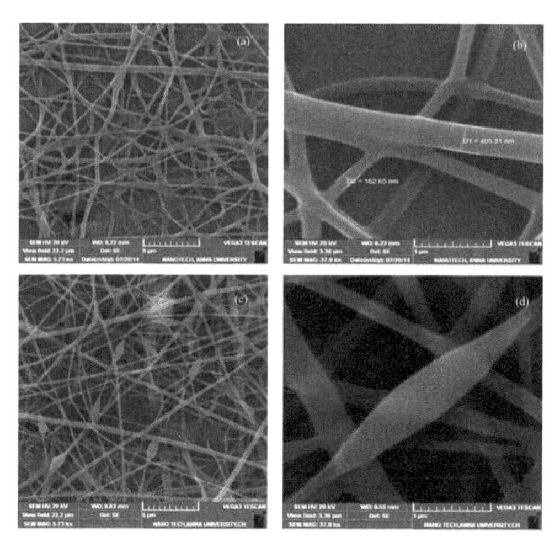

FIGURE 9.18 SEM image of papain-loaded polyvinyl alcohol (PVA) nanofiber mats with different concentration of papain: (a, b) 30 wt% (c, d) 50 wt% (b, d)]. (Reprinted with permission from [38].)

9.5.1.4 POROSITY MEASUREMENTS

Porosity and fiber diameter distribution are the most important factors that determine the performance of nanofibers. The porosity of the nanofiber is the measure of pore volume of nanofibrous membrane to total volume of the nanofibrous membrane [40]. The techniques available for characterization of porous materials are mercury porosimetry, capillary flow porosimetry, and adsorption condensation/Brunauer–Emmett–Teller (BET).

Mercury porosimetry is a familiar method to determine the pore size distribution in the nanofibrous materials. This technique is based on immersing the scaffold in a nonwetting liquid (mercury) and applying the external pressure on it. The porosity is measured by the amount of pressure applied to mercury intrusion into pore spaces [41] The differences in porosities of polylactide (PLA) (86.6%), PVA, PCL (77.4%) nanofibers showed variations in the release of incorporated molecules [35] BET is used to measure the specific surface area of a sample and pore-size distribution. This information is used to predict the dissolution rate, as this rate is proportional to the specific surface area. Thus, the surface area can be used to predict bioavailability and determine the performance of the product.

9.5.1.5 FIELD-EMISSION SCANNING ELECTRON MICROSCOPE

Field-emission scanning electron microscope (FESEM) uses a beam of electrons to observe small specimens on the surface of cells and nanofibers. The FESEM is used to get specific information about topography, surface characteristics, and elemental composition. In FESEMs, the SE detectors are capable of imaging 1–5 nm features at high voltage (15–30 kV) and 5–10 nm features at low voltage (1–5 kV). FESEMs are capable of providing high-quality images.

9.5.1.6 ENVIRONMENTAL SCANNING ELECTRON MICROSCOPE

Environmental scanning electron microscopy (ESEM) is operated in "wet" mode, in which both nonconductive and conductive samples can be analyzed. It is employed in examining the biological and industrial materials with an electron beam in a high-chamber pressure atmosphere of water vapor. Additional sample preparations are not required and specimens can be analyzed without destruction. Dynamic experiments are also possible in the ESEM, such as drying or crystallization.

Liu et al. [42] had demonstrated a new three-dimensional (3D) cell culture substrate that is built by electrospun PVDF polymeric nanofiber bundles. The morphology of cardiomyocytes on PVDF nanofibers was characterized using ESEM. The random nanofiber bundles had pentagonal-shaped randomly distributed cardiomyocytes. But the aligned nanofiber bundle had adhered aligned cardiomyocytes and the interface of a single cardiomyocyte and parallel nanofibers was observed using ESEM.

9.5.1.7 *TRANSMISSION ELECTRON MICROSCOPE*

TEM is a technique used to examine very small structures less than 1 nm. It provides the information about the nature and structure of the material. Whereas SEM provides information only about the morphology of a material. The beam of electrons transmitted on to specimen depends on density, composition, and uniformity of material being examined.

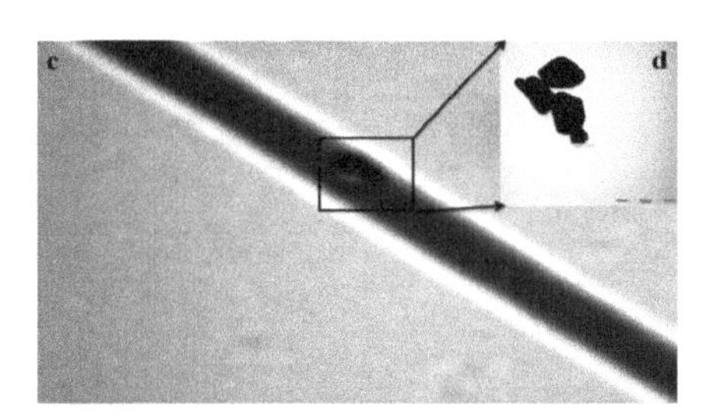

FIGURE 9.19 TEM images (c) Mo-PCL nanofibers (d) MoO_3 nanoparticles. (Reprinted with permission from [43]. © 2018 Elsevier.)

The TEM (Figure 9.19) had clearly shown the MoO_3 nanoparticles (NPs) incorporated in PCL nanofiber scaffolds. The average diameter of the control NPs was 150 nm, whereas the PCL fibers containing the NPs varied from 300 to 600 nm [43].

9.5.1.8 *ATOMIC FORCE MICROSCOPY—SURFACE ROUGHNESS*

AFM is an analytical technique used to analyze the surface of the nanofibers down to atomic resolution and generate images with high resolution. This technique is based on the interatomic forces and is a relatively nondestructive technique as compared to SEM and TEM. It is also capable of 3D mapping of the surface. Based on the tip modification, electric, magnetic, and chemical properties of the materials can be measured

The copolymer poly(L-lactic acid)-*co*-poly (ε-caprolactone) (PLACL), silk fibroin (SF), and aloe vera (AV) were fabricated to produce biocomposite nanofibrous scaffolds for cardiac tissue engineering. The surface morphologies of PLACL/SF and PLACL/SF/AV nanofibers were observed using a tapping mode (silicon tip) on the AFM [44]. The surface morphology of PLACL was identified as the most rough surface with average roughness of

131 nm (Figure 9.20a), PLACL/SF showed 86 nm of average roughness (Figure 9.20b), and PLACL/SF/AV showed 54.7 nm of average roughness (Figure 9.20c). Hence, they were able to suggest that the rough surface would benefit cell adhesion and proliferation.

AFM has also been employed in characterizing core/shell nanofibers. Chen et al. [45] had observed the surface morphologies of thermoplastic polyurethane (TPU), collagen, and TPU/collagen using AFM. Using a height mode on the AFM, the morphology analysis has shown that core/shell collagen/TPU nanofiber blend had a rough surface and collagen–TPU exhibited smooth surface. The rough surface of nanofibers promotes the cell adhesion and growth on nanofibers.

9.5.2 CHEMICAL CHARACTERIZATION

The approaches for designing the nanofibrous scaffolds for tissue engineering and drug delivery systems require the knowledge of elemental compositions and the chemical interaction in the material. Characterizing the internal architecture of nanofibrous materials can be performed using various methods, such as elemental analysis, FTIR, XRD, X-ray photoelectron spectroscopy (XPS), and so on.

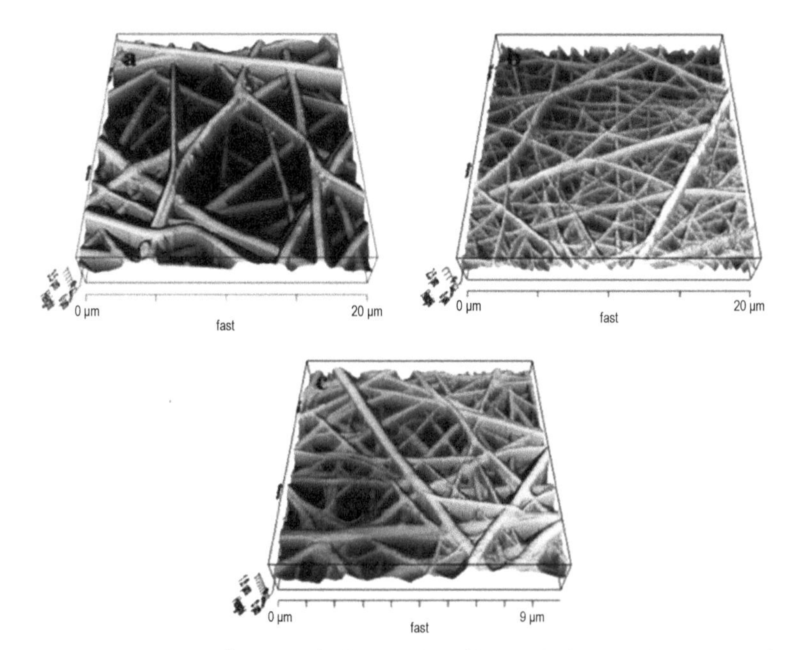

FIGURE 9.20 3D AFM images of (a) PLACL, (b) PLACL/SF, and (c) PLACL/SF/AV nanofibers. (Reprinted with permission from [44].) © 2014 Elsevier.

9.5.2.1 FOURIER TRANSFORM INFRARED SPECTROSCOPY—FUNCTIONAL GROUP DETECTION

FTIR spectroscopy is an analytical technique used to identify organic, polymeric, inorganic materials, or other constituents in the sample. When infrared radiation (IR) passes through a sample, some radiation gets absorbed by the sample and some are transmitted. The signal received at the detector produces a pattern representing the molecular fingerprint of the sample. Each molecule has a unique chemical structure which gives different spectral fingerprint. Absorption bands from 4000 cm^{-1} to 1500 cm^{-1} are due to the presence of various functional groups and the range of 1500 cm^{-1} to 400 cm^{-1} is known as the fingerprint region. Absorption pattern in the fingerprint region is highly specific to the material.

Rychter et al. [46] had designed the drug/polymer composite using electrospinning method. The different concentrations of cilostazol (CIL) drug (6.25, 12.50, and 18.75%) were incorporated into the biodegradable PCL fibers (Figure 9.21). The presence of CIL in PCL nanofibrous mat and crystallinity of CIL after encapsulation were confirmed using FTIR spectra.

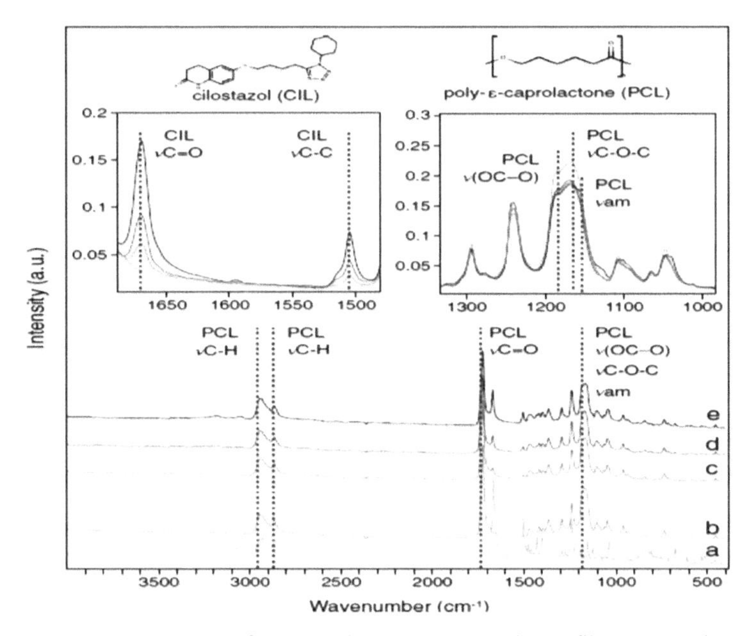

FIGURE 9.21 FTIR spectra of CIL powder (a), as prepared PCL fibrous mats (b), and PCL fibrous mats with 6.25% (c); 12.50% (d); 18.75% (e) of CIL. (Reprinted with permission from [46]. © 2018 Springer.)

9.5.2.2 ATTENUATED TOTAL REFLECTANCE-FOURIER TRANSFORM INFRARED SPECTROSCOPY ANALYSIS

ATR is used to measure the surface composition of very small materials. The IR spectra obtained is based on the internal reflection of material on the crystal. It is applied to the samples requiring intense penetration such as thick biological materials, coatings, viscous liquids, and free-flowing aqueous solutions. The depth of penetration is around 1 or 2 μm depending on the ATR crystal material.

9.5.2.3 CONTACT ANGLE MEASUREMENTS

Contact angle is the method used to measure the wettability of a surface or fibrous material. The angle created by the liquid when comes in contact with the solid material is termed as the contact angle. The contact angles of the membrane are evaluated by drop shape analysis and the modified Washburn method. The contact angle less than 90° is considered as the liquid wets the surface and contact angle greater than 90° as nonwetting with the liquid. The contact angle is strongly influenced by the surface texture such as pores and surface roughness. The wettability is most importantly studied to characterize the surface morphology of the different polymers in blends. It was reported that the studies have shown that the nanocomposites with hydrophilic nature and increased wettability have an ability to enhance bone regeneration [47]

9.5.2.4 X-RAY DIFFRACTION—CRYSTALLINITY PROPERTIES

XRD is a nondestructive technique for identification and quantification of crystalline materials. It provides in-depth information on structural analysis, crystallinity, and orientations (texture). XRD diffractogram analysis is based on the intensity as a function of diffraction angle. The XRD pattern generates a unique fingerprint of crystals in the material. The accuracy of the methods mainly depends on the number of compounds embedded in the sample; a smaller number reveals high accuracy.

9.5.2.5 X-RAY PHOTOELECTRON SPECTROSCOPY—FUNCTIONAL GROUP QUANTIFICATION

The surface chemical composition of the nanofibers was evaluated by XPS. Its application in diverse range of materials makes it a versatile tool for studying surface modification. XPS, otherwise known as electron spectroscopy for chemical analysis is based on the photoelectric effect. XPS is surface-sensitive because the X-rays used are of low intensity (1500 eV) and its average depth of analysis is around 5 nm. XPS system works under high vacuum. Hence, the samples must be high-vacuum compatible and thickness of the sample to be analyzed must be lesser than 13 mm.

Rezaei et al. [48] revealed that the surface chemistry of the nanofibers was altered upon plasma treatment. XPS can be used to analyze the chemical composition and elemental states of the peptide-tagged nanofiber. XPS being an effective tool to evaluate the elements present in the surface has been extensively used for identifying the presence of peptides tagged to the NP and hence used for the characterization of NPs and nanofibers. Characterization of transactivator of transcription (TAT) peptide conjugated to the poly(L-lactide) (PLA) NPs were determined by XPS [49] The presence of N_2 peak in the peptide conjugated NPs confirms the conjugation of TAT peptide since PLA doesn't contain any nitrogen group.

9.5.2.6 DIFFERENTIAL SCANNING CALORIMETER—THERMAL ANALYSIS

Differential scanning calorimeter (DSC) is a powerful and versatile thermal analyzer that allows for property measurements on a broad variety of materials from −150 to 600°C. The DSC measures the rate of heat flow and the transition of material with respect to the temperature and time. DSC is used widely for examining polymers to check their composition. In DSC, thermograms are used to analyze the polymer's heat capacity, melting and boiling temperature, glass transitions (T_g), composition of the material, oxidative/thermal stability, reaction kinetics, and purity of materials.

DSC thermogram [41] has shown that the inclusion of chitosan/polyethylene oxide (Cs/PEO) leads to increase in the T_g temperature of the composite nanofibers compared to polyurethane (PU) nanofibers, where the Cs incorporation increases the hydrophilic properties of the composite by acting as a plasticizer for PU nanofibers. The core–shell PU/Cs shows similar degradation curve to that of PU nanofibers except for the mass loss. Hence, the core–shell PU/Cs nanofibers were suggested to be potential platform for the bioactive scaffolds with further biological tests.

9.5.2.7 THERMOGRAVIMETRIC ANALYSIS—THERMAL ANALYSIS

Thermogravimetric analysis (TGA) measures the changes in physical and chemical properties of material as a function of increasing temperature (with constant heating rate), or as a function of time (with constant temperature and/or constant mass loss). The most significant applications of TGA is the assessment of the compositional analysis of polymeric blends, thermal stabilities, oxidative stabilities, estimation of product lifetimes, decomposition kinetics, effects of reactive atmospheres on materials, filler content of materials, moisture, and volatiles content.

DSC and TGA were used to analyze the thermal properties of the PLLA/PHBV (poly (L-lactic acid) and poly(3-hyroxybutyrate-*co*-3-hydroxyvalerate)) fibrous mats [50] The crystallinity of PHBV is affected when the weight ratio of PHBV varies in the blend whereas the crystallinity of PLLA remains unaffected at any blend ratio of the polymer. Feng et al. [51] demonstrated that the combination of PHBV with PLLA improves the biocompatibility and wettability of electrospun polymer scaffolds compared to PLLA nanofiber and it can be used for tissue engineering.

The PCL and Mo-PCL nanofiber were analyzed using DSC. The melting point of PCL was about 59.55 °C whereas Mo-PCL had exhibited melting point at 62.11 °C (Figure 9.22). Hence, it was demonstrated that the marginal shift in the peak was due to the incorporation of very minimal amount of NPs into the PCL scaffold. Similarly, when PCL and Mo-PCL were subjected to TGA, there was weight loss of about 2%–5% from 50 °C to 380 °C. The weight loss of PCL was observed at 384 °C whereas the Mo-PCL nanofibers had shown denaturation at 391 °C (Figure 9.23). The slight increase in the thermal degradation temperature between PCL and

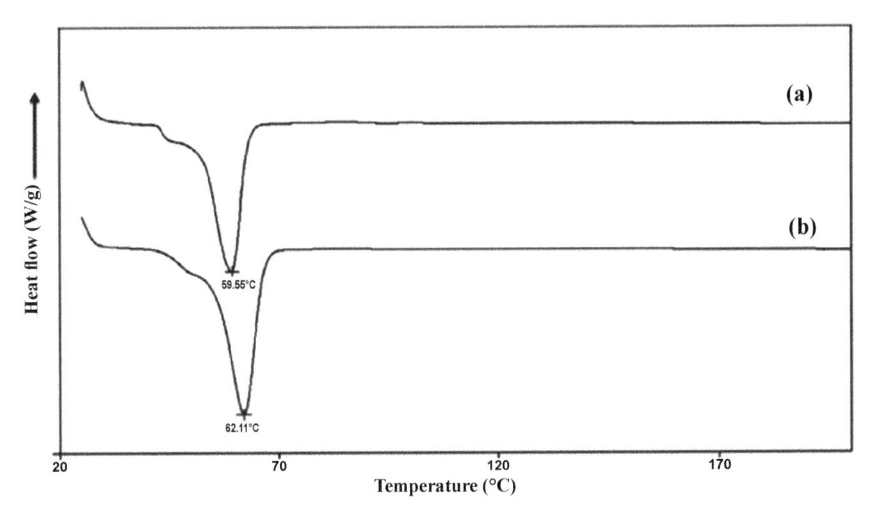

FIGURE 9.22 DSC thermogram (a) PCL (b) Mo-PCL nanofibers. (Reprinted with permission from [43]. © 2018 Elsevier.)

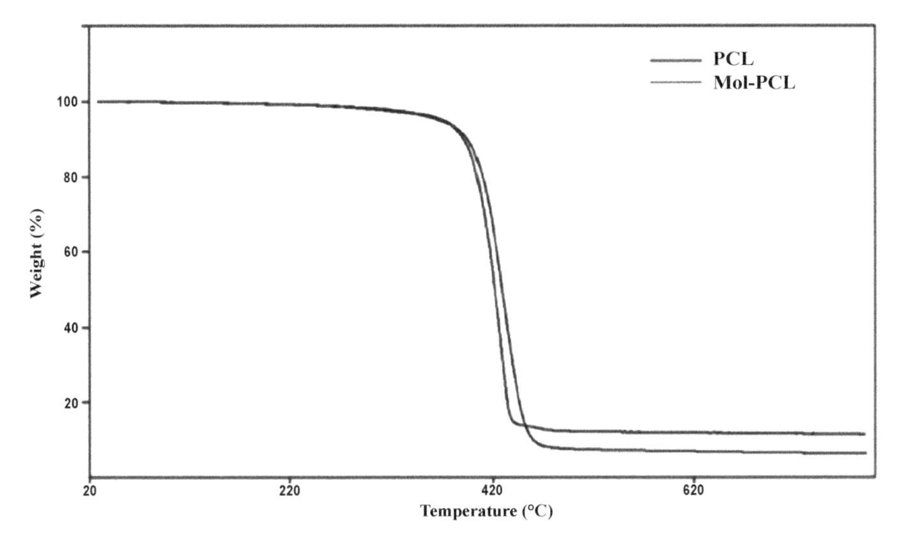

FIGURE 9.23 TGA of (a) PCL (b) Mo-PCL nanofibers. (Reprinted with permission from [43]. © 2018 Elsevier.)

Mo-PCL nanofibers was due to the incorporation of the NPs in the polymeric scaffold [43].

9.5.2.8 *RAMAN SPECTROSCOPY*

Raman spectroscopy is a spectroscopic analysis used to provide the detailed information on molecular vibrations and spatial distribution of the chemical content within the samples. It is based upon the interaction of light with the chemical bonds within a sample. Raman spectroscopy is a noncontact and nondestructive chemical analysis. In-depth analysis of transparent samples is made using a confocal optical system. It has a high spatial resolution in the order of 0.5–1 μm. It takes only 10 ms to 1 s exposure to get a Raman spectrum.

Sfakis et al. [52] had produced biodegradable/biocompatible PGS/PLGA core/shell nanofibers and explored their core/shell chemical structure and morphology using Raman spectroscopic mapping. They characterized PGS/PLGA composite fiber using confocal Raman microscopy, which combines confocal imaging with Raman spectroscopy. Further, they used singular value decomposition analysis for identification of different chemical regions in the image, and observed the organization of the fiber components which was proposed to be the substrate for tissue engineering.

Janani et al. [43] had shown the significant signals of molybdenum in Raman spectra and incorporation of molybdenum NPs in PCL scaffold.

The Mo-PCL peaks were observed at 994 cm^{-p} and 800 cm^{-a}, indicating the presence of MoO_3 NPs with PCL nanofibers (Figure 9.24).

9.5.3 MECHANICAL CHARACTERIZATION

The biomaterial or nanofibrous scaffold should mechanically match the host tissue to enhance the delivery system [53] (drug delivery, cell therapy). The deformation or permanent failures of the nanofibers are due to the strains and stresses from the surrounding media such as local density, degree of crystallinity, and arrangement of polymer molecules. The activity of the nanofibers is enhanced by controlling and optimizing its tensile strength. Tensile strength is the ability of a material to withstand at maximum stressed condition or before it breaks under tension.

Young's modulus is used to quantify the stiffness of the material where the material is stretched under unit stress and the deformation is elastic (i.e., material returns to the original state or slight change occurs when the stress is removed). It evaluates the elasticity of materials, which is the relation between the deformation of a material and the power needed to deform it [54]

The aligned nanofibers exhibit high strength, high modulus, and significantly increase the efficiency of the fibers. The misalignment will greatly reduce the mechanical properties of nanofibers. In addition,

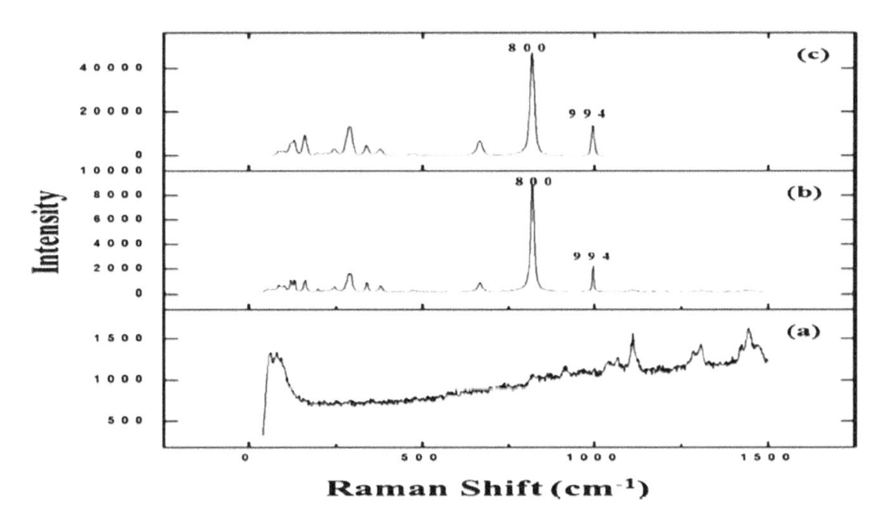

FIGURE 9.24 Raman spectra of (a) Control PCL nanofibers, (b) Mol-PCL nanofibers, and (c) MoO3 nanoparticles. (Reprinted with permission from [43]. © 2018 Elsevier.)

it was proposed that the increase in mechanical properties was mainly related to confined molecular orientation of amorphous regions with decreasing nanofiber diameter.

9.5.4 BIOLOGICAL CHARACTERIZATION

Understanding the cellular responses and their behavior to nanofiber composites in biological systems is required. The interaction properties and the activity of nanofibers are enhanced by characterizing them in possible biological aspect and they play an important role in tissue engineering applications.

Demonstrating the interaction of electrospun nanofibers, incorporation of biological factors, effects of surface modifications with cells such as chondrocytes, osteoblasts[55], neural cells[56], fibroblasts, keratinocytes, stem cells [57] and so on, and evaluating their biocompatibility are important for their biological applications. Electrospun nanofibers mimic the structure of original ECM; therefore, they create a favorable geometrical environment for the cells to attach, migrate, and proliferate [58]

9.5.4.1 CYTOCOMPATIBILITY ANALYSIS OF THE SCAFFOLDS

9.5.4.1.1 Cell Proliferation Assay

Biocompatibility assays are performed to study the cell adhesion, survival, migration, and proliferation with response to various stimuli and their effects on cell morphology and growth. Assays for cell proliferation are used to monitor the number of cells over time, cellular divisions, metabolic activity, or DNA synthesis. The cytotoxicity of the fabricated scaffolds is assessed to check its biocompatibility and its effect on physiological process of regeneration mechanism. This evaluates whether scaffold has any toxic effect on cells. Proliferation of cells is indicative of the influence of scaffold to elicit proliferative response in cells [32]

9.5.4.1.2 Metabolic Proliferation Assays

Assays that measure metabolic activity are suitable for analyzing proliferation, viability, and cytotoxicity. The reduction of tetrazolium salts such as MTT (3-(4,5-dimethylthiazolyl-2)-2,5-diphenyltetrazolium bromide), MTS, XTT (2,3-bis-(2-methoxy-4-nitro-5-sulfophenyl)-2H-tetrazolium-5-carboxanilide), WST-1 to colored formazan compounds occurs in metabolically active cells. The cells exposed to toxic compounds will have decreased cell viability whereas the active compounds show the increased metabolic activity indicating cell proliferation.

9.5.4.1.3 Fluorescence Dye Proliferation Assays

Live and dead double staining can be used for simultaneous fluorescence staining of viable and dead cells. Acridine orange is an intercalating dye that stains all kind of cells (live and dead) and emits green fluorescence. Propidium iodine (PI) cannot pass through a viable cell membrane. It stains the nucleus of the dead cells by passing through the disrupted cell membrane and emits red fluorescence. Therefore, all live cells fluoresce green and all dead cells fluoresce red. The cells stained with acridine orange (AO) and PI quench and emit red fluorescence. Similarly, calcein—a highly lipophilic and cell membrane permeable dye—and PI simultaneously monitor both viable and dead cells with a fluorescence microscope.

Janani et al. [43] had shown the toxicity of fabricate PCL and Mol-PCL nanofibers of keratinocytes (HaCaT) compared to epidermoid cancer cells (A431) using AO/PI staining. Hence, it was proposed that the fabricated Mol-PCL nano constructs as a good tunable targeted delivery system (Figure 9.25).

9.5.4.1.4 Angiogenic Ability of Scaffold

Angiogenesis, the formation of new blood capillaries from pre-existing capillaries, marks an important step in regeneration. An ideal scaffold for regeneration should enhance the cell–cell communication/cell connectivity, thereby to promote sprout formation, tube formation, and neovascularization. The new capillaries maintain the blood flow

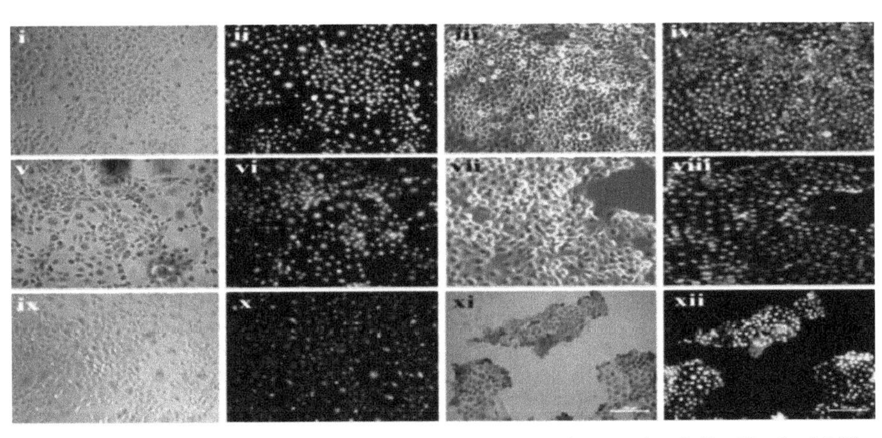

FIGURE 9.25 The phase contrast and fluorescence micrograph of live/dead viability analysis (AO/PI staining) of HaCaT (i, ii, v, vi, ix, x) and A431 (iii, iv, vii, viii, xi, xii) cells. (Reprinted with permission from [43]. © 2018 Elsevier.)

of the proliferating cells and ensure adequate supply of nutrients, toxin removal, as well as exchange of gases. The nanofibrous scaffold used for the delivery of specific growth factors, drugs, or other bioactives for regeneration purpose should provide the suitable microenvironment for cell adherence and proliferation. The angiogenic potential of the nanofibers can be confirmed using in-*vivo* chick chorioallantoic membrane assay and rat/chick aortic arch assay [59]

Yoo et al. [60] had investigated the effect of dimethyloxalylglycine (DMOG)-embedded poly(ε-caprolactone) fiber (PCLF/DMOG) on odontoblastic differentiation. The PCLF/DMOG-treated dentin slices have shown the high cellularity of human dental pulp-derived cells, enhanced host cell recruitment, and angiogenesis, with PCLF/DMOG fibers.

9.5.4.1.5 Hemocompatibility of the Scaffold

Hemolysis is the lysis of red blood cells due to hemolytic reagents such as drugs or toxins. The release of the cytoplasmic contents, in particular adenosine diphosphate, recruits the assembly of blood platelets which accelerates blood clotting. The scaffold should possess hemocompatible property as it would be in contact with blood vessels and damaged tissues. To understand the hemocompatibility of the fabricated scaffold and the bioactives incorporated, in vitro blood compatibility assay is carried out and the percentage of hemolysis when the blood comes in contact with nanofibrous scaffold is determined [61]

9.5.4.1.6 Cell Adhesion

Cell adhesion is the adherence of the cells to the ECM or a specific area, where the growth and survival of the cells is essential for cell–cell communication. The cell adhesion assay determines the ability of a specific type of cell to adhere to a specific adhesive substrate and also tests the sensitivity of the cell–substrate interaction [59] The cell–cell interactions and cell–scaffolding interactions can be visualized using confocal fluorescence microscope. Keller et al. [62] has investigated the capacity of osteoblasts cultivated as 3D microtissues. When human primary osteoblast microtissues are seeded onto the scaffold, cells exhibit osteoblast migration along the nanofibers. The cells on fibrous scaffolds have exhibited normal cell shapes and integrated well with surrounding fibers [63] The studies have also shown that electrospun fibrous scaffolds support the attachment and the proliferation of cells. Most studies have shown that the performance of the cells is improved by their interaction with nanofiber composites. Therefore, they were proposed to be the suitable substrate for tissue regeneration and artificial implants. Shoba et al. [59] had shown the adhesion, migration, and proliferation of rat cardio myoblast H9c2 cells on PCL/GE NF and Sal B-MAAP PCL/GE NF (Figure 9.26). Hence, they proved that bioactive-incorporated scaffold influenced good adherence of cells with better morphology, increased proliferation, and differentiation for regeneration of complete damaged myocardium.

9.5.4.1.7 Cell Migration Assay (In Vitro Wound Assay)

Cell migration is the movement of cells from one area to the other promoting cell differentiation and regeneration in response to various chemical signals. The scratch assay is a popular method for the study of cell migration in vitro. Scratch wound assay is to assess the ability of the scaffold to re-establish the tissue integrity and restore the function of cells around the injured site (Figure 9.27). The basic steps involve creation of a scratch

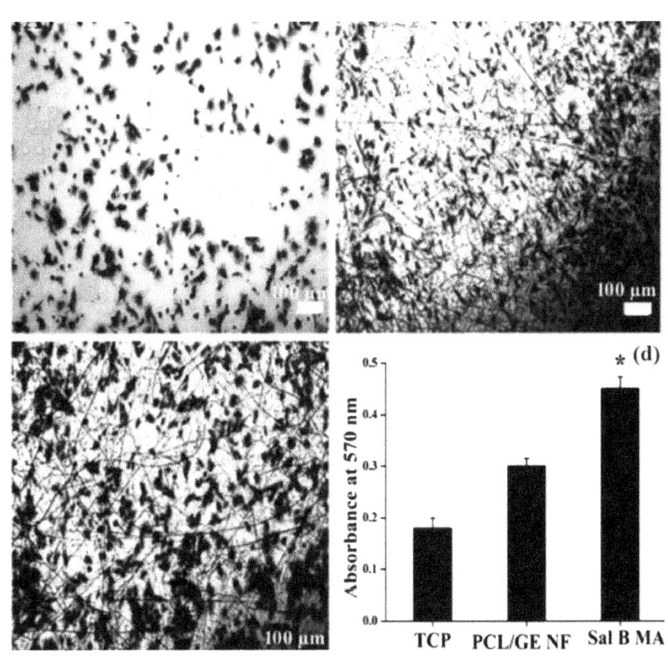

FIGURE 9.26 Phase contrast images of adherent cells on (a) tissue culture plate (TCP), (b) PCL poly(ε-caprolactone) (PCL)/gelatin (GE) nanofiber, (c) salvianolic acid B (Sal B)-Br PCL/GE NF, and (d) quantitative analysis of the cell adhesion potential of the fabricated scaffold. (Reprinted with permission from [59]. © 2018 Elsevier.)

(gap) on a monolayer of confluent cells. The cells are stimulated and activated by various signaling molecules to initiate migration in the damaged site, and cells on the edge of the scratch will migrate to close the gap and to establish cell–cell connection. The cells migrate from random or two-dimensional layered matrixes, since they do not possess the ability to grow in 3D orientations [64].

9.5.4.1.8 Flow Cytometry (Fluorescence-activated Cell Sorting)

Flow cytometry is a laser-based technique to count and profile cells simultaneously in a heterogeneous mixture of biological sample. Sorting of biological samples is depending on forward scattering (FSC), side scattering (SSC), and fluorescence activity of individual cell. On the

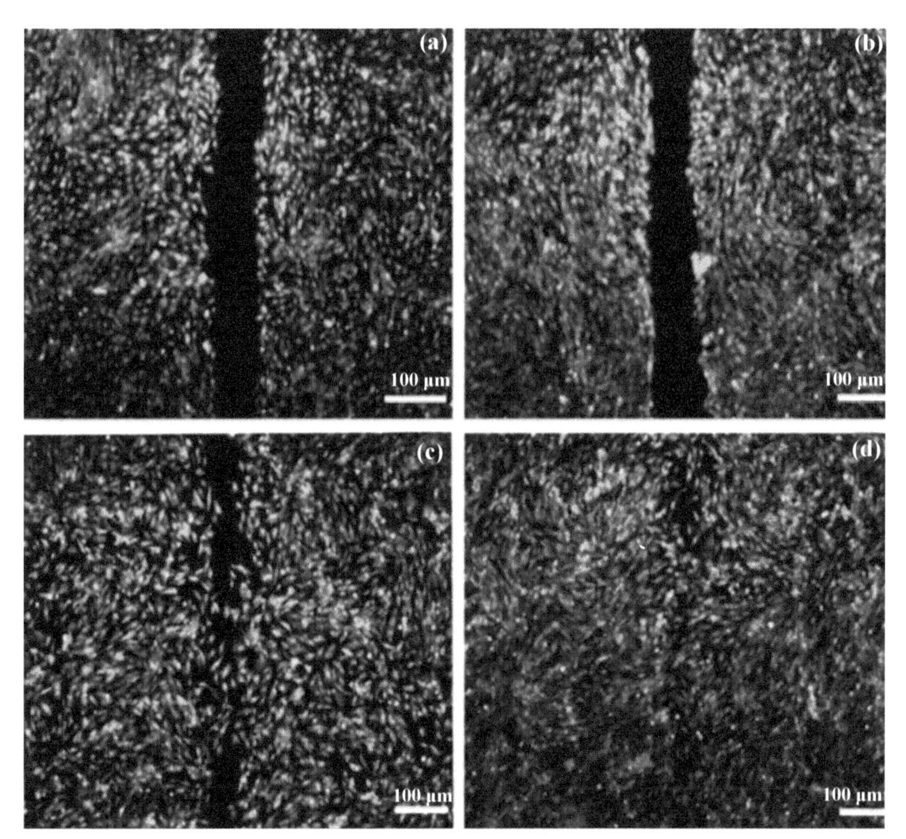

FIGURE 9.27 Fluorescence microscopic image revealing migration pattern of H9c2 cells (a, c) PCL/GE NF treated cells at 0 and 8 hours. (b, d) Sal B-MAAP PCL/GE NF treated cells at 0 and 8 hour. (Reprinted with permission from [59]. © 2018 Elsevier.)

basis of FSC and SSC, the amount of cells at different stages of the cell cycle, highly specific antibodies labeled with fluorescent dyes are sorted and analyzed. In FACS, FSC monitors the size of healthy and dead cell and SSC monitors the granularity of cell. The presence of the healthy cells at higher concentration will be indicated by FSC value whereas the cell death (apoptosis) is indicated through low FSC and SSC value.

Pereirae et al. [65] had analyzed the fractions of the living and dead fibroblast cells using flow cytometry. They have also shown that the proliferation and attachment of fibroblast cells on the CNFs were low and determined the rate of apoptosis. Similarly, Zhang et al. [66] have studied the antitumor activity of PLA nanofibers loaded with 5-fluorouracil (5-Flu) and oxaliplatin (O). FACS analysis indicated the effective cell (HCT-8 cells) growth inhibition than the combination of pure oxaliplatin and 5-Flu. Hence, it was proven that drug-loaded electrospun nanofibers can be used as an implantable local drug delivery system suitable for preventing local tumor recurrence against colon rectal cancer.

9.5.4.1.9 Degradability of the Scaffolds

The scaffolds used for regeneration application should have a capacity of degradation in a controlled manner and nonimmunogenic to the system. The degradation rate of the polymer depends on the tissue to be regenerated and at the same time, degradation of the scaffold should not be toxic to the cells directly or through production of undesirable byproducts, which may cause inflammatory reactions [67] For example, the degradation of scaffold should be slow in case of skeletal muscle regeneration, as it should maintain the mechanical properties to native skin. In addition, it should not evoke any immunological reaction and pathogenic infections which could hamper the healing process [68]

9.6 CONCLUSION

Thus, the speed, accuracy, and process precision of the architecture form the basis for any successful and functional nanomaterial design which would entirely depend on the concerted efforts from combination of scientists, engineers, biochemists, doctors, material physicists, and so on. This is even more stupendous for revolutionizing the entire healthcare industry that demands not only site-specificity but also targeted efficiency to overcome undesirable consequences hindering the biological functions. Therefore, a sound knowledge of the functional requisites rely on molecular pathology, genotypic, and phenotypic characteristics which are of major concern while designing any nanomaterial. Further the design should also overcome the hurdles and challenges involved in the bulk preparation, stability, shelf life, and so on, coercing an incessant need for designing newer synthetic and analytical tools for marketing strategies. This would thus lead to a continuous indispensable and excited research platform based on the drawbacks of the hitherto known material designs.

KEYWORDS

- **nanofibers**
- **bicomponent fibers**
- **electrospinning**
- **physicochemical characterizations**
- **nanofiber fabrication**

REFERENCES

1. Silva, G. A. Introduction to Nanotechnology and Its Applications to Medicine. *Surg. Neurol.* 2004, *61* (3), 216–220. https://doi.org/10.1016/j.surneu.2003.09.036.
2. Roco, M. C.; Bainbridge, W. S. Societal Implications of Nanoscience and Nanotechnology: Maximizing Human Benefit. *J. Nanoparticle Res.* 2005, *7* (1), 1–13. https://doi.org/10.1007/s11051-004-2336-5.
3. Grzybowski, B. A.; Huck, W. T. S. The Nanotechnology of Life-Inspired Systems. *Nat. Nanotechnol.* 2016, *11* (7), 585–592. https://doi.org/10.1038/nnano.2016.116.
4. Gajanan, K.; Tijare, S. N. Applications of Nanomaterials. *Mater. Today Proc.* 2018, *5* (1), 1093–1096. https://doi.org/10.1016/j.matpr.2017.11.187.
5. Bhatia, S. *Natural Polymer Drug Delivery Systems*; 2016. https://doi.org/10.1007/978-3-319-41129-3.
6. Kulinets, I. *Biomaterials and Their Applications in Medicine*; Woodhead Publishing Limited, 2014. https://doi.org/10.1533/9780857099204.1.
7. Fitzgerald, W. E.; Knudsen, J. P. Mixed-Stream Spinning of Bicomponent Fibers1. *Text. Res. J.* 1967, *37* (6), 447–453. https://doi.org/10.1177/004051756703700602.
8. Nakata, K.; Fujii, K.; Ohkoshi, Y.; Gotoh, Y.; Nagura, M.; Numata, M.; Kamiyama, M. Poly(Ethylene Terephthalate) Nanofibers Made by Sea-Island-Type Conjugated Melt Spinning and Laser-Heated Flow Drawing. *Macromol. Rapid Commun.* 2007, *28* (6), 792–795. https://doi.org/10.1002/marc.200600624.
9. Ma, P. X.; Zhang R. Synthetic Nano-Scale Fibrous Extracellular Matrix. *Nano-Fiber Matrix* 1998, *46* (1), 60–72.
10. Feng, L.; Li, S.; Li, H.; Zhai, J.; Song, Y.; Jiang, L.; Zhu, D. Super-Hydrophobic Surface of Aligned Polyacrylonitrile Nanofibers. *Angew. Chem. Int. Ed.* 2002, *41* (7), 1221–1223.
11. Li, X.; Tian, S.; Ping, Y.; Kim, D. H.; W. K. One-Step Route to the Fabrication of Highly Porous Polyaniline Nanofiber Films by Using PS-b-PVP Diblock Copolymers as Templates. *Langmuir.* 2005, *21* (15), 9393–9397. https://doi.org/10.1021/la0514009.
12. Feng, J.; Li, J.; Lv, W.; Xu, H.; Yang, H.; Yan, W. Synthesis of Polypyrrole Nano-Fibers with Hierarchical Structure and Its Adsorption Property of Acid

Red G from Aqueous Solution. *Synth. Met.* 2014, *191*, 66–73. https://doi.org/10.1016/j.synthmet.2014.02.013.
13. Ondarcuhu, T.; Joachim, C. Drawing a Single Nanofibre over Hundreds of Microns Related Content Combing a Nanofibre in a Nanojunction. 1998, *42*, 215.
14. Ren, L.; Ozisik, R.; Kotha, S. P. Rapid and Efficient Fabrication of Multilevel Structured Silica Micro-/Nanofibers by Centrifugal Jet Spinning. *J. Colloid. Interface Sci.* 2014, *425*, 136–142. https://doi.org/10.1016/j.jcis.2014.03.039.
15. Padron, S.; Fuentes, A.; Caruntu, D.; Lozano, K. Experimental Study of Nanofiber Production through Forcespinning. *J. Appl. Phys.* 2013, *113* (2), 024318. https://doi.org/10.1063/1.4769886.
16. Weitz, R. T.; Harnau, L.; Rauschenbach, S.; Burghard, M.; Kern, K. Polymer Nanofibers via Nozzle-Free Centrifugal Spinning. *Nano Lett.* 2008, *8* (4), 1187–1191. https://doi.org/10.1021/nl080124q.
17. Ducree, J.; Haeberle, S.; Lutz, S.; Pausch, S.; Stetten, F. V; Zengerle, R. The Centrifugal Microfluidic Bio-Disk Platform. *J. Micromech. Microeng.* 2007, *17*, S103–S115. https://doi.org/10.1088/0960-1317/17/7/S07.
18. Lu, Y.; Li, Y.; Zhang, S.; Xu, G.; Fu, K.; Lee, H.; Zhang, X. Parameter Study and Characterization for Polyacrylonitrile Nanofibers Fabricated via Centrifugal Spinning Process. *Eur. Polym. J.* 2013, *49* (12), 3834–3845. https://doi.org/10.1016/j.eurpolymj.2013.09.017.
19. Beachley, V.; Wen, X. Polymer Nanofibrous Structures: Fabrication, Biofunctionalization, and Cell Interactions. *Prog. Polym. Sci.* 2010, *35* (7), 868–892. https://doi.org/10.1016/j.progpolymsci.2010.03.003.
20. Rollings, D. E.; Tsoi, S.; Sit, J. C.; Veinot, J. G. C.; Uni, V.; Canada, T. G.; Engineering, C.; Uni, V.; Tg, A. Formation and Aqueous Surface Wettability of Polysiloxane Nanofibers Prepared via Surface Initiated, Vapor-Phase Polymerization of Organotrichlorosilanes. *Langmuir.* 2007, *23*, 5275–5278.
21. Raghavan, B.; Soto, H.; Lozano, K. Fabrication of Melt Spun Polypropylene Nanofibers by Forcespinning. *J. Eng. Fiber. Fabr.* 2013, *8* (1), 52–60.
22. Bhardwaj, N.; Kundu, S. C. Electrospinning: A Fascinating Fiber Fabrication Technique. *Biotechnol. Adv.* 2010, *28* (3), 325–347. https://doi.org/10.1016/j.biotechadv.2010.01.004.
23. Doshi, J.; Reneker, D. H. Electrospinning Process and Applications of Electrospun Fibers. *J. Electrostat.* 1995, *35*(2–3), 151–160, ISSN 0304-3886, https://doi.org/10.1016/0304-3886(95)00041-8.
24. Theron, A.; Zussman, E.; Yarin, A. L. Electrostatic Field-Assisted Alignment of Electrospun Nanofibres. *Nanotechnology*. 2001, *12* (3), 384–390. https://doi.org/10.1088/0957-4484/12/3/329.
25. Wei, Q. F.; Gao, W. D.; Hou, D. Y.; Wang, X. Q. Surface Modification of Polymer Nanofibres by Plasma Treatment. *Appl. Surf. Sci.* 2005, *245* (1–4), 16–20. https://doi.org/10.1016/j.apsusc.2004.10.013.
26. Liu, Z.; Jia, L.; Yan, Z.; Bai, L. Plasma-Treated Electrospun Nanofibers as a Template for the Electrostatic Assembly of Silver Nanoparticles. *New J. Chem.* 2018, *42* (13), 11185–11191. https://doi.org/10.1039/c8nj01151f.
27. Wei, X.; Xia, Z.; Wong, S-.C.; Baji, A. Modelling of Mechanical Properties of Electrospun Nanofibre Network. *Int. J. Exp. Comput. Biomech.* 2009,

1 (1), 45. https://doi.org/10.1504/ijecb.2009.022858.
28. Başaran, İ.; Oral, A. Grafting of Poly(ε-Caprolactone) on Electrospun Gelatin Nanofiber through Surface-Initiated Ring-Opening Polymerization. *Int. J. Polym. Mater. Polym. Biomater.* 2018, *67* (18), 1051–1058. https://doi.org/10.1080/00914037.2017.1417287.
29. Tao, D.; Wei, Q.; Cai, Y.; Xu, Q.; Sun, L. Functionalization of Polyamide 6 Nanofibers by Electroless Deposition of Copper. *J. Coatings Technol. Res.* 2008, *5* (3), 399–403. https://doi.org/10.1007/s11998-008-9118-4.
30. Díaz Silva, N.; Salas, B. V.; Nedev, N.; Alvarez, M. C.; Rull, J. M. B.; Zlatev, R.; Stoytcheva, M. Synthesis of Carbon Nanofibers with Maghemite via a Modified Sol-Gel Technique. *J. Nanomater.* 2017, *2017*. https://doi.org/10.1155/2017/5794312.
31. Chen, W. S.; Guo, L. Y.; Tang, C. C.; Tsai, C. K.; Huang, H. H.; Chin, T. Y.; Yang, M. L.; Chen-Yang, Y. W. The Effect of Laminin Surface Modification of Electrospun Silica Nanofiber Substrate on Neuronal Tissue Engineering. *Nanomaterials.* 2018, *8* (3). https://doi.org/10.3390/nano8030165.
32. Satish, A.; Korrapati, P. S. Strategic Design of Peptide-decorated Aligned Nanofibers Impregnated with Triiodothyronine for Neural Regeneration. *J. Tissue Eng. Regen. Med.* 2019. https://doi.org/10.1002/term.2822.
33. Ladd, M. R.; Hill, T. K.; Yoo, J. J.; Jin, S. Electrospun Nanofibers in Tissue Engineering. *Nanofibers—Prod. Prop. Funct. Appl.* 2012. https://doi.org/10.5772/24095.
34. Subramanian, A.; Krishnan, U. M.; Sethuraman, S. Fabrication of Uniaxially Aligned 3D Electrospun Scaffolds for Neural Regeneration. *Biomed. Mater.* 2011, *6* (2). https://doi.org/10.1088/1748-6041/6/2/025004.
35. Hrib, J; Sirc,J.; Hobzoval, R.; Hampejsova, Z.; Bosakova, Z.; Munzarova, M.; Michalek, J. Nanofibers for Drug Delivery—Incorporation and Release of Model Molecules, Influence of Molecular Weight and Polymer Structure. *Beilstein J. Nanotechnol.* 2015, *6*, 1939–1945. https://doi.org/10.3762/bjnano.6.198.
36. Chaudhary, L. S.; Ghatmale, P. R.; Chavan, S. S.; Patil, R. N. Characterization Technique of Nanofibers. *Int. J. Sci. Res. Dev.* 2016, *4* (03), 2035–2041.
37. Asmatulu, R.; Yildirim, M. B.; Khan, W. S.; Adeniji, A.; Wamocha, H. Nanofiber Fabrication and Characterization for the Engineering Education. *2007 ASEE Midwest Regional Conference*, September 19–21, 2007, Wichita, KS, 9 pages.
38. Shoba, E.; Lakra, R.; Kiran, M. S.; Korrapati, P. S. Design and Development of Papain-Urea Loaded PVA Nanofibers for Wound Debridement. *RSC Adv.* 2014, *4*, 60209–60215. https://doi.org/10.1039/c4ra10239h.
39. Qi, J.; Huang, Z.; Wang, Y.; Mani, M. P.; Jaganathan, S. K. Development and Blood Compatibility Assessment of Electrospun Polyvinyl Alcohol Blended with Metallocene Polyethylene and Plectranthus Amboinicus (PVA/MPE/PA) for Bone Tissue Engineering. *Int. J. Nanomedicine.* 2018, *13*, 2777–2788. https://doi.org/10.2147/ijn.s151242.
40. Sreedhara, S. S.; Tata, N. R. A Novel Method for Measurement of Porosity in Nanofiber Mat Using Pycnometer in Filtration. *J. Eng. Fiber. Fabr.* 2013, *8* (4). https://doi.org/10.1177/155892501300800408.

41. Maleknia, L.; Dilamian, M.; Pilehrood, M. K.; Sadeghi-Aliabadi, H.; Hekmati, A. H. Preparation, Process Optimization and Characterization of Core-Shell Polyurethane/Chitosan Nanofibers as a Potential Platform for Bioactive Scaffolds. *Res. Pharm. Sci.* 2018, *13* (3), 273–282. https://doi.org/10.4103/1735-5362.228957.
42. Liu, X.; Xu, S.; Kuang, X.; Wang, X. 3D Cardiac Cell Culture on Nanofiber Bundle Substrates for the Investigation of Cell Morphology and Contraction. *Micromachines.* 2017, *8* (5). https://doi.org/10.3390/mi8050147.
43. Janani, I.; Lakra, R.; Syamala, M.; Kiran, P.; Sai. K. Selectivity and Sensitivity of Molybdenum Oxide-Polycaprolactone Nanofiber Composites on Skin Cancer: Preliminary in-Vitro and in-Vivo Implications. *J. Trace Elem. Med. Biol.* 2018, *49* (April), 60–71. https://doi.org/10.1016/j.jtemb.2018.04.033.
44. Bhaarathy, V.; Venugopal, J.; Gandhimathi, C.; Ponpandian, N.; Mangalaraj, D.; Ramakrishna, S. Biologically Improved Nanofibrous Scaffolds for Cardiac Tissue Engineering. *Mater. Sci. Eng. C.* 2014, *44*, 268–277. https://doi.org/10.1016/j.msec.2014.08.018.
45. Chen, R; Qiu, L; Ke, Q; He, C; Mo, X. Electrospinning Thermoplastic Polyurethane-Contained Collagen Nanofibers for Tissue-Engineering Applications. *J Biomater Sci Polym Ed.* 2009, *20* (11), 1513–1536.
46. Rychter, M.; Baranowska-Korczyc, A.; Milanowski, B.; Jarek, M.; Maciejewska, B. M.; Coy, E. L.; Lulek, J. Cilostazol-Loaded Poly(ε-Caprolactone) Electrospun Drug Delivery System for Cardiovascular Applications. *Pharm. Res.* 2018, *35* (32). https://doi.org/10.1007/s11095-017-2314-0.
47. Abdal-hay, A.; Khalil, K. A.; Hamdy, A. S.; Al-Jassir, F. F. Fabrication of Highly Porous Biodegradable Biomimetic Nanocomposite as Advanced Bone Tissue Scaffold. *Arab. J. Chem.* 2017, *10* (2), 240–252. https://doi.org/10.1016/j.arabjc.2016.09.021.
48. Rezaei, F.; Gorbanev, Y.; Chys, M.; Nikiforov, A.; Van Hulle, S. W. H.; Cos, P.; Bogaerts, A.; De Geyter, N. Investigation of Plasma-Induced Chemistry in Organic Solutions for Enhanced Electrospun PLA Nanofibers. *Plasma Process. Polym.* 2018, *15* (6), 1–18. https://doi.org/10.1002/ppap.201700226.
49. Rao, K. S.; Reddy, M. K.; Horning, J. L.; Labhasetwar, V. TAT-Conjugated Nanoparticles for the CNS Delivery of Anti-HIV Drugs. *Biomaterials.* 2009, *29* (33), 4429–4438. https://doi.org/10.1016/j.biomaterials.2008.08.004. TAT-conjugated.
50. Wagner, A.; Poursorkhabi, V.; Mohanty, A. K.; Misra, M. Analysis of Porous Electrospun Fibers from Poly(l-Lactic Acid)/Poly(3-Hydroxybutyrate-Co-3-Hydroxyvalerate) Blends. *ACS Sustain. Chem. Eng.* 2014, *2* (8), 1976–1982. https://doi.org/10.1021/sc5000495.
51. Feng, S.; Shen, X.; Fu, Z.; Shao, M. Preparation and Characterization of Gelatin–Poly(L-Lactic) Acid/Poly(Hydroxybutyrate-Co-Hydroxyvalerate) Composite Nanofibrous Scaffolds. *J. Macromol. Sci. Part B.* 2011, *50* (9), 1705–1713.
52. Sfakis, L.; Kamaldinov, T.; Khmaladze, A.; Hosseini, Z. F.; Nelson, D. A.; Larsen, M.; Castracane, J. Mesenchymal Cells Affect Salivary Epithelial Cell Morphology on PGS/PLGA Core/Shell Nanofibers. *Int. J. Mol. Sci.* 2018, *19* (4), 1–15. https://doi.org/10.3390/ijms19041031.

53. Xu, B.; Li, Y.; Fang, X.; Thouas, G. A.; Cook, W. D.; Newgreen, D. F.; Chen, Q. Mechanically Tissue-like Elastomeric Polymers and Their Potential as a Vehicle to Deliver Functional Cardiomyocytes. *J. Mech. Behav. Biomed. Mater.* 2013, *28*, 354–365. https://doi.org/10.1016/j.jmbbm.2013.06.005.
54. Hidetoshi Matsumoto; Tanioka, A. Functionality in Electrospun Nanofibrous Membranes Based on Fiber's Size, Surface Area, and Molecular Orientation. *Membranes (Basel).* 2011, *1* (3), 249–264. https://doi.org/10.3390/membranes1030249.
55. Chocholata, P.; Kulda, V.; Babuska, V. Fabrication of Scaffolds for Bone-Tissue Regeneration. *Materials (Basel).* 2019, *12* (568), 1–25. https://doi.org/10.3390/ma12040568.
56. Wang, Y.; Tan, H.; Hui, X. Biomaterial Scaffolds in Regenerative Therapy of the Central Nervous System. *Biomed Res. Int.* 2018, *2018*, 7848901.
57. Wolf, M. T.; Dearth, C. L.; Sonnenberg, S. B.; Loboa, E. G.; Badylak, S. F. Naturally Derived and Synthetic Scaffolds for Skeletal Muscle Reconstruction. *Adv Drug Deliv Rev.* 2018, *84*, 208–221. https://doi.org/10.1016/j.addr.2014.08.011.
58. Chen, F.-M.; Liu, X. Advancing Biomaterials of Human Origin for Tissue Engineering. *Prog Polym Sci.* 2016, *1* (53), 86–168. https://doi.org/10.1016/j.progpolymsci.2015.02.004.
59. Shoba, E.; Lakra, R.; Kiran, M. S.; Korapatti, P. S. Strategic Design of Cardiac Mimetic Core-Shell Nanofibrous Scaffold Impregnated with Salvianolic Acid B and Magnesium L-Ascorbic Acid 2 Phosphate for Myoblast Differentiation. *Mater. Sci. Eng. C.* 2018, *90*, 131–147. https://doi.org/10.1016/j.msec.2018.04.056.
60. Yoo, Y.; Oh, J.; Zhang, Q. Dimethyloxalylglycine-Embedded Poly(ε-Caprolactone) Fiber Meshes Promote Odontoblastic Differentiation of Human Dental Pulp-Derived Cells. *J. Endod.* 2018, *44* (1), 98-103.e1. https://doi.org/10.1016/j.joen.2017.09.002.
61. Poornima, B.; Korrapati, P. S. Fabrication of Chitosan-Polycaprolactone Composite Nanofibrous Scaffold for Simultaneous Delivery of Ferulic Acid and Resveratrol. *Carbohydr. Polym.* 2017, *157*. https://doi.org/10.1016/j.carbpol.2016.11.056.
62. Keller, L.; Idoux-Gillet, Y.; Wagner, Q.; Eap, S.; Brasse, D.; Schwinté, P.; Arruebo, M.; Benkirane-Jessel, N. Nanoengineered Implant as a New Platform for Regenerative Nanomedicine Using 3D Well-Organized Human Cell Spheroids. *Int. J. Nanomedicine.* 2017, *12*, 447–457. https://doi.org/10.2147/IJN.S116749.
63. Beachley, V.; Katsanevakis, E.; Zhang, N.; Wen, X. A Novel Method to Precisely Assemble Loose Nanofiber Structures for Regenerative Medicine Applications. *Adv. Healthc. Mater.* 2013, *2*, 343–351. https://doi.org/10.1002/adhm.201200125.
64. Villarreal-Gómez, L. J.; Vera-Graziano, R.; Vega-Ríos, M. R.; Pineda-Camacho, J. L.; Almanza-Reyes, H.; Mier-Maldonado, P.A.; Cornejo-Bravo, J. M. Biocompatibility Evaluation of Electrospun Scaffolds of Poly (L-Lactide) with Pure and Grafted Hydroxyapatite. *J. Mex. Chem. Soc.* 2014, *58* (4), 435–443. https://doi.org/10.29356/jmcs.v58i4.53.
65. Pereira, M. M.; Raposo, N. R. B.; Brayner, R.; Teixeira, E. M.; Oliveira, V.; Quintao, C. C. R.; Camargo, L. S. A.; Mattoso, L. H. C.; Brandao, H. Cytotoxicity and Expression of Genes Involved in the Cellular Stress Response and Apoptosis in

Mammalian Fibroblast Exposed to Cotton Cellulose Nanofibers. *Nanotechnology.* 2013, *25*. https://doi.org/10.1088/0957-4484/24/7/075103.

66. Zhang, J.; Wang, X.; Liu, T.; Liu, S.; Jing, X. Antitumor Activity of Electrospun Polylactide Nanofibers Loaded with 5-Fluorouracil and Oxaliplatin against Colorectal Cancer Antitumor Activity of Electrospun Polylactide Nanofibers Loaded with 5-Fluorouracil and Oxaliplatin against Colorectal Cancer. *Drug Deliv.* 2016, *23* (3), 784–790. https://doi.org/10.3109/10717544.2014.916768.
67. Patel, H.; Minal, B.; Ganga, S. Biodegradable Polymer Scaffold for Tissue Engineering. *Trends Biomater. Artif. Organs.* 2011, *25* (1), 20–29.
68. Vasita, R.; Katti, D. S. Nanofibers and Their Applications in Tissue Engineering. *Int. J. Nanomedicine.* 2006, *1* (1), 15–30.
69. Hooper, J.P. Centrifugal Spinneret. U.S.A: 1500931, 1924.

PART II

Nanotechnology Approaches to Tissue Engineering

CHAPTER 10

Hydrogel Systems for Tissue Engineering

SENTHILKUMAR MUTHUSAMY, SHINY VELAYUDHAN, and
P. R. ANIL KUMAR*

Biomedical Technology Wing, Sree Chitra Tirunal Institute for Biomedical Sciences and Technology, Thiruvananthapuram, Kerala 695 012, India

Corresponding author. E-mail: anilkumarpr@sctimst.ac.in

ABSTRACT

Hydrogels are having potential applications in tissue engineering for regenerating lost tissues either by acting as a scaffold for intrinsic cell growth or by acting as a reservoir to deliver required biomolecules and cells in the damaged area. Enormous number of biocompatible materials have been studied as a hydrogel and classified based on several factors. Here we described elaborately about the hydrogel classification and factors influencing hydrogel properties in the context with tissue engineering. We also provided methods for physical and biological characterization of hydrogel along with their use as a bioink in 3D/4D bioprinting.

10.1 INTRODUCTION

The regeneration ability of human body is limited based on tissue or cell type and the surrounding micro-environment. Regenerative approach with tissue engineering significantly supports regaining the tissue loss and current research is progressing toward making human tissue and organs in-vitro for medical applications [1]. Key components involved in tissue engineering are; (1) the cells of interest, (2) the biomaterial as a scaffold, (3) adequate nourishment with active environment for cell expansion or differentiation [2]. The cell source for tissue engineering can be chosen depending on the target tissue requirement. Stem cells such as bone marrow stem cells, mesenchymal stem cells (MSC), and induced pluripotent stem cells are some of the other sources that have differentiation potential toward multicellular lineage. Whereas, the complex organ can be made using different types of primary cells with or without stem cell [3–5]. Biomaterials are the natural or synthetic substances that make a platform with direct or indirect contact to the cells and also act as a delivery system to the site of implantation. The ability of gas, nutrient, and growth factor exchange by the biomaterial could facilitate the cell function, proliferation, and differentiation. Besides, biomaterials used in tissue engineering should exhibit biocompatibility, capable of biodegradation, or bioresorption in the host [6–8]. Based on their properties, biomaterials are fabricated into scaffold for tissue engineering applications using various physical and chemical processes such as self-assembly, solvent casting, freeze-drying, electrospinning, and others [9]. Initially, additive manufacturing with the process of creating prototypes that has the ability of making complex geometric structures became sophisticated in mechanical industries and commonly named as three dimensional (3D) printing. As native 3D architecture of the tissues strongly influences the cellular fate in both physical and functional, researches developed novel strategies using this 3D printing technology with several biomaterials and cells for tissue engineering [10–13]. Recently, 4D printing technology has been evolved by applying mathematical modeling along with time in 3D printing. This would help to fabricate dynamic structures with shape variations by different external stimuli on smart biomaterials [14, 15]. Bioinks are the important component in 3D/4D bioprinting, which provide mechanical and/or biological support for the cells. Various scaffold-based (hydrogels, decellularized matrix, and microcarriers) and scaffold-free (cell aggregates and spheroids) have been developed as bioink [16–18]. Currently, hydrogels are widely used as bioink that can hold several fold water or biological fluids to mimic physiochemical properties of tissues [19]. This chapter focuses on the hydrogels, a building block in tissue engineering and 3D/4D bioprinting applications.

10.2 DEFINITIONS

10.2.1 TISSUE ENGINEERING

Tissue engineering is an interdisciplinary process that aims to create native tissue-like structure in-vitro to restore or enhance function of injured or diseased tissue in-vivo [20–23]. For example, the cells are embedded into the biocompatible scaffold and transferred to the target site, where the scaffold can be degraded and integrated into host with formation of new tissue or eliminated through blood stream.

10.2.2 BIOMATERIALS

Biomaterials are nondrug substance or combination of substances derived from biological (e.g., Collagen) or synthetic materials (e.g., ceramic), which can be used to augment tissue function or replace damaged tissue at least in part. Each biomaterial has its own physical and chemical properties, those makes them useful in wide range of medical applications especially in regenerative medicine and tissue engineering [7, 24, 25].

10.2.3 HYDROGEL

Hydrogels are 3D colloidal hydrophilic polymer network either physically or chemically crosslinked, which filled with water within macromolecular space. As a good biocompatible scaffold, hydrogels are widely used as cell-laden biomaterial in tissue engineering [9, 26, 27]. New generation hydrogels have the potential for responding to external stimuli those named as "smart hydrogels" [28].

10.2.4 3D/4D BIOPRINTING

3D bioprinting is a fabrication of complex shape using the desired living cells and high-dense biomaterial (bioink) layer by layer accurately with the help of computer-aided design. Combination of smart biomaterial that responds to time-dependent external stimuli along with 3D bioprinting gives fourth dimension to biomaterial [29–32].

10.2.5 BIOINK

Hydrogels are hydrophilic, can encapsulate cells/bioactive molecules and mimic many of the characteristics of natural extracellular matrix (ECM), making them an attractive cell delivery vehicle. They have good porosity for diffusion of oxygen, nutrients, and metabolites; can be processed under mild cell-friendly conditions; and produce little to no irritation. When hydrogels are loaded with living cells, they become bioink, and are the "raw material" of bioprinting processes. The biomaterial formulation used to print the living cells is called bioink. Printability, mechanical integrity, stability, and functionality are some important features of bioink in 3D bioprinting [33, 34]. The mono or multicellular spheroids also used as bioink for scaffold-free tissue engineering, in which, the cells in the aggregates forms self-assembly with cell–cell interaction or their own ECM synthesis [35, 36].

10.3 HYDROGEL FORMULATION

A myriad of physical and chemical mechanisms such as ionic, hydrophilic, hydrophobic, and chemical crosslinking are involved in the hydrogel formation when the polymer dispersed in an aqueous medium. Upon selection of suitable material for hydrogel preparation, the polymer chains are crosslinked by physical, chemical, or radiation that enables them to swell without dissolving while retaining their three-dimensional structure. The intrinsic property of the material is a major factor that limits hydrogel formation in physical method whereas chemical crosslinking can be controlled dynamically. Certain natural and synthetic polymers such as gelatin, agarose, and polyacrylic acid (PAA) form hydrogel in response to temperature change. The alteration in the solubility at a given temperature renders the gel formation. These thermoresponsive properties may be altered by the addition/changing copolymer or molecular weight of the polymer [37]. Ionic bonding like chelation and electrostatic interaction like interaction of two charged molecules also play role in the formation of hydrogel spontaneously. For example, presence of divalent ions such as Ca^{2+} and Ba^{2+} facilitates gelation of G-blocks present in the alginate through chelation and the presence of polyelectrolytes in the medium allows to form poly(L-lysine)/PAA complex [38, 39].

Majority of natural and synthetic polymers possess active moieties on their backbone or side chains, which allows to chemically crosslink to obtain hydrogels. Over the decades several reactions such as carbodiimide formation, aldehyde complementation, radical polymerization, enzyme-based linking, irradiation, and click chemistry are developed to prepare hydrogels [40, 41]. Each method has its own advantages and limitations and now combinational methods have been developing to improve the hydrogel formulations with excellent physicochemical and mechanical properties such as printable bioink, injectable and self-healing gel for tissue engineering, and other regenerative therapeutic applications [42]. The sol–gel transition (solution phase to hydrogel phase) has to be considered as critical during the formulation of hydrogel for tissue engineering application. The function of encapsulated cells or bioactivity of embedded molecules may be compromised if the monomer, solvent, polymerization initiator, or temperature is having significant effect during this sol–gel transition. Therefore approaches that achieve minimal effect on cells and bioactive molecules should be developed during hydrogel preparation.

10.4 HYDROGEL FOR TISSUE ENGINEERING

"PubMed search" revealed the tremendous increase in hydrogel since year 2000 (Figure 10.1a and b). Hydrogels are employed in tissue engineering applications such as space-filling, wound healing, delivering cells or bioactive molecules, and to provide 3D structure where the cells could organize to develop desired tissue (Figures 10.2 and 10.3). In tissue engineering, it is expected that a hydrogel should maintain the rate of cellular proliferation and function throughout life of the hydrogel. Therefore a critical design of the hydrogel has to be considered between the transitions from degradation to healing or new tissue formation. Since, the required materials and their properties will be quite different depending on the tissue of interest and the specific application. Both physical and chemical cues of the materials have significant impact on the cellular functions such as adhesion, proliferation, migration, and differentiation. Numerous hydrogels are being employed as a scaffold materials, which are either natural or synthetic and their structural integrity depends on their crosslinking. Hydrogels used in tissue engineering applications are typically biodegradable, which processed under mild conditions and have properties that recapitulate tissue or ECM architecture.

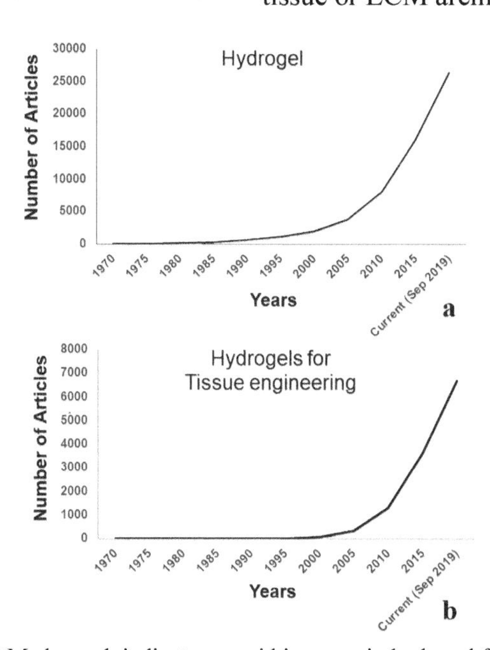

FIGURE 10.1 PubMed search indicates a rapid increase in hydrogel formulation and their use in tissue engineering.

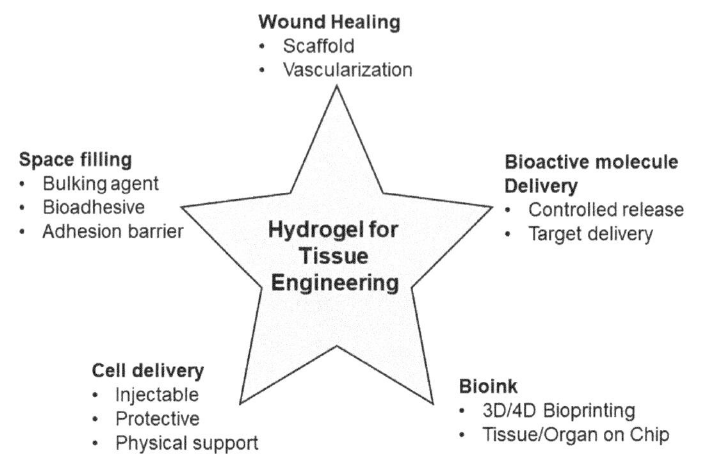

FIGURE 10.2 Applications of the hydrogel in tissue engineering.

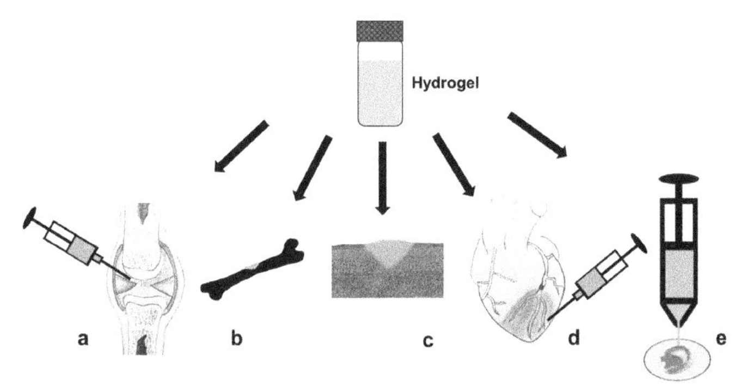

FIGURE 10.3 Examples of hydrogel-based tissue regeneration. Repair of cartilage (a), bone (b), skin (c), infarcted myocardium (d), and bioprinting of artificial organs (e).

10.4.1 HYDROGELS FOR SPACE FILLING

Certain hydrogels have the ability to maintain structural integrity and desired volume for a given period, which is employed as a filling material that could provide bulking effect, prevent adhesions, and act as bioadhesives (Figure 10.4a). It is important for a hydrogel to keep a desired volume and structural integrity for the intended period. Modulating the filling hydrogel to present stimuli that directs the growth and formation of a desired tissue on the defective site upon delivery is equally important. The cellular adhesion to the hydrogel directly influences their proliferation, migration, and organization. Some natural hydrogels like ECM proteins possess the adhesion moieties such as RGD (arginine–glycine–aspartic acid) peptide sequence for cell attachment whereas in other hydrogels, incorporation of this RGD peptide have shown to favor the binding of cells including fibroblasts, endothelial cells, smooth muscle cells, osteoblasts, and chondrocytes [43, 44].

As hydrogels are hydrophilic polymers with interconnected networks by physical and chemical bonds, controlling their properties such as polymer concentration, the distance between molecular interactions in the network could alter the hydrogel stiffness, swelling, and pore size, which significantly impact cell spreading and vascularization in newly developing tissue. Alginate, chitosan, and collagen-based hydrogels have been shown as potential bulking agents. Yang et al. [45] formulated cellulose nanocrystals-reinforced injectable hydrogels that could fill in irregular cavities and shapes, which act as a promising bone tissue engineering material [45]. Using crystal templating technique, Zawko and Schmidt fabricated nanostructured hyaluronic acid (HA) scaffolds with space-filling dendritic pore networks and fibrillar microtopography that suggested to be applicable as regenerative patches for skin and other tissues [46]. Ma et al. [47] developed a hydrogel with human-like collagen and chitosan crosslinked with dialdehyde starch that showed enhanced biological stability and excellent biocompatibility, where they concluded for using them as potentially promising hydrogel for skin patch scaffolds, wrinkle treatments, and tissue cavity fillers.

Other than space-filling, synthetic hydrogel such as polyethylene glycol (PEG) and natural hydrogel such as fibrin works well as a bioadhesive. As an FDA-approved agent, PEG-based hydrogels meet mechanical and biological requirements for applications as a sealant [48]. Fibrin is a natural hemostatic agent forms a gel in the presence of Ca^{2+} via the cross-linking of fibrinogen that promotes wound healing and commercial products (e.g., Tissucol®) are currently available as tissue sealants. Hydrogels composed of chitosan are now used as biological adhesives in surgical procedures to seal small wounds to improve the effectiveness of wound dressings [49]. In certain cases, the postoperative adhesions are unavoidable especially during abdominal surgery, which cause serious complications such as severe pain and bowel obstruction. Hydrogels have been used as barriers to prevent postoperative adhesion formation. For example, mitomycin C-loaded crosslinked HA, PEG-poly(ε-caprolactone) (PCL) composite and hydroxybutyl chitosan hydrogels have been shown effective in reducing postoperative adhesion formation [50].

10.4.2 HYDROGELS FOR BIOACTIVE MOLECULES DELIVERY

Delivering the bioactive molecules at the target site could not only improve the rate of regeneration but also reduce the cost and side effects. For example, delivery of most extensively studied vascular endothelial growth factor (VEGF), systemically to promote local angiogenesis, may lead to serious side-effects including neovascularization of off-target tissues and tumor growth [51, 52]. Hydrogels can act as scaffolds to stabilize and deliver these drugs and bioactive molecules (Figure 10.4b). In-vitro and in-vivo delivery of VEGF from ionically crosslinked alginate hydrogels and glutaraldehyde-crosslinked collagen sponges has been demonstrated efficient angiogenesis around the implanted site [53, 54]. Similarly basic fibroblast growth factor (bFGF) release from heparin alginate hydrogels has been shown to enhance angiogenesis and bone morphogenic protein-2 release from photo-crosslinked poly(lactic acid) (PLA)–PEG and PLA–dexamethasone (DX)–PEG systems has been shown to promote osteogenesis [55, 56]. Gelatin-based hydrogels have been investigated in broad range of biomolecule delivery applications [57–59]. The formation of PEO–cl–PEI nanogel by cationic network between poly(ethylene oxide) (PEO) and polyethyleneimine (PEI) augments the interaction of oligonucleotides, which found to transport of oligonucleotides across the gastrointestinal epithelium and

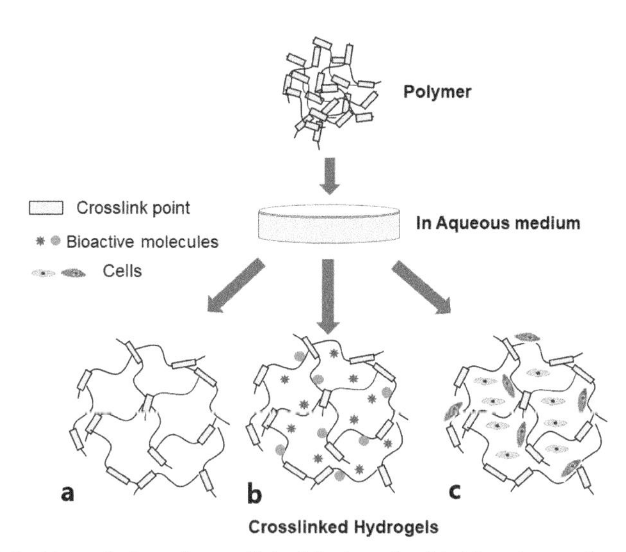

FIGURE 10.4 Formulation of crosslinked hydrogel with bioactive molecules and cells either encapsulated or bound to the polymer.

even the blood–brain barrier [60]. Needham et al. [61] demonstrated bPEI-HA conjugate zwitterionic gene delivery vectors those had significant transfection efficiency.

10.4.3 HYDROGEL AS CELL CARRIER

Cell-based therapeutic research is in pipeline for regenerative medicine. Cell-laden hydrogels are particularly attractive as scaffolding materials that harbor cells and deliver them at defective sites (Figure 10.4c). For tissue engineering applications, hydrogels retain the encapsulated cells by providing biological and physical supports. The hydrogels interconnected micropores permit nutrient and gas transport as well as waste removal from the encapsulated cells. Hydrogels can also protect the encapsulated cells from the host immune systems [62]. Various natural hydrogels such as alginate, collagen, and synthetic hydrogels like PEG and hybrid hydrogels are employed on the encapsulation of cells. Hernandez et al. [63] formulated alginate-HA composite hydrogel, which promoted MSC growth and released therapeutic proteins. Microencapsulated MSCs in methacrylated HA showed enhanced chondrogenic differentiation [64]. Siltanen et al. [65] encapsulated embryonic stem cells in heparin/PEG microgel using a microfluidic technique and showed enhanced spheroid formation and differentiation. Chitosan/gelatin thermosensitive hydrogel showed sustained release of adipose-derived stem cells that maintained survival, which can be used for angiogenesis applications [66]. Ekerdt et al. [67] used hydrogel comprised of HA and poly(*N*-isopropolyacrylamide), which has thermo-responsive properties that readily enable mixing of cells at low temperatures and encapsulation within the hydrogel upon elevation to 37 °C. This system showed human pluripotent stem cells recovery and expansion over long cell culture periods while maintaining cell pluripotency. Encapsulation of neural stem cells in the click-chemistry-based polyethylene glycol-based hydrogels showed in-vitro differentiation to primary neurons over two weeks without addition of differentiation factor [68].

The important properties required for hydrogel formulation to use in tissue engineering applications and their classification therefrom are discussed below.

10.5 HYDROGEL PROPERTIES

Significant transfer from the 2D to 3D tissue culture has been accomplished in modern tissue engineering and regenerative medicine. Hydrogel is the major component in the 3D culture especially in modern computer-aided 3D bioprinting. Normally, the polymer chains in biopolymers swell in the water, hence called “hydrogel.” Hence the ECM provides skeletal structure to the tissue that supports the cells to maintain their integrity and functions, hydrogel that closely mimics the ECM could provide a better environment for new tissue formation in tissue engineering. The physiochemical, mechanical, and biological properties of the materials used to prepare hydrogel directly could influence the morphology, cell–cell interaction, and gene expression. Except hard tissue like bone and teeth, these hydrogels can be manipulated physio-chemically to recapitulate the elastic modulus of soft tissues in our body. Factors influencing the hydrogel properties are:

10.5.1 WATER CONTENT

The hydrophilic nature of the polymeric network in the hydrogel facilitates the absorption of water and other biological fluids. This water-absorbing ability gives density to hydrogels that determines their physiochemical characteristics and function. Increasing the density of hydrogel decreases the interpolymeric pore size thereby reduces molecule diffusion rate [69]. Reduction in the dextran diffusion was observed when the concentration of collagen increased in collagen hydrogel encapsulated with hematopoietic stem cells and niche cells [70]. Changes in the alginate–chitosan hydrogel density showed difference in metabolic activity of bone marrow MSC [71]. Berkovitch and Seliktar [72] demonstrated that the dorsal root ganglion cell outgrowth was highly correlated with the density of different hydrogel compositions containing PEG, ECM proteins (fibrinogen, gelatin), and albumin.

10.5.2 SURFACE PROPERTIES

The surface properties of the hydrogel play a vital role in cell adhesion, proliferation, and function. Relatively large accessible surface area facilitates more cell attachment. The encapsulated cells and polymer material interact based on their surface properties and the surrounding microenvironment [73, 74]. Either the micronutrients in the cell-laden hydrogel or the incorporation of growth factors (such as bFGF, VEGF, and EGF), hormones (like insulin), RGD peptide, and proteins (like collagen and fibronectin) improve the surface characteristics including biocompatibility of the hydrogel [75, 76]. Increasing surface charge of the polymer network has been shown to promote cell attachment and positively charged surfaces showed to induce differentiation of stem cells [77–81]. The polymer chain in the hydrogels has functional groups, such as –OH, $–NH_2$, –COOH, $–SO_3H$, and $–CONH_2$, which provide affinity and binding characteristics to cells as well as the surrounding molecules. In the aqueous media, the negatively charged functional groups like $–SO_3$ and –COO chain in the polymer adsorb metal ions through electrostatic interactions [82].

10.5.3 POROSITY

The interconnected pores within the hydrogel play an essential role in the transfer of nutrients, gas exchange, cell growth, penetration as well as neovascularization [83]. The pore characteristics such as shape, size distribution, roughness and interconnectivity are important to prepare a hydrogel. It has been demonstrated that various pore size influences the activity of encapsulated cells. For example, pore size of 20 μm enhance hepatocyte growth, 200–350 μm increase osteoconduction ability and 20–125 μm augment the regeneration of adult mammalian skin [84–86].

10.5.4 MECHANICAL PROPERTIES

The mechanical properties of hydrogels have been shown to significantly affect the nature of encapsulated cells. As the isometric tension of the ECM in various tissues varies depending on the tissue, it is important to design hydrogel close to tissue specific mechanical characteristics. It has been noted that changes in the mechanical parameters such as stress and tension on ECM alters the gene expression in the cells [87]. Crosslinking is one of the important parameter that control the mechanical compliance of hydrogel and the changes in the crosslinking density has been reported to change the cell growth and morphology [88]. It was also noted that the stiffness of the hydrogel influences the differentiation of MSC [89]. The influence of stiffness and elasticity was further evaluated in the differentiation of stem cell into muscle, regulation of liver progenitor cell, and expansion of hematopoietic stem cell [90–92].

10.5.5 BIODEGRADATION

Biodegradability of hydrogel is a major factor in tissue engineering, which ensures cell proliferation, distribution, and new tissue formation upon use. In tissue engineering, depending on the application, hydrogel can be prepared as completely biodegraded within certain period (e.g., skin graft) or partial biodegradable (e.g., cartilage and corneal graft). Hence the critical design of the hydrogels involves modulating their biodegradation rate by varying properties and crosslinking ability [93]. Studies showed that the distribution of the cells were actively proliferating and distributed in biodegradable hydrogel whereas cells were grouped as cluster in nonbiodegradable hydrogel [94]. Many natural and synthetic polymers such as collagen, chitosan, alginate, PLA, PCL, and their combinations have been used to make biodegradable hydrogel [95].

10.5.6 BIOCOMPATIBILITY

Biocompatibility was defined as "the ability of a biomaterial to perform with an appropriate host response in the specific application" [96]. In general, a specific material becomes biocompatible when it shows negligible or no damage to specific tissue or physiological systems like blood, interstitial fluids, and immune cells. The key to make the hydrogel for tissue engineering involves the understanding of biocompatibility by which mechanisms the cells interact with the biomaterial and analyze the consequences of cell-material interaction [97]. Other than the material per se, the chemical components used to polymerize or crosslink the hydrogel may cause damage to the cells or tissue. For example, Irgacure, a photoinitiator that forms free radicles has been shown to cause cell death even at lower concentration. So caution measures have to be taken to remove the unreacted hazard from the hydrogel before application for in-vivo tissue regeneration.

10.5.7 VASCULARIZATION

Vascularization is a major concern in tissue engineering especially in a hydrogel with larger construct. The vasculature has complex structure that varies between tissues. Based on its application in desired tissue, the hydrogel with suitable porosity allows adequate gas, nutrient, and waste exchange through perfusion. Though the mechanical strength of the hydrogel is low compared to the hard biomaterial, it efficiently allows the formation of new vessels (neovascularization). Few natural biomaterials, such as collagen, attract the cells that secrete growth factors essential for vascularization [98]. Other approaches such as incorporation of vasculogenic/angiogenic growth factors that enhance vasculature by neighboring cells, addition of endothelial cells into the hydrogel that directly initiate vascularization were found to be improving vascularization [99]. To date, many natural and synthetic hydrogel are used to engineer vascularized tissue constructs by co-culturing cells with endothelial cells and other mesenchymal cells [100].

10.6 HYDROGEL CLASSIFICATION

Hydrogels have been classified in many ways. Figure 10.5 illustrates the broad classification of hydrogels that are detailed in the following sections.

10.6.1 BASED ON ORIGIN

Hydrogels have been classified into three groups based on the origin of polymers used—natural, synthetic, and hybrid.

10.6.1.1 NATURAL POLYMERS

Natural polymers from the components of mammalian ECM such as collagen, gelatin, silk, fibrin, HA, and elastin and nonmammalian polymers such as chitosan, alginate, and agarose are used in hydrogel for tissue engineering and regenerative medicine application.

The organized triple helix structure in the collagen provides tensile strength and the presence of integrin-binding sites allows adherent of cells and enhance their function [101–104]. The biodegradation of collagen is through matrix metalloproteinases and the degradation rate can be altered by crosslinking [105]. Gelatin is a denatured collagen, which is thermoresponsive undergoes sol–gel transition below 35 °C. Like collagen, gelatin can be crosslinked to maintain gel state at physiological temperature [106, 107]. The catalytic cleavage of fibronectin by thrombin gives fibrin, which is used as a hydrogel in tissue engineering and hemostatic application [108]. Fibrin hydrogels have been found to support neuronal, vascular regeneration, and stem cell differentiation [109–112]. Unmodified fibrin hydrogel tends to degrade faster and losses its weight due to in-vitro remodeling. This can be

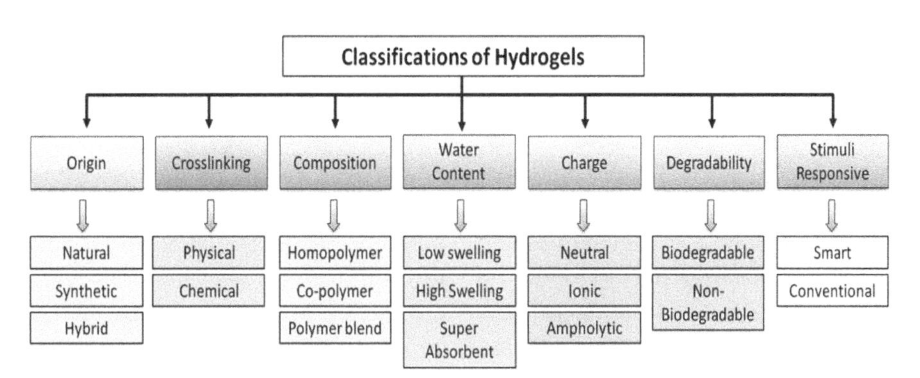

FIGURE 10.5 Classifications of hydrogels based on various factors.

eliminated by modification of fibrin. PEGylation of fibrin has shown to control the degradation and maintain the 3D morphogenesis longer time [113]. Hyaluronic acid or hyaluronan (HA) is a polysaccharide consists repeated units of D-glucuronic acid and *N*-acetyl-D-glucosamine, which is an important component of connective tissues, cartilage, skin, and synovial fluid. The biocompatibility and tunable properties allow modifying HA to form hydrogel that are used in a myriad of applications [114]. Elastin is another ECM protein from the tropoelastin, which is a major component of elastic fibers that impart mechanical strength and elasticity to almost all connective tissues. Due to increase in crosslinks, elastin is poor soluble and other forms of elastin such as tropoelastin, α-elastin, and elastin-like polypeptides are developed [115–118]. Incorporation of exogenous elastin into other biopolymers has been shown to support cell growth and maintain the stem cell phenotype. As a surgical adhesive, photo-crosslinkable methacrylat tropoelastin provided mechanical properties that required to repair elastic tissues including lung, vasculature, and skin [119]. Silk fibroin hydrogel has been recently received great attention in tissue engineering due to retention of a large amount of water in interconnected pores. This 3D formation provides the physiologically stable condition and microenvironment for cell survival [120, 121]. Silk fibroin showed slower degradation compared to other natural proteins due to high crystallinity. Incorporation of collagen and fibronectin has been shown to increase the cell adhesion property of silk proteins [122, 123].

Chitosan is partial deacetylated chitin from exoskeletons of crustaceans, some insects, and fungi [124, 125]. Chitosan is prone to degrade by lysozyme, which makes this biodegradable [126]. The rate of deacetylation denotes the rate of degradation, cell biocompatibility, and cell adhesion [127]. Chitosan hydrogels have been used in the regeneration of various tissues like cartilage, bone, skin, and nerves [128–131]. Alginate is a natural polysaccharide from brown seaweed that composed of L-guluronate and D-mannuronate. Alginate undergoes sol–gel transformation in the presence of divalent cations such as calcium in aqueous solutions, which makes precise control over the gelation for cell encapsulation [132]. Agarose is a neutral polysaccharide from red algae that undergoes sol–gel transition while cooling [133]. The concentration, porosity, and mechanical properties of agarose are interconnected those can be optimized based on applications in tissue engineering. The stability and porous agarose hydrogel facilitate cell mobility that has been found to provide structural integrity for muscle function and culture chondrocytes for cartilage repair [134–136].

10.6.1.2 SYNTHETIC POLYMERS

Synthetic polymers are highly versatile materials with physical and chemical properties that can be easily manipulated. Conversely, synthetic polymers are not bioactive and less biocompatible compared to natural polymers. Hence most of the synthetic hydrogels function as a passive scaffold for cells. Various synthetic polymers have been used to make hydrogels for applications in tissue engineering and regenerative medicine.

A wide spectrum of synthetic materials used for hydrogels fabrication are PLA-derived polymers, PEG-derivatives, poly(glycolic acid) and copolymers like PAA, PEO, polyvinyl alcohol (PVA), 2-hydroxyethyl methacrylate and 2-hydroxypropyl methacrylate. The PLA–PEG-based hydrogels are amphiphilic (contains both hydrophilic and hydrophobic groups), which are used in various biomedical applications [8, 137]. The crosslinked PLA–PEG hydrogel encapsulated with osteoinductive growth factor showed influence in differentiation and mineralization [138]. Microchanneled PEG hydrogel cylinders with bFGF have been shown to increase vascularization and adipogenicity of bone marrow mesenchymal stem cells (BMSCs) [139]. Further, it has been shown that PLGA micro-channels allow the perfusion of oxygen and nutrients for vascular migration and alignment [140].

10.6.1.3 HYBRID POLYMERS

As described earlier natural polymers possess active structural components that directly regulate cellular response and synthetic polymers exhibit precise structure with strength, combination of synthetic and natural polymers to form hybrid hydrogels has become another approach to make bioactive hydrogel for tissue engineering applications. For example, chondrocytes encapsulated in collagen mimic peptide (CMP)–PEG hybrid hydrogels showed better outcome in cartilage repair compared to PEG hydrogel alone, where PEG provided inert polymeric backbone and CMP alleviated biological function [141]. The graphene–chitosan hybrid hydrogel showed improved conductivity, mechanical properties, and increased proliferation of fibroblast [142]. Collagen-*N*-isopropylacrylamide hydrogel promoted corneal cell and nerve growth. Further implantation of this hybrid hydrogel showed rapid recruitment of epithelial and stromal cells on implanted pig cornea [143].

An osteoconductive alginate–PVA–HA hydrogel formulation showed intact for 14 days and potentially could improve the bone formation [144].

10.6.2 BASED ON WATER CONTENT

The water uptake and swelling capacity of the polymer changes depends on the hydrophobic and hydrophilic moieties present. Based on the water content, polymers can be divided into (1) low swelling hydrogel that absorbs water around 20%–50%, (2) high swelling hydrogel that absorbs water around 50%–90%, and (3) superabsorbent hydrogel that absorbs water greater than 100% [145]. Increase in the water content increases the permeability of the polymer, especially the biocompatibility of synthetic polymers increases upon more water uptake capability. Though high water containing hydrogels resemble tissue, it weakens the mechanical properties that affect the stability of the polymer. The swelling property of the hydrogel is indirectly proportional to the amount of crosslinking [146].

10.6.3 BASED ON CROSSLINKING

The physiochemical and functional properties of most hydrogels can be influenced by altering the degree of crosslinking. Normally the crosslinking of the hydrogel is achieved by either physical or chemical methods [147]. The physical crosslinking occurs through hydrogen bonding, hydrophobic interaction, polymer chain aggregation and complexion, and crystallization [148–151]. The chemical crosslinking is done by covalent crosslinking between the polymer chains or by the use of chemical linkers. The chemical crosslinking is irreversible due to changes in the structural configuration whereas physical crosslinking is reversible because of changes in the conformation in water [152–154]. However, dual crosslinking (both physical and chemical) has been employed to overcome the disadvantages of using sole physical or chemical crosslinking [155]. High energy radiation such as gamma (γ) or electron beam radiations also used to polymerize the materials in hydrogel preparation [28, 32].

10.6.4 BASED ON COMPOSITION

Based on the composition of the hydrogel, it can be further classified into homopolymer, copolymer, and multipolymer hydrogel. Homopolymer hydrogels are polymeric network structures from single polymer units, whereas copolymer hydrogels consist of monomer units of two polymers, in which at least one is hydrophilic to render it to swell. Many natural polymers such as collagen, gelatin, alginate, and chitosan and the synthetic polymers like PEG, PAA, and polyhydroxyethyl methacrylate are in use for tissue engineering applications either mono or copolymer hydrogels [156–158]. Three or more polymers (polymer blends) offer mechanical strength that overcomes the limitations of homopolymer and copolymer hydrogels [159, 160].

10.6.5 BASED ON IONIC CHARGE

Hydrogels are classified into three groups based on the presence or absence of electrical charge units. (1) Neutral (no charge in their backbone or side chain), (2) ionic (cationic that possess positively charged groups like amines group; anionic that possess negatively charged group like carboxylic acid), and (3) ampholytic that carry positive and negative charge on the same polymer chain. Neutral hydrogel can swell in aqueous medium solely based on water–polymer interaction. Swelling capability of ionic hydrogels is directly influenced by the pH of the aqueous medium. The cationic hydrogel swells superior at lower pH, whereas the anionic hydrogel swells better at higher pH. Charges in the ampholytic hydrogels are balanced at isoelectric point and a slight change in pH can change the overall ionic properties of hydrogels [161–164].

10.6.6 BASED ON BIODEGRADATION

The rate of biodegradation is critical for the anticipated hydrogel life depends on their need in specific tissue engineering applications. Normally based on the biodegradation ability the hydrogels can be classified as biodegradable and nonbiodegradable. In case of regenerative medicine, the distribution and proliferation of encapsulated cells were found to be better in the biodegradable hydrogel rather than the nonbiodegradable hydrogel [165, 166]. It is highly desirable to ensure that the rate of biodegradation coincides with new tissue regeneration at the site of use. In addition, it has been shown that the proteolytic cleavage of the biodegradable hydrogel enhances the proliferation and ECM synthesis of encapsulated cell [167]. In general, biodegradability of the nonbiodegradable hydrogels are alleviated by the incorporation of degradable components such as ECM proteins (collagen, fibronectin, and HA) and polysaccharides (alginate and chitosan) into the polymer backbone [168].

10.6.7 BASED ON STIMULI RESPONSE

From the past decade, hydrogels with advanced characteristics have been prepared for various biomedical applications named as "smart hydrogels." Though the preparation and characterization of smart hydrogels are similar to conventional hydrogels, in response to various external stimuli the smart hydrogels exhibit unusual changes in their swelling behavior and mechanical characteristics. The physical stimuli such as temperature, light, mechanical stress, pressure, and chemical stimuli including pH, ionic strength of aqueous medium and the surrounding environment or the combination of both physical and chemical stimuli have been applied in smart hydrogel systems [169–173].

10.7 HYDROGEL CHARACTERIZATION

10.7.1 SCANNING ELECTRON MICROSCOPY

The polymer networks in the hydrogel make porous structures that allow vascularization, cell entrapment or attachment, and influences the proliferation and differentiation in tissue regeneration [174]. The morphological characteristics of hydrogels can be analyzed by using scanning electron microscopy. Mostly, the sputter coating of the freeze-dried hydrogel is accomplished by gold and the 2D images can be taken at higher magnifications. It determines the surface topology such as pore size and interconnected pore structure of building block of the hydrogel before as well as after encapsulating the cells or drugs [175, 176].

10.7.2 ATOMIC FORCE MICROSCOPY

The advantage of atomic force microscopy (AFM) is that it can work well in ambient air and in a liquid environment. AFM is a high-resolution scanning probe microscopy, which provides 3D surface profile of the hydrogel. In AFM, the oscillation amplitude is decreased upon increased interaction of hydrogel surface and the tip of the probe [177, 178].

10.7.3 SWELLING RATIO

In hydrogel, the polymer chains interact with the solvent molecule and expand to the fully solvated state. The swelling characteristics are important for hydrogels for the use of tissue engineering applications. The degree of hydration can be measured by determining the amount of solvent in the hydrogel. The gravimetric technique is the most common method to analyze the hydration degree of hydrogel. In which, the wet weight and dry weight of the hydrogel composite can be measured using an electronic balance and the swelling ratio calculated as follows:

$$Q = [(W_w - W_d)/W_d]$$

where Q is the swelling ratio of the hydrogel, W_w denotes the wet weight, and W_d is the dry weight of the hydrogel [179].

10.7.4 FOURIER TRANSFORM INFRARED SPECTROSCOPY

The functional groups present in the polymer plays a significant role in the property of hydrogel. Both mechanical and functional properties of the hydrogel can be improved by modification of functional groups. The presence of functional groups either in naïve or modified state can be analyzed using Fourier transform infrared spectroscopy. The changes in the vibration of functional groups under infrared are transformed into peaks at a certain wavelength and represented in the graphical form [180, 181].

10.7.5 MECHANICAL CHARACTERIZATION

When force is applied to hydrogel, it deforms depending on the mechanical property, size, and direction. The mechanical properties are affected by the degree of swelling, polymer composition, and crosslinking density [159, 182]. Most common methods used to determine the mechanical properties of hydrogels are tensile testing (strip or ring extensiometry) and compression test. In tensile testing, tensile force is applied to the strips or ring material held between grips. From the shear–strain chart obtained between the applied force and elongation of the material, the mechanical properties of the hydrogels can be identified (including Young's modulus, yield strength, and ultimate tensile strength). Viscoelastic nature of a hydrogel material can also be analyzed by elongating the material strip to a particular length and examining the stress relaxation response over time at a constant strain. In a compression test, the specimen is loaded between two plates and subjected to compression by application of force. Pressure is applied to the hydrogel surface, the stress–strain curves can be calculated from the axial force and displacement using a theoretical model. The main advantage of compression test over extensiometry is that it does not limit the hydrogel geometry to strips or rings [183].

10.7.6 RHEOLOGICAL CHARACTERIZATION

Rheological properties of bioink polymers are evaluated to understand the yield stress, shear dependent viscosity, and recovery behavior [184]. Bioink passes through relaxed and applied pressure phase during bioprinting. The viscoelastic behavior in the relaxed phase can be characterized by shear stress rheometry. The yield stress of the hydrogel can be determined from the shear stress ramping values. In the viscosity plot, the intersection point of the tangent from the plateau region and the region where the viscosity drops signify the yield stress. This information is very essential for the hydrogel to be qualified as a boink for extrusion-based bioprinting. When viscosity is plotted over shear stress, the point at which the hydrogel starts flowing represents the relaxed viscosity. Rheological properties are dependent on the properties of hydrogels like their association, entanglement, and crosslinking [185]. All viscous polymers follow the law $G' \sim \omega^2$ and $G'' \sim \omega^2$; where G' and G'' represent shear storage modulus and shear loss modulus, respectively, while ω represents the angular velocity. In most cases, elasticity dominates ($G' > G''$) which corresponds to Maxwell-type behavior having single relaxation time and this behavior increases with increase in concentration [186]. The G' and G'' of the crosslinked hydrogel dispersions are mostly independent of oscillation frequency.

10.7.7 OTHER TECHNIQUES

The hydrogels can be assessed to confirm the crosslinked network structures by using thermogravimetric and sol–gel analysis [187]. The amount of free and bound water in the hydrogel can be analyzed quantitatively using differential scanning calorimetry (DSC) and nuclear magnetic resonance (NMR). In DSC, the frozen hydrogel is analyzed by endotherm measurement where it assumed frozen water as free water. The bound water can be obtained from the difference between total water and free water [188]. The interchange of water molecules between free and bound states is measured by proton NMR [189].

10.7.8 DEGRADATION RATE

The degradation rate of hydrogels can be analyzed in-vitro by incubating in phosphate buffer saline at 37 °C and at predetermined intervals. Simple measurement of the weight of hydrogel or analyzing the degraded end product of the material leached into the buffer will tell us the degradation rate against time. The degradation of natural hydrogel can be assessed by incubating with specific enzymes (e.g., Gelatinase for gelatin-based hydrogel) at physiological concentrations for different time points. In-vivo degradation rate can also be achieved by implanting the hydrogel material in animal models and analyze at different time intervals [190].

10.7.9 BIOLOGICAL EVALUATION

Biological evaluation of biomaterials includes interaction with blood, cells, and tissues. It can be broadly classified as in-vitro and in-vivo evaluation. In-vitro evaluation is economical and rapid method that gives information on biological interaction with material. In-vivo evaluation is done on animals that model the biological interaction that might be encountered when the material is used for clinical application in humans. International guidelines such as ISO10993-5 and American Society for Testing and Materials are available for the biological evaluation of hydrogels [191–193]. The following section briefly describes the in-vitro cytotoxicity evaluation of biomaterials. Readers are advised to refer to other references for in-vivo and hemocompatibility tests [194].

10.7.9.1 CYTOTOXICITY

The primary and the mandate test that a hydrogel proposed for biomedical applications including 3D bioprinting would be the cytotoxicity testing. Generally, the bioink formulation before adding the cells would be advisable to be tested for the biological response of cells toward materials. The cell material interactions of hydrogels can be broadly classified as cytotoxicity and cytocompatibility. The cytotoxicity evaluates whether the biomaterial is causing toxicity to cells and cell death. The primary methods that are done for evaluating material induced cell toxicity that are most recommended as per ISO 10993-5 are direct contact, agar diffusion, and test on the extract of materials. Appropriate test method will be chosen to collect meaningful data for the evaluation of biocompatibility. Cytotoxicity evaluation is done along with positive and negative control materials. A positive control will give a confirmed severe cytotoxicity response whereas the negative control will show a repeated noncytotoxicity.

10.7.9.1.1 Test by Direct Contact

Direct contact test is a qualitative analysis that gives first-hand information on cytotoxicity of material toward a used characterized cell type. Usually fibroblast cell line (ATCC, NTC clone L929) is used in the form of a subconfluent monolayer. Material of size 10% to the cellular area will be kept on the cell layer and incubated at physiological conditions for not less than 24 h at 37±1 °C. Cell morphology, lysis, vacuolization, and detached are evaluated to grade cytotoxicity from 0 to 4 (0—none, 1—sight, 2—mild, 3—moderate, and 4—severe cytotoxicity).

10.7.9.1.2 Agar Diffusion

In agar diffusion assay, culture medium from a subconfluent monolayer of cells is replaced with culture medium containing 2% agar. Once the agar gets solidified, test and control materials are placed on it and incubated at physiological conditions for not less than 24 h at 37±1 °C. This allows diffusion of molecules in differently through the agar and will give indication of possible toxicant from the material based on its molecular weight.

10.7.9.1.3 Test on Extract

Another screening test for biomaterial is the test by elusion method, where the cells are exposed to culture medium or buffer that has been incubated with test material. Different dilutions of the material extract shall give direct information on the possible toxicant that could elute from the biomaterial. The toxicity shall be evaluated and graded based on the percentage cell damage to a particular concentration of extract, that is, 0—No damage; 1—70% good cells; 2—50% good cells; 3—20%–50% good cells; and 4—<20% good cells. The test also has an advantage of quantitative analysis of a cell-based assays such as MTT (3-(4,5-dimethylthiazol-2-yl)-2,5-diphenyltetrazolium bromide), CCK-8 (Cell Counting Kit) or LDH (lactate dehydrogenase), which can be done after exposing the cells to the extracts.

10.7.9.2 CYTOCOMPATIBILITY

Cytocompatibility is a battery of biological analysis that gives comprehensive information on cell-material interaction for a particular application. Hydrogels are used along with the cells or without cells. When hydrogel is intended to use with cells, one should look for cell encapsulation, followed by analysis of biological parameters like viability, proliferation, migration, and cell function. The hydrogel proposed to be used without cells shall be analyzed for cell adhesion on the finished material. Cell lines of particular cell type are preferred for initial studies and primary isolated cells from the tissue of application would be ideal for functional analysis.

10.7.9.3 TEST CULTURE METHOD

Mostly the biological evaluation of hydrogels is done on cell monolayer or in cell encapsulated form. However, certain analysis requires dynamic culture conditions, for example, hydrogels for cartilage repair might reflect the tissue functions when cultured in bioreactor with cyclic loading and deloading forces. Continuous perfusion bioreactor would be essential for efficient mass transfer in the case of hydrogels for bioartificial liver.

1.8 HYDROGEL AS BIOINK

The hydrogel proposed as bioink is expected to show certain specific characteristics such as biocompatibility, tissue remodeling, biodegradation, mass transfer efficiency, gelation, viscoelasticity, printability, and shear thinning property [195]. The bioink used in the 3D bioprinting processes should be designed for application in a particular tissue and it demands the general mandatory characters such as biocompatibility and tissue remodeling. Based on the tissue type, the hydrogel for bioink should be carefully selected.

1.8.1 SPECIFICITY

When considering hydrogels in context of bioprinting, one of the foremost characteristic features that it should possess is printability. The printability of hydrogels to a large extent depends on (1) the modality of bioprinting and (2) the material's viscous and shear-thinning properties. These two properties are especially important to the nozzle-based inkjet and extrusion printing methods [34]. Viscosity is the material's resistance to flow and in the context of bioprinting, this resistance is measured in terms of pressure. For example, inkjet printing, thermal or acoustic pulsations are created within the cell-laden suspension to achieve sufficient air pressure which in turn propels the cell suspension through the printing nozzle in a "drop-on-demand" fashion. Inkjet printers deposit very small volumes (<100 pL) of cell suspensions and require hydrogels in their precursor (uncrosslinked), low-viscosity state (ideally below 0.1 Pa s). Extrusion-based bioprinting, on the other hand, hydrogel is deposited as continuous filaments by applying pressure in a pneumatic, piston-driven, or screw manner. The hydrogels employed in this modality are significantly more viscous ($30\text{-}6\times10^6$ m Pa s) and require higher pressures to extrude through nozzles [196, 197].

1.8.2 GELATION AND CROSSLINKING OF HYDROGEL FOR BIOINK

In order for a hydrogel to be applied to bioprinting, it must have (1) a relatively quick gelation time such that subsequent layers can be immediately deposited on top,

(2) a noncytotoxic mechanism of crosslinking, and (3) the crosslinks formed should be stable in an aqueous environment at physiologic temperature and pH [198]. Generally, two methods of crosslinking are used (1) physical crosslinking (which includes ionic interactions and hydrogen bonding) and (2) chemical crosslinking (which forms covalent bonds secondary to photo-initiation or enzyme catalysis). There are thus four major classes of hydrogels in terms of crosslinking: (1) thermo-sensitive hydrogels, (2) ionically crosslinking hydrogels, (3) enzymatically crosslinking hydrogels, and (4) photo-polymerizable hydrogels. Thermo-sensitive hydrogels are formed secondary to conformational changes in their constituent monomers at a certain temperature. This temperature is called the critical solution temperature. Depending on the hydrogel, it can have a lower or upper critical solution temperature, above which or below an insoluble semisolid gel forms. For example, gelatin has an upper critical solution temperature of around 30 °C. This means that gelatin solutions are soluble liquids above 30 °C and insoluble gels below 30 °C. The solution-to-gel transition is usually rather gradual depending on distance from the critical solution temperature. Other examples of thermo-sensitive hydrogels are collagen I, agarose, and Pluronic®.

10.9 CONCLUSION

Hydrogels have distinct properties like high water content with controllable swelling behavior using external stimuli, which makes them attractive for pharmaceutical biomedical applications. One can customize hydrogel as 2D, 3D, and 4D with various fabrication techniques. In this review, we attempt to provide an overview of hydrogels, their classification, and characterization with special attention as bioink. Scientific development and evaluation of hydrogels for regenerative medicine have progressed tremendously in this decade. Future development of hydrogels with modern technology could make it a valuable asset in tissue engineering. More research is going on to develop novel neo tissue formation using hydrogels that activates regeneration of tissues by providing the 3D structural framework and the cell adhesion cues that support cell adhesion, migration, proliferation, and differentiation.

KEYWORDS

- **hydrogel**
- **biomaterial**
- **bioink**
- **tissue engineering**
- **polymers**

REFERENCES

1. Atala, A. Advances in Tissue and Organ Replacement. *Curr. Stem Cell Res. Ther.* **2008**, *3(1)*, 21–31.
2. Utech, S.; Boccaccini, A. R. A Review of Hydrogel-Based Composites for Biomedical Applications: Enhancement of Hydrogel Properties by Addition of Rigid Inorganic Fillers. *J. Mater. Sci.* **2006**, *51(1)*, 271–310.
3. Fisher, M. B.; Mauck, R. L. Tissue Engineering and Regenerative Medicine: Recent Innovations and the Transition to Translation. *Tissue Eng. Part B Rev.* **2013**, *19(1)*, 1–13.
4. Mao, A. S.; Mooney, D. J. Regenerative Medicine: Current Therapies and Future Directions. *Proc. Natl. Acad. Sci. U. S. A.* **2015**, *112(47)*, 14452–14459.
5. Witten, C. M.; McFarland, R. D.; Simek, S. L. Concise Review: The U.S. Food and Drug Administration and Regenerative Medicine. *Stem Cells Transl. Med.* **2015**, *4(12)*, 1495–1499.
6. Lee, E. J.; Kasper, F. K.; Mikos, A. G. Biomaterials for Tissue Engineering. *Ann. Biomed. Eng.* **2014**, *42(2)*, 323–337.
7. Nair, L. S.; Laurencin, C. T. Polymers as Biomaterials for Tissue Engineering and Controlled Drug Delivery. *Adv. Biochem. Eng. Biotechnol.* **2006**. *102*, 47–90.
8. O'Brien, F. J.; Harley, B. A. Editorial. Tissue Engineering. *J. Mech. Behav. Biomed. Mater.* **2012**, *11*, 1–2.
9. El-Sherbiny, I. M.; Yacoub, M. H. Hydrogel Scaffolds for Tissue Engineering: Progress and Challenges. *Glob. Cardiol. Sci. Pract.* **2013**, *2013(3)*, 316–342.
10. Gao, G.; Cui, X. Three-Dimensional Bioprinting in Tissue Engineering and Regenerative Medicine. *Biotechnol. Lett.* **2006**, *38(2)*, 203–11.
11. Mandrycky, C., et al. 3D Bioprinting for Engineering Complex Tissues. Biotechnol. Adv. **2006**, *34(4)*, 422–434.
12. Ozbolat, I. T.; Peng, W.; Ozbolat, V. Application Areas of 3D Bioprinting. *Drug Discov. Today.* **2006**, *21(8)*, 1257–1271.
13. Sears, N. A., et al. A Review of Three-Dimensional Printing in Tissue Engineering. *Tissue Eng. Part B Rev.* **2006**, *22(4)*, 298–310.
14. Mitchell, A., et al. Additive Manufacturing—A Review of 4D Printing and Future Applications. *Addit. Manuf.* **2018**, *24*, 606–626.
15. Momeni, F., et al. A Review of 4D Printing. Mater. Des. **2017**, *122*, 42–79.
16. Hospodiuk, M., et al. The Bioink: A Comprehensive Review on Bioprintable Materials. *Biotechnol. Adv.* **2017**, *35(2)*, 217–239.
17. Jakus, A. E.; Rutz, A. L.; Shah, R. N. Advancing the Field of 3D Biomaterial Printing. *Biomed. Mater.* **2006**, *11(1)*, 014102.
18. Mironov, V., et al. Organ Printing: Tissue Spheroids as Building Blocks. *Biomaterials*. **2009**, *30(12)*, 2164–2174.
19. Vashist, A.; Ahmad, S. Hydrogels in Tissue Engineering: Scope and Applications. *Curr. Pharm. Biotechnol.* **2015**, *16(7)*, 606–620.
20. Griffith, L. G.; Naughton, G. Tissue Engineering—Current Challenges and Expanding Opportunities. *Science*. **2002**, *295*(5557), 1009–1014.
21. Langer, R.; Vacanti, J. P. Tissue Engineering. *Science*. **1993**, *260*(5110), 920–926.
22. Tibbitt, M. W., et al. Progress in Material Design for Biomedical Applications. Proc. Natl. Acad. Sci. U. S. A. **2015**, *112(47)*, 14444–14451.
23. Lanza, R. P.; Langer, R. S.; Vacanti, J. *Principles of tissue engineering*. 3rd ed. Elsevier Academic Press: Amsterdam; Boston, 2007; xxvii, 1307 p.
24. Tabata, Y. Biomaterial Technology for Tissue Engineering Applications. *J. R. Soc. Interface.* **2009**, *6*, S311–S324.

25. Williams, D. F., *The Williams dictionary of biomaterials*. Liverpool University Press: Liverpool, 1999; xvii, 343 p.
26. Slaughter, B. V., et al. Hydrogels in Regenerative Medicine. *Adv. Mater.* **2009**, *21*(32–33), 3307–3329.
27. Suntornnond, R., et al. A Highly Printable and Biocompatible Hydrogel Composite for Direct Printing of Soft and Perfusable Vasculature-Like Structures. *Sci. Rep.* **2017**, *7*(1), 16902.
28. Zhou, Y., et al. Fully Printed Flexible Smart Hybrid Hydrogels. *Adv. Funct. Mater.* **2018**, *28(9)*, 1705365.
29. Choi, J., et al. 4D Printing Technology: A Review. *3D Print. Addit. Manuf.* **2015**, *2(4)*, 159–167.
30. Ge, Q.; Qi, H. J.; Dunn, M. L. Active Materials by Four-Dimension Printing. *Appl. Phys. Lett.* **2013**, *103(13)*, 131901.
31. Ji, S.; Guvendiren, M. Recent Advances in Bioink Design for 3D Bioprinting of Tissues and Organs. *Front. Bioeng. Biotechnol.* **2017**, *5*, 23.
32. Zhou, Y., et al., From 3D to 4D Printing: Approaches and Typical Applications. *J. Mech. Sci. Technol.* **2015**, *29(10)*, 4281–4288.
33. Gungor-Ozkerim, P. S., et al. Bioinks for 3D Bioprinting: An Overview. *Biomater. Sci.* **2018**, *6(5)*, 915–946.
34. Hölzl, K., et al. Bioink Properties Before, During and After 3D Bioprinting. *Biofabrication*, **2006**, *8(3)*, 032002.
35. Jakab, K., et al. Engineering Biological Structures of Prescribed Shape Using Self-Assembling Multicellular Systems. *Proc. Natl. Acad. Sci. U. S. A.* **2004**, *101(9)*, 2864–2869.
36. Norotte, C., et al. Scaffold-Free Vascular Tissue Engineering using Bioprinting. *Biomaterials*. **2009**, *30(30)*, 5910–5917.
37. Djabourov, M.; Leblond, J.; Papon, P. Gelation of Aqueous Gelatin Solutions .2. Rheology of the Sol-Gel Transition. *J. Phys.* **1988**. *49(2)*, 333–343.
38. Leijten, J., et al. Advancing Tissue Engineering: A Tale of Nano-, Micro-, and Macroscale Integration. *Small.* **2006**, *12(16)*, 2130–2145.
39. Braccini, I.; Perez, S. Molecular Basis of Ca^{2+}-Induced Gelation in Alginates and Pectins: The Egg-Box Model Revisited. Biomacromolecules. **2001**, *2(4)*, 1089–1096.
40. Azagarsamy, M. A.; Anseth, K. S. Bioorthogonal Click Chemistry: An Indispensable Tool to Create Multifaceted Cell Culture Scaffolds. *ACS Macro Lett.* **2013**, *2(1)*, 5–9.
41. Hennink, W. E.; van Nostrum, C. F. Novel Crosslinking Methods to Design Hydrogels. *Adv. Drug Delivery Rev.* **2012**, *64*, 223–236.
42. Li, J. Y., et al. *Hybrid Hydrogels with Extremely High Stiffness and Toughness. ACS Macro Lett.* **2014**, *3(6)*, 520–523.
43. Ma, P. X. Biomimetic Materials for Tissue Engineering. *Adv. Drug Delivery Rev.* **2008**, *60(2)*, 184–198.
44. Shin, H.; Jo, S.; Mikos, A. G. Biomimetic Materials for Tissue Engineering. *Biomaterials.* **2003**, *24(24)*, 4353–4364.
45. Yang, J.; Xu, F. Synergistic Reinforcing Mechanisms in Cellulose Nanofibrils Composite Hydrogels: Interfacial Dynamics, Energy Dissipation, and Damage Resistance. *Biomacromolecules*. **2017**, *18(8)*, 2623–2632.
46. Zawko, S. A.; Schmidt, C. E. *Crystal Templating Dendritic Pore Networks and Fibrillar Microstructure into Hydrogels. Acta Biomater.* **2010**, *6(7)*, 2415–2421.
47. Ma, X. X., et al. A Novel Chitosan-Collagen-Based Hydrogel for Use as a Dermal Filler: Initial In Vitro and In

Vivo Investigations. *J. Mater. Chem. B.* **2014**, *2(18)*, 2749–2763.
48. Li, Y. L.; Rodrigues, J.; Tomas, H. Injectable and Biodegradable Hydrogels: Gelation, Biodegradation and Biomedical Applications. *Chem. Soc. Rev.* **2012**, *41(6)*, 2193–2221.
49. Ishihara, M. Photocrosslinkable Chitosan Hydrogel as a Wound Dressing and a Biological Adhesive. *Trends Glycosci. Glycotechnol.* **2002**, *14(80)*, 331–341.
50. Liu, Y. C., et al. Crosslinked Hyaluronan Hydrogels Containing Mitomycin C Reduce Postoperative Abdominal Adhesions. *Fertil. Steril.* **2005**, *83*, 1275–1283.
51. Simons, M., et al. Clinical Trials in Coronary Angiogenesis: Issues, Problems, Consensus— An Expert Panel Summary. *Circulation.* **2000**, *102(11)*, E73–E86.
52. Epstein, S. E., et al. Therapeutic Interventions for Enhancing Collateral Development by Administration of Growth Factors: Basic Principles, Early Results and Potential Hazards. *Cardiovasc. Res.* **2001**, *49(3)*, 532–542.
53. Peters, M. C., et al. Release from Alginate Enhances the Biological Activity of Vascular Endothelial Growth Factor. *J. Biomater. Sci.-Polym. Ed.* **1998**, *9(12)*, 1267–1278.
54. Tabata, Y., et al. Controlled Release of Vascular Endothelial Growth Factor by Use of Collagen Hydrogels. *J. Biomater. Sci.-Polym. Ed.* **2000**, *11(9)*, 915–930.
55. Tanihara, M., et al. Sustained Release of Basic Fibroblast Growth Factor and Angiogenesis in a Novel Covalently Crosslinked Gel of Heparin and Alginate. *J. Biomed. Mater. Res.* **2001**, *56(2)*, 216–21.
56. Saito, N., et al. A Biodegradable Polymer as a Cytokine Delivery System for Inducing Bone Formation. *Nat. Biotechnol.* **2001**, *19(4)*, 332–335.
57. Kim, K., et al. Osteochondral Tissue Regeneration Using a Bilayered Composite Hydrogel with Modulating Dual Growth Factor Release Kinetics in a Rabbit Model. *J. Controll. Release.* **2013**, *168(2)*, 166–178.
58. Wang, H. N., et al. Combined Delivery of BMP-2 and bFGF from Nanostructured Colloidal Gelatin Gels and its Effect on Bone Regeneration In Vivo. *J. Controll. Release.* **2013**, *166(2)*, 172–181.
59. Kimura, Y.; Tabata, Y. Controlled Release of Stromal-Cell-Derived Factor-1 from Gelatin Hydrogels Enhances Angiogenesis. *J. Biomater. Sci.-Polym. Ed.* **2010**, *21(1)*, 37–51.
60. Vinogradov, S. V.; Bronich, T. K.; Kabanov, A. V. Nanosized Cationic Hydrogels for Drug Delivery: Preparation, Properties and Interactions with Cells. *Adv. Drug Deliv. Rev.* **2002**, *54(1)*, 135–147.
61. Needham, C.J., et al. Engineering a Polymeric Gene Delivery Vector Based on Poly(ethylenimine) and Hyaluronic Acid. *Biomacromolecules.* **2012**, *13(5)*, 1429–1437.
62. Guilak, F., et al. Control of Stem Cell Fate by Physical Interactions with the Extracellular Matrix. *Cell Stem Cell.* **2009**, *5(1)*, 17–26.
63. Canibano-Hernandez, A., et al. Alginate Microcapsules Incorporating Hyaluronic Acid Recreate Closer in Vivo Environment for Mesenchymal Stem Cells. *Mol. Pharm.* **2017**, *14(7)*, 2390–2399.
64. Chung, C.; Burdick, J. A. Influence of Three-Dimensional Hyaluronic Acid Microenvironments on Mesenchymal Stem Cell Chondrogenesis. *Tissue Eng. Part A.* **2009**, *15(2)*, 243–254.
65. Siltanen, C., et al. Microfluidic Fabrication of Bioactive Microgels

for Rapid Formation and Enhanced Differentiation of Stem Cell Spheroids. *Acta Biomater.* **2006**, *34*, 125–132.

66. Cheng, N.C., et al. Sustained Release of Adipose-Derived Stem Cells by Thermosensitive Chitosan/Gelatin Hydrogel for Therapeutic Angiogenesis. *Acta Biomater.* **2017**, *51*, 258–267.
67. Ekerdt, B.L., et al. Thermoreversible Hyaluronic Acid-PNIPAAm Hydrogel Systems for 3D Stem Cell Culture. *Adv. Healthcare Mater.* **2018**, *7(12)*.
68. Li, H., et al. Neural Stem Cell Encapsulation and Differentiation in Strain Promoted Crosslinked Polyethylene Glycol-Based Hydrogels. *J. Biomater. Appl.* **2018**, *32(9)*, 1222–1230.
69. Drury, J. L.; Mooney, D. J. Hydrogels for Tissue Engineering: Scaffold Design Variables and Applications. *Biomaterials.* **2003**, *24(24)*, 4337–51.
70. Mahadik, S. P., et al. Regulating Dynamic Signaling Between Hematopoietic Stem Cells and Niche Cells Via a Hydrogel Matrix. *Biomaterials.* **2017**, *125*, 54–64.
71. Kuhn, P. T., et al. Anti-Microbial Biopolymer Hydrogel Scaffolds for Stem Cell Encapsulation. *Polymers.* **2017**, *9(4)*.
72. Berkovitch, Y.; Seliktar, D. Semi-Synthetic Hydrogel Composition and Stiffness Regulate Neuronal Morphogenesis. *Int. J. Pharm.* **2017**, *523(2)*, 545–555.
73. Chang, H. Y., et al. Effect of Surface Potential on Epithelial Cell Adhesion, Proliferation and Morphology. *Coll. Surf. B.* **2006**, *141*, 179–186.
74. Ventre, M.; Causa, F.; Netti, P. A. Determinants of Cell-Material Crosstalk at the Interface: Towards Engineering of Cell Instructive Materials. *J. R. Soc. Interface.* **2012**, *9(74)*, 2017–2032.
75. Elbert, D. L.; Hubbell, J. A. Surface Treatments of Polymers for Biocompatibility. *Annu. Rev. Mater. Sci.* **1996**, *26*, 365–394.
76. Hersel, U.; Dahmen, C.; Kessler, H. RGD Modified Polymers: Biomaterials for Stimulated Cell Adhesion and Beyond. *Biomaterials.* **2003**, *24(24)*, 4385–4415.
77. De Luca, I., et al. Positively Charged Polymers Modulate the Fate of Human Mesenchymal Stromal Cells via EphrinB2/EphB4 Signaling. *Stem Cell Res.* **2006**, *17(2)*, 248–255.
78. De Rosa, M., et al. Cationic Polyelectrolyte Hydrogel Fosters Fibroblast Spreading, Proliferation, and Extracellular Matrix Production: Implications for Tissue Engineering. *J. Cell. Physiol.* **2004**, *198(1)*, 133–143.
79. Schneider, G.B., et al. The Effect of Hydrogel Charge Density on Cell Attachment. *Biomaterials.* **2004**, *25(15)*, 3023–3028.
80. Tan, F., et al. Charge Density is More Important Than Charge Polarity in Enhancing Osteoblast-Like Cell Attachment on Poly(Ethylene Glycol)-Diacrylate Hydrogel. *Mater. Sci. Eng. C.* **2017**, *76*, 330–339.
81. Zhang, W., et al. Upregulation of BMSCs Osteogenesis by Positively-Charged Tertiary Amines on Polymeric Implants via Charge/iNOS Signaling Pathway. *Sci. Rep.* **2015**, *5*, 9369.
82. Abou Taleb, M.; Ismail, S.; El-Kelesh, N. Radiation Synthesis and Characterization of Polyvinyl Alcohol/Methacrylic Acid-Gelatin Hydrogel for Vitro Drug Delivery. *J. Macromol. Sci. Part A.* **2009**, *46(2)*, 170–178.
83. Yang, S. F., et al. The Design of Scaffolds for Use in Tissue Engineering. Part 1. Traditional Factors. *Tissue Eng.* **2001**, *7(6)*, 679–689.
84. Brauker, J. H., et al. Neovascularization of Synthetic Membranes Directed

by Membrane Microarchitecture. *J. Biomed. Mater. Res.* **1995**, *29(12)*, 1517–1524.
85. Whang, K., et al. Engineering Bone Regeneration with Bioabsorbable Scaffolds with Novel Microarchitecture. *Tissue Eng.* **1999**, *5(1)*, 35–51.
86. Yannas, I. V., et al. Synthesis and Characterization of a Model Extracellular-Matrix That Induces Partial Regeneration of Adult Mammalian Skin. *Proc. Natl. Acad. Sci. U. S. A.* **1989**, *86(3)*, 933–937.
87. Ingber, D. E., Cellular Mechanotransduction: Putting All the Pieces Together Again. *FASEB J.* **2006**, *20(7)*, 811–827.
88. Bryant, S. J., et al. Crosslinking Density Influences the Morphology of Chondrocytes Photoencapsulated in PEG Hydrogels During the Application of Compressive Strain. *J. Orthop. Res.* **2004**, *22(5)*, 1143–1149.
89. Engler, A.J., et al. Matrix Elasticity Directs Stem Cell Lineage Specification. *Cell.* **2006**, *126(4)*, 677–689.
90. Chaudhuri, T., et al. Preparation of Collagen-Coated Gels That Maximize In Vitro Myogenesis of Stem Cells by Matching the Lateral Elasticity of In Vivo Muscle. *Methods Mol. Biol.* **2010**, *621*, 185–202.
91. Holst, J., et al. Substrate Elasticity Provides Mechanical Signals for the Expansion of Hemopoietic Stem and Progenitor Cells. *Nat. Biotechnol.* **2010**, *28(10)*, 1123–1128.
92. Lozoya, O. A., et al. Regulation of Hepatic Stem/Progenitor Phenotype by Microenvironment Stiffness in Hydrogel Models of the Human Liver Stem Cell Niche. *Biomaterials*. **2011**, *32(30)*, 7389–402.
93. Lutolf, M. P. Spotlight on Hydrogels. *Nat. Mater.* **2009**, *8(6)*, 451–453.
94. Deshmukh, M., et al. Biodegradable Poly(Ethylene Glycol) Hydrogels based on a Self-Elimination Degradation Mechanism. *Biomaterials*. **2010**, *31(26)*, 6675–6684.
95. Lee, J.; Cuddihy, M. J.; Kotov, N. A. Three-Dimensional Cell Culture Matrices: State of the Art. *Tissue Eng. Part B.* **2008**, *14(1)*, 61–86.
96. Naahidi, S., et al. Biocompatibility of Engineered Nanoparticles for Drug Delivery. *J. Control Release* **2013**, *166(2)*, 182–194.
97. Williams, D. F., On the Mechanisms of Biocompatibility. *Biomaterials*. **2008**, *29(20)*, 2941–2953.
98. Allen, P.; Melero-Martin, J.; Bischoff, J. Type I Collagen, Fibrin and PuraMatrix Matrices Provide Permissive Environments for Human Endothelial and Mesenchymal Progenitor Cells to Form Neovascular Networks. *J. Tissue Eng. Regener. Med.* **2011**, *5(4)*, E74–E86.
99. Koike, N., et al. Tissue Engineering: Creation of Long-Lasting Blood Vessels. *Nature*. **2004**, *428*(6979), 138–139.
100. Caliari, S. R.; Burdick, J. A. A Practical Guide to Hydrogels for Cell Culture. *Nat. Methods*. **2006**, *13(5)*, 405–414.
101. Fratzl, P., et al. Fibrillar Structure and Mechanical Properties of Collagen. *J. Struct. Biol.* **1998**, *122*(1–2), 119–122.
102. Heino, J. The Collagen Receptor Integrins have Distinct Ligand Recognition and Signaling Functions. *Matrix Biol.* **2000**, *19(4)*, 319–323.
103. Walters, B. D.; Stegemann, J. P. Strategies for Directing the Structure and Function of Three-Dimensional Collagen Biomaterials Across Length Scales. *Acta Biomater.* **2014**, *10(4)*, 1488–1501.
104. White, D. J., et al. The Collagen Receptor Subfamily of the Integrins.

Int. *J. Biochem. Cell Biol.* **2004**, *36(8)*, 1405–1410.
105. Kuijpers, A. J., et al. Characterization of the Network Structure of Carbodiimide Cross-Linked Gelatin Gels. *Macromolecules.* **1999**, *32(10)*, 3325–3333.
106. Mahony, O., et al. Silica-Gelatin Hybrids with Tailorable Degradation and Mechanical Properties for Tissue Regeneration. *Adv. Funct. Mater.* **2010**, *20(22)*, 3835–3845.
107. Van den Bulcke, A. I., et al. Structural and Rheological Properties of Methacrylamide Modified Gelatin Hydrogels. *Biomacromolecules.* **2000**, *1(1)*, 31–38.
108. Shaikh, F. M., et al. Fibrin: a Natural Biodegradable Scaffold in Vascular Tissue Engineering. *Cells Tissues Organs.* **2008**, *188(4)*, 333–346.
109. Bensaid, W., et al. A Biodegradable Fibrin Scaffold for Mesenchymal Stem Cell Transplantation. *Biomaterials.* **2003**, *24(14)*, 2497–2502.
110. Johnson, P. J., et al. Controlled Release of Neurotrophin-3 and Platelet-Derived Growth Factor From Fibrin Scaffolds Containing Neural Progenitor Cells Enhances Survival and Differentiation Into Neurons in a Subacute Model of SCI. *Cell Transplant.* **2010**, *19(1)*, 89–101.
111. Rowe, S. L.; Lee, S.; Stegemann, J. P. Influence of Thrombin Concentration on the Mechanical and Morphological Properties of Cell-Seeded Fibrin Hydrogels. *Acta Biomater.* **2007**, *3(1)*, 59–67.
112. Willerth, S. M., et al. Optimization of Fibrin Scaffolds for Differentiation of Murine Embryonic Stem Cells into Neural Lineage Cells. *Biomaterials.* **2006**, *27(36)*, 5990–6003.
113. Zhang, G., et al. A PEGylated Fibrin Patch for Mesenchymal Stem Cell Delivery. *Tissue Eng.* **2006**, *12(1)*, 9–19.
114. Burdick, J. A.; Prestwich, G. D. Hyaluronic Acid Hydrogels for Biomedical Applications. *Adv. Mater.* **2011**, *23(12)*, H41–H56.
115. Broekelmann, T. J., et al. Tropoelastin Interacts with cell-Surface Glycosaminoglycans via its COOH-Terminal Domain. *J. Biol. Chem.* **2005**, *280(49)*, 40939–40947.
116. Girotti, A., et al. Design and Bioproduction of a Recombinant Multi(Bio)Functional Elastin-Like Protein Polymer Containing Cell Adhesion Sequences for Tissue Engineering Purposes. *J. Mater. Sci.-Mater. Med.* **2004**, *15(4)*, 479–484.
117. Leach, J. B., et al. Crosslinked Alpha-Elastin Biomaterials: Towards a Processable Elastin Mimetic Scaffold. *Acta Biomater.* **2005**, *1(2)*, 155–164.
118. Mithieux, S. M.; Wise, S. G.; Weiss, A. S. Tropoelastin—A Multifaceted Naturally Smart Material. *Adv. Drug Delivery Rev.* **2013**, *65(4)*, 421–428.
119. Weiss, A. S.; Khademhosseini, A. Engineering a Highly Elastic Human Protein-Based Sealant for Surgical Applications. *Sci Transl Med.* **2017**, *9*(410), eaai7466
120. Davis, N. E., et al. Enhanced Function of Pancreatic Islets Co-Encapsulated with ECM Proteins and Mesenchymal Stromal Cells in a Silk Hydrogel. *Biomaterials.* **2012**, *33(28)*, 6691–6697.
121. Wang, X. Q., et al. Sonication-Induced Gelation of Silk Fibroin for Cell Encapsulation. *Biomaterials.* **2008**, *29(8)*, 1054–1064.
122. Leal-Egana, A.; Scheibel, T. Silk-Based Materials for Biomedical Applications. *Biotechnol. Appl. Biochem.* **2010**, *55*, 155–167.
123. Yanagisawa, S., et al. Improving Cell-Adhesive Properties of Recombinant

Bombyx Mori Silk by Incorporation of Collagen or Fibronectin Derived Peptides Produced by Transgenic Silkworms. *Biomacromolecules.* **2007**, *8(11)*, 3487–3492.
124. Merzendorfer, H. The Cellular Basis of Chitin Synthesis in Fungi and Insects: Common Principles and Differences. *Eur. J. Cell Biol.* **2011**, *90(9)*, 759–769.
125. Khor, E.; Lim, L. Y. Implantable Applications of Chitin and Chitosan. *Biomaterials.* **2003**, *24(13)*, 2339–2349.
126. Pangburn, S. H.; Trescony, P. V.; Heller, J. Lysozyme Degradation of Partially Deacetylated Chitin, its Films and Hydrogels. *Biomaterials.* **1982**, *3(2)*, 105–8.
127. Freier, T., et al. Controlling Cell Adhesion and Degradation of Chitosan Films by N-acetylation. *Biomaterials.* **2005**, *26(29)*, 5872–8.
128. Cao, L.Y., et al. Bone Regeneration Using Photocrosslinked Hydrogel Incorporating rhBMP-2 Loaded 2-N, 6-O-Sulfated Chitosan Nanoparticles. *Biomaterials.* **2014**, *35(9)*, 2730–2742.
129. Gnavi, S., et al. The Use of Chitosan-Based Scaffolds to Enhance Regeneration in the Nervous System. *Int. Rev. Neurobiol.* **2013**, *109*, 1–62.
130. Miguel, S. P., et al. Thermoresponsive Chitosan-Agarose Hydrogel for Skin Regeneration. *Carbohydr. Polym.* **2014**, *111*, 366–373.
131. Mirahmadi, F., et al. Enhanced Mechanical Properties of Thermosensitive Chitosan Hydrogel by Silk Fibers for Cartilage Tissue Engineering. *Mater. Sci. Eng. C-Mater. Biol. Appl.* **2013**, *33(8)*, 4786–4794.
132. Andersen, T.; Auk-Emblem, P.; Dornish, M. 3D Cell Culture in Alginate Hydrogels. *Microarrays (Basel).* **2015**, *4(2)*, 133–161.
133. Fu, X. T.; Kim, S. M. Agarase: Review of Major Sources, Categories, Purification Method, Enzyme Characteristics and Applications. *Marine Drugs.* **2010**, *8(1)*, 200–218.
134. Carroll, S. L.; Klein, M. G.; Schneider, M. F. Calcium Transients in Intact Rat Skeletal-Muscle Fibers in Agarose-Gel. *Am. J. Physiol.-Cell Physiol.* **1995**, *269(1)*, C28-C34.
135. Guaccio, A., et al. Oxygen Consumption of Chondrocytes in Agarose and Collagen Gels: A Comparative Analysis. *Biomaterials.* **2008**, *29(10)*, 1484–1493.
136. Gunja, N. J., et al. Effects of Agarose Mould Compliance and Surface Roughness on Self-Assembled Meniscus-Shaped Constructs. *J. Tissue Eng. Regener. Med.* **2009**, *3(7)*, 521–530.
137. Reis, R. L.; Cohn, D. Polymer Based Systems on Tissue Engineering, Replacement, and Regeneration. *NATO Science Series Series II, Mathematics, Physics, and Chemistry.* Kluwer Academic Publishers: Boston, MA, USA, 2002; xii, 422 p.
138. Burdick, J. A.; Anseth, K. S. Photoencapsulation of Osteoblasts in Injectable RGD-Modified PEG Hydrogels for Bone Tissue Engineering. *Biomaterials.* **2002**, *23(22)*, 4315–4323.
139. Stosich, M. S., et al. Vascularized Adipose Tissue Grafts from Human Mesenchymal Stem Cells with Bioactive Cues and Microchannel Conduits. *Tissue Eng.* **2007**, *13(12)*, 2881–2890.
140. Kaully, T., et al. Vascularization-The Conduit to Viable Engineered Tissues. *Tissue Eng. Part B-Rev.* **2009**, *15(2)*, 159–169.
141. Lee, H. J., et al. Collagen Mimetic Peptide-Conjugated Photopolymerizable PEG Hydrogel. *Biomaterials.* **2006**, *27(30)*, 5268–5276.

142. Sayyar, S., et al. Processable Conducting Graphene/Chitosan Hydrogels for Tissue Engineering. *J. Mater. Chem. B.* **2015**, *3(3)*, 481–490.
143. Li, F. F., et al. Cellular and Nerve Regeneration Within a Biosynthetic Extracellular Matrix for Corneal Transplantation. *Proc. Natl. Acad. Sci. U. S. A.* **2003**, *100(26)*, 15346–15351.
144. Bendtsen, S. T.; Quinnell, S. P.; Wei, M. Development of a Novel Alginate-Polyvinyl Alcohol-Hydroxyapatite Hydrogel for 3D Bioprinting Bone Tissue Engineered Scaffolds. *J. Biomed. Mater. Res. Part A.* **2017**, *105(5)*, 1457–1468.
145. Peppas, N. A., et al. Hydrogels in Pharmaceutical Formulations. *Eur. J. Pharm. Biopharm.* **2000**, *50(1)*, 27–46.
146. Allen, P. E. M.; Bennett, D. J.; Williams, D. R. G. Water in Methacrylates .1. Sorption and Desorption Properties of Poly(2-Hydroxyethyl Methacrylate-Co-Glycol Dimethacrylate) Networks. *Eur. Polym. J.* **1992**, *28(4)*, 347–352.
147. Hennink, W. E.; van Nostrum, C. F. Novel Crosslinking Methods to Design Hydrogels. *Adv. Drug Delivery Rev.* **2002**, *54(1)*, 13–36.
148. Gacesa, P. Alginates. *Carbohydr. Polym.* **1988**, *8(3)*, 161–182.
149. Liu, Y., et al. Physically Crosslinked Composite Hydrogels of PVA with Natural Macromolecules: Structure, Mechanical Properties, and Endothelial Cell Compatibility. *J. Biomed. Mater. Res. B Appl. Biomater.* **2009**, *90(2)*, 492–502.
150. Willcox, P. J., et al. Microstructure of Poly(Vinyl Alcohol) Hydrogels Produced by Freeze/Thaw Cycling. *J. Polym. Sci. Part B-Polym. Phys.* **1999**, *37(24)*, 3438–3454.
151. Zhang, Y. X., et al. pH Switching On-Off Semi-IPN Hydrogel Based on Cross-Linked Poly(Acrylamide-Co-Acrylic Acid) and Linear Polyallyamine. *Polymer.* **2005**, *46(18)*, 7695–7700.
152. Hu, B. H.; Su, J.; Messersmith, P. B. Hydrogels Cross-Linked by Native Chemical Ligation. *Biomacromolecules.* **2009**, *10(8)*, 2194–2200.
153. Lee, C. R.; Grodzinsky, A. J.; Spector, M. The Effects of Cross-Linking of Collagen-Glycosaminoglycan Scaffolds on Compressive Stiffness, Chondrocyte-Mediated Contraction, Proliferation and Biosynthesis. *Biomaterials.* **2001**, *22(23)*, 3145–3154.
154. Prestwich, G. D., et al. Controlled Chemical Modification of Hyaluronic Acid: Synthesis, Applications, and Biodegradation of Hydrazide Derivatives. *J. Controll. Release.* **1998**, *53*(1–3), 93–103.
155. Cong, H. P.; Wang, P.; Yu, S. H. Stretchable and Self-Healing Graphene Oxide-Polymer Composite Hydrogels: A Dual-Network Design. *Chem. Mater.* **2013**, *25(16)*, 3357–3362.
156. Benamer, S., et al. Synthesis and Characterisation of Hydrogels Based on Poly(Vinyl Pyrrolidone). *Nucl. Instrum. Meth. Phys. Res. Sect. B-Beam Interact. Mater. Atoms.* **2006**, *248(2)*, 284–290.
157. Brovold, M., et al. Naturally-Derived Biomaterials for Tissue Engineering Applications. *Adv. Exp. Med. Biol.* **2018**, *1077*, 421–449.
158. Iizawa, T., et al. Synthesis of Porous Poly(N-Isopropylacrylamide) Gel Beads by Sedimentation Polymerization and their Morphology. *J. Appl. Polym. Sci.* **2007**, *104(2)*, 842–850.
159. Jaiswal, M.; Koul, V. Assessment of Multicomponent Hydrogel Scaffolds of Poly(Acrylic Acid-2-Hydroxy Ethyl Methacrylate)/Gelatin for Tissue Engineering Applications. *J. Biomater. Appl.* **2013**, *27(7)*, 848–861.

160. Jia, X. Q.; Kiick, K. L. Hybrid Multicomponent Hydrogels for Tissue Engineering. *Macromol. Biosci.* **2009**, *9(2)*, 140–156.
161. Dubrovskii, S. A.; Rakova, G. V. Elastic and Osmotic Behavior and Network Imperfections of Nonionic and Weakly Ionized Acrylamide-Based Hydrogels. *Macromolecules.* **1997**, *30(24)*, 7478–7486.
162. Kabiri, K., et al. Superabsorbent Hydrogel Composites and Nanocomposites: A Review. *Polym. Compos.* **2011**, *32(2)*, 277–289.
163. Kopecek, J.; Yang, J. Y. Revie—Hydrogels as Smart Biomaterials. *Polym. Int.* **2007**, *56(9)*, 1078–1098.
164. Singhal, R.; Gupta, K. A Review: Tailor-made Hydrogel Structures(Classifications and Synthesis Parameters). *Polym.-Plast. Technol. Eng.* **2006**, *55(1)*, 54–70.
165. Elisseeff, J., et al. Photoencapsulation of Chondrocytes in Poly(Ethylene Oxide)-Based Semi-Interpenetrating Networks. *J. Biomed. Mater. Res.* **2000**, *51(2)*, 164–171.
166. Suggs, L. J.; Mikos, A. G. Development of Poly(Propylene Fumarate-Co-Ethylene Glycol) as an Injectable Carrier for Endothelial Cells. *Cell Transplant.* **1999**, *8(4)*, 345–350.
167. Nguyen, K.T.; West, J. L. Photopolymerizable Hydrogels for Tissue Engineering Applications. *Biomaterials*. **2002**, *23(22)*, 4307–14.
168. Lutolf, M. P.; Hubbell, J. A. Synthetic Biomaterials as Instructive Extracellular Microenvironments for Morphogenesis in Tissue Engineering. *Nat. Biotechnol.* **2005**, *23(1)*, 47–55.
169. Gao, Y., et al. Preparation and Characteristic of Electric Stimuli Responsive Hydrogel Composed of Polyvinyl Alcohol/Poly (Sodium Maleate-Co-Sodium Acrylate). *J. Appl. Polym. Sci.* **2008**, *107(1)*, 391–395.
170. Kim, S. J., et al. Self-Oscillatory Actuation at Constant DC Voltage with pH-Sensitive Chitosan/Polyaniline Hydrogel Blend. *Chem. Mater.* **2006**, *18(24)*, 5805–5809.
171. Kim, S.J., et al. Electrical Behavior of Polymer Hydrogel Composed of Poly(Vinyl Alcohol)-Hyaluronic Acid in Solution. *Biosens. Bioelectron.* **2004**, *19(6)*, 531–536.
172. Suzuki, A., Light-Induced Phase-Transition of Poly(N-Isopropylacrylamide-Co-Chlorophyllin) Gels. *J. Intell. Mater. Syst. Struct.* **1994**, *5(1)*, 112–116.
173. Wang, D.; Hao, J. C. Multiple-Stimulus-Responsive Hydrogels of Cationic Surfactants and Azoic Salt Mixtures. *Coll. Polym. Sci.* **2013**, *291(12)*, 2935–2946.
174. Annabi, N., et al. Controlling the Porosity and Microarchitecture of Hydrogels for Tissue Engineering. *Tissue Eng. Part B-Rev.* **2010**, *16(4)*, 371–383.
175. El Fray, M., et al. Morphology Assessment of Chemically Modified and Cryostructured Poly(Vinyl Alcohol) Hydrogel. *Eur. Polym. J.* **2007**, *43(5)*, 2035–2040.
176. Omidian, H.; Rocca, J. G.; Park, K. Advances in Superporous Hydrogels. *J. Controll. Release*. **2005**, *102(1)*, 3–12.
177. Alarcon, C. D. H.; Pennadam, S.; Alexander, C. Stimuli Responsive Polymers for Biomedical Applications. *Chem. Soc. Rev.* **2005**, *34(3)*, 276–285.
178. Geisse, N. A. AFM and Combined Optical Techniques. *Mater. Today.* **2009**, *12*(7–8), 40–45.
179. Tang, C., et al. Swelling Behavior and Biocompatibility of Carbopol-Containing Superporous Hydrogel Composites. *J. Appl. Polym. Sci.* **2007**, *104(5)*, 2785–2791.
180. Kumar, A., et al. Smart Polymers: Physical Forms and Bioengineering

Applications. *Progr. Polym. Sci.* **2007**, *32(10)*, 1205–1237.
181. Mansur, H. S.; Orefice, R. L.; Mansur, A. A. P. Characterization of Poly(Vinyl Alcohol)/Poly(Ethylene Glycol) Hydrogels and PVA-Derived Hybrids by Small-Angle X-Ray Scattering and FTIR Spectroscopy. *Polymer.* **2004**, *45(21)*, 7193–7202.
182. Kazakia, G. J., et al. Effects of In Vitro Bone Formation on the Mechanical Properties of a Trabeculated Hydroxyapatite Bone Substitute. *J. Biomed. Mater. Res. Part A.* **2006**, *77A(4)*, 688–699.
183. Svensson, A., et al. Bacterial Cellulose as a Potential Scaffold for Tissue Engineering of Cartilage. *Biomaterials.* **2005**, *26(4)*, 419–431.
184. Paxton, N., et al. Proposal to Assess Printability of Bioinks for Extrusion-Based Bioprinting and Evaluation of Rheological Properties Governing Bioprintability. *Biofabrication.* **2017**, *9(4)*, 044107.
185. Kheirandish, S., et al. Shear and Elongational Flow Behavior of Acrylic Thickener Solutions Part I: Effect of Intermolecular Aggregation. *Rheol. Acta.* **2008**, *47(9)*, 999–1013.
186. Weng, L. H.; Chen, X. M.; Chen, W. L. Rheological Characterization of In Situ Crosslinkable Hydrogels Formulated from Oxidized Dextran and N-Carboxyethyl Chitosan. *Biomacromolecules.* **2007**, *8(4)*, 1109–1115.
187. Janik, I., et al. Radiation Crosslinking and Scission Parameters for Poly(Vinyl Methyl Ether) in Aqueous Solution. *Nucl. Instrum. Methods Phys. Res. Sect. B-Beam Interact. Mater. Atoms.* **2003**, *208*, 374–379.
188. Hoffman, D. M. Dynamic Mechanical Signatures of a Polyester-Urethane and Plastic-Bonded Explosives Based on this Polymer. *J. Appl. Polym. Sci.* **2002**, *83(5)*, 1009–1024.
189. Capitani, D.; Crescenzi, V.; Segre, A. L. Water in Hydrogels. An NMR Study of Water/Polymer Interactions in Weakly Cross-Linked Chitosan Networks. *Macromolecules.* **2001**, *34*(12), 4136–4144.
190. Shahriari, D., et al. Characterizing the Degradation of Alginate Hydrogel for Use in Multilumen Scaffolds for Spinal Cord Repair. *J. Biomed. Mater. Res. Part A.* **2006**, *104(3)*, 611–619.
191. ASTM F895–11, *Standard Test Method for Agar Diffusion Cell Culture Screening for Cytotoxicity*. ASTM International: West Conshohocken, PA, USA, 2016.
192. ASTM F813–07, *Standard Practice for Direct Contact Cell Culture Evaluation of Materials for Medical Devices*. ASTM International: West Conshohocken, PA, USA, 2016.
193. ISO 10993–5, *Biological Evaluation of Medical Devices. Part 5:*, in *Tests for in Vitro Cytotoxicity.* International Organization for Standardization: Geneva, Switzerland, 2009; p. 3.
194. Ratner, B.D., et al. *Biomaterials Science: An Introduction to Materials in Medicine*. Elsevier Science, 2004.
195. Gopinathan, J.; Noh, I. Recent Trends in Bioinks for 3D Printing. *Biomater. Res.* **2018**, *22*, 11–11.
196. Derakhshanfar, S., et al. 3D Bioprinting for Biomedical Devices and Tissue Engineering: A Review of Recent Trends and Advances. *Bioact. Mater.* **2018**, *3(2)*, 144–156.
197. Gopinathan, J.; Noh, I. Recent Trends in Bioinks for 3D Printing. *Biomater. Res.* **2018**, *22*, 11.
198. Parak, A., et al. Functionalizing Bioinks for 3D Bioprinting Applications. *Drug Discov. Today*. **2019**, *24(1)*, 198–205.

CHAPTER 11

Injectable Hydrogel Scaffolds for Neural Tissue Engineering Applications

JOSNA JOSEPH[1,4*], ANNIE ABRAHAM[1], and MYRON SPECTOR[2,3]

[1]*Advanced Centre for Tissue Engineering, Department of Biochemistry, University of Kerala, Thiruvananthapuram, India*

[2]*Tissue Engineering/Regenerative Medicine, VA Boston Healthcare System, Boston, MA, USA*

[3]*Department of Orthopedics, Brigham and Women's Hospital, Harvard Medical School, Boston, MA, USA*

[4]*Department of Clinical Immunology & Rheumatology, Christian Medical College, Vellore, India*

[*]*Corresponding author. E-mail: josna.joseph@fulbrightmail.org*

ABSTRACT

Neural injuries and hereditarily or sporadically acquired neurodegenerative diseases leading to life-debilitating conditions necessitate the search for novel therapeutic approaches overcoming the absence of regeneration potential of the central nervous system. Also, the search for alternate in vitro testing systems that merged with the pursuit to examine cellular interactions in a three-dimensional (3D) in vivo milieu had led to the development of 3D cell culture systems. Here, the focus will primarily be on presenting a brief overview of the various neural injuries and neurodegenerative diseases, the existing treatment options, emerging research areas, and models to address this enormous challenge.

11.1 INTRODUCTION

This chapter outlines the various injuries and disorders of central nervous system (CNS) that necessitate the search for devising the goals of neuroregeneration, which includes exogenous cell transplantation and the approach of recruiting endogenous neural stem progenitors to the site of neural injury. The various biomaterials currently used for CNS repair with emphasis on injectable hydrogels, including their types and desirable characteristics, also need to be presented. The role of these same hydrogels as 3D cell culture systems will also be covered. Finally, the usage of various growth factors/neurotrophins/phytochemicals for neural regeneration, creation of biomatrix-oriented neuro-stimulative environment, and a methodical approach to the same are included.

11.2 INJURIES OF THE CNS

11.2.1 TRAUMATIC BRAIN INJURY

The statistics of the Centers for Disease Control and Prevention [1] reveals that the occurrence rate of traumatic brain injury (TBI) in the United States is 577 in 100,000 (1.7 million cases/year), and the mortality rate is 17.6 in 100,000. Another systemic review depicts the incidence rate of TBI as 262/100,000 and the mortality rate as 10.5/100,000 in Europe, with and falls and road traffic accidents being the most common causes [2]. TBI falls into different categories like (1) closed, where the physical damage incurred does not involve the cranial space or (2) penetrating, when if a foreign object physically invades the cranial space. Existing treatment for TBI focuses on the supportive measures that preserve the remaining healthy tissue after injury rather than the regeneration of damaged tissue. The treatment methods include reducing the intracranial pressure by placing a catheter into the ventricles or by the usage of diuretics to reduce the fluid in the system. Other remedial measures include seizure control by the administration of benzodiazepines and surgery.

11.2.2 STROKE

Stroke results in serious long-term disability for millions of patients worldwide. Brain stroke occurs when a blood clot blocks an artery or if a blood vessel ruptures, leading to disruption in the blood flow to the brain. Of the two subtypes, ischemic and hemorrhagic stroke, latter is associated with significantly higher morbidity and mortality and mostly accounts for 8%–18% of all the strokes [3]. Further, based on the location of the ruptured vessel and bleeding, hemorrhagic strokes can occur as subarachnoid hemorrhage or intracerebral hemorrhage.

Ischemic stroke is predominant of all the strokes and affects almost 800,000 people in the United States alone. The major event underlying ischemic stroke is the obstruction of blood vessels to brain, creating a blockage in the supply of oxygen and nutrients, often leading to injury or death for the affected brain tissue. Considering the cause of the vessel occlusion, ischemic strokes can further be grouped into thrombotic or embolic strokes.

11.2.3 SPINAL CORD INJURY

Another grave neural damage situation is the injury caused to the spinal cord. There are approximately 4,500,000 individuals affected with spinal cord injury (SCI) in the United States and an estimate of 11,000 new cases are expected to be diagnosed each year [4]. SPI could be categorized into primary and secondary injury: primary being the initial mechanical trauma and secondary the subsequent cellular and biochemical reactions that lead to further damage. The causal factors of SPI are (1) blunt impact, (2) compression, and (3) penetrating trauma. Blunt impacts could lead to concussion, contusion, laceration, transection, or intraparenchymal hemorrhage.

Primary and secondary injuries to the spinal cord could lead to cell necrosis and subsequent cell loss, whereas penetrating injuries culminate in scarring and tethering of the cord [5]. Demyelination occurs following the death of oligodendrocytes, causing conduction deficits [6]. In contusion injuries, a cystic cavity surrounded by an astrocytic scar is formed following the tissue loss and this scar does inhibit axons to grow across the cavity [7], hindering the axonal regeneration (Figure 11.1).

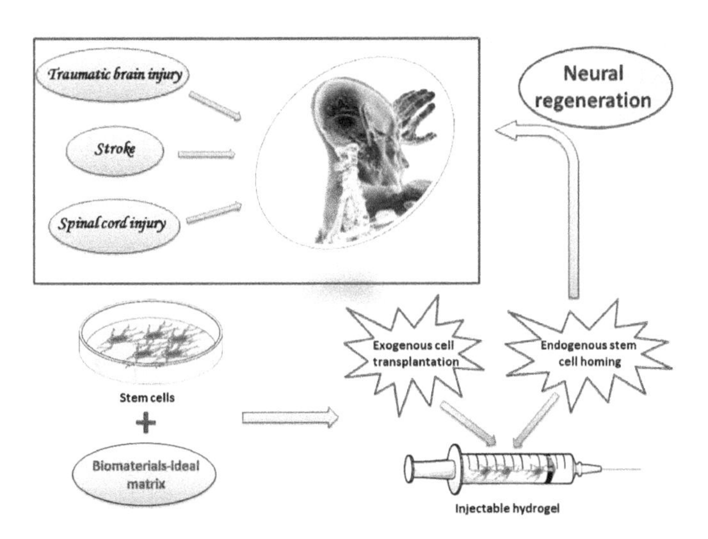

FIGURE 11.1 Illustration of the post-injury sequel in spinal cord injury.

The existing treatment regimen for SCI includes the methyl prednisolone administration, surgery, hypothermia, intensive multisystem medical management, and rehabilitative care.

11.2.4 POST-INJURY CHANGES

Although the nature of the neural injury may vary significantly from hemorrhage, ischemia to trauma, the subsequent disturbance to the homeostatic state of the complex environment of brain may trigger a parallel sequelae of events. This include excitotoxicity (excess release of excitatory amino acid neurotransmitters) [8] and ionic imbalance, increased production of reactive oxygen species, necrosis and apoptosis, subsequent inflammation and tissue loss by atrophy and liquefactive necrosis, demyelination, release of growth inhibitory molecules inhibiting axonal regeneration and neurite outgrowth, and glial scar formation. These injury-associated changes in the molecular and cellular homeostasis could result in chronic neurobehavioral alterations, including cognitive deficits, changes in personality, and increased rates of psychiatric illness [9]. The common denominators to injuries to the brain and spinal cord are cavitary lesions filled with fluid or necrotic tissue. There is no extracellular matrix (ECM) within these lesions which can serve as a stroma/framework to support the infiltration by host progenitor cells from the surrounding tissue. Moreover, the lesions contained within otherwise healthy tissue are possible. These challenging features of the lesions represent opportunities for the development of injectable hydrogels.

11.3 NEUROREGENERATION: GOALS

The current treatment modalities to combat these ailments include pharmacotherapy, prevention, and rehabilitation but they fail to restore the injury-induced loss of function. Even after the administration of tissue plasminogen, the only medical-approved therapy for stroke, the involved patients are left with lifelong motor, sensory, and cognitive disabilities. This is due to the existence of damaged brain tissue and hence the need for neuronal repair. The hallmarks of these CNS defects, which include small lesions volumes surrounded by healthy tissue, necessitate a biomaterial matrix capable of being delivered by injection.

Neuroregeneration in the CNS and SPI differs considerably; wherein, the former involves the regrowth or repair of the nervous tissue by generating new neurons, axons, synapses, and glial cells and in latter, it comprises of elongation of existing axons, sprouting and growth of new axons from neural cell soma, remyelination, plasticity among surviving connections, and functional recovery. SPI includes nerve cell death within the spinal cord, disruption of nerve pathways, and demyelination disrupting the signal transduction. Neural stem cells (NSCs) could aid as a neuron-rich substrate to the injured spinal cord segments, possibly promoting and supporting repair, regeneration, and reorganization. Current treatment for SCI is the direct administration of NSCs resulting in partial recovery of motor functions.

The presence of stem cells, neural and glial precursor cells in the subventricular zone (SVZ), and subgranular zone (SGZ) of dentate gyrus in the adult brain and their ability migrate to cerebral cortex, CA1 region of the hippocampus and the striatum, offer newer therapeutic avenue in neuroregeneration research. Despite the presence of these cells with regenerative potential, neuroregeneration in the adult brain is delayed due to the inhibitory action of the glial cells and a challenging extracellular environment that counteracts the myelin and neuronal repair [10]

The major considerations of biomatrix-oriented neuroregeneration include minimizing cell death and inflammation after scaffold implantation by choosing ideal biocompatible materials, controlling drug release over an appropriate time course to prevent multiple surgeries or injections, and adapting minimally invasive procedures to limit further damage to healthy brain tissue while preserving the integrity of the blood–brain barrier [11]

Homing of endogenous stem and progenitor cells to biological sites of injury or the delivery of such cells from exogenous sources are promising approaches in regenerative medicine. Such cells can replenish degenerated cells in the lesion through their differentiation to specific cell lineages and/or facilitate a regenerative response through their immunoregulation of the injury/diseased site. These approaches have potential applications in the treatment of millions of patients suffering from CNS disorders including retinal disease as well as stroke and SPI, where the common factor is the loss of functional cells and depleted capability of spontaneous/intrinsic cellular repair and proliferation following an injury or disease.

Select neuroregenerative approaches are focused on (1) growth-factor-oriented targeting of regeneration associated genes to stimulate axonal sprouting and growth [12], (2) silencing the effects of inhibitory glycoproteins and proteoglycans, (3) replenishing brain cells from exogenous sources, and (4) endogenous origins.

11.3.1 EXOGENOUS CELL TRANSPLANTATION

Cell transplantation from exogenous source is a warranting strategy for neuroregeneration. Over the years, different cell sources have been isolated and administered to substitute neuronal loss resulted by an injury. This include neurons [13] embryonic stem cells [14], NSCs, [15] bone-marrow-derived mesenchymal stem cells [16], adipose-derived stem cells [17], umbilical cord blood-derived stem cells [18], and peripheral blood-derived mono nuclear cells [19]. NSCs are defined by their ability to self-renew and differentiate into multiple cell types including neurons, astrocytes, and oligodendrocytes, and hence can serve as a source for cell replacement for damaged neurons. Transplantation of exogenous stem cells to substitute lost cells might lead to suppression of post-injury immune responses, and thereby create a conducive environment for neuroregeneration [20].

In this context, encapsulated NSC is a potential treatment strategy and the currently used biopolymers for NSC encapsulation are mainly hyaluronan, collagen, heparin and fibronectin, poly(lactic-*co*-glycolic acid, and so on. Though injectable hydrogels like Matrigel and self-assembling peptide hydrogels are used for the delivery of NSCs, they often lack mechanical stability and desired dissolution profiles. A recent class of biopolymers, gelatin-hydroxyphenyl propionic acid utilizing enzyme-mediated oxidative coupling has shown its potential for conductive growth, proliferation, and differentiation of NSCs [21]. Incorporation of neural chemokines like stromal cell derived factor (SDF-1α) into these hydrogels [22] has enhanced their regeneration potential as there is involvement of both transplanted and endogenous neural stem progenitor cells. Moreover, with their in situ gelling capacity, they could re-establish structure to the damaged tissue and could act as active matrices that match the mechanical properties of surrounding neural tissues.

11.3.2 ENDOGENOUS REGENERATIVE APPROACHES

The alternate approach for neurogenesis is the endogenous regeneration, exploiting the resident stem cell progenitors/stem cells in the brain itself. Adult neurogenesis involves endogenous stem cells originating from the two zones in the brain, that is, SVZ, located along the lateral wall of ventricle and SGZ, which is positioned between the dentate gyrus and the hilus of the hippocampus. Two prominent barriers to endogenous regeneration against an injury are the loss of chemokine expression and disappearance of stroma resultant to tissue loss in the brain lesion, leading to lesser endogenous regenerative responses. Also, the challenge in analyzing the regeneration potential in in vivo models is manifold and it involves delineation of endogenous progenitor cells from the transplanted NSCs. Hence, tracking the route of endogenous progenitor migration will yield promising insights into nerve wound healing.

Even after the administration of tissue plasminogen, the only medical approved therapy for stroke, the affected patients are left with lifelong motor, sensory, and cognitive disabilities. This is due to the existence of damaged brain tissue and hence, the need for neuronal repair. One of the critical elements of these therapeutic approaches is the biomaterial matrix to enable endogenous cell infiltration of the defect or to serve as the delivery vehicle for exogenous cells. The hallmarks of these CNS defects, which include small lesions volumes surrounded by healthy tissue, necessitate a biomaterial matrix capable of being delivered by injection. The direct injection of stem/progenitor cells into the peri-infarct cortex has the disadvantage of damaging the newly regenerating tissue that recovers after stroke [23]. Moreover, the necrotic core of infarcted tissue resulting from an acute ischemic stroke, an ideal homing environment with a viable matrix, and required growth factors for the transplanted cells to help them regenerate and reorganize the damaged tissue [24].

11.4 BIOMATERIALS FOR NEURAL REPAIR AND REGENERATION

The design of scaffolds should consider including the creation of a viable, biocompatible environment that allows cellular infiltration and restoration of damaged neuronal connections. Further, the scaffolds should also provide accurate cues for promoting nerve regeneration in a controlled, localized manner [25]. Other considerations while selecting the scaffold include the ability to conform to the dimensions of the implantation site and maintenance of its appropriate form after implantation, matching to the mechanical behavior of surrounding tissue, nonimmunogenicity, and methods of presterilization. The ideal materials employed for neuroregeneration application include synthetic materials like polyethylene glycol (PEG) [26]/polyethylene oxide, poly(ethylene-*co*-vinylacetate [27, 28], poly(glycolic acid)/poly(lactic acid)/poly(lactic-*co*-glycolic acid), poly(2-hydroxyethyl methacrylate) 29, poly(2-hydroxyethyl methacrylate-*co*-methyl methacrylate), polypyrrole, [30] and natural materials like alginate [31–33]/ agarose, [34] chitosan/methylcellulose/nitrocellulose [35–37], collagen, [38] dextran, [39] fibrin, fibronectin, [40] starch, elastin, hyaluronic acid (HA), [41] and so on. ECM constitutes 10%–20% of CNS and has inductive, tissue regenerative properties. Effective scaffold should mimic the architecture, chemical composition, and mechanical properties of natural ECM, niche for endogenous stem cell homing. Here lies the advantage of employing hydrogels, which are highly cross-linked, water insoluble polymer networks for tissue engineering applications. The first successful clinical application of hydrogel was represented by the development of soft hydrogel contact lenses by Wichterle in 1961 [42] Hydrogels which possess the ligands for the integrins of the target cells can create an environment where NSCs could perform their normal behaviors such as proliferation, differentiation, and migration. Pure synthetic hydrogels lack many similarities to natural neural tissue. Further, many biological studies indicate that nanoparticle addition to hydrogels might improve their topography, mechanical properties, electroconductivity, and biological functions. Hence, the most promising biomaterials which meet the requirements of CNS tissue engineering seem to be injectable thermosensitive hydrogels loaded with specific micro- and nanoparticles.

11.5 INJECTABLE HYDROGEL MATRICES

Injectable hydrogels have wide application in tissue engineering and drug delivery applications due to ease of administration, site-specific shape formation providing good linkage, and interface between gel and host tissue. Various drug moieties and cells could be mixed to the polymer precursor solution to form, making them handy tools of a surgeon. For an ideal injectable system, it should meet several criteria [43] as follows:

(1) Low viscosity of the precursor solution allowing homogenous dispersion of drug molecule/ cells;
(2) Precursor solution should flow under pressure;
(3) Optimal gelation time avoiding clogging of the injecting needle or overflowing from the defect site to the surrounding tissues;
(4) Gelation reaction should not produce any temperature changes causing damage to adjacent tissue;
(5) Should attain sufficient mechanical stability in short time and maintain the integrity and strength for long time;

(6) Precursor solution and degradation products need to be biocompatible;
(7) Precursor solution should be sterilizable; and
(8) Controlled degradation rate

11.5.1 TYPES OF INJECTABLE HYDROGELS

Injectable hydrogels could be broadly classified as physical and chemical gels based on their nature of cross-linking. Physical gels also called as "smart hydrogels" are formed in response to changes in environment such as temperature, pH, and ionic changes. Examples include poly(N-isopropylacrylamide (poly(NIPAAm))-based copolymers, poly(ethylene oxide)/poly(propylene oixide) copolymers, PEG polyester copolymers, poly organophospazene derivatives, chitosan, alginate, gellan gum hydrogels, and so on.

11.5.2 METHODOLOGIES OF HYDROGEL PREPARATION

Whereas chemical gels are synthesized by the cross-linking methods like Schiff base formation, free-radical polymerization (redox polymerization, photopolymerization), Michael-type addition reaction, disulphide cross-linking, Genipin cross-linking, Click reaction chemistry, enzymatic cross-linking using the enzymes transglutaminases, peroxidases, tyrosinase, laccase, lysyl oxidase, phosphatase, metalloproteinases, DNA ligase, polymerase, and so on [44]. The factors influencing hydrogel formation are multifactorial, including the concentration of the reactants, temperature, pH, presence of cross-linking agents/catalysts/salts in the reaction mixture, and so on.

11.5.3 MODIFICATIONS AND SURFACE FUNCTIONALIZATION OF HYDROGELS

Several modifications have been carried out by various groups for obtaining optimum results. PEG-diacrylamide was synthesized by reacting acrylamide groups with thiol groups, and peptides containing a single thiol group were coupled to the PEG-diacrylamide in aqueous solution at room temperature [45]. In a similar way, reactive star-shaped poly(ethylene glycol) prepolymers (Star PEG) were modified with linear RGD peptides (gRGDsc) in different concentrations [46]. Danielle et al. [47] modified Heparin with methacrylate groups, copolymerized with dimethacrylated poly(ethylene glycol), and analyzed as a localized delivery vehicle for bFGF and synthetic ECM for the differentiation of hMSCs. Another approach is to incorporate growth factor/drug moieties to polymers. Gomez et al. [48] immobilized nerve growth factor (NGF) onto the electrically conducting polymer polypyrrole (PPy). NGF was immobilized using an intermediate linker provided by a layer of polyallylamine conjugated to an arylazido functional group, via activation of the azido groups by UV light, and NGF was fixed to the substrate. Also, the drug dexamethasone was stored in a conducting polymer coating (PEDOT/Dex) and selectively deposited on the electrode sites of neural probes [49] and active-controlled release in a chronic in vivo model was observed. Another group [50] has tried cross-linking of collagen using *N*-(3-dimethylaminopropyl)-*N*′-ethylcarbodiimide (EDC) and *N*-hydroxysuccinimide (NHS) to yield a material containing 14 free primary amino groups per 1000 amino acid residues (E/N14C) noncytotoxic cross-linked collagen substrate for endothelial cell seeding which was investigated. Hydrogels could also be functionalized using various methods. Polylysine-functionalized and thermoresponsive chitosan hydrogel was synthesized [51] for neural tissue engineering applications and two gelatin-based hydrogels cross-linked with EDC/NHS were fabricated for corneal tissue engineering [52]. Fibrin gels were formed with incorporated peptides like laminin and N-cadherin alone and in combination at concentrations up to 8.2 mol peptide per mole of fibrinogen [53]. A defined HA-based hydrogel cross-linked with matrix metalloproteinase cleavable peptides and modified with multiphoton labile nitrodibenzofuran was synthesized and epidermal growth factor (EGF) in gradients was photochemically immobilized [54].

Another interesting research evaluates histologically the effect of administration of HA hydrogels containing ECM derived from mouse embryonic stem cell-derived astrocytes on following SCI in rat [55]. Similarly, hyaluronan-methylfuran hydrogels [56], capable of encapsulating live cells, were prepared via Diels–Alder chemistry.

11.5.4 BIODEGRADATION AND BIOCOMPATIBILITY OF HYDROGELS

Hydrogels degrade either in bulk or by local diffusion. Local/surface dissolution is advantageous for drug delivery, as it can warranty a steady release over a long time. Hydrogel degradation can occur in different ways such as hydrolysis, enzymatic reaction, disentanglement of the network, or by environmental triggers [44]. Degradation rate is directly proportional to the rigidity and extent of cross-linking in the case of chemical gels, whereas physical gels can undergo revert back and undergo dissolution by changes in physical conditions or applications of stress [57]

Due to their high water absorbing ability, hydrogels are flexible and very similar to biological soft tissues and contributes to minimal irritation of surrounding tissues. Also, hydrogels exhibit low interfacial free energy, when in contact with body fluids, leading to reduced adsorption of proteins, and thus yielding better biocompatibility. Thus, biocompatibility means compatibility of the hydrogel and its degradation products with the immune system, the ability of them being metabolized into harmless products which could be excreted out of the body.

11.5.5 CHARACTERIZATION OF HYDROGELS

Hydrogels could be characterized physicochemically using various techniques such as solubility analysis, swelling properties, wettability, presence of chemical groups by FTIR spectroscopy, surface topography, and electrical conductivity by scanning electron microscopy, molecular distribution, and light scattering by gel permeation chromatography, yield of cross-linking and degradation by sol–gel analysis, gel point, and storage modulus by rheology analysis, the amount of free and bound water by differential scanning calorimetry and proton NMR, confirmation of cross-linking by thermogravimetric analysis, and X-ray diffraction analysis. The detailed principle, methods, and analysis are represented elsewhere [58].

Biological characterization of cells seeded or encapsulated in hydrogels could be carried out by immunofluorescent staining of specific cellular/cytoskeleton markers, the cellular proliferation by live–dead staining methods (Calcein AM-Ethidium homodimer staining) [59], migration by phase contrast microscopy, and activation by adapting molecular biology/proteomic techniques [60, 61].

11.6 THREE-DIMENSIONAL CELL CULTURE SYSTEMS

The need for alternate in vitro testing systems is substantiated by the concept of 3Rs [62] in animal research proposed by Russell and Burch [62], which advocates the replacement, reduction, and refinement of animal usage in biomedical research. International Co-operation on Alternative test Methods is an international organization established by the European Union Reference Laboratory for Alternatives to Animal Testing, the Japanese Center for the Validation of Alternative Methods, and Health Canada with the objective of establishing international cooperation necessary to ensure the development and practice of alternative test methods. The hanging drop method devised by Ross Granville Harrison (1870–1959) could be considered as the primitive form of three-dimensional (3D) culture system. Tissue microenvironment and 3D culturing techniques for cancer research was first proposed in the early 1980s by Mina Bissell, as a pioneer initiative. The comparative advantages and disadvantages between two-dimensional (2D) and 3D cell culture systems are represented in Table 11.1.

TABLE 11.1 Comparison between Two-Dimensional and Three-Dimensional Cell Culture Systems

2D	3D
Advantages	**Advantages**
• It is inexpensive	• Much better biomimetic tissue models
• It is well established	• Interaction between different types of cells
• There is a lot of comparative literature	• Establishment of barrier tissues
• It is what most researchers and scientists understand	• Better simulation of conditions in a living organism
• Easier cell observation and measurement	• Reduces use of animal models
Disadvantages	**Disadvantages**
• They are not representative of real cell environments	• Techniques are cumbersome and time-consuming
• Lack of predictivity	• Challenges in microscopy and measurement
• Issues caused by the growth media and expansion of cells	• Getting oxygen and other essential nutrients to the right place

Filter well inserts were one of the first technologies that adapted a 3D exposure of cells to a substrate by allowing all membrane sides to interact with the environment. Another type of 3D cell culture systems are microcarriers. The curved surface of the microcarrier is the simplest 3D substrate for cell culture. Microcarriers are small spheres, typically less than 500 mm in diameter, whose enormous surface area of up to 500 cm^2/g can culture large numbers of cells in small volumes, for example, cytoline, GEMTM, and so on. In addition to these, there are several ECM-based hydrogels mentioned in Table 11.2, available commercially. These culture matrices are used for differentiation studies of stem cells in a 3D milieu.

TABLE 11.2 Commercially Available Matrices for Neuronal Differentiation Studies

Cell Type	3D Model	Culture Substrate and Matrix
Human embryonic stem cells (HES)	Differentiation to smooth muscle, neurons	Polymer scaffold with ECM coating
Human embryonic stem cells	Differentiation to neural precursors	Hyaluronic acid matrix
Mouse embryonic stem cells (MES)	Differentiation to neurons	Collagen I scaffold
Mouse embryonic stem cells	Differentiation to neurons and astrocytes	Fibrin scaffold
Human neural progenitor stem cells (NPC)	Differentiation to neuronal cells	Corning Puramatrix (RADA-16) scaffold
Human neural progenitor stem cells	Differentiation to neuronal cells	Corning Puramatrix ± functionalized scaffold (laminin I)

11.7 NEURO-REGENERATIVE GROWTH FACTORS/TROPHINS/ PHYTOCHEMICALS

Several research groups worldwide are involved in active research employing different growth factor/ mitogens to elicit desired cellular response, mainly endogenous regenerative response. In addition to the several growth factors reported to have neurogenic effect, there are several phytochemicals used in the traditional medicine, which are treasure houses having wide range of biochemical and functional activities. Hydrogels loaded with phytomodulators as drug delivery and tissue engineering scaffolds is a comparatively newer approach. Arai et al. [63] reported the neurogenin-2 mediated NSC differentiation potential of phytochemicals from Butea Superba. Studies report that Ginkgo biloba extract could promote proliferation of endogenous NSCs in vascular dementia model in rats [64]. Also, Amalaki Rasayana (a preparation of Amla fruits-*Phyllanthus emblica*) has efficiently demonstrated reduction in DNA damage in brain cells proving its genomic stability in neurons and astrocytes [65]. Other neurogenic phytochemicals include Baicalin, from roots of *Scutellaria baicalensis*, [66] epimedium flavonoids [67], natural cerebrolysin, [68] Ecdyterone, [69], Xeihuo decoction, [70] Acanthopanax, Angelica, Rhodioloa, Ganoderma spore polygala, Gardenia, Astragaloside, Ginsenoside Rg1, Panax Notoginseng saponins, and so on [71]. These phytochemicals act as inducers of neurogenic differentiation in mesenchymal stem cells and the proliferation and survival of NSCs.

11.8 BIOMATRIX-INTEGRATED NEURO SIMULATIVE ENVIRONMENT—IN VIVO MODELS OF NEUROREGENERATION

Imitola et al. [72] demonstrated for the first time that chemokines like SDF-1a released during injury from neuronal cells will direct the migration of human and mouse NSCs to areas of injury in mice. Similar studies with neurogenic growth factors (SDF-1alpha, EGF) have been carried out by Lim et al. [60] where they have observed the chemotactic migration of NSCs to the site of injury. In another study, Macaya et al. [61] employed growth factor laden collagen-genipin (naturally occurring irridoid compound from plant origin) hydrogels loaded with FGF-laden lipid microtubules which could cross-link in vivo and become matrices for endogenous recruitment of NSCs to SPI site affecting motor function recovery. Tsai et al. [29] after complete spinal cord transection at T8, either filled pHEMA-MMA channels with matrices like collagen, fibrin, Matrigel, methylcellulose, or placed smaller pHEMA-MMA tubes within a larger pHEMA-MMA channel, then implanted into adult Sprague Dawley rats. This was carried out to probe whether the presence and type of matrix contained within synthetic hydrogel guidance channels were affecting the quantity and origin of axons that regenerate after complete spinal cord transection. In addition, several in vivo models for neural regeneration have been studied worldwide and discussed elsewhere [73].

11.9 CONCLUSION

The various challenges faced in the neuroregenerative therapy and the emerging research methodologies involving the usage of suitable biomatrices and natural and synthetic trophic factors are discussed in this chapter. Additionally, the development of 3D cell culture systems that help in the better understanding of complex in vivo cellular interactions is presented in this overview. Though research has been evolving in the in vivo simulating 3D cell culture systems, this chapter attempts to present in coherence the need for 3D culture systems, the most suitable matrices and growth factors/phytochemicals utilized for this purpose, and their potential applications with specific emphasis to neural tissue regeneration.

ACKNOWLEDGMENTS

The first author acknowledges the financial support of KSCSTE-YIPB project. The authors are grateful for the technical help in graphical illustration rendered by Ms Gayathri Sundar and Mr Vineeth CA.

KEYWORDS

- **neural regeneration**
- **tissue engineering**
- **injectable hydrogels**
- **three-dimensional cell culture system**

REFERENCES

1. Faul, M.; Xu, L.; Wald, M.; Coronado, V. Traumatic brain injury in the United States: emergency department visits, hospitalizations, and deaths 2002–2006. Atlanta, GA: Centers for Disease Control and Prevention, 2010.
2. Peeters, W.; van den Brande, R.; Polinder, S.; Brazinova, A.; Steyerberg, E.W.; Lingsma, H.F.; et al. Epidemiology of traumatic brain injury in Europe. *Acta Neurochir. (Wien)*. 2015, 157, 1683–1696.
3. Feigin, V.L.; Lawes, C.M.; Bennett, D.A.; Anderson, C.S. Stroke epidemiology: a review of population-based studies of incidence, prevalence, and case-fatality in the late 20th century. *Lancet Neurol*. 2003, 2, 43–53.
4. National Spinal Cord Injury Statistical Center Fact Sheet. National Spinal Cord Injury Center, 2006.
5. Hulsebosch, C.E. Recent advances in pathophysiology and treatment of spinal cord injury. *Adv. Physiol. Educ.* 2002, 26, 238–255.
6. Barami, K.; Diaz, F.G. Cellular transplantation and spinal cord injury. *Neurosurgery*. 2000, 47, 691–700.
7. Houle, J.D.; Tessler, A. Repair of chronic spinal cord injury. *Exp. Neurol*. 2003, 182, 247–260.
8. Bullock, R.; Zauner, A.; Woodward, J.J.; Myseros, J.; Choi, S.C; Ward, J.D.; et al. Factors affecting excitatory amino acid release following severe human head injury. *J Neurosurg*. 1998, 89(4), 507–518.
9. McAllister, T.W. Neurobehavioral sequelae of traumatic brain injury: evaluation and management. *World Psychiatry*. 2008, 7(1), 3–10.
10. Bardia, N.; Waubant, E. Neurodegeneration and remyelination in multiple sclerosis. In *Multiple Sclerosis—A Mechanistic View*, Minagar, A.; Ed; Academic Press: Cambridge, USA, 2016; p 311.
11. Willerth, S.M.; Sakiyama-Elbert, S.E. Approaches to neural tissue engineering using scaffolds for drug delivery. *Adv. Drug Deliv. Rev*. 2007, 59(4–5), 325–338.
12. Wosnick, J.H.; Baumann, M.D.; Shoichet, M.S. Tissue therapy: Central nervous system. In *Principles of Regenerative Medicine*, 2008; pp 1248–1269.
13. Zhang, C.; Saatman, K.E.; Royo, N.C.; Soltesz, K.M.; Millard, M.; Schouten, J.W.; et al. Delayed transplantation of human neurons following brain injury in rats: a long-term graft survival and behavior study. *J. Neurotrauma*. 2005, 22, 1456–1474.
14. Nonaka, M.; Yoshikawa, M.; Nishimura, F.; Yokota, H.; Kimura, H.; Hirabayashi, H.; et al. Intraventricular transplantation of embryonic stem cell-derived neural stem cells in intracerebral rats. *Neurol Res*. 2004, 26, 265–272.
15. Wang, Z.; Cui, C.; Li, Q.; Zhou, S.; Fu, J.; Wang, X.; et al. Intracerebral transplantation of foetal neural stem cells improves brain dysfunction

induced by intra cerebral haemorrhage stroke in mice. *J. Cell. Mol. Med.* 2011, 15, 2624–2633.
16. Bao, X.J.; Liu, F.; Lu, S.; Han, Q.; Feng, M.; Wei, J.J; et al. Transplantation of Flk-1+ human bone marrow-derived mesenchymal stem cells promotes behavioural recovery and anti-inflammatory and angiogenesis effects in an intracerebral hemorrhage rat model. *Int J. Mol. Med.* 2013, 31, 1087–1096.
17. Yang, K.L.; Lee, J.T.; Pang, C.Y.; Lee, T.Y.; Chen, S.P.; Liew, H.K.; et al. Human adipose-derived stem cells for the treatment of intracerebral hemorrhage in rats via femoral intravenous injection. *Cell. Mol. Biol. Lett.* 2012, 17, 376–392.
18. Liao, W.; Zhong, J.; Yu, J.; Xie, J.; Liu, Y.; Du, L.; et al. Therapeutic benefit of human umbilical cord derived mesenchymal stromal cells in intracerebral hemorrhage rat: implications of antiinflammation and angiogenesis. *Cell. Physiol. Biochem.* 2009, 24, 307–316.
19. Tara, S.; Krishnan, L.K. Bioengineered fibrin based niche to direct outgrowth of circulating progenitors into neuron-like cells for potential use in cellular therapy. *J. Neural. Eng.* 2015, 12, 036011.
20. Kokaia, Z.; Martino, G.; Schwartz, M.; Lindvall, O. Cross-talk between neural stem cells and immune cells: the key to better brain repair? *Nat. Neurosci.* 2012, 15, 1078–1087.
21. Yan, M.; Dai, H.; Ding, T.; Dai, A.; Zhang, F.; Yu, L.; et al. Effects of dexmedetomidine on the release of glial cell line-derived neurotrophic factor from rat astrocyte cells. *Neurochem. Int.* 2011, 58, 549–557.
22. Ashton, R.S.; Conway, A.; Pangarkar, C.; Bergen, J.; Lim K.; Shah, P.; et al. Astrocytes regulate adult hippocampal neurogenesis through ephrin-B signaling. *Nat. Neurosci.* 2012, 15, 1399–1406.
23. Moshayedi, P.; Carmichael, S.T. Hyaluronan, neural stem cells and tissue reconstruction after acute ischemic stroke. *Biomatter*. 2013, 3(1), e23863.
24. Fisher, S.A.; Tam, R.Y.; Stoichet, M. Tissue mimetics: engineered hydrogel matrices provide biomimetic environments for cell growth. *Tissue Eng. A*. 2014, 20(5, 6), 1–4.
25. Sakiyama-Elbert, S.E.; Hubbell, J.A. Controlled release of nerve growth factor from a heparin-containing fibrin-based cell ingrowth matrix. *J. Control Release*. 2000, 69(1), 149–158.
26. Peppas, N.A.; Keys, K.B.; Torres-Lugo, M.; Lowman, A.M. Poly(ethylene glycol)-containing hydrogels in drug delivery. *J. Control Release*. 1999, 62(1–2), 81–87.
27. Barras, F.M.; Pasche, P.; Bouche, N.; Aebischer, P.; Zurn, A.D. Glial cell line-derived neurotrophic factor released by synthetic guidance channels promotes facial nerve regeneration in the rat. *J. Neurosci. Res.* 2002, 70(6), 746–755.
28. Hoffman, D.; Wahlberg, L.; Aebischer, P. NGF released from a polymer matrix prevents loss of ChAT expression in basal forebrain neurons following a fimbria-fornix lesion. *Exp. Neurol.* 1990, 110(1), 39–44.
29. Tsai, E.C.; Dalton, P.D.; Shoichet, M.S.; Tator, C.H. Matrix inclusion within synthetic hydrogel guidance channels improves specific supraspinal and local axonal regeneration after complete spinal cord transection. *Biomaterials.* 2006, 27(3), 519–533.
30. Richardson, R.T.; Thompson, B.; Moulton, S.; Newbold, C.; Lum, M.G.; Cameron, A.; et al. The effect of polypyrrole with incorporated neurotrophin-3 on the promotion

of neurite outgrowth from auditory neurons. *Biomaterials.* 2007, 28(3), 513–523.
31. Kataoka, K.; Suzuki, Y.; Kitada, M.; Hashimoto, T.; Chou, H.; Bai, H.; et al. Alginate enhances elongation of early regenerating axons in spinal cord of young rats. *Tissue Eng.* 2004, 10(3–4), 493–504.
32. Kataoka, K.; Suzuki, Y.; Kitada, M.; Ohnishi, K.; Suzuki, K.; Tanihara, M.; et al. Alginate, a bioresorbable material derived from brown seaweed, enhances elongation of amputated axons of spinal cord in infant rats. *J. Biomed. Mater. Res.* 2001, 54(3), 373–384.
33. Prang, P.; Muller, R.; Eljaouhari, A.; Heckmann, K.; Kunz, W.; Weber, T.; et al. The promotion of oriented axonal regrowth in the injured spinal cord by alginate based anisotropic capillary hydrogels. *Biomaterials.* 2006, 27(19), 3560–3569.
34. Balgude, A.P.; Yu, X.; Szymanski, A.; Bellamkonda, R.V. Agarose gel stiffness determines rate of DRG neurite extension in 3D cultures. *Biomaterials.* 2001, 22(10), 1077–1084.
35. Bellamkonda, R.; Ranieri, J.P.; Bouche, N.; Aebischer, P. Hydrogel-based three-dimensional matrix for neural cells. *J. Biomed. Mater. Res.* 1995, 29(5), 663–671.
36. Tate, M.C.; Shear, D.A.; Hoffman, S.W.; Stein, D.G.; LaPlaca, M.C. Biocompatibility of methylcellulose based constructs designed for intracerebral gelation following experimental traumatic brain injury. *Biomaterials.* 2001, 22(10), 1113–1123.
37. Crompton, K.E.; Goud, J.D.; Bellamkonda, R.V.; Gengenbach, T.R.; Finkelstein, D.I.; Horne, M.K. Polylysine-functionalised thermoresponsive chitosan hydrogel for neural tissue engineering. *Biomaterials.* 2007, 28(3), 441–449.
38. Wissink, M.J.; Beernink, R.; Pieper, J.S.; Poot, A.A.; Engbers, G.H.; Beugeling, T. Binding and release of basic fibroblast growth factor from heparinized collagen matrices. *Biomaterials.* 2001, 22(16), 2291–2299.
39. Levesque, S.G.; Shoichet, M.S. Synthesis of cell-adhesive dextran hydrogels and macroporous scaffolds. *Biomaterials.* 2006, 27(30), 5277–5285.
40. Phillips, J.B.; King, V.R.; Ward, Z.; Porter, R.A; Priestley, J.V; Brown, R.A. Fluid shear in viscous fibronectin gels allows aggregation of fibrous materials for CNS tissue engineering. *Biomaterials.* 2004, 25(14), 2769–2779.
41. Gupta, D.; Tator, C.H.; Shoichet, M.S. Fast-gelling injectable blend of hyaluronan and methylcellulose for intrathecal, localized delivery to the injured spinal cord. *Biomaterials.* 2006, 27(11), 2370–2379.
42. Wichterle, O.; Lim, D. Hydrophilic gels for biological use. *Nature.* 1960, 185(4706), 117–118.
43. Balakrishnan, B.; Jayakrishnan, A. Injectable hydrogels for biomedical applications. In *Injectable Hydrogels for Regenerative Engineering*; Lakshmi, S.N., Ed.; Imperial College Press, World Scientific Publishing: Singapore, 2016.
44. Lakshmi S. Ed. *Injectable Hydrogels for Regenerative Engineering*. Imperial College Press, World Scientific Publishing: Singapore, 2006.
45. Donald, E.; Hubbell, J.A. Conjugate addition reactions combined with free-radical cross-linking for the design of materials for tissue engineering. *Biomacromolecules.* 2001, 2, 430–441.
46. Groll, J.; Fiedler, J.; Engelhard, E.; Ameringer, T.; Tugulu, S.; Klok, H.A.; et al. A novel star PEG–derived surface

coating for specific cell adhesion. J. Biomed. Mater. Res. A. 2005, 74(4), 607–617.
47. Danielle S.W. Benoit, Kristi S. Anseth, Heparin functionalized PEG gels that modulate protein adsorption for hMSC adhesion and differentiation. *Acta Biomaterialia*. 2005, 1(4), 461–470.
48. Gomez, N.; Schmidt, C.E. Nerve growth factor-immobilized polypyrrole: bioactive electrically conducting polymer for enhanced neurite extension. *J. Biomed. Mater. Res Part A*. 2007, 81(1), 135–149. https://doi.org/10.1002/jbm.a.31047.
49. Boehler, C.; Kleber, C.; Martini, N.; Xie, Y.; Dryg, I.; Stieglitz, T.; et al. Actively controlled release of dexamethasone from neural microelectrodes in a chronic in vivo study. *Biomaterials*. 2017, 129, 176–187.
50. Wissink, M.J.B.; Beernink, R.; Pieper, J.S.; Poot, A.A.; Engbers, G.H.M.; Beugeling, T.; et al. Immobilization of heparin to EDC/NHS-crosslinked collagen. Characterization and in vitro evaluation. *Biomaterials*. 2001, 22(2), 151–163.
51. Crompton, K.E.; Goud, J.D.; Bellamkonda, R.V.; Gengenbach, T.R.; Finkelstein, D.I.; Horne, M.K.; et al. Polylysine-functionalised thermoresponsive chitosan hydrogel for neural tissue engineering. *Biomaterials*. 2007, 28(3), 441–449.
52. Freier, T.; Koh, H.S.; Kazazian, K.; Shoichet, M.S. N-acetylation of chitosan films. Controlling cell adhesion and degradation of chitosan films by N-acetylation. *Biomaterials*. 2005, 26(29), 5872–5878.
53. Goodarzi, H.; Jadidi, K.; Pourmotabed, S.; Sharifi, E.; Aghamollaei, H. Preparation and *in vitro* characterization of cross-linked collagen–gelatin hydrogel using EDC/NHS for corneal tissue engineering applications. *Int. J. Biol. Macromol*. 2019, 126, 620–632.
54. Fisher, S.A.; Tam, R.Y.; Fokina, A.; Mahmoodi, M.M.; Distefano, M.D.; Shoichet, M.S. Photo-immobilized EGF chemical gradients differentially impact breast cancer cell invasion and drug response in defined 3D hydrogels. *Biomaterials*. 2018, 178, 751–766.
55. Thompson, R.E.; Pardieck, J.; Smith, L.; Kenny, P.; Crawford, L.; Shoichet, M.; et al. Effect of hyaluronic acid hydrogels containing astrocyte-derived extracellular matrix and/or V2a interneurons on histologic outcomes following spinal cord injury. *Biomaterials*. 2018, 162, 208–223. https://doi.org/10.1016/j.biomaterials.2018.02.0.
56. Smith, L.J.; Taimoory, S.M.; Tam, R.Y.; Baker, A.E.G.; Mohammad, N.B.; Trant, J.F.; et al. Diels-Alder click-cross-linked hydrogels with increased reactivity enable 3D cell encapsulation. *Biomacromolecules*. 2018, 19(3), 926–935. https://doi.org/10.1021/acs.biomac.7b01715.
57. Hennink, W.E.; Nostrum.C.F.V. Novel cross-linking methods to design hydrogels. *Adv. Drug Deliver. Rev.* 2002, 54(1), 107–113.
58. Gulrez, S.K.H.; Al-Assaf, S.; Phillips, G.O. Hydrogels: methods of preparation, characterisation and applications. In *Progress in Molecular and Environmental Bioengineering—From Analysis and Modeling to Technology applications*; Carpi, A. Eds; IntechOpen, 2011.
59. Joseph, J.; Hayakawa, K.; Kurisawa, M.; Wang, L.; Niu, W.; Spector, M. Astrocyte migration in injectable gelatin-hydroxyphenyl propionic acid matrices for neuronal guidance in spinal cord injury. *Front. Bioeng Biotechnol*, 2016. Conference Abstract: 10th World Biomaterials

Congress. https://doi.org/10.3389/conf.FBIOE.2016.01.01911
60. Lim, T.C.; Rokkapavanar, S.; Toh, W.S.; Spector, M. Chemotactic recruitment of adult neural progenitor cells into multifunctional hydrogels providing sustained SDF-1 release and compatible structural support. FASEB J. 2012, 27(3), 1023–1033.
61. Macaya, D.J.; Hayakawa, K.; Arai, K.; Spector, M. Astrocyte infiltration into injectable collagen-based hydrogels containing FGF-2 to treat spinal cord injury. Biomaterials. 2013, 34(14), 3591–3602.
62. Russell, W.M.S.; Burch, R.L. *The Principles of Humane Experimental Technique*. Methuen: London, 1959; pp 3–4, 137–139.
63. Arai, M.A.; Koryudzu, K.; Koyano, T.; Kowithayakorn, T.; Ishibashi, M. Naturally occurring Ngn2 promoter activators from Butea Superba. *Mol. Biosyst.* 2013, 9, 2489–2497.
64. Wang, J.W.; Chen, W.; Wang, Y.L. A Ginkgo biloba extract promotes proliferation of endogenous neural stem cells in vascular dementia rats. *Neural. Regen. Res.* 2013, 8, 1655–1662.
65. Swain, U.; Sindhu, K.K.; Boda, U.; Pothani, S.; Giridharan, N.V.; Raghunath, M. Studies on the molecular correlates of genomic stability in rat brain cells following Amalakirasayana therapy. *Mech. Ageing Dev*. 2012, 133, 112–117.
66. Ming, L.; Sze-Ting, C.; Kam-Sze, T.; Pang-Chui, S.; Kwok-Fai, L. DNA microarray expression analysis of baicalin-induced differentiation of C17.2 neural stem cells. *Chem. Biol. Chem*. 2012, 13, 1286–1290.
67. Ruiqin, Y.; Lan, Z.; Xiaoli, L.; Lin, L. Effects of epimedium flavonoids on proliferation and differentiation of neural stem cells *in vitro*. *Neurol. Res.* 2010, 32, 736–742.
68. Yinghong, L.; Zhengzhi, W.; Ming, L.; Xiuqin, J.; Min, Y.; Manyin, C. Gene expression in rat mesenchymal stem cells following treatment with natural cerebrolysin-containing serum: validation of a whole genome microarray technique. *Neural Regen. Res.* 2010, 5, 424–432.
69. Weihua, C.; Hua, F.; Zhi, C.; Mengxiong, Z.; Gang, Z.; Jiankai, L. Chinese herbal medicine derived ecdysterone for inducing differentiation of mammalian neural stem cells to neuronal cells. *Faming Zhuanli Shenqing.* 2008, CN101280291, A 20081008.
70. Peng-cheng, Z.; Xing-qun, L.; Qing-hua, L.; Wen-fang, L.; Qing-er, L.; Jie-Kun, L.; et al. Effects of Xiehuo Bushen Decoction on survival and differentiation of transplanted neural stem cells in brain of rats with intracerebral hemorrhage. *Zhongxiyi Jiehe Xuebao*. 2008, 6, 626–631.
71. Yin-Chu, S.; Qiang, L.; Chun-EX.; Xin, N.; Xiao-Hui, X.; Chang-Yuan, Y. Chinese herbs and their active ingredients for activating xue (blood) promote the proliferation and differentiation of neural stem cells and mesenchymal stem cells. *Chin. Med.* 2014, 9, 13.
72. Imitola, J.; Raddassi, K.; Park, K.I.; Mueller, F.J.; Nieto, M.; Teng, Y.D; et al. Directed migration of neural stem cells to sites of CNS injury by the stromal cell-derived factor 1alpha/CXC chemokine receptor 4 pathway. *Proc. Natl. Acad. Sci*. 2004, 101, 18117–18122.
73. Grinsell, D.; Keating, C.P. Peripheral nerve reconstruction after injury: a review of clinical and experimental therapies. *BioMed Res. Int.* 2014, 13, 698256.

CHAPTER 12

Nanoceria: A Rare-Earth Nanoparticle for Tissue Engineering Applications

SHIV DUTT PUROHIT, HEMANT SINGH, INDU YADAV, SAKCHI BHUSHAN, and NARAYAN CHANDRA MISHRA*

Department of Polymer and Process Engineering, Indian Institute of Technology Roorkee, Saharanpur Campus, Uttar Pradesh, 247001, India

**Corresponding author. E-mail: narayan.mishra@pe.iitr.ac.in*

ABSTRACT

Nowadays, nanoceria has gained significant research thrust, owing to its extensive biomedical applications, which include free radical scavenging, tissue engineering, anticancer activity, drug delivery systems, and antibacterial properties. Nanoceria could switch between the Ce(III) and Ce(IV) oxidation states. This produces the oxygen defects in the lattice, which result in free radical scavenging properties. Free radical scavenging property of nanoceria changes with the pH: at basic or physiological pH, it acts as an antioxidant, while at acidic pH it acts as a pro-oxidant. These free radical scavenging and antibacterial properties of nanoceria make it a potential material for tissue engineering. In this chapter, synthesis, properties, and application, in soft and hard tissue engineering of nanoceria have been discussed. Further, characterization techniques and factors influencing the nanoceria synthesis have been discussed briefly. The outcome of this chapter will develop the basic perceptive about various synthesis methods, properties, and tissue engineering applications of nanoceria, which might, sequentially, produce novel approaches for the treatment of various health disorders.

12.1 INTRODUCTION

Recently, the progress in nanotechnology has garnered research interest globally. Thus, nanoparticles are being used in various areas, that is, industrial, environmental, and medical fields. As compared to bulk materials, nanoparticles possess diverse mechanical, electronic, and magnetic properties owing to their distinctive small particle size, higher aspect ratio, and morphology [1]. There are various kinds of nanoparticles based on their physical and chemical nature such as carbon-based nanostructures (graphene, graphene oxide, fullerenes, and carbon nanotubes), quantum dots (CdSe, CdTe), metallic nanoparticles (Au, Ag), and metal oxides (CeO_2, ZnO_2, TiO_2) [2]. The exceptional physicochemical and biological properties of nanoparticles have attracted biomedical scientists to use it in diverse fields like tissue engineering (TE), biodetection of pathogens, drug delivery, detection of proteins, diagnostics, cancer treatment, separation and purification of biological molecules [3].

Nowadays, metal oxide nanoparticles are being used extensively in biomedical field owing to their various oxidation states. Among various metal oxide nanoparticles, cerium oxide nanoparticles (Nanoceria) draw significant research thrust due to their exceptional properties, that is, free radical scavenging, antioxidant, and antibacterial property.

Cerium, a rare-earth metal, belonging to the lanthanide series, is currently used in diverse industrial applications such as fuel cells, catalysis, and microelectronics [4]. Cerium has two redox states, Ce(III) and Ce(IV) which form two types of cerium oxide (ceria), that is, Ce_2O_3 and CeO_2.

Nanoceria possesses oxygen defect due to its large surface-to-volume ratio, which aids in free radical scavenging [5]. Additionally, various properties of therapeutic use such as antioxidant, antibacterial, anti-inflammatory, antiapoptotic, and angiogenic activity make nanoceria an ideal candidate for its use in TE applications [6–11]. Nanoceria, alone or in combination with other biomaterials, had been employed to repair/regeneration of various tissues, for example, skin, bone, heart, and nerve [12–15]. Thus nanoceria has become an important material of research and for finding its new applications.

This chapter includes various methods for synthesizing nanoceria; properties of nanoceria and its applications in various tissue engineering fields. The unique properties and state-of-the-art applications of nanoceria discussed in this chapter will motivate the researchers and scientists to further explore the applicability of nanoceria in diverse biomedical fields.

12.2 SYNTHESIS METHODS OF NANOCERIA

There are several methods available for the synthesis of nanoceria particles, which is depicted in Figure 12.1, discussed briefly in the following [16].

12.2.1 PRECIPITATION METHOD

It is the most suitable method skilled by various scientists for the development of nanoceria particles [17–26]. In this method, $Ce(NO_3)_3 \cdot 6H_2O$ is dissolved in double-distilled water (DDW), and for adjusting the pH in the range 2.5–3, liquid ammonia is added dropwise. Thereafter, NH_4HCO_3 solution is added dropwise to the above solution and stirred for 75 min at 90 °C temperature. Then the mixture is stirred for another 30 min at 90 °C for the reaction to complete. Finally, the resulting precipitate is washed multiple times with DDW and ethanol alternatively, followed by drying in the oven at 80 °C. Afterward, dried powders were calcined at 800 °C to obtain nanoceria particles.

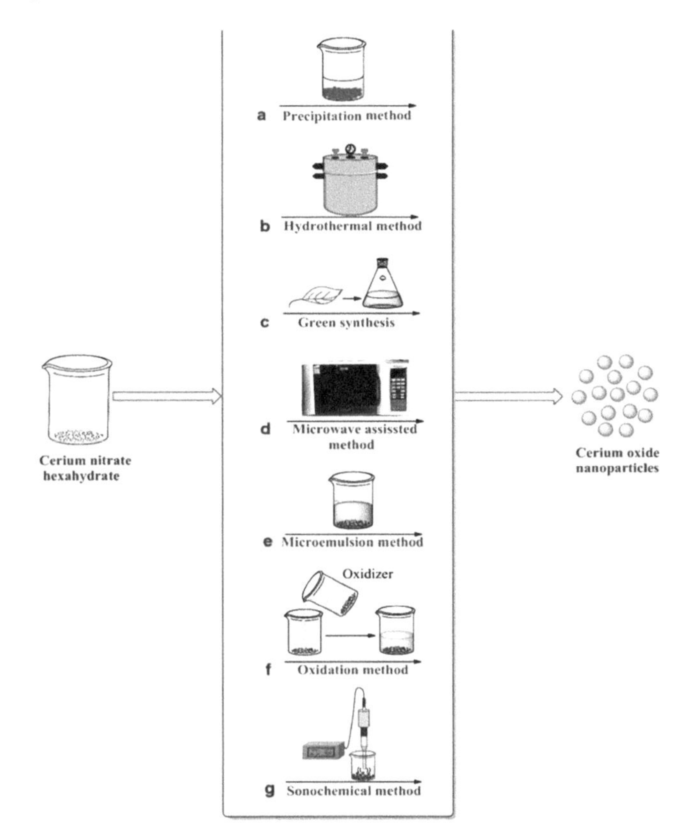

FIGURE 12.1 Different methods for the synthesis of nanoceria. (a) Precipitation method, (b) hydrothermal method, (c) green synthesis, (d) microwave-assisted method, (e) microemulsion method, (f) oxidation method, and (g) sonochemical method.

Source: Reprinted from [16]. (http://creativecommons.org/licenses/by/4.0/

12.2.2 HYDROTHERMAL METHOD

Apart from the precipitation method, hydrothermal method is also commonly used for the synthesis of nanoceria particles. Firstly in 2002, hydrothermal method-based nanoceria preparation was reported by Masui et al. [27]. In this study, a mixture of 1.0 M aqueous solutions of citric acid and cerium chloride (1:1) was added to 3.0 M ammonia water. Then, the mixture was stirred at 50 °C for 24 h to obtain a dark brown transparent liquid. Thereafter, the above synthesized solution was transferred into a Teflon vessel and heated at 80 °C for 24 h to improve the crystallinity of the particles. Further, the above-mentioned solution underwent centrifugation (2.1 × 104 rpm for 24 h) to obtain nanoparticle pellet. Finally, the particles were washed with DDW and methanol by ultrasonic agitation, followed by freeze drying of the particles to obtain fine nanoparticles.

Furthermore, various scientists also synthesized the nanoceria by hydrothermal method with different novel approaches such as; Renu et al. [28] reported the spherical nanoceria with diameter 100–200 nm in the absence of citric acid. Mai et al. [29] reported the synthesis of nanoceria with three different morphologies (polyhedron; size: 11.5 ± 1.8 nm, nanorods; size: 9.6 ± 1.2 nm, and nanocubes; size: 36.1 ± 7.1 nm) using cerium nitrate hexahydrate and sodium hydroxide by altering the temperature of the reaction and the base concentration. Different morphological nanoceria have different shape-dependent oxygen storage capacity, like nanorods and nanocubes store the oxygen on their surface in bulk, unlike nano-polyhedron. Further, Patil et al. [30] developed positively charged nanoceria with sizes 8–10 nm from cerium nitrate hexahydrate and ammonium hydroxide. Zou et al. [31] reported the surfactant (polyvinyl pyrrolidone)-assisted hydrothermal technique for the synthesis of conductive nanoceria (spherical; size: 5–10 nm) from cerium nitrate hexahydrate and PVP as precursors. Sutradhar et al. [32] used different precursors, like ceric ammonium nitrate and ammonium carbonate. The developed nanoceria was spherical in shape with diameters 3–4 nm, possessing high surface area, a large number of oxygen vacancies, and showed perfect catalytic activity.

12.2.3 GREEN SYNTHESIS

Presently, scientists showed much emphasis on Green synthesis process for nanoparticle development to avoid the harmful effects caused by chemical methods due to toxic chemicals. This technique allows easy and environment-friendly procedure. The main methodologies include in the following:

12.2.3.1 NUTRIENT-MEDIATED SYNTHESIS

In this method, nanoceria (plate shape: avg. dia.: 6–30 nm) was synthesized by Maensiri et al. [33] using cerium acetate hydrate ($Ce(CH_3CO_2)_3{\cdot}xH_2O$) with freshly extracted egg white as precursors. In brief, 40 mL deionized water was added to the 60 mL egg white under vigorous stirring to obtain the homogeneous mixture. Then, 0.1 M of $(CH_3CO_2)_3Ce{\cdot}xH_2O$ was added dropwise to the above-synthesized solution. Further, this mixture is stirred vigorously for 2 h, here the egg white entraps the cerium ions and produces a gel-like matrix. Thereafter, this gel-like matrix was dried at 80 °C and the obtained dried matrix was crushed to get nanoceria powder. Later, Kargar et al. [34] have also synthesized nanoceria from cerium nitrate hexahydrate and fresh egg white.

12.2.3.2 PLANT-MEDIATED SYNTHESIS

In this method, Priya et al. [35] synthesized the nanoceria by using aloe vera leaf extract. In brief, to obtain the aloe vera extract, the leaves of Aloe barbadensis miller plant were harvested and washed thoroughly, followed by spooning off the gel to the sterile beaker. On the other hand, a transparent solution was prepared by adding 40 mL of distilled water to 0.1 M of cerium(III) nitrate hexahydrate. This solution was added to 10 mL of the aloe vera leaf extract. Then the mixture underwent stirring for 30 min, followed by heat treatment at 80 °C to dry the mixture and the dried powder was obtained. Finally, the dried powder was calcinated for 2 h at 600 °C to achieve the nanoceria. Several other scientists also synthesized nanoceria by using different plant extracts such as *Acalypha indica* leaf extract [36], *Olea europaea* leaf extract [37] and Hibiscus sabdariffa flower extract [38].

12.2.3.3 FUNGUS-MEDIATED SYNTHESIS

For fungus-mediated synthesis of nanoceria, Munusamy et al. [39] developed this method. In this method, *Curvularia lunata* was inoculated in Czapek-Dox-Broth medium and kept at 37 °C temperature and 120 rpm for 72 h in an incubator shaker. Then, the produced fungal mycelium was filtered by Whatman filter paper No.1 and stored at room temperature. Subsequently, 0.1 M $CeCl_3{\cdot}7H_2O$ was added to 100 mL of above-synthesized filtrate and the mixture was kept on a stirrer at 80 °C for 6 h. Finally, the obtained yellow precipitate was washed with DDW and calcined at 400 °C for 2 h to get the nanoceria powder.

12.2.3.4 POLYMER-MEDIATED SYNTHESIS

This method was used by various scientists to prepare nanoceria [40–42]. Patil et al. [43] used pectin, cerium nitrate, and ammonia solution as precursors to synthesize nanoceria (spherical; size: ≤40 nm). Briefly, a 0.2 wt% aqueous solution of pectin was prepared and then 50 mL of 0.5 M cerium nitrate solution was added dropwise to this solution, while the solution is kept at 60 °C under vigorous stirring. Further, the excess amount of ammonia solution was added dropwise until the pH of the solution reached to 10. The solution was allowed to stir for an hour. Initially, the color of the solution was light yellow which later changed into yellow with increase in ammonia concentration. The yellow-colored precipitate obtained was centrifuged and washed several times with acetone and subsequently with water to make it free from nitrate, ammonia, and organic impurities. The washed precipitate was dried at 60 °C for 12 h, which was then annealed at 400 °C for 4 h.

12.2.4 MICROWAVE-ASSISTED METHOD

In this method, the microwave directly is coupled with the reactant and solvent available in the reaction mixture: this makes a quick transfer of energy resulting in a rapid rise in temperature of the mixture [44]. This technique provides high a yield in a short time. Wang et al. [45] firstly reported the preparation of nanoceria (diameter 2–3 nm) by microwave-assisted (power-650 W) technique by using ammonium cerium nitrate, hexamethylenetetramine, and polyethylene glycol as precursors. In brief, an appropriate amount of $(NH_4)_2Ce(NO_3)_6$, hexamethylenetetramine and poly(ethylene glycol) (PEG, 19,000) were introduced into 100 mL distilled water to give final concentrations of 0.02 mol L^{-1} $(NH_4)_2Ce(NO_3)_6$, 0.1 mol L^{-1} hexamethylenetetramine and 0.5 wt.% PEG. Then the mixture was placed in the microwave reflux system and refluxed under ambient air for 10 min. The microwave oven followed a working cycle of 9 s on and 21 s off (30% power). After cooling to room temperature, the precipitate was centrifuged, washed with distilled water and acetone repeatedly, and dried in air. The final products were collected as nanoceria.

12.2.5 MICROEMULSION METHOD

Here, in general, two immiscible solvents, a surfactant and co-surfactant are mixed and stirred, which creates emulsion. Patil et al. [46] synthesized nanoceria by stirring the mixture of cerium nitrate hexahydrate, ammonium hydroxide, toluene, bis(2-ethylhexyl) sulfosuccinate, and water. Briefly, sodium bis(2-ethylhexyl) sulfosuccinate dissolved in toluene, where 2.5 mL of 0.1 mol/L aqueous cerium nitrate solution was added and stirred for 45 min. After that ammonium hydroxide solution was added drop-by-drop. After 1 h, the reaction mixture was separated into two layers. Toluene containing nanoceria form the upper layer while water is the lower layer. Nanoceria settled down at the bottom due to gravity, which separated and dried.

Further, Sathyamurthy et al. [47] synthesized polyhedral nanoceria by stirring the mixture of cerium nitrate hexahydrate, ammonium hydroxide, cetyltrimethylammonium bromide (surfactant), 1-butanol (co-surfactant), *N*-octane (oil), and water (aqueous phase).

12.2.6 OXIDATION METHOD

Lee et al. [48] synthesized spherical nanoceria particles in water by oxidizing cerium ions with hydrogen peroxide: ammonium hydroxide was used for mineralization. Briefly, hydrogen peroxide (concentration: 10–20 times the concentration of Ce(III) ion) is added to 0.1 M $Ce(NO_3)_3.6H_2O$ aqueous solution and stirred in an ice bath, where deep orange color was observed at a high concentration of hydrogen peroxide. Ammonia solution (3 M), after 10 min of incubation, was added to the reaction mixture while stirring at 3000 rpm. On addition of ammonia, a violent reaction occurred changing the color of the solution from deep orange to dark brown. The precipitate formed was aged for 1 day which turned into yellow after aging. Neutral pH was maintained by washing the precipitate with ethanol. The wet precipitates were taken in a Teflon vessel which is placed in a pressure vessel. In the sealed condition, the vessel heated at 200 °C for 6 h in a thermostatic oven. Finally, nanoceria particles (precipitates) were rewashed several times with ethanol and dried at 80 °C for 24 h.

Several other methods have also been applied by different scientists to develop nanoceria [17, 22, 49–52].

12.2.7 SONOCHEMICAL METHOD

In this method, ultrasound waves of high intensity cause acoustic cavitations (i.e., bubble formation, growth, and collapsing), resulting in high temperature and pressure in a chemical reaction. The first nanoceria synthesis by sonochemical technique was reported by Yin et al. [53]. In brief, 0.434 g $Ce(NO_3)_3{\cdot}9H_2O$ was dissolved in 50 mL doubly distilled water, and 0.11 g azodicarbonamide was added to the aqueous solution of $Ce(NO_3)_3$. This is followed by the addition of different amounts of additives. The solution was sonicated at room temperature for 3 h by a high-intensity ultrasonic probe. During sonication, the temperature of the reaction mixture raised to ~80 °C. The suspension was then centrifuged, and the precipitate was washed with distilled water and ethanol and then dried under vacuum at room temperature: a yellow fine powder (nanoceria) was obtained. They reported that mixing of additives decreased the particle size, while the absence of additives causes agglomeration in nanoceria particles. Furthermore, Dutta et al. [54] reported the development of flower-type nanoceria by using cerium nitrate hexahydrate, tetraethylene glycol, and ammonium hydroxide. Later, other scientists also synthesized nanoceria by using cerium nitrate hexahydrate [55, 56].

12.3 FACTORS AFFECTING THE SYNTHESIS OF NANOCERIA

It had been observed that the size and shape of the nanoceria particles depend on various factors, for example, synthesis method, precursors selected, pH of the solution during synthesis, temperature of the reaction, and aging time. These are discussed briefly in the following.

12.3.1 SYNTHESIS METHOD

There are many methods available for the synthesis of the nanoceraia as described above. Every method has specific pros and cons. Size and shape of the synthesized nanoceria largely depend on the precursors used for the synthesis and synthesis method. For instance, the size of nanoceria, synthesized by coprecipitation method, ranges from 15 to 40 nm, the while size of nanoceria, by hydrothermal method, ranges from 3 to 5 nm.

12.3.2 PH

pH is also an essential aspect that influences the synthesis of nanoceria. Size of nanoceria can be controlled by varying the pH. The study conducted by Alpaslan et al. [57] revealed that the size of nanoceria was found to be smaller (15.5 nm) in diameter at basic pH (9.0), while diameter increases (24 nm) in acidic pH (6.0).

12.3.3 TEMPERATURE

Reaction temperature is an important factor which influences the shape of the nanoceria. Sakthivel et al. [58] observed the phase boundaries to shift from nanoparticles to nanorods and nanocubes at higher temperatures with a low concentration of sodium hydroxide. The gradual increase in temperature showed that the formation of nanorod and nanocube is mainly due to an increase in temperature. When the temperature is raised above 180 °C, the nanorods are converted to nanocubes.

12.3.4 TIME

The quality and size of nanoceria significantly depend on the aging time. Kuchibhatla et al. [59] demonstrated that aging time is the function of the oxidation state. Their study shows that when nanoceria solution was aged for 1 day then Ce(IV) oxidation state is dominant, while at the aging time of 1 week, obtained nanoceria gains both the oxidation states (Ce(III) and Ce(IV)) significantly.

12.4 CHARACTERIZATION TECHNIQUES

Several techniques have been widely used to characterize the nanoceria. Some of them are described below.

12.4.1 NANOPARTICLE FORMATION ANALYSIS

Formation of nanoceria is primarily analyzed by UV–visible spectroscopy [43]. In this technique, plasmon resonance is measured by evaluating the collective oscillations of conduction band electrons. Formation of nanoceria is confirmed by a well-defined sharp and strong absorbance peak located around 320 nm.

12.4.2 MORPHOLOGY AND PARTICLE SIZE

Morphology and particle size are key parameters for the nanoceria. These parameters could be determined by microscopic techniques such as scanning electron microscopy (SEM), transmission electron microscopy (TEM), and atomic force microscopy (AFM), which are described briefly in the following.

12.4.2.1 SEM

SEM provides a visual depiction of the nanoceria. For SEM analysis, the nanoceria suspension is drop casted and dried onto a glass slide, and sputter coated with a conductive metal (i.e., gold and gold/palladium alloy) [35, 60]. Next, a high-energy electrons beam is aimed at the sample to generate a topographical image.

12.4.2.2 TEM

TEM is a widely used technique, which determines the size and morphology of nanoceria [35, 60]. For TEM analysis nanoceria suspension is drop casted on carbon-coated copper grids and allowed to dry. After that, a monochromatic beam of electrons penetrate the sample and is projected onto a viewing screen to generate an image. By this technique, a near-atomic crystallographic structure of the sample can be visualized.

12.4.2.3 AFM

AFM produces three-dimensional images so that the height and volume of the nanoceria particles can be evaluated [60]. AFM can produce quantitative information regarding individual nanoparticles and groups of particles (i.e., size, morphology, and surface texture) with the help of software-based image processing. In this technique, a probe tip hovers over the surface of the sample in contact mode or tapping mode. The key advantage of AFM is its ability to image nonconducting samples without any specific treatment.

12.4.3 FOURIER-TRANSFORM INFRARED (FT-IR) SPECTROSCOPY

FT-IR spectroscopy is conducted to identify the functional groups present on nanoparticles [35]. The FT-IR spectra show a fingerprint region, which consists of stretching and vibrations bonds present in the nanoceria. Since each type of nanoparticle contains a unique combination of atoms, we can identify functional groups present inside the nanoparticles based on the FT-IR spectra. For obtaining the FT-IR spectra, the nanoceria is pelletized with potassium bromide (KBr) and the IR is passed through the pellet to obtain the IR spectra.

12.4.4 CRYSTALLINITY ANALYSIS

X-ray diffraction spectroscopy is used to evaluate the crystallinity of synthesized nanoceria [35, 60]. This technique is used to identify and quantitatively examine various crystalline forms. To accomplish this, the structure and lattice parameters of the powder specimen are analyzed by measuring the angle of diffraction, when X-ray beams are made incident on them.

12.4.5 RAMAN SPECTROSCOPY

Raman spectroscopy is a nondestructive chemical analysis technique, based on inelastic scattering of monochromatic light, usually from a laser source: this is used to understand about the chemical structure; phase and polymorphy; molecular interactions; crystallinity; purity and the defect levels associated with the nanoceria [33, 60].

12.4.6 NUCLEAR MAGNETIC RESONANCE SPECTROSCOPY

Nuclear magnetic resonance spectroscopy is a more advanced and convenient technique that noninvasively provides unique information regarding the interactions between the molecules of the nanoceria at an atomic level as well as a method to determine the size of nanoparticles in solution [43].

12.5 PROPERTIES OF NANOCERIA

12.5.1 PHYSIOCHEMICAL PROPERTIES

The physiochemical properties (i.e., size, morphology, surface chemistry, agglomeration, etc.) of nanoceria depend largely on the synthesis method, which eventually impacts its biological properties [61]. The size range for nanoceria varies from 3 to 70 nm based on various synthesis methods. Nanoceria has a large number of surface defects due to oxygen vacancies at the surface of its lattice [62]. The grain size is an important factor for the reactivity of the nanoceria: the lattice of nanoceria grows as the particle becomes smaller. This increase in the lattice decreases reabsorption and oxygen release [58]: this indicates that nanoceria with smaller crystalline size stabilizes less Ce^{4+} ions on their surface as compared to Ce^{3+} [62]. To evaluate the effect of synthesis methods on morphology, Sakthive and colleagues synthesized nanoceria, nanocubes, and nanorods by using the hydrothermal method with varying NaOH concentration, reaction time, and reaction temperature [63]. They exhibited that the synthesis methods can affect the crystalline size, lattice constant, oxygen vacancies, photoluminescence intensity, and bandgap of nanoceria. Their study also showed that the nanocubes of cerium oxide exhibit a reduced catalytic activity as compared to the nanoparticles and nanorods. Thus it could be stated that various synthesis methods of nanoceria allow tailoring the surface properties of the nanoceria, and therefore, its effect on biological systems. Furthermore, the addition of a surfactant or polymer during the synthesis of the nanoceria improves its dispersion. This improvement results in the decreased agglomeration in water or biological fluids, which could have an impact on the in vitro/in vivo cell interactions [64].

12.5.2 BIOLOGICAL PROPERTIES

Nanoceria exhibits exceptional antioxidant properties by inhibiting the oxidation of other molecules from reactive oxygen species (ROS) due to its free radical scavenging [65]. This ability is one of the most important features of nanoceria which protects the cells in an in-vivo environment [66, 67]. Nanoceria can remove free radicals by switching between the CeO_2 and Ce_2O_3 oxidative state and creating "reactive sites" in their structure. In this regard, Rubio and coworkers [68] evaluated the antioxidant property of nanoceria by adding an oxidative stress-inducing agent ($KBrO_3$) and they found a significant reduction in intracellular production of ROS. Further, Ciofani and associates [69] quantitatively evaluated the transcriptional profile of genes involved in oxidative stress and antioxidant defence: the results indicated that the nanoceria act as strong ROS scavengers as well as can modulate, downregulate, and upregulate genes involved in natural cell defenses.

Nanoceria can downregulate the mRNA expression of the key molecules involved in apoptosis including Bak1, caspase-8, and cytochrome C [70]. Nanoceria also can increase mitochondrial membrane potential and decrease the level of lactate dehydrogenase, both of which are essential for showing antiapoptotic activity [71]. But, the molecular mechanisms involved in antiapoptotic effects of nanoceria are not studied properly.

The antibacterial effects of nanoceria mostly depend on particle size, pH, and dose [72, 73]. Recently, Farias et al. [74], in their review, showed that nanoceria has antibacterial activity at size less than 54 nm on various bacterial strains including *Escherichia coli*, *Bacillus subtilis*, *Staphylococcus aureus*, *Streptococcus pneumoniae*, *Pseudomonas aeruginosa*. To evaluate the effect of pH, studies conducted by Alpaslan et al. [57] depict that nanoceria was significantly more effective at basic pH values (pH 9) for Gram-negative and positive strains than at acidic pH values (pH 6). In another study, Unnithan and colleagues [75] fabricated nanoceria doped electrospun antibacterial composite mats consisting of polyurethane, cellulose acetate, and zein. Results of antibacterial studies showed increased antibacterial properties with an increase in the nanoceria concentration. Moreover, the incubation time was determined as another important factor regarding the antibacterial activity of nanoceria particles.

Nanoceria does not induce any significant inflammation, even at high doses (50 μg/mL) [76]. This anti-inflammatory property of nanoceria aids for its potential use for TE and regenerative medicine [77]. The effects of nanoceria on molecules involved in inflammation revealed that it can upregulate anti-inflammatory interleukins IL-6 and IL-13 in vitro [78]. Also, the expression level of IL-10 was increased in the lungs of C57BL/6 mice instilled with 100 μg of CeO_2 [78]. Nanoceria also exhibits angiogenesis properties, which is a significant criterion toward repairing and regeneration of the damaged tissues through the formation of new vessels. To evaluate the angiogenesis properties of nanoceria, Das et al. [79] studied the relationships between physicochemical properties of nanoceria and angiogenesis. Their study depicts that the angiogenetic potential of nanoceria is dependent on the Ce^{3+}/Ce^{4+}, and surface of nanoceria regulates intracellular oxygen: which leads to efficient induction of angiogenesis. However, it should be noted that nanoceria particles can also inhibit angiogenesis depending on the microenvironment. Nanoceria at high concentrations has produced antiangiogenic effects [80]. For example, the proliferation of human umbilical vein endothelial cells is reduced by nanoceria at concentrations greater than 8.6 mg/mL [81]. It has also been shown that the antiangiogenic effect is more significant when nanoceria are functionalized with heparin [82].

12.6 APPLICATIONS OF NANOCERIA FOR TISSUE ENGINEERING

Nanoceria plays an important role in various types of tissue engineering, which are discussed in the following.

12.6.1 WOUND HEALING AND SKIN TISSUE ENGINEERING

The aptness of nanoceria for wound healing and skin regeneration has been evaluated by many researchers [83–85]. In this regard, Popov and coworkers [86] depicted that citrate-stabilized nanoceria can expedite the proliferation of primary mouse embryonic fibroblasts. This study demonstrated that nanoceria could stimulate cell proliferation to different extents in a wide range of concentrations (10^{-3}–10^{-9} M), while 10^{-7} M being the most effective concentration for this type of cell. Their study further revealed that stimulation for cell proliferation is being regulated by decreasing intracellular ROS during the lag phase of cell growth along with the modification of the expression level of major antioxidant enzymes such as SOD1, SOD2, glutathione reductase, glutathione peroxidase, and catalase.

To expedite the wound healing process, coordination between the various cells involved in wound repair is necessary. In this aspect, Chigurupati and coworkers examined the effects of nanoceria on the proliferation of fibroblasts, keratinocytes, and vascular endothelial cells for healing of cutaneous wounds [87]. They performed in vitro and in vivo experiments: in vitro studies evaluated the cell migration and proliferation after treatment with 500 nM, 1, and 10 mM of nanoceria, and in vivo studies were conducted on male C57BL/6 mice to evaluate the healing process of full-thickness wounds. For this purpose, 10 mL of a 10 mM nanoceria suspension was applied at the wound on a daily basis and the results revealed that topical application of nanoceria speeded up the wound healing process in mice by enhancing the cell proliferation and migration. Additionally, this study also reported that nanoceria is capable to infiltrate into the damaged site of the wounded tissue and reduce damage due to cellular membranes and proteins by scavenging ROS and, in turn, protect regenerated tissue.

In other studies, it has been revealed that ROS production increases at the wound sites, which results in activating a set of deleterious measures such as inflammation, fibrotic scarring, and alteration in cellular senescence [88]. This increased ROS concentration can be reduced by nanoceria-based drugs/scaffolds, which will enhance wound healing as well as reduce inflammation.

Recently, Wu and his group developed a highly versatile ROS-scavenging tissue adhesive nanocomposite by fixing the ultra-small ceria nanocrystals onto the surface of uniform mesoporous silica nanoparticles and evaluated its potential for its use in wound healing and skin regeneration applications [89]. The results obtained by in vitro as well as in vivo studies show that the silica–ceria nanocomposites have robust adhesion strength and demonstrate remarkable ROS-scavenging effects. These properties aid in the speedy recovery of a cutaneous wound and promote tissue regeneration at the damaged site. Wu and colleagues stated that their strategy may not be limited to cutaneous wound healing but it could also potentially be applicable in various types of wound repair and regeneration strategies, where both tissue adhesive ability and ROS-scavenging activity are the prime objective. In 2017, a study conducted by Naseri-Nosar et al. utilized the ROS scavenging property of the nanoceria and developed a prospective wound dressing material [13]. In this study, they incorporated different concentration of nanoceria (1.5, 3 and 6% [w/v]) to the polycaprolactone and gelatin (1:1 [w/w]) electrospun films. Preliminary results such as cell proliferation revealed that electrospun films containing 1.5% nanoceria could be the best suited for wound healing. Further, in in vivo studies, the film containing 1.5% nanoceria has been presumed as the optimum wound dressing material and this film was employed on the full-thickness excisional wounds in Wistar rats. Results revealed that the wounds treated with the nanoceria containing dressing achieved complete wound closure on day 14 as compared to nearly 63% closure with the sterile gauze.

Recently, Carlos et al. [90] have used nanoceria and its free-radical scavenging properties to heal the diabetic wounds. In their study, miR-146a micro-RNA was conjugated to different concentrations of nanoceria (1, 10, 100, or 1000 ng) and the effect of nanoceria-miR146a conjugate on diabetic wounds were evaluated. Results of this study revealed that a 100-ng dose of nanoceria-miR146a improved diabetic wound healing and did not impair the biomechanical properties of the skin post-healing.

The studies explained above established that free radical scavenging property of nanoceria makes it ideal for its use in wound healing and skin regeneration applications.

12.6.2 BONE TISSUE ENGINEERING

Based on the exceptional biological properties of nanoceria, researchers investigated its suitability in bone TE applications. Ball and coworkers assessed the biocompatibility of porous ceria foams for orthopedic tissue engineering [91]. In their study, the porous ceria scaffolds were fabricated via direct foaming approach, and further, their biocompatibility and ability to scavenge free radicals had been evaluated, which depicted that the ceria foams do not induce any cytotoxicity on mouse osteoblast (7F2) cell line and no significant inflammatory response was generated by human monocyticleukemia cells. Karakoti and colleagues have conducted a comparative study, which investigated the osteogenic potential of nanoceria containing bioactive glass-based scaffolds and the scaffolds without nanoceria [92]. Their in vitro studies showed that nanoceria-containing scaffolds do not induce any cytotoxicity to the human mesenchymal stem cells. Further, increased osteogenic differentiation and collagen production were observed in the nanoceria containing scaffolds as compared to the scaffolds without nanoceria.

In bone repair/regeneration, angiogenesis, that is, the formation of vascular structures plays a vital role [93]. In this regard, Xiang et al. have taken benefit of this intrinsic property associated with the nanoceria [94]. In their study, the cancellous bone was immersed in the 2 wt% poly-L-lactic acid solution, after that different concentration (upto 10 wt%) of nanoceria was added to the above solutions. Further, these mixtures were undergone freeze drying to obtain nanocomposite scaffolds. The results of the study depicted that the scaffold containing 5 wt% nanoceria is best suited for cell adhesion and proliferation. It was also revealed that the interaction between nanoceria and the cell membrane of mesenchymal stem cells could activate the available calcium channels, which leads to an increase in intracellular free Ca^{2+} ions. This increase in calcium ions can improve the stability of hypoxia-inducible factor 1-alpha and in turn results in the high expression of vascular endothelial growth factor.

In 2018, Robin et al. [95] exploited the redox properties of nanoceria to promote cell adhesion and angiogenesis, which is the key for bone regeneration. They fabricated nanoceria incorporated electrospun polycaprolactone nano-composite scaffolds. They observed that nanoceria-containing scaffolds aid in cell adhesion and angio-genesis. Gene-expression studies revealed that angiogenesis-related factors such as vascular endothelial growth factor and hypoxia-inducible factor 1-alpha were upregulated.

As described in various litera-ture [96–98], deposition of surface nanoceria coatings on nonbioactive orthopedic implants is a remarkable strategy to enhance the osteointe-gration of the prosthesis with host bone tissue. This can be attained by providing good osteogenic responses and decreased inflam-matory response. In this aspect, Li and coworkers developed nanoceria containing calcium silicate coatings by deploying the plasma spraying method [99]. And, they assessed the responses of bone-marrow-derived mesenchymal stem cells (BMSCs) and macrophages (RAW264.7 cell line) on addition of nanoceria. Their results depicted increased cell adhesion, proliferation, mineralized nodule formation, and infy alkaline phosphatase activity in BMSCs with increasing nanoceria content in the coatings. Furthermore, it has been observed that calcium silicate coat-ings with a high content of nanoceria can downregulate the gene expres-sions of proinflammatory (M1) markers (TNF-α, IL-6, and CCR7) and can upregulate the expressions of anti-inflammatory (M2) markers (IL-1ra, IL-10, and CD206).

The above-mentioned studies proved that the use of nanoceria is a game-changer with respect to enhancement of angiogenesis, cell adhesion, and cell proliferation, Hence nanoceria can be a potential material for bone tissue engineering.

12.6.3 CARDIAC TISSUE ENGINEERING

It has been observed by clinicians that during myocardial infarction, ROS are generated in ischemic surround-ings of heart tissue. This enhanced production of ROS can limit the cell–cell or cell–matrix adhesions, which result in the impairment of the integrity of transplanted cells (e.g., mesenchymal stem cells) used for cardiac regeneration [100]. Further, cardiomyocytes and the vasculature of the heart are at risk of damages due to oxidative stress [101]. Hence, significant attention has been given to apply nanoceria, owing to its ROS-scavenging properties, to overcome these problems,. In this respect, Pagliari and coworkers depicted that nanoceria is capable of protecting cardiac progenitor cells from oxidative stress [83]. Their study revealed that cells, treated with different concentrations (5, 10, and 50 μg/mL) of nanoceria are viable for at least 7 days without showing any inhibitory effect on cell growth and function, against H_2O_2-induced cytotoxicity. In another study, Mandoli et al. [102] developed a hybrid 2D polymeric-ceramic (PLGA–nanoceria) nanocomposites, which induce the growth of aligned murine-derived cardiac stem cells. In this study, the nanoceria concentrations and orientation strongly affected the mechanical, topographical, and biological properties of the nanocomposite. Moreover, cells, cultured on the hybrid nanocomposite for 1, 3, and 6 days, revealed enhanced cardiac stem cell proliferation for the nanoceria-loaded scaffolds. This suggests that antioxidative activity of nanoceria is preserved after the integration of nanoceria in the biopolymeric matrix.

In addition to in vitro studies, the clinical potential of nanoceria for cardiac repair/regeneration applications has been evaluated in terms of in vivo studies. In this regard, Kolli and coworkers reported that nanoceria can reduce the effect of right ventricular hypertrophy following 4 weeks of monocrotaline-induced pulmonary arterial hypertension [103]. In this study, nanoceria (0.1 mg/kg) was injected in the tail vein of male sprague dawley rats. The results depicted that nanoceria treatment reduces the occurrence of monocrotaline-induced pulmonary arterial hypertension, and right ventricular remodeling in the rats by diminishing oxidative stress. Moreover, the results also confirmed a reduction in Bax/Bcl2 ratio, serum inflammatory markers, and caspase-3 activation, which in turn aids in controlling cardiomyocyte apoptosis.

Based on the above studies, it could be stated that ROS scavenging property of nanoceria plays an important role in cardiac tissue engineering applications.

12.6.4 NERVE TISSUE ENGINEERING

For the treatment of nervous system disorders, tissue engineering-based strategies have gained significant importance [104, 105]. In this perspective, Arya and his group reported that nanoceria promotes nerve generation and diminishes the effect of hypoxia-induced memory impairment through aMPK-PKC-cBP signaling cascade [106]. In this study, polyethylene glycol-coated nanoceria (PEG-nanoceria) was injected intraperitoneally in adult male sprague dawley rats at a concentration of 5 μg/kg of body weight. The results showed a substantial decrease in oxidative stress and associated damage during hypoxia exposure. Furthermore, the results of flow cytometric, microscopic, and histological studies revealed increased nerve generation and the augmented hippocampus neuronal survival. The molecular tests depicted that PEG-nanoceria triggered the adenosine monophosphate kinase-protein kinase C-cAMP response element-binding protein (AMPK-PKC-CBP) protein pathway which in turn promotes the nerve generation. Other important molecular pathways in terms of neuronal survival have also been stimulated by nanoceria. In this regard, D'Angelo and coworkers reported that nanoceria modulates brain-derived neurotrophic factor pathway, which promotes neuronal survival in a human Alzheimer's disease model [107]. They assessed the signal transduction pathways involved in neuronal survival, neurotrophins modulation, and neuronal death in control and nanoceria-treated SH-SY5Y human neuroblastoma cells. The results depicted that nanoceria neutralizes neuronal death by increasing TrkB, brain-derived neurotrophic factor, and p75NTR levels, and promote cell survival by PI3K/Akt signaling pathway. Based on the above encouraging studies, Marino et al. recently developed electrospun nanocomposite scaffolds of highly aligned gelatin and nanoceria for nerve tissue engineering [108]. Further, the effects of the fabricated nanocomposite scaffolds were investigated on neuron-like SH-SY5Y cells. The results of this study show that the nanocomposite scaffolds aid in the growth and differentiation of neuronal cells, with improved neuronal phenotype as a result of strong topographical cues and antioxidant properties as depicted in Figure 12.2.

A latest in vivo study conducted by Kim and his group has shown that nanoceria could play a very important role in the treatment of spinal cord injury (SCI) [109]. In their study, nanoceria, with a wide range of doses (250–2000 μg/mL), was injected to a contused SCI model of rats, and then the functional behaviors and inflammatory responses were monitored up to 8 weeks. Results revealed that nanoceria at doses less than 1000 μg/mL significantly reduced the inducible nitric oxide synthase level and inflammatory responses of cells. Moreover, the use of nanoceria improved the locomotor functional recovery. Thus the study suggests the potential use of nanoceria in the treatment of SCI as an alternative tool to other

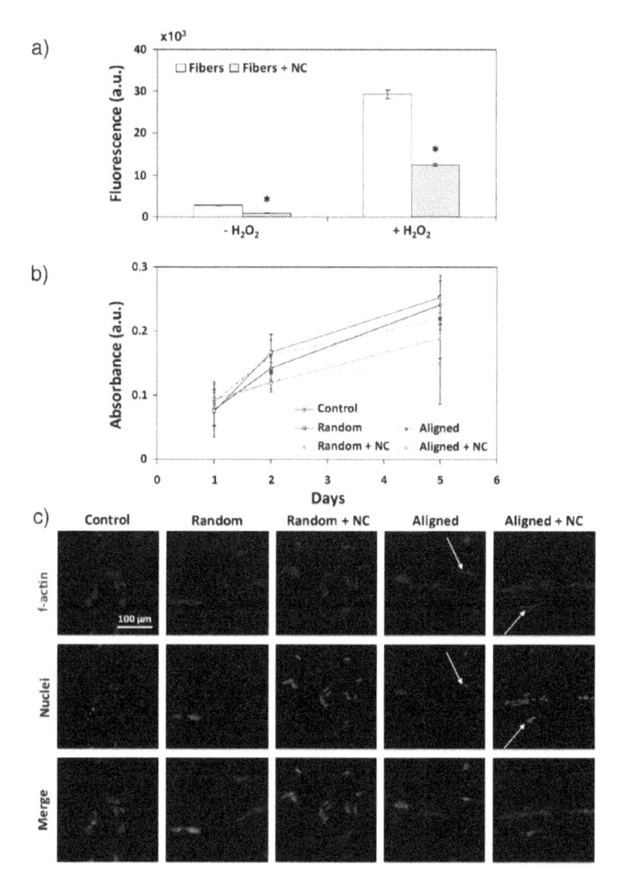

FIGURE 12.2 NC (Nanoceria)-loaded gelatin scaffolds reduce ROS levels and sustain cell proliferation. (a) ROS levels in the cells grown on gelatin vs gelatin/NC composite scaffolds in presence or absence of an oxidative insult (50 μM H_2O_2 for 30 min). (b) WST-1 proliferation measurements of cells grown on the different substrates at days 1, 2, and 5 of culture. (c) Staining of f-actin and nuclei of SH-SY5Y cells in proliferation. White arrows indicate polarized cells on anisotropic substrates ("aligned" and "aligned + NC"). * $p < 0.05$.

Source: Reprinted with permission from Ref. [108]. © 2017 Elsevier.

conventional approaches such as drug therapy.

Hence, nanoceria, owing to its remarkable properties such as antioxidant, anti-inflammatory, and strong topographical cues, could be a potential candidate for nerve tissue engineering applications.

12.7 CONCLUSION AND PERSPECTIVE OF NANOCERIA

Nanoceria is a comparatively new biomaterial than other vastly used biomaterials such as gelatin, collagen, alginate, hydroxyapatite, bioglass, chitosan, and so forth, which have been used for decades in tissue engineering applications. Nanoceria has attracted the scientific community in the mid-2000s for its use in tissue engineering applications. Since then nanoceria has been used in a variety of tissue regeneration applications, including the regeneration of soft tissues (e.g., skin, heart, and peripheral nerve) as well as hard tissue (e.g., bone).

A plenty of potential biomaterials have been developed for use in tissue engineering applications, and they possess many of the properties which are present in the nanoceria. But, nanoceria has certain limitations, for example, it cannot be processed to give the desired shape like polymers and bioactive glasses. Hence, nanoceria might not be sustainable to use it alone and it requires to be incorporated with some polymeric biomaterials. When nanoceria is incorporated in some polymeric biomaterial, it can tailor the physicochemical and biological properties depending on the specific application required. It is to mention that nanoceria, at high concentration, could induce cytotoxicity, which is a limitation associated with all nanomaterials.

The strength of nanoceria is its intrinsic free-radical scavenging property, which in turn yields in the following remarkable properties:

- Nanoceria depicts an antibacterial effect, which is not related with the release of antibacterial ions that could cause cytotoxicity to healthy cells.
- Nanoceria does not induce any significant inflammation, even at high doses (50 μg/mL).
- Nanoceria could protect cells from oxidative stress owing to its antioxidant property.
- Nanoceria can modulate the gene expression associated with regeneration of skin, bone, and cardiac tissues.
- Nanoceria could also affect the signal transduction pathways associated with nerve regeneration.

Thus, nanoceria could prove to be a potential biomaterial for its use in wound healing, skin, bone, cardiac, and nerve tissue engineering.

12.8 PRESENT CHALLENGES AND FUTURE RESEARCH OF NANOCERIA

In current societal, clinical, and biomedical scenario, there are many challenges as depicted in the following:

- Finding the way to treat the various types of cancer, which is ever increasing since its inception.

- Finding novel strategies to combat antibiotic-resistant bacterial strains, which have been considered as global threat by the World Health Organization.
- Globally, major cause of death is cardiac diseases and there is an urgent need to find potential treatment for cardiac diseases.
- The development of therapeutic cues for the treatment of problems associated with nerve degeneration.

The noteworthy properties of nanoceria could hold a great potential for finding the solutions of the above-mentioned challenges, and we believe nanoceria will provide a significant input in developing effective therapeutic approaches for treating the ever-augmenting pathological states in our society, which in turn will improve the quality of life of mankind.

KEYWORDS

- **nanoceria**
- **scaffold**
- **bone**
- **biomaterials**
- **tissue engineering**

REFERENCES

1. Fard, J. K., Jafari, S., & Eghbal, M. A. A review of molecular mechanisms involved in toxicity of nanoparticles. Advanced Pharmaceutical Bulletin, 5(4), 447 (2015).
2. Ju-Nam, Y., & Lead, J. R. Manufactured nanoparticles: an overview of their chemistry, interactions and potential environmental implications. Science of the Total Environment, 400(1–3), 396–414 (2008).
3. Baalousha, M., Le Coustumer, P., Jones, I., & Lead, J. R. Characterisation of structural and surface speciation of representative commercially available cerium oxide nanoparticles. Environmental Chemistry, 7(4), 377–385 (2010).
4. Pirmohamed, T., Dowding, J. M., Singh, S., Wasserman, B., Heckert, E., Karakoti, A. S., & Self, W. T. Nanoceria exhibit redox state-dependent catalase mimetic activity. Chemical Communications, 46(16), 2736–2738 (2010).
5. Sahu, T., Singh Bisht, S., Ranjan Das, K., & Kerkar, S. Nanoceria: synthesis and biomedical applications. Current Nanoscience, 9(5), 588–593 (2013).
6. Kyosseva, S. V., Chen, L., Seal, S., & McGinnis, J. F. Nanoceria inhibit expression of genes associated with inflammation and angiogenesis in the retina of Vldlr null mice. Experimental Eye Research, 116, 63–74 (2013).
7. Sardesai, N. P., Andreescu, D., & Andreescu, S. Electroanalytical evaluation of antioxidant activity of cerium oxide nanoparticles by nanoparticle collisions at microelectrodes. Journal of the American Chemical Society, 135(45), 16770–16773 (2013).
8. Shah, V., Shah, S., Shah, H., Rispoli, F. J., McDonnell, K. T., Workeneh, S., & Seal, S. Antibacterial activity of polymer coated cerium oxide nanoparticles. PLoS One, 7(10), e47827 (2012).

9. Fahy, N., Farrell, E., Ritter, T., Ryan, A. E., & Murphy, J. M. Immune modulation to improve tissue engineering outcomes for cartilage repair in the osteoarthritic joint. Tissue Engineering Part B: Reviews, 21(1), 55–66 (2014).
10. Stegen, S., van Gastel, N., & Carmeliet, G. Bringing new life to damaged bone: the importance of angiogenesis in bone repair and regeneration. Bone, 70, 19–27 (2015).
11. Zhang, Q., Ge, K., Ren, H., Zhang, C., & Zhang, J. Effects of cerium oxide nanoparticles on the proliferation, osteogenic differentiation and adipogenic differentiation of primary mouse bone marrow stromal cells in vitro. Journal of Nanoscience and Nanotechnology, 15(9), 6444–6451 (2015).
12. Naseri-Nosar, M., Farzamfar, S., Sahrapeyma, H., Ghorbani, S., Bastami, F., Vaez, A., & Salehi, M. Cerium oxide nanoparticle-containing poly (ε-caprolactone)/gelatin electrospun film as a potential wound dressing material: in vitro and in vivo evaluation. Materials Science and Engineering: C, 81, 366–372 (2017).
13. Pagliari, F., Mandoli, C., Forte, G., Magnani, E., Pagliari, S., Nardone, G., & Traversa, E. Cerium oxide nanoparticles protect cardiac progenitor cells from oxidative stress. ACS Nano, 6(5), 3767–3775 (2012).
14. Arya, A., Gangwar, A., Singh, S. K., Roy, M., Das, M., Sethy, N. K., & Bhargava, K. Cerium oxide nanoparticles promote neurogenesis and abrogate hypoxia-induced memory impairment through AMPK–PKC–CBP signaling cascade. International Journal of Nanomedicine, 11, 1159 (2016).
15. Thakur, N., Manna, P., & Das, J. Synthesis and biomedical applications of nanoceria, a redox active nanoparticle. Journal of Nanobiotechnology, 17(1), 84 (2019).
16. Kamruddin, M., Ajikumar, P. K., Nithya, R., Tyagi, A. K., & Raj, B. Synthesis of nanocrystalline ceria by thermal decomposition and soft-chemistry methods. Scripta Materialia, 50(4), 417–422 (2004).
17. Du, N., Zhang, H., Chen, B., Ma, X., & Yang, D. Ligand-free self-assembly of ceria nanocrystals into nanorods by oriented attachment at low temperature. The Journal of Physical Chemistry C, 111(34), 12677–12680 (2007).
18. Farahmandjou, M., Zarinkamar, M., & Firoozabadi, T. P. Synthesis of Cerium Oxide (CeO_2) nanoparticles using simple CO-precipitation method. Revista Mexicana de Física, 62(5), 496–499 (2016).
19. Sulthana, S., Banerjee, T., Kallu, J., Vuppala, S. R., Heckert, B., Naz, S., & Santra, S. Combination therapy of NSCLC using Hsp90 inhibitor and doxorubicin carrying functional nanoceria. Molecular Pharmaceutics, 14(3), 875–884 (2017).
20. Kalashnikova, I., Mazar, J., Neal, C. J., Rosado, A. L., Das, S., Westmoreland, T. J., & Seal, S. Nanoparticle delivery of curcumin induces cellular hypoxia and ROS-mediated apoptosis via modulation of Bcl-2/Bax in human neuroblastoma. Nanoscale, 9(29), 10375–10387 (2017).
21. Perez, J. M., Asati, A., Nath, S., & Kaittanis, C. Synthesis of biocompatible dextran-coated nanoceria with pH-dependent antioxidant properties. Small, 4(5), 552–556 (2008).
22. Kumar, A., Babu, S., Karakoti, A. S., Schulte, A., & Seal, S. Luminescence properties of europium-doped cerium oxide nanoparticles: role of vacancy

and oxidation states. Langmuir, 25(18), 10998–11007 (2009).
23. Q.L. Zhang, Z.M. Yang, & B.J. Ding, Synthesis of cerium oxide nanoparticles by the precipitation method, Materials Science Forum, 610–613, 233–238 (2009).
24. Zhang, Q. L., Yang, Z. M., & Ding, B. J. Synthesis of cerium oxide nanoparticles by the precipitation method. Materials Science Forum, 610, 233–238 (2009).
25. Rahdar, A., Aliahmad, M., Hajinezhad, M. R., & Samani, M. Xanthan gum-stabilized nano-ceria: Green chemistry based synthesis, characterization, study of biochemical alterations induced by intraperitoneal doses of nanoparticles in rat. Journal of Molecular Structure, 1173, 166–172 (2018).
26. Masui, T., Hirai, H., Imanaka, N., Adachi, G., Sakata, T., & Mori, H. Synthesis of cerium oxide nanoparticles by hydrothermal crystallization with citric acid. Journal of Materials Science Letters, 21(6), 489–491 (2002).
27. Renu, G., Rani, V. V., Nair, S. V., Subramanian, K. R. V., & Lakshmanan, V. K. Development of cerium oxide nanoparticles and its cytotoxicity in prostate cancer cells. Advanced Science Letters, 6(1), 17–25 (2012).
28. Mai, H. X., Sun, L. D., Zhang, Y. W., Si, R., Feng, W., Zhang, H. P., & Yan, C. H. Shape-selective synthesis and oxygen storage behavior of ceria nanopolyhedra, nanorods, and nanocubes. The Journal of Physical Chemistry B, 109(51), 24380–24385 (2005).
29. Patil, S., Sandberg, A., Heckert, E., Self, W., & Seal, S. Protein adsorption and cellular uptake of cerium oxide nanoparticles as a function of zeta potential. Biomaterials, 28(31), 4600–4607 (2007).
30. Zhou, F., Zhao, X., Xu, H., & Yuan, C. CeO_2 spherical crystallites: synthesis, formation mechanism, size control, and electrochemical property study. The Journal of Physical Chemistry C, 111(4), 1651–1657 (2007).
31. Sutradhar, N., Sinhamahapatra, A., Pahari, S., Jayachandran, M., Subramanian, B., Bajaj, H. C., & Panda, A. B. Facile low-temperature synthesis of ceria and samarium-doped ceria nanoparticles and catalytic allylic oxidation of cyclohexene. The Journal of Physical Chemistry C, 115(15), 7628–7637 (2011).
32. Maensiri, S., Masingboon, C., Laokul, P., Jareonboon, W., Promarak, V., Anderson, P. L., & Seraphin, S. Egg white synthesis and photoluminescence of plate-like clusters of CeO_2 nanoparticles. Crystal Growth & Design, 7(5), 950–955 (2007).
33. Kargar, H., Ghazavi, H., & Darroudi, M. Size-controlled and bio-directed synthesis of ceria nanopowders and their in vitro cytotoxicity effects. Ceramics International, 41(3), 4123–4128 (2015).
34. Priya, G. S., Kanneganti, A., Kumar, K. A., Rao, K. V., & Bykkam, S. Biosynthesis of Cerium oxide nanoparticles using Aloe barbadensis miller gel. International Journal of Scientific and Research Publications, 4(6), 199–224 (2014).
35. Kannan, S. K. & Sundrarajan, M. A green approach for the synthesis of a cerium oxide nanoparticle: characterization and antibacterial activity. International Journal of Nanoscience, 13(03), 1450018 (2014).
36. Maqbool, Q., Nazar, M., Naz, S., Hussain, T., Jabeen, N., Kausar, R., & Jan, T. Antimicrobial potential of green synthesized CeO_2 nanoparticles from *Olea europaea*

leaf extract. International Journal of Nanomedicine, 11, 5015 (2016).
37. Thovhogi, N., Diallo, A., Gurib-Fakim, A., & Maaza, M. Nanoparticles green synthesis by Hibiscus sabdariffa flower extract: main physical properties. Journal of Alloys and Compounds, 647, 392–396 (2015).
38. Munusamy, S., Bhakyaraj, K., Vijayalakshmi, L., Stephen, A., & Narayanan, V. Synthesis and characterization of cerium oxide nanoparticles using *Curvularia lunata* and their antibacterial properties. International Journal of Innovative Research in Science and Engineering, 2(1), 318–323 (2014).
39. Kargar, H., Ghasemi, F., & Darroudi, M. Bioorganic polymer-based synthesis of cerium oxide nanoparticles and their cell viability assays. Ceramics International, 41(1), 1589–1594 (2015).
40. Darroudi, M., Sarani, M., Oskuee, R. K., Zak, A. K., & Amiri, M. S. Nanoceria: gum mediated synthesis and in vitro viability assay. Ceramics International, 40(2), 2863–2868 (2014).
41. Darroudi, M., Sarani, M., Oskuee, R. K., Zak, A. K., Hosseini, H. A., & Gholami, L. Green synthesis and evaluation of metabolic activity of starch mediated nanoceria. Ceramics International, 40(1), 2041–2045 (2014).
42. Patil, S. N., Paradeshi, J. S., Chaudhari, P. B., Mishra, S. J., & Chaudhari, B. L. Bio-therapeutic potential and cytotoxicity assessment of pectin-mediated synthesized nanostructured cerium oxide. Applied Biochemistry and Biotechnology, 180(4), 638–654 (2016).
43. Hayes, B. L. Recent advances in microwave-assisted synthesis. Aldrichimica Acta, 37(2), 66–76(2004).
44. Wang, H., Zhu, J. J., Zhu, J. M., Liao, X. H., Xu, S., Ding, T., & Chen, H. Y. Preparation of nanocrystalline ceria particles by sonochemical and microwave assisted heating methods. Physical Chemistry Chemical Physics, 4(15), 3794–3799 (2002).
45. Patil, S., Kuiry, S. C., Seal, S., & Vanfleet, R. Synthesis of nanocrystalline ceria particles for high temperature oxidation resistant coating. Journal of Nanoparticle Research, 4(5), 433–438 (2002).
46. Sathyamurthy, S., Leonard, K. J., Dabestani, R. T., & Paranthaman, M. P. Reverse micellar synthesis of cerium oxide nanoparticles. Nanotechnology, 16(9), (2005).
47. Lee, J. S. & Choi, S. C. Crystallization behavior of nano-ceria powders by hydrothermal synthesis using a mixture of H_2O_2 and NH_4OH. Materials Letters, 58(3–4), 390–393(2004).
48. Karakoti, A. S., Singh, S., Kumar, A., Malinska, M., Kuchibhatla, S. V., Wozniak, K., & Seal, S. PEGylated nanoceria as radical scavenger with tunable redox chemistry. Journal of the American Chemical Society, 131(40), 14144–14145 (2009).
49. Alili, L., Sack, M., von Montfort, C., Giri, S., Das, S., Carroll, K. S., & Brenneisen, P. Downregulation of tumor growth and invasion by redox-active nanoparticles. Antioxidants & Redox Signaling, 19(8), 765–778 (2013).
50. Karakoti, A. S., Monteiro-Riviere, N. A., Aggarwal, R., Davis, J. P., Narayan, R. J., Self, W. T., & Seal, S. Nanoceria as antioxidant: synthesis and biomedical applications. JOM, 60(3), 33–37 (2008).
51. Pierscionek, B. K., Li, Y., Yasseen, A. A., Colhoun, L. M., Schachar, R. A., & Chen, W. Nanoceria have no genotoxic

effect on human lens epithelial cells. Nanotechnology, 21(3), 035102 (2009).
52. Yin, L., Wang, Y., Pang, G., Koltypin, Y., & Gedanken, A. Sonochemical synthesis of cerium oxide nanoparticles—effect of additives and quantum size effect. Journal of Colloid and Interface Science, 246(1), 78–84 (2002).
53. Dutta, D. P., Manoj, N., & Tyagi, A. K. White light emission from sonochemically synthesized rare earth doped ceria nanophosphors. Journal of Luminescence, 131(8), 1807–1812 (2011).
54. Agawane, S. M., & Nagarkar, J. M. Nano ceria catalyzed synthesis of α-aminophosphonates under ultrasonication. Tetrahedron Letters, 52(27), 3499–3504 (2011).
55. Choudhury, B., & Choudhury, A. Ce^{3+} and oxygen vacancy mediated tuning of structural and optical properties of CeO_2 nanoparticles. Materials Chemistry and Physics, 131(3), 666–671(2012).
56. Lew, A., Krutzik, P. O., Hart, M. E., & Chamberlin, A. R. Increasing rates of reaction: microwave-assisted organic synthesis for combinatorial chemistry. Journal of Combinatorial Chemistry, 4(2), 95–105 (2002).
57. Alpaslan, E., Geilich, B. M., Yazici, H., & Webster, T. J. pH-controlled cerium oxide nanoparticle inhibition of both gram-positive and gram-negative bacteria growth. Scientific Reports, 7, 45859 (2017).
58. Sakthivel, T., Das, S., Kumar, A., Reid, D. L., Gupta, A., Sayle, D. C., & Seal, S. Morphological phase diagram of biocatalytically active ceria nanostructures as a function of processing variables and their properties. ChemPlusChem, 78(12), 1446–1455 (2013).
59. Kuchibhatla, S. V., Karakoti, A. S., Baer, D. R., Samudrala, S., Engelhard, M. H., Amonette, J. E., & Seal, S. Influence of aging and environment on nanoparticle chemistry: implication to confinement effects in nanoceria. The Journal of Physical Chemistry C, 116(26), 14108–14114 (2012).
60. Salame, P. H., Pawade, V. B., & Bhanvase, B. A. Characterization tools and techniques for nanomaterials. In Nanomaterials for Green Energy; Bhanvase, B. A., Pawade, V. B., Dhoble, S. J., Sonawane, S. H., & Muthupandian, A.K. (Eds), Elsevier, 83–111 (2018).
61. Deshpande, S., Patil, S., Kuchibhatla, S. V., & Seal, S. Size dependency variation in lattice parameter and valency states in nanocrystalline cerium oxide. Applied Physics Letters, 87(13), 133113 (2005).
62. Reed, K., Cormack, A., Kulkarni, A., Mayton, M., Sayle, D., Klaessig, F., & Stadler, B. Exploring the properties and applications of nanoceria: is there still plenty of room at the bottom? Environmental Science: Nano, 1(5), 390–405 (2014).
63. Lynch, I. & Dawson, K. A. Protein-nanoparticle interactions. Nano Today, 3(1–2), 40–47 (2008).
64. Grulke, E., Reed, K., Beck, M., Huang, X., Cormack, A., & Seal, S. Nanoceria: factors affecting its pro-and anti-oxidant properties. Environmental Science: Nano, 1(5), 429–444 (2014).
65. Chen S., Hou Y., Cheng G., Zhang C., Wang S, & Zhang J. Cerium oxide nanoparticles protect endothelial cells from apoptosis induced by oxidative stress. Biological Trace Element Research, 154(1), 156–166 (2013).
66. Colon, J., Hsieh, N., Ferguson, A., Kupelian, P., Seal, S., Jenkins, D. W., & Baker, C. H. Cerium oxide nanoparticles protect gastrointestinal epithelium from

radiation-induced damage by reduction of reactive oxygen species and upregulation of superoxide dismutase 2. Nanomedicine: Nanotechnology, Biology and Medicine, 6(5), 698–705 (2010).
67. Nelson B. C., Johnson M. E., Walker M. L., Riley K. R., & Sims C. M. Antioxidant cerium oxide nanoparticles in biology and medicine. Antioxidants, 5(2), 15 (2016).
68. Rubio, L., Annangi, B., Vila, L., Hernández, A., & Marcos, R. Antioxidant and anti-genotoxic properties of cerium oxide nanoparticles in a pulmonary-like cell system. Archives of Toxicology, 90(2), 269–278 (2016).
69. Ciofani G., Genchi G. G., Mazzolai B., & Mattoli V. Transcriptional profile of genes involved in oxidative stress and antioxidant defense in PC12 cells following treatment with cerium oxide nanoparticles. Biochimica et Biophysica Acta, 1840(1), 495–506 (2014).
70. Krishnamoorthy K., Veerapandian M., Zhang L.-H., Yun K., & Kim S. J. Surface chemistry of cerium oxide nanocubes: Toxicity against pathogenic bacteria and their mechanistic study. Journal of Industrial and Engineering Chemistry, 20(5), 3513–3517 (2014).
71. Colon J., Hsieh N., Ferguson A., et al. Cerium oxide nanoparticles protect gastrointestinal epithelium from radiation-induced damage by reduction of reactive oxygen species and upregulation of superoxide dismutase 2. Nanomedicine, 6(5), 698–705 (2010).
72. Fahy N., Farrell E., Ritter T., Ryan A. E., & Murphy J. M. Immune modulation to improve tissue engineering outcomes for cartilage repair in the osteoarthritic joint. Tissue Engineering Part B Reviews, 21(1), 55–66 (2014).
73. Kartsonakis I. A., Liatsi P., Daniilidis I., & Kordas G. Synthesis, characterization, and antibacterial action of hollow ceria nanospheres with/without a conductive polymer coating. Journal of the American Ceramic Society, 91(2), 372–378 (2008).
74. Farias, I. A. P., Santos, C. C. L. D., & Sampaio, F. C. Antimicrobial activity of cerium oxide nanoparticles on opportunistic microorganisms: a systematic review. BioMed Research International, 2018, 14 (2018).
75. Unnithan A. R., Sasikala A. R. K., Sathishkumar Y., Lee Y. S., Park C. H., & Kim C. S. Nanoceria doped electrospun antibacterial composite mats for potential biomedical applications. Ceramics International, 40(8), 12003–12012 (2014).
76. Ngobili TA & Daniele MA. Nanoparticles and direct immunosuppression. Experimental Biology and Medicine, 241(10), 1064–1073 (2016).
77. Wingard, C. J., Walters, D. M., Cathey, B. L., Hilderbrand, S. C., Katwa, P., Lin, S., & Brown, J. M. Mast cells contribute to altered vascular reactivity and ischemia-reperfusion injury following cerium oxide nanoparticle instillation. Nanotoxicology, 5(4), 531–545 (2011).
78. Gojova A., Lee J.-T., Jung H. S., Guo B., Barakat A. I., & Kennedy I. M. Effect of cerium oxide nanoparticles on inflammation in vascular endothelial cells. Inhalation Toxicology, 21(sup1), 123–130 (2009).
79. Das, S., Singh, S., Dowding, J. M., Oommen, S., Kumar, A., Sayle, T. X., & Self, W. T. The induction of angiogenesis by cerium oxide nanoparticles through the modulation of oxygen in intracellular

environments. Biomaterials, 33(31), 7746–7755 (2012).
80. Saghiri M. A., Orangi J., Asatourian A., Sorenson C. M., & Sheibani N. Functional role of inorganic trace elements in angiogenesis part III:(Ti, Li, Ce, As, Hg, Va, Nb and Pb). Critical Reviews in Oncology/Hematolgy, 98, 290–301 (2016).
81. Dowding, J. M., Das, S., Kumar, A., Dosani, T., McCormack, R., Gupta, A., & Self, W. T. Cellular interaction and toxicity depend on physicochemical properties and surface modification of redox-active nanomaterials. ACS Nano, 7(6), 4855–4868 (2013).
82. Lord MS, Tsoi B, Gunawan C, Teoh WY, Amal R, & Whitelock JM. Anti-angiogenic activity of heparin functionalised cerium oxide nanoparticles. Biomaterials, 34(34), 8808–8818 (2013).
83. Pagliari, F., Mandoli, C., Forte, G., Magnani, E., Pagliari, S., Nardone, G., & Traversa, E. Cerium oxide nanoparticles protect cardiac progenitor cells from oxidative stress. ACS Nanomaterials, 6(5), 3767–3775 (2012).
84. Wu, H., Li, F., Wang, S., Lu, J., Li, J., Du, Y., & Ling, D. Ceria nanocrystals decorated mesoporous silica nanoparticle based ROS-scavenging tissue adhesive for highly efficient regenerative wound healing. Biomaterials, 151, 66–77 (2018).
85. Davan, R., Prasad, R. G. S. V., Jakka, V. S., Aparna, R. S. L., Phani, A. R., Jacob, B., & Raju, D. B. Cerium oxide nanoparticles promotes wound healing activity in in-vivo animal model. Journal of Bionanoscience, 6(2), 78–83 (2012).
86. Popov, A. L., Popova, N. R., Selezneva, I. I., Akkizov, A. Y., & Ivanov, V. K. Cerium oxide nanoparticles stimulate proliferation of primary mouse embryonic fibroblasts in vitro. Materials Science and Engineering: C, 68, 406–413 (2016).
87. Chigurupati, S., Mughal, M. R., Okun, E., Das, S., Kumar, A., McCaffery, M., & Mattson, M. P. Effects of cerium oxide nanoparticles on the growth of keratinocytes, fibroblasts and vascular endothelial cells in cutaneous wound healing. Biomaterials, 34(9), 2194–2201 (2013).
88. Davalli P, Mitic T, Caporali A, Lauriola A, D'Arca D. ROS, cell senescence, and novel molecular mechanisms in aging and age-related diseases. Oxidative Medicine and Cellular Longevity, 2016, 3565127 (2016).
89. Wu, H., Li, F., Wang, S., Lu, J., Li, J., Du, Y., & Ling, D. Ceria nanocrystals decorated mesoporous silica nanoparticle based ROS-scavenging tissue adhesive for highly efficient regenerative wound healing. Biomaterials, 151, 66–77 (2018).
90. Zgheib, C., Hilton, S.A., Dewberry, L.C., Hodges, M.M., Ghatak, S., Xu, J., Singh, S., Roy, S., Sen, C.K., Seal, S., & Liechty, K.W. Use of cerium oxide nanoparticles conjugated with MicroRNA-146a to correct the diabetic wound healing impairment. Journal of the American College of Surgeons, 228(1), 107–115 (2019).
91. Ball J. P., Mound B. A., Monsalve A. G., Nino J. C., Allen J. B. Biocompatibility evaluation of porous ceria foams for orthopedic tissue engineering. Journal Biomedical Materials Research Part A, 103(1), 8–15 (2015).
92. Karakoti, A. S., Tsigkou, O., Yue, S., Lee, P. D., Stevens, M. M., Jones, J. R., & Seal, S. Rare earth oxides as nanoadditives in 3-D nanocomposite scaffolds for bone regeneration. Journal

of Materials Chemistry, 20(40), 8912–8919 (2010).
93. Kargozar, S., Hashemian, S. J., Soleimani, M., Milan, P. B., Askari, M., Khalaj, V., & Mozafari, M. Acceleration of bone regeneration in bioactive glass/gelatin composite scaffolds seeded with bone marrow-derived mesenchymal stem cells over-expressing bone morphogenetic protein-7. Materials Science and Engineering: C, 75, 688–698 (2017).
94. Xiang, J., Li, J., He, J., Tang, X., Dou, C., Cao, Z., & Dong, S. Cerium oxide nanoparticle modified scaffold interface enhances vascularization of bone grafts by activating calcium channel of mesenchymal stem cells. ACS Applied Materials & Interfaces, 8(7), 4489–4499 (2016).
95. Augustine, R., Dalvi, Y. B., Dan, P., George, N., Helle, D., Varghese, R., & Sandhyarani. Nanoceria can act as the cues for angiogenesis in tissue-engineering scaffolds: toward next-generation in situ tissue engineering. ACS Biomaterials Science & Engineering, 4(12), 4338–4353 (2018).
96. Baino F & Vitale-Brovarone C. Feasibility of glass–ceramic coatings on alumina prosthetic implants by airbrush spraying method. Ceramics International, 41(2), 2150–2159 (2015).
97. Chen, Q., Baino, F., Pugno, N. M., & Vitale-Brovarone, C. Bonding strength of glass-ceramic trabecular-like coatings to ceramic substrates for prosthetic applications. Materials Science and Engineering: C, 33(3), 1530–1538 (2013).
98. Vitale-Brovarone C., Baino F., Tallia F., Gervasio C., & Vern´e E. Bioactive glass-derived trabecular coating: a smart solution for enhancing osteointegration of prosthetic elements. Journal of Materials Science: Materials in Medicine, 23(10), 2369–2380 (2012).
99. Li K., Yu J., Xie Y., You M., Huang L., & Zheng X. The effects of cerium oxide incorporation in calcium silicate coating on bone mesenchymal stem cell and macrophage responses. Biological Trace Element Research, 177(1), 148–158 (2017).
100. Song, H., Cha, M. J., Song, B. W., Kim, I. K., Chang, W., Lim, S., & Jang, Y. Reactive oxygen species inhibit adhesion of mesenchymal stem cells implanted into ischemic myocardium via interference of focal adhesion complex. Stem Cells, 28(3), 555–563 (2010).
101. Ilkun, O. & Boudina, S. Cardiac dysfunction and oxidative stress in the metabolic syndrome: an update on antioxidant therapies. Current Pharmaceutical Design, 19(27), 4806–4817 (2013).
102. Mandoli, C., Pagliari, F., Pagliari, S., Forte, G., Di Nardo, P., Licoccia, S., & Traversa, E. Stem cell aligned growth induced by CeO_2 nanoparticles in PLGA scaffolds with improved bioactivity for regenerative medicine. Advanced Functional Materials, 20(10), 1617–1624 (2010).
103. Kolli, M. B., Manne, N. D., Para, R., Nalabotu, S. K., Nandyala, G., Shokuhfar, T., & Dornon, L. Cerium oxide nanoparticles attenuate monocrotaline induced right ventricular hypertrophy following pulmonary arterial hypertension. Biomaterials, 35(37), 9951–9962 (2014).
104. Orive, G., Anitua, E., Pedraz, J. L., & Emerich, D. F. Biomaterials for promoting brain protection, repair and regeneration. Nature Reviews Neuroscience, 10(9), 682 (2009).

105. Naz S., Beach J., Heckert B., et al. Cerium oxide nanoparticles: a 'radical' approach to neurodegenerative disease treatment. Nanomedicine 12(5), 545–553 (2017).
106. Arya, A., Gangwar, A., Singh, S. K., Roy, M., Das, M., Sethy, N. K., & Bhargava, K. Cerium oxide nanoparticles promote neurogenesis and abrogate hypoxia-induced memory impairment through AMPK–PKC–CBP signaling cascade. International Journal of Nanomedicine, 11, 1159 (2016).
107. D'Angelo, B., Santucci, S., Benedetti, E., Di Loreto, S., Phani, R. A., Falone, S., & Cimini, A. Cerium oxide nanoparticles trigger neuronal survival in a human Alzheimer disease model by modulating BDNF pathway. Current Nanoscience, 5(2), 167–176 (2009).
108. Marino, A., Tonda-Turo, C., De Pasquale, D., Ruini, F., Genchi, G., Nitti, S., & Ciofani, G. Gelatin/nanoceria nanocomposite fibers as antioxidant scaffolds for neuronal regeneration. Biochimica et Biophysica Acta (BBA)-General Subjects, 1861(2), 386–395 (2017).
109. Kim, J. W., Mahapatra, C., Hong, J. Y., Kim, M. S., Leong, K. W., Kim, H. W., & Hyun, J. K. Functional recovery of contused spinal cord in rat with the injection of optimal-dosed cerium oxide nanoparticles. Advanced Science, 4(10), 1700034 (2017).

CHAPTER 13

Micro- and Nanostructured Biomaterials for Bone Tissue Engineering

AJITA JINDAL[1], JAYDEEP BHATTACHARYA[1], and RANJITA GHOSH MOULICK[2,*]

[1]*School of Biotechnology, Jawaharlal Nehru University, New Delhi, India*

[2]*Amity Institute of Integrative Sciences & Health, Amity University, Gurgaon, Haryana, India*

**Corresponding author. E-mail: ranjita.ghoshmoulick@gmail.com/ rgmoulick@ggn.amity.edu*

ABSTRACT

Bone tissue engineering (BTE) techniques are aimed at the repair, regeneration, and restoration of damaged bone by utilizing cells in combination with biomaterials and growth factors. Ceramic, metallic, polymeric biomaterials, and their composites augment the natural inherent capacity of bone tissue to regenerate itself as a result they have found wide-scale application in BTE. Over the years, chemical and topographical modification of material surface has emerged as a powerful tool for regulating cellular properties such as adhesion, morphology, attachment, proliferation, and differentiation via alteration of physical attributes and chemical composition of engineered biomaterials. Consequently, the recent trend in biofabrication is to impart micro-and nanofeatures to the implant material that closely mimic the intrinsic composition and topography of the natural bone. A majority of emerging material fabrication methods are therefore focused on achieving high precision and control at the micro- and nanoscale. This chapter provides an overview of the bone tissue regeneration process and discusses a broad classification of biomaterials along with their unique properties towards BTE applications. We have also discussed the role of material properties in regulating biological responses in terms of cellular attachment, proliferation, and differentiation. Furthermore, the chapter also briefly sheds light on the three major fabrication techniques (i.e., lithography, electrospinning, and 3D bioprinting) for the development of micro and nanostructured constructs.

13.1 TISSUE ENGINEERING: CURRENT STATE OF ART

Tissue engineering unifies cells, scaffolds, signaling molecules, and other biochemical factors to replace, regenerate, and restore a damaged tissue. Engineered biomaterials for scaffold preparation have various chemical and topographical surfaces which influence the cellular activities such as determination of cell shape, adhesion, proliferation, fate, and so on. Therefore, the potential to promote such cellular functions can be determined by the quality of biomaterials. Till date, various techniques, biomaterials, and a wide range of cell types have been explored to fabricate, monitor, and preserve the artificial tissues and organs. Advanced biomaterials having desired mechanical, chemical, and biological properties are currently under investigation for easy adaptation of complex tissue environment. These biomaterials need to be biocompatible and their fabrication should be simple and cost effective. A drastic progress in the understanding of cellular cues, tissue architectural, and gene editing technologies together with improved immunomodulation have strongly affected the current innovations [1].

13.2 BONE TISSUE ENGINEERING: AN OVERVIEW

13.2.1 STRUCTURE OF BONE TISSUE

Bone is a dynamic and vascularized tissue which possesses the intrinsic capacity to repair, regenerate, and remodel itself. It plays a vital role by facilitating locomotion, serving as a reservoir for minerals, and providing a skeletal framework for support and protection to the internal organs [2]. At the time of birth, a human body has around 270 bones. During the development period, these bones undergo remodeling and some of them fuse together resulting in a total of 206 bones in an adult human. Femur (thigh-bone) and stapes (middle ear) are the largest and the smallest bones in humans, respectively.

Architecturally, bone can be widely divided into two types: (1) compact bone and (2) cancellous bone.

Compact or cortical bone constitutes around 80% of the bone mass. It is present at the outer region, is dense and hard, and provides mechanical strength. The outermost layer of bone covering the compact bone is called periosteum. Cancellous, trabecular, or spongy bone, on the other hand, is porous and comparatively less dense, lighter, and more flexible [3]. The newly formed bone is organic, unmineralized, and is known as osteoid. It is majorly composed of type-I collagen and other proteins. Osteoid undergoes mineralization process involving deposition of minerals such as hydroxyapatite (HA) ($Ca_{10}(PO_4)_6(OH)_2$) leading to maturation and hardening of the bone. Based on composition, bone is a composite with inorganic and organic components. About 70% of the bone weight is due to inorganic components which are mainly calcium (Ca) and phosphorus (P) present as hydroxyapatite (HA) crystals. Other minerals present are sodium (Na), potassium (K), magnesium (Mg), fluoride (F), chlorine (Cl), and carbonate (CO_3^{2-}). Trace amounts of silicon (Si), strontium (Sr), iron (Fe), zinc (Zn), and copper (Cu) are also reported [4]. The organic component constitutes about 35% of the bone weight and is majorly composed of type-I collagen, other proteins and water (Figure 13.1). The inorganic components of the bone act as a reservoir for mineral ions and are also responsible for mechanical strength and stiffness of the bone due to mineralization. Whereas, the organic part of the bone is made up of proteins and imparts flexibility to the bone [5, 6].

Bone consists of various structural components ranging from macroscale to subnanoscale. As mentioned earlier, bone is a composite macrostructure which comprises of cancellous and cortical regions. Each of these regions has microstructures, namely trabeculae (cancellous) and osteons (cortical). These structures are made up of self-assembled collagen fibers. Individual

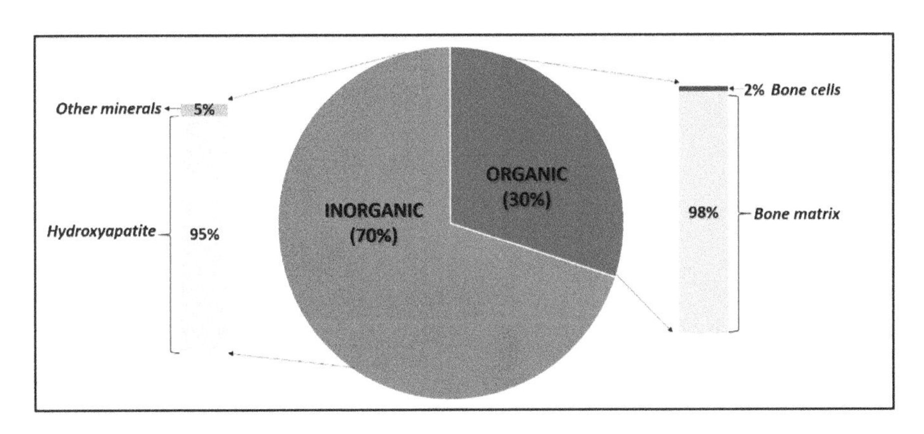

FIGURE 13.1 Chemical composition of the bone.

collagen proteins are around 200 nm in length and 2–3 nm in width and they self-assemble to form collagen fibrils (~500 nm in diameter). The gaps between individual collagen molecules are occupied by calcium phosphate (hydroxyapatite like) nanocrystals which are 2–3 nm in diameter (Figure 13.2) [7, 8].

13.2.2 TYPES OF CELLS INVOLVED

Osteoprogenitor cells are bone stem cells that are derived from mesenchymal stem cells. These cells possess the ability to differentiate and give rise to osteoblast and osteocyte cells. Osteoblasts are bone-forming cells that arise from mesenchymal stem cell lineage and play a pivotal role in osteogenesis. They also secrete bone matrix, collagen, and other proteins such as osteocalcin and osteopontin. They also produce hydroxyapatite which is responsible for mineralization of the bone matrix. Osteoclasts are multinucleated and large cells which resorb the mineralized bone by secreting enzymes that digest the bone matrix. These cells originate from hematopoietic lineage. By facilitating the bone resorption, osteoclasts help in maintaining the levels of calcium and other minerals. Osteocyte cells are mature osteoblast cells that lose the ability to secrete matrix and are integrated into the newly formed bone (osteoid) which is calcified over the time. They are responsible for cell-to-cell communications and other homeostasis functions. The inactive bone surface which is not undergoing the remodeling process during the resting state is covered by bone-lining cells. The various bone cells orchestrate the remodeling process by regulating

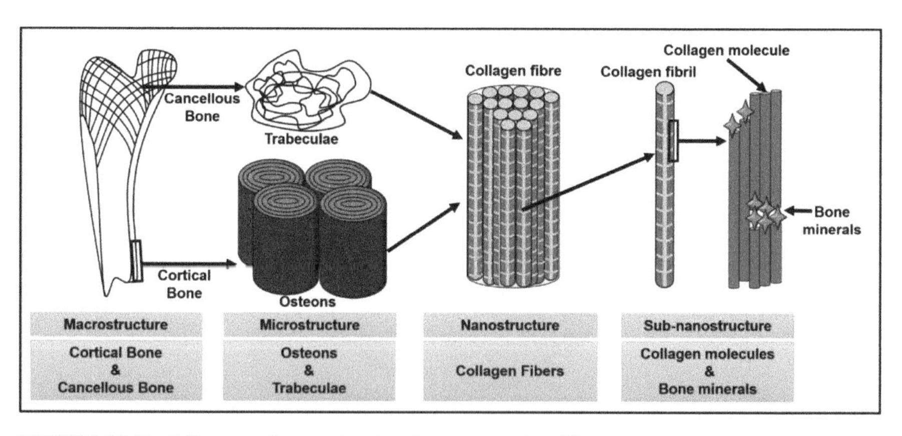

FIGURE 13.2 Micro- and nanostructured components of bone.

the balance between bone formation and resorption (Figure 13.3). These processes are crucial for formation of new bone, and maintaining the strength and mineral homeostasis [9–12].

13.2.3 BONE REGENERATION PROCESS

The process of bone formation by cells is termed as osteogenesis or ossification. There are two types of processes that can lead to formation of new bone—intramembranous and endochondral ossification. Intramembranous ossification involves formation of new bone within the existing mesenchymal connective tissue. Endochondral ossification is a comparatively much complex process which begins with initial formation of cartilage by chondrocytes and subsequent removal and replacement by new bone. Natural healing and repair of bone is majorly endochondral type of ossification [13, 14]. The repair process is gradual and sequential involving induction, hematoma, formation of soft callus, ossification, and remodelling [15, 16].

During an impact leading to a bone fracture, the surrounding blood vessels are damaged and ruptured. Within hours, clot (hematoma) formation takes place followed by infiltration of the immune cells and resultant inflammation. Over the next few days, the inflammation subsides and simultaneously mesenchymal progenitor cells are recruited. These progenitor cells differentiate into chondrocytes and facilitate formation of a soft (cartilaginous) callus. Soft callus consists of new blood vessels, inner callus (spongy bone), and external callus (cartilage). Gradually, the soft callus is replaced by endochondral bone which matures into woven bone and further into hard callus owing to osteoblast activity. Remodeling of the healed bone is governed by osteoclast and osteoblast activity. At the end of the process, the fractured

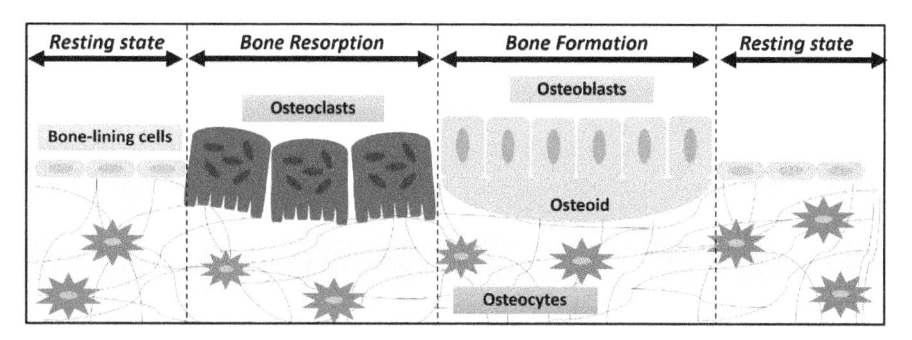

FIGURE 13.3 Types of cells involved in bone remodeling process.

bone is entirely repaired and its original shape is restored [16].

Osteoconduction refers to the process by which an osteoconductive material serves as a structural scaffolding or a matrix upon which the new bone can be laid. Osteoconductive materials are porous in nature to allow formation of new blood vessels (angiogenesis) and bone formation (osteogenesis). Osteoinduction is the process of recruitment of osteoprogenitor cells and the ability to stimulate their differentiation toward osteogenic lineages in order to induce bone formation. It is a phenomenon which is involved in natural bone repair process. The formation of a chemical bond between the implant material and the living bone is termed as osseointegration [17].

13.3 BIOMATERIALS FOR BONE TISSUE REGENERATION

Even though bone possess the ability to regenerate and repair, defects larger than a critical size require therapeutic intervention to facilitate and accelerate the natural healing process [18]. Bone grafting is a process which involves replacement of the damaged or missing bone tissue to repair fractures that are large, complex and unable to heal naturally. As the healing process advances, the graft material is gradually replaced by new bone growth [19–21]. An ideal grafting material should be biocompatible and thus should not elicit any adverse immunological reactions when implanted within the patient's body. Higher the biocompatibility of the graft, lower are the chances of rejection due to adverse responses of the patient's immune system. During the course of repair process, the graft should also be able to replace the fundamental functions of damaged tissue till successful integration of the implant within the newly formed native bone takes place [22–24]. Table 13.1 depicts various types of bone grafts along with their sources and properties.

TABLE 13.1 Types of Bone Grafts and Their Properties

Type of Bone Graft	Source		Biocompatibility	Chances of Rejection
Autograft	Patient		High	Low
Allograft	Other donor (same species)		Moderate	Moderate
Xenograft	Other donor (other species)		Low	High
Alloplast	Synthetic	Metallic	Low	High
		Inorganic	Moderate	Moderate
		Organic	High	Low

Autologous bone grafts (autografts) are harvested from a healthy site and implanted at the damage/defect site of the same patient. Autografts display osteoinduction, osteoconduction as well as osteogenesis. As these grafts are extracted from and inserted within the same individual, they have extremely low risk of rejection. Major drawback of this technique is that it results in multiple surgical sites and in case of large defect sites, the donor tissue is insufficient. Allografts or homografts are obtained from different individuals of the same species. These grafts are most commonly retrieved from cadavers and implanted after processing. However, due to lack of viable cells, allografts display only osteoconductive nature and also pose the risk of pathogen transmission. Xenografts are implanted from one species into the other. Bovine implants are most commonly used for this purpose. Due to differences in the genetic makeup of the donor and the patient, xenografts lack biocompatibility, and are prone to rejection by the patient's immune system. Alloplasts are synthetic implants with mechanical strength and properties similar to the natural bone. These include metallic, inorganic, and organic constructs. Among the alloplasts, metallic implants have the highest chances of rejection due to their low biocompatibility, nonresorbable nature and toxicity [22, 25].

A biomaterial can be defined as any natural or synthetic construct which can be implanted within a biological system to augment, repair, or replace a living tissue. In medical applications, biomaterials are commonly used for either a therapeutic or a diagnostic purpose [26–28]. When implanted within a living system, the ability of a material to react directly with the surrounding tissues and form a bond (Bioactive fixation) is termed as bioactivity. Tissue engineering exploits the use of biomaterials to repair a damaged tissue while replacing its function within the body. Such biomaterials when used as implants have various promising properties such a porous nature, excellent biocompatibility, and resorbable nature. Bioactivity of a material can be defined as the ability to elicit a biological response from its surroundings when implanted into a biological system [29, 30].

13.3.1 METALS

Metallic biomaterials are mostly exploited because of their high mechanical strength, stiffness, wear resistance, ductility, and inertness. However, they mostly lack biological properties such as biocompatibility, bioactivity, and osteoconduction [31, 32]. Several studies have been carried out over the years to improve the biofunctionalities of such implants via various surface modifications. Enhancement in the surface roughness and resistance to corrosion have been major areas of improvement in metallic implants to improve attachment on cells and prevent implant-associated toxicity, respectively [33–35]. Coating of the implants with ceramics and polymers have also shown to improve the bioactivity and biocompatibility of various implants [36–38]. Some of the problems associated with the metallic implants are in vivo corrosion, debris generation, lacking bioactivity, localized inflammation or necrosis, adverse allergic and/or immunological reactions, nonbiodegradable nature, toxicity, and rejection [39, 40]. Due to these significant limitations, metallic implants have mostly been replaced by more compatible biomaterials such as ceramics and polymers. However, for fractures occurring at the load-bearing sites it is essential for the implant to have good mechanical strength and durability; thus, for such cases, metal implants still remain the most suitable candidate.

13.3.2 CERAMICS

Ceramics are inorganic compounds which exhibit numerous promising properties toward their use in vivo. Hardness, bioactivity, biocompatibility, biodegradability, resistance to abrasion, resistance to fatigue, and in some cases osteoinductive properties have led to a wide-scale use of these materials for various in vivo applications [41]. Biocompatible class of ceramics are termed as bioceramics and are widely used as implant coatings, bone fillers, surgical cements, scaffolds, composite materials, and so on [42]. These materials when used for biological applications can be further classified into inert bioceramics like zirconia which are nonresorbable in nature and noninert bioceramics, which are biodegradable and resorbable in nature like calcium phosphates. Due to their biodegradability, the noninert ceramics are gradually replaced within the body as the new bone is formed [42]. Coating of metal implants with certain ceramics has proved beneficial in reducing the adverse inflammatory responses and reducing chances of implant rejection [43–45]. Porous implants composed of bioceramics have the ability to serve as an osteoconductive scaffold material by allowing infiltration of osteogenic cells and proteins, attachment of cells within the material, and supporting formation of new blood vessels and bone ingrowth (Biological fixation) [46]. Some of the bioceramics such as hydroxyapatite-based composites and bioglasses also exhibit osteoinductive properties by releasing ionic products which elicit biological responses stimulating bone formation [47–49]. However, brittleness of the bioceramics still remains a major limitation for their applications [50]. To address this issue, over the years, bioceramics have been incorporated with various reinforcement materials (metals and/or polymers) resulting in hybrid composites with improved tensile strength and durability [51–54].

13.3.3 POLYMERS

Natural polymers such as collagen and cellulose are derived from animal or plant tissues. They occur naturally and are similar to the natural components of the native bone and extracellular matrix. Therefore, they serve as a promising biomaterial which have lower immunogenic properties, fewer side-effects, and possess excellent bioactivity with inherent biological cues that promote cellular attachment and facilitate the healing process even in the form of an implant [55, 56]. Polymers have also shown biodegradability resulting into degradation products which are resorbable in vivo. Additionally, due to their fibrous nature, polymers can easily be used to create porous scaffolds which provide an interconnected mesh-like structure where the cells can attach, proliferate, and form new bone [58]. These scaffolds support bone formation not only at the surface but also within the implant. As the new bone forms (Figure 13.4), the scaffolds are degraded and ultimately completely replaced by the native bone [59]. However, natural polymers when used as a biomaterial for implants have some serious limitations such as rigidity, poor stability, and high degradation rates which result in failure of grafts [60].

Synthetic polymers provide a range of biomaterials with tailored properties. The physical and chemical properties of synthetic polymers can be tuned by various methods ranging from modification during

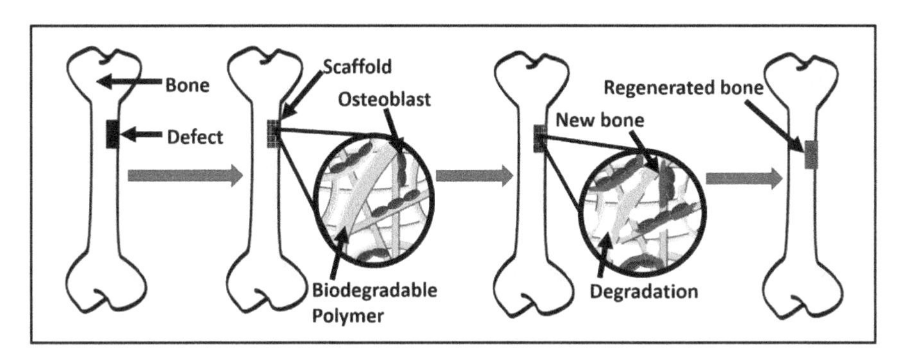

FIGURE 13.4 Schematic representation of bone repair process using polymeric scaffolds.

synthesis to surface modification, resulting in more advanced biomaterials with controllable degradation rates, higher in vivo stability, flexibility, and improved mechanical strength. However, due to synthetic nature of these materials, they are immunogenic in nature and lack desirable biological properties such as bioactivity, biocompatibility, osteoinductivity, and osteoconductivity [61–63]. Therefore, combination of natural and synthetic components can help to create a hybrid biomaterial which can be used to overcome limitations. This has resulted in an extensive research and interest in the field of composite materials and polymer blends which merge together the unique properties of individual components to create a material with desirable and superior properties. Various polymeric blends of both natural and synthetic polymers have shown excellent features as a biomaterial when used as coating, fillers, scaffolds, implants, and so on [64–66]. Table 13.2 compares the biological properties of various biomaterials for bone tissue engineering.

13.4 FABRICATION TECHNIQUES

Cell–biomaterials interaction within a scaffold is important for various cellular functions. An ideal scaffold should not only provide mechanical support but also exhibit proper topography, surface chemistry, interconnections, or pores to promote cellular activities [67]. Studies also suggested that cell cytoskeletal organization is an important event during attachment; it reflects the mechanosensitivity of the cells, in turn the force balance. A microenvironment resembling the elasticity of tissue enhances the interaction between the biomaterials and the cells. In addition, engineering at the nano level sharpens the surface features, aspect ratios, geometries and also helps in focal adhesion. As bone contains arrangement of components in the dimension from macroscale (cortical bone with porous cancellous bone) to nanoscale (collagen fiber); therefore, a variety of processing technologies to obtain the best topographical dimensions with exact mechanical strength, stiffness, and toughness are being tried. Nanostructures in

TABLE 13.2 Comparison of Biological Properties of Biomaterials

	Metals	Ceramics	Polymers
Bioactive	–	√	√
Osteoconductive	–	–	√
Osteoinductive	–	√	√
Biodegradable	–	√	√
Mechanical strength	√	–	–

the form of nanopatterns, nanocomposites, and nanofibers are being investigated in order to match such architectural design of bone [68]. The techniques through which these nano-engineered substrates are fabricated include electrospinning of fibers, three dimensional (3D) bioprinting, different lithography-based methods, and so on.

13.4.1 LITHOGRAPHY-BASED METHODS

Lithography-based fabrication methods are of prime importance to create artificially designed micro and nano environment that support cell and tissue growth. Lithography technique uses a light (optical lithography), electron (electron-beam lithography [EBL]), or ion (ion-beam lithography) source to write on a sample with computer-aided design (CAD). A variety of surface landscapes, for example, grooves, ridges, and wells, could be patterned well with help of these techniques [68]. UV lithography, the optical one, is the standard technique that uses a photo resist and a mask to draw patterns. But EBL is the most powerful technique when structures of few nanometers are required [69]. Using EBL, a large array of nanopits of 120 nm diameter and 100 nm deep were patterned. It was observed that nanopits fabricated in a random manner promoted cell adhesion than an ordered nanopits [70]. Direct laser writing lithography uses ultrafast pulsed laser to create 3D nanostructures and it was demonstrated that by varying the condition of laser processing technique, cell attractive and cell repellent areas could be generated [71]. Two photon lithography has also been helpful in development of nanostructures from nanocomposites mimicking portion of spongy bone [72]. Another very simple and low-cost lithographic technique that has been tested enormously is soft lithography. In this technique, designs or patterns from an elastic stamp is transferred on a metal or solid substrate. By utilizing this procedure, a variety of patterns has been generated [73, 74]. Stereo lithography is also a popular 3D printing technology commonly used for plastic models using layer-by-layer deposition. This technology is also being used for designing scaffolds resembling bone structure [75, 76]. A report showed that glass-ceramic scaffolds with suitable compressive strength could be fabricated by taking help from this technique [77]. Nowadays, lithography-based additive manufacturing technology is becoming favorite to fabricate structures made from bioglass [78]

13.4.2 ELECTROSPINNING

Electrospinning is a simple and a favorite technique to produce ultra-thin fibers from a wide variety of materials such as polymers, composites, ceramics, and so on, for scaffold fabrication. It uses an electric field to overcome the surface tension of a polymer solution or a molten substance to shoot it out of a fine needle. When the solvent of the solution evaporates, the fibers remains. Fibers of extreme thin diameters (micro to nano), with additives and tunable properties could be electrospun through this technique [79]. Since nanofiber materials became best choices for bone tissue engineering due to their properties similar to the natural fibrillar extra cellular matrices, so a broad range of natural like silk, collagen, chitosan, gelatin, and so on, and synthetic polymers (polycaprolactone [PCL], polylactic acid, polyglycolic acid, etc.) have been spun using this technique [80]. In recent achievements, electrospun fibers of composites have shown improved mechanical strength and supported osteoblast cell growth [81]. On the other hand, when a PCL nanofiber was spun using innovative approach of freeze drying, a high number of pores could be seen in the scaffolds ideal for interconnections [82]. Therefore, implementation of new strategies and other methods like crosslinking [83] and surface chemistries [84] have also enhanced the mechanical strength of these fibers. In addition, a variety of recent advancements reflected that electrospinning will remain a favorite technique for development of new strategies in bone tissue engineering as they are easy to apply here in this technique [85].

13.4.3 3D BIOPRINTING

3D printing is one of the major innovations in the current times. The area of regenerative medicines is progressing rapidly as a result of this innovation. This technology involves bio-ink materials to be printed in a layer-by-layer fashion to build a 3D structure governed by a CAD. It has shown potential to develop a complex tissue construct from a cell-laden biomaterial [86]. Among variety of other bio-ink materials, polymeric hydrogels have drawn immense attention for their properties. Chemically crosslinked (photo, enzyme, di-sulphide) hydrogels or using peptide crosslinker, the mechanical property of bio-ink hydrogel could be improved to match the extracellular matrix [87]. One of the current investigations showed formation of stable human scale tissue structure of any shape using mechanically improved hydrogel [88]. Scope of using a variety of material to be used as bio-ink is under investigation [89].

13.5 CONTROLLING CELLULAR BEHAVIOR VIA BIOMATERIALS

During the fracture repair, when an implant is placed in vivo, the implant surface interacts with the cells from the surrounding tissues. Studies have indicated that the structural, chemical, and physical features of biomaterial are capable of altering the behavior of surrounding cells and tissue through various biological processes [90, 91]. An understanding of these interactions provides important information related to the influence of the biomaterial on altering the cellular behavior. This information allows strategic modifications of the biomaterial to facilitate cellular functions which promote the healing process while minimizing the undesirable biological effects such as adverse immunological responses. Besides the chemical signaling, the morphological changes experienced by the cells like cell shape and cytoskeletal tension initiate multiple cascades leading to altered cellular expressions [91].

13.5.1 MORPHOLOGY, ATTACHMENT, AND ADHESION

As soon as the implant surface comes in contact with the surrounding microenvironment, various proteins present in the biological systems interact with the surface leading to protein adsorption on the implant [92]. This process depends highly on the physical attributes of implants such as roughness, topography, surface charge, wettability, and so on [93–95]. Micro- and nanoscale surface modifications enhance the roughness as well as increase the available surface area for adsorption of proteins present in the serum. Protein adsorption is followed by cellular attachment on the implant [96]. Amount as well as type of protein adsorption on the surface regulates the cellular attachment process. In a study, albumin adsorption did not show any significant difference in osteoblast attachment, however, adsorption of fibronectin resulted in better cellular attachment. It was also demonstrated that the cellular morphology was independent of the type of protein adsorbed [92]. The extracellular matrix proteins interact with the cell-surface receptors (integrins) to stimulate and regulate the process of cellular adhesion [97]. Culturing of osteoblast cells on nanofiber surfaces has also shown reduced spreading behavior, enhanced expression of sialoproteins when compared to flat surfaces [98]. Enhanced osteoblast attachment has been reported in polymeric implants coated with metals such as titanium when compared to the uncoated polymers [99]. Additionally, features like nanopits and nanogrooves have also been reported to exhibit decreased adhesion and spreading of cells resulting in higher motility of cells while simultaneously promoting differentiation toward osteoblastic lineage [100, 101]. Interestingly, several studies have demonstrated that cells grown on ordered nanotopographies show poor or no cellular adhesion while cells on random disordered nanotopographies show better adhesion and more osteoblast-like morphology [102, 103].

Some of the common techniques of creating micro- and nano features on metal implants are acid etching, plasma spraying, electrochemical anodization, lithography, and coating with biomaterials. Surface functionalization of the implants and subsequent attachment of various bioactive substances is also another popular approach to modulate interactions between the implant and its surroundings [104, 105].

13.5.2 PROLIFERATION

Smooth surfaces support much more rapid spreading, attachment, and proliferation of osteoblast cells [106]. Studies have reported that nanotopologies facilitate selective adhesion of cells around the implant favoring the cellular proliferation and accelerating the healing process [107]. Metallic implants when coated with bioactive materials such as ceramics, polymers, and/or proteins enable increased osteoblast proliferation [108–111]. Ceramics and their dissolution products facilitate the proliferation and differentiation of osteogenic cells by enhancing the expression of osteoblast-related genes [112–115]. Recently, it was reported that electrospun chitosan nanofibers with flat surfaces and nano-diameters enhance the surface area and porosity of the scaffolds. These nanofibrous scaffolds support cellular attachment while promoting osteoblast proliferation and maturation more effectively than the chitosan films [116]. Additionally, polymeric biomaterials with crystalline and rigid surfaces promote fibroblastic growth whereas the amorphous and comparably flexible surfaces show much better osteoblastic growth [117].

13.5.3 DIFFERENTIATION

Changes in the cellular morphology, attachment, and adhesion behavior have shown to closely affect the commitment of precursor cells toward osteogenic differentiation process [118]. Much enhanced differentiation and matrix mineralization properties have been observed for cells grown on rough metal surfaces [119, 120]. When cultured on surfaces with microstructures, osteoblast cells stimulate an osteogenic microenvironment which facilitates the bone formation [121, 122]. Mesenchymal cells when grown on microstructured titanium implant surfaces display osteoblast-like characteristics which were sufficient to promote differentiation of mesenchymal stem cells toward osteoblastic cell lineage [122]. Ceramics which are bioactive in nature have the capacity to form apatite layer on the surface when in contact with the biological surroundings. Due to this bioactivity, ceramic biomaterials promote differentiation toward osteoblastic lineage [123–125]. Also, culturing of precursor cells on nanofibrous surfaces exhibited enhanced osteogenic differentiation when compared to smoother counterpart of the same material [126]. This behavior has been observed in various types of bioactive and bioinert materials, indicating that the fibrous structures of dimension less than 1 μm are responsible for stimulating the osteoinductive behavior independent of the material composition. Therefore, surface micro- and nanotopography can stimulate the differentiation of progenitor cells toward osteoblastic lineage.

13.6 CONCLUSIONS

In case of a damage, bone has the ability to repair itself. However, when the damage or defect is beyond the natural regenerative capacity of the bone, a surgical intervention is necessary to introduce biomaterial at the site of damage to aid the natural healing process. A biomaterial can be defined as any natural or synthetic construct which can be implanted within a biological system to augment, repair, or replace a living tissue. An ideal biomaterial to be used as an implant for bone tissue repair should possess biocompatibility, bioactivity, biodegradability, osteoconductivity, osteoinductivity, and mechanical properties. Common biomaterials applied for bone tissue engineering are metals, bioceramics, and polymers. Each class of these biomaterial has unique properties along with some limitations restricting their wide-scale application as implant material. Therefore, composite materials consisting of a blend of different biomaterials provide an effective solution to fabricate customized biomaterials with all desirable properties. Several studies have reported control over the cellular properties such as spreading, attachment, proliferation, and differentiation via modifying physical and chemical attributes of the implant material. Incorporation of micro- and nano features on within the implant imparts biomimetic properties due to presence of similar micro- and nanoscale components in the natural bone. Mimicking the native composition and topography of the bone serves as a useful tool in studying the implant interactions and manipulating the cellular responses toward it. Henceforth, recent emerging technologies provide precise control over fabrication of versatile biomaterials with state-of-the-art features to aid the process of tissue repair and regeneration.

KEYWORDS

- **nanomaterials**
- **bone tissue engineering**
- **biomaterials**
- **material fabrication**

REFERENCES

1. Chen, F.-M., and X. Liu. "Advancing biomaterials of human origin for tissue engineering." Progress in Polymer Science 53 (2016): 86–168.
2. Marks Jr, S. C., and S. N. Popoff. "Bone cell biology: the regulation of development, structure, and function in the skeleton." American Journal of Anatomy 183(1), (1988): 1–44.
3. Bilezikian, J. P., L. G. Raisz, and T. J. Martin, eds. Principles of bone biology. Academic Press, 2008.
4. Goto, T., and K. Sasaki. "Effects of trace elements in fish bones on crystal characteristics of hydroxyapatite obtained by calcination." Ceramics International 40(7), (2014): 10777–10785.
5. Quelch, K. J., et al. "Chemical composition of human bone." Archives of Oral Biology 28(8), (1983): 665–674.
6. Gong, J. K., J. S. Arnold, and S. H. Cohn. "Composition of trabecular and cortical bone." The Anatomical Record 149(3), (1964): 325–331.
7. Kane, R., and P. X. Ma. "Mimicking the nanostructure of bone matrix to regenerate bone." Materials Today 16(11), (2013): 418–423.
8. Olszta, M. J., et al. "Bone structure and formation: a new perspective." Materials Science and Engineering: R: Reports 58(3–5), (2007): 77–116.
9. Sommerfeldt, D., and C. Rubin. "Biology of bone and how it orchestrates the form and function of the skeleton." European Spine Journal 10(2), (2001): S86-S95.
10. Bonnet, D. "Biology of human bone marrow stem cells." Clinical and Experimental Medicine 3(3), (2003): 140–149.
11. Reddi, A. H. "Cell biology and biochemistry of endochondral bone development." Collagen and Related Research 1(2), (1981): 209–226.
12. Aarden, E. M., P. J. Nijweide, and E. H. Burger. "Function of osteocytes in bone." Journal of Cellular Biochemistry 55(3), (1994): 287–299.
13. Ortega, N., D. J. Behonick, and Z. Werb. "Matrix remodeling during endochondral ossification." Trends in Cell Biology 14(2), (2004): 86–93.
14. Percival, C. J., and J. T. Richtsmeier. "Angiogenesis and intramembranous osteogenesis." Developmental Dynamics 242(8), (2013): 909–922.
15. Taylor, D., J. G. Hazenberg, and T. C. Lee. "Living with cracks: damage and repair in human bone." Nature Materials 6(4), (2007): 263.
16. Einhorn, T. A., and L. C. Gerstenfeld. "Fracture healing: mechanisms and interventions." Nature Reviews Rheumatology 11(1), (2015): 45.
17. Albrektsson, T., and C. Johansson. "Osteoinduction, osteoconduction and osseointegration." European Spine Journal 10(2), (2001): S96-S101.
18. Kalfas, I. H. "Principles of bone healing." Neurosurgical Focus 10(4), (2001): 1–4.
19. Einhorn, T. A. "The cell and molecular biology of fracture healing." Clinical Orthopaedics and Related Research 355 (1998): S7-S21.
20. Marsell, R., and T. A. Einhorn. "The biology of fracture healing." Injury 42(6), (2011): 551–555.
21. Phillips, A. M. "Overview of the fracture healing cascade." Injury 36(3), (2005): S5-S7.
22. Bauer, T. W., and G. F. Muschler. "Bone graft materials: an overview of the basic science." Clinical Orthopaedics and Related Research 371 (2000): 10–27.
23. De Boer, H. H. "The history of bone grafts." Clinical Orthopaedics and Related Research 226 (1988): 292–298.

24. Chase, S. W., and C. H. Herndon. "The fate of autogenous and homogenous bone grafts: a historical review." JBJS 37(4), (1955): 809–841.
25. Machibya, F. M., Y. Zhuang, and J. Chen. "Comparing properties of bone regeneration materials (BRMs) for optimal clinical outcome in dentistry: a review of the current literature." Research & Reviews: A Journal of Dentistry 4 (4) (2016): 10–16.
26. Hubbell, J. A. "Biomaterials in tissue engineering." Bio/technology 13(6), (1995): 565.
27. Kim, B.-S., C. E. Baez, and A. Atala. "Biomaterials for tissue engineering." World Journal of Urology 18(1), (2000): 2–9.
28. O'brien, F. J. "Biomaterials & scaffolds for tissue engineering." Materials Today 14(3), (2011): 88–95.
29. Hench, L. L., and H. A. Paschall. "Direct chemical bond of bioactive glass-ceramic materials to bone and muscle." Journal of Biomedical Materials Research 7(3), (1973): 25–42.
30. Shirtliff, V. J., and L. L. Hench. "Bioactive materials for tissue engineering, regeneration and repair." Journal of Materials Science 38(23), (2003): 4697–4707.
31. Hermawan, H., D. Ramdan, and J. R. P. Djuansjah. "Metals for biomedical applications." Biomedical Engineering-From Theory to Applications. InTech, 2011.
32. Srivastav, A. "An overview of metallic biomaterials for bone support and replacement." Biomedical Engineering, Trends in Materials Science. IntechOpen, 2011.
33. Hazan, R., R. Brener, and U. Oron. "Bone growth to metal implants is regulated by their surface chemical properties." Biomaterials 14(8), (1993): 570–574.
34. Asri, R. I. M., et al. "Corrosion and surface modification on biocompatible metals: a review." Materials Science and Engineering C 77 (2017): 1261–1274.
35. Puleo, D. A., and A. Nanci. "Understanding and controlling the bone–implant interface." Biomaterials 20(23–24), (1999): 2311–2321.
36. McEntire, B. J., et al. "Ceramics and ceramic coatings in orthopaedics." Journal of the European Ceramic Society 35(16), (2015): 4327–4369.
37. Wijesinghe, W. P. S. L., et al. "Preparation of bone-implants by coating hydroxyapatite nanoparticles on self-formed titanium dioxide thin-layers on titanium metal surfaces." Materials Science and Engineering C 63 (2016): 172–184.
38. Wong, H. M., et al. "A biodegradable polymer-based coating to control the performance of magnesium alloy orthopaedic implants." Biomaterials 31(8), (2010): 2084–2096.
39. Kamachimudali, U., T. M. Sridhar, and B. Raj. "Corrosion of bio implants." Sadhana 28(3–4), (2003): 601–637.
40. Sansone, V., D. Pagani, and M. Melato. "The effects on bone cells of metal ions released from orthopaedic implants. A review." Clinical Cases in Mineral and Bone Metabolism 10(1), (2013): 34.
41. Raucci, M. G., D. Giugliano, and L. Ambrosio. "Fundamental properties of bioceramics and biocomposites." Handbook of Bioceramics and Biocomposites, 2016: 35–58.
42. Ravaglioli, A., and A. Krajewski. Bioceramics: Materials·properties·app lications. Springer Science & Business Media, 1991.
43. Hayashi, K., et al. "Evaluation of metal implants coated with several types of ceramics as biomaterials." Journal of Biomedical Materials Research 23(11), (1989): 1247–1259.

44. Sollazzo, V., et al. "Zirconium oxide coating improves implant osseointegration *in vivo*." Dental Materials 24(3), (2008): 357–361.
45. Sola, A., et al. "Bioactive glass coatings: a review." Surface Engineering 27(8), (2011): 560–572.
46. Hench, L. L. "Bioceramics and the future." Ceramics and Society, 1995: 101–120.
47. Cao, W., and L. L. Hench. "Bioactive materials." Ceramics International 22(6), (1996): 493–507.
48. Zhu, H., et al. "Nanostructural insights into the dissolution behavior of Sr-doped hydroxyapatite." Journal of the European Ceramic Society 38(16), (2018): 5554–5562.
49. Hoppe, A., N. S. Güldal, and A. R. Boccaccini. "A review of the biological response to ionic dissolution products from bioactive glasses and glass-ceramics." Biomaterials 32(11), (2011): 2757–2774.
50. Doremus, R. H. "Bioceramics." Journal of Materials Science 27(2), (1992): 285–297.
51. Tavassoli, H., et al. "Incorporation of nanoalumina improves mechanical properties and osteogenesis of hydroxyapatite bioceramics." ACS Biomaterials Science & Engineering 4(4), (2018): 1324–1336.
52. Dorozhkin, S. "Calcium orthophosphate-containing biocomposites and hybrid biomaterials for biomedical applications." Journal of Functional Biomaterials 6(3), (2015): 708–832.
53. Dorozhkin, S., and T. Ajaal. "Toughening of porous bioceramic scaffolds by bioresorbable polymeric coatings." Proceedings of the Institution of Mechanical Engineers, Part H: Journal of Engineering in Medicine 223(4), (2009): 459–470.
54. Martínez-Vázquez, F. J., et al. "Improving the compressive strength of bioceramic robocast scaffolds by polymer infiltration." Acta Biomaterialia 6(11), (2010): 4361–4368.
55. Sonia, T. A., and C. P. Sharma. "An overview of natural polymers for oral insulin delivery." Drug Discovery Today 17(13–14) (2012): 784–792.
56. Dang, J. M., and K. W. Leong. "Natural polymers for gene delivery and tissue engineering." Advanced Drug Delivery Reviews 58(4), (2006): 487–499.
57. Agrawal, C. M., and R. B. Ray. "Biodegradable polymeric scaffolds for musculoskeletal tissue engineering." Journal of Biomedical Materials Research: An Official Journal of The Society for Biomaterials, The Japanese Society for Biomaterials, and The Australian Society for Biomaterials and the Korean Society for Biomaterials 55(2), (2001): 141–150.
58. Liu, X., and P. X. Ma. "Polymeric scaffolds for bone tissue engineering." Annals of Biomedical Engineering 32(3), (2004): 477–486.
59. Holland, T. A., and A. G. Mikos. "Biodegradable polymeric scaffolds. Improvements in bone tissue engineering through controlled drug delivery." Tissue Engineering I. Springer, 2005: 161–185.
60. Jayakumar, R., et al. "Sulfated chitin and chitosan as novel biomaterials." International Journal of Biological Macromolecules 40(3), (2007): 175–181.
61. Middleton, J. C., and A. J. Tipton. "Synthetic biodegradable polymers as orthopedic devices." Biomaterials 21(23), (2000): 2335–2346.
62. Gunatillake, P. A., and R. Adhikari. "Biodegradable synthetic polymers for tissue engineering." European Cells and Materials 5(1), (2003): 1–16.

63. Lutolf, M. P., and J. A. Hubbell. "Synthetic biomaterials as instructive extracellular microenvironments for morphogenesis in tissue engineering." Nature Biotechnology 23(1), (2005): 47.
64. Coenen, A. M. J., et al. "Elastic materials for tissue engineering applications: Natural, synthetic, and hybrid polymers." Acta Biomaterialia 79 (2018): 60–82.
65. Jindal, A., et al. "Mesoporous zinc silicate bio-composite: Preparation, characterization and in vitro evaluation." Microporous and Mesoporous Materials 277 (2019): 124–131.
66. Lei, Bo, et al. "Hybrid polymer biomaterials for bone tissue regeneration." Frontiers of Medicine (2018): 1–13.
67. Tuzlakoglu, K., et al. "Nano-and micro-fiber combined scaffolds: a new architecture for bone tissue engineering." Journal of Materials Science: Materials in Medicine 16(12), (2005): 1099–1104.
68. Gong, Tao, et al. "Nanomaterials and bone regeneration." Bone Research 3 (2015): 15029.
69. Martinez, E., et al. "Effects of artificial micro-and nano-structured surfaces on cell behaviour." Annals of Anatomy-Anatomischer Anzeiger 191(1), (2009): 126–135.
70. Dalby, M. J., et al. "The control of human mesenchymal cell differentiation using nanoscale symmetry and disorder." Nature Materials 6(12), (2007): 997.
71. Yiannakou, C., et al. "Cell patterning via laser micro/nano structured silicon surfaces." Biofabrication 9(2), (2017): 025024.
72. Marino, A., et al. "Two-photon lithography of 3D nanocomposite piezoelectric scaffolds for cell stimulation." ACS Applied Materials & Interfaces 7(46), (2015): 25574–25579.
73. Kiyama, R., et al. "Micro patterning of hydroxyapatite by soft lithography on hydrogels for selective osteoconduction." Acta Biomaterialia 81 (2018): 60–69.
74. Moraes, C., et al. "Supersoft lithography: candy-based fabrication of soft silicone microstructures." Lab on a Chip 15(18), (2015): 3760–3765.
75. Mačiulaitis, J., et al. "Preclinical study of SZ2080 material 3D microstructured scaffolds for cartilage tissue engineering made by femtosecond direct laser writing lithography." Biofabrication 7(1), (2015): 015015.
76. Kumaresan, T., et al. "Conceptual design and fabrication of porous structured scaffold for tissue engineering applications." Biomedical Research 26 (4) (2015): S42–S48.
77. Elsayed, H., et al. "Comparative analysis of wollastonite-diopside glass-ceramic structures fabricated via stereo-lithography." Advanced Engineering Materials 21(6), (2019): 1801160.
78. Tesavibul, P., et al. "Processing of 45S5 Bioglass® by lithography-based additive manufacturing." Materials Letters 74 (2012): 81–84.
79. Li, D., and Y. Xia. "Electrospinning of nanofibers: reinventing the wheel?." Advanced Materials 16(14), (2004): 1151–1170.
80. Khajavi, R., M. Abbasipour, and A. Bahador. "Electrospun biodegradable nanofibers scaffolds for bone tissue engineering." Journal of Applied Polymer Science 133(3), (2016).
81. Shao, W., et al. "Coaxial electrospun aligned tussah silk fibroin nanostructured fiber scaffolds embedded with hydroxyapatite–tussah silk fibroin nanoparticles for bone tissue engineering." Materials Science and Engineering: C 58 (2016): 342–351.

82. Xu, T., et al. "Electrospun polycaprolactone 3D nanofibrous scaffold with interconnected and hierarchically structured pores for bone tissue engineering." Advanced Healthcare Materials 4(15), (2015): 2238–2246.
83. Luo, X., et al. "Study on structure, mechanical property and cell cytocompatibility of electrospun collagen nanofibers crosslinked by common agents." International Journal of Biological Macromolecules 113 (2018): 476–486.
84. Kouhi, M., et al. "Bredigite reinforced electrospun nanofibers for bone tissue engineering." Materials Today: Proceedings 7 (2019): 449–454.
85. Sankar, S., C. S. Sharma, and S. N. Rath. "Enhanced osteodifferentiation of MSC spheroids on patterned electrospun fiber mats-An advanced 3D double strategy for bone tissue regeneration." Materials Science and Engineering: C 94 (2019): 703–712.
86. Murphy, S. V., and A. Atala. "3D bioprinting of tissues and organs." Nature Biotechnology 32(8), (2014): 773.
87. Park, K. M., et al. "Tissue engineering and regenerative medicine 2017: A year in review." Tissue Engineering Part B: Reviews 24(5), (2018): 327–344.
88. Kang, H.-W., et al. "A 3D bioprinting system to produce human-scale tissue constructs with structural integrity." Nature Biotechnology 34(3), (2016): 312.
89. Mandrycky, C., et al. "3D bioprinting for engineering complex tissues." Biotechnology Advances 34(4), (2016): 422–434.
90. Lim, J. Y., and H. J. Donahue. "Cell sensing and response to micro-and nanostructured surfaces produced by chemical and topographic patterning." Tissue Engineering 13(8), (2007): 1879–1891.
91. McBeath, R., et al. "Cell shape, cytoskeletal tension, and RhoA regulate stem cell lineage commitment." Developmental Cell 6(4), (2004): 483–495.
92. Yang, Y., R. Cavin, and J. L. Ong. "Protein adsorption on titanium surfaces and their effect on osteoblast attachment." Journal of Biomedical Materials Research Part A: An Official Journal of The Society for Biomaterials, The Japanese Society for Biomaterials, and The Australian Society for Biomaterials and the Korean Society for Biomaterials 67(1), (2003): 344–349.
93. Hartvig, R. A., et al. "Protein adsorption at charged surfaces: the role of electrostatic interactions and interfacial charge regulation." Langmuir 27(6), (2011): 2634–2643.
94. Jia, S., et al. "Enhanced hydrophilicity and protein adsorption of titanium surface by sodium bicarbonate solution." Journal of Nanomaterials 2015 (2015): 5.
95. Wang, Z., Y. Yan, and L. Qiao. "Protein adsorption on implant metals with various deformed surfaces." Colloids and Surfaces B: Biointerfaces 156 (2017): 62–70.
96. Howlett, C. R., et al. "Mechanism of initial attachment of cells derived from human bone to commonly used prosthetic materials during cell culture." Biomaterials 15(3), (1994): 213–222.
97. Stuiver, I., and T. E. O'Toole. "Regulation of integrin function and cellular adhesion." Stem Cells 13(3), (1995): 250–262.
98. Hu, J., X. Liu, and P. X. Ma. "Induction of osteoblast differentiation phenotype on poly (L-lactic acid) nanofibrous matrix." Biomaterials 29(28), (2008): 3815–3821.
99. Yao, C., D. Storey, and T. J. Webster. "Nanostructured metal coatings on polymers increase osteoblast

attachment." International Journal of Nanomedicine 2(3), (2007): 487.

100. Biggs, M. J. P., et al. "The effects of nanoscale pits on primary human osteoblast adhesion formation and cellular spreading." Journal of Materials Science: Materials in Medicine 18(2), (2007): 399–404.
101. Biggs, M. J. P., et al. "Regulation of implant surface cell adhesion: characterization and quantification of S-phase primary osteoblast adhesions on biomimetic nanoscale substrates." Journal of Orthopaedic Research 25(2), (2007): 273–282.
102. Dalby, M. J., et al. "The control of human mesenchymal cell differentiation using nanoscale symmetry and disorder." Nature Materials 6(12), (2007): 997.
103. Aminuddin, N. I., et al. "Osteoblast and stem cell response to nanoscale topographies: a review." Science and Technology of Advanced Materials 17(1), (2016): 698–714.
104. Beutner, R., et al. "Biological nano-functionalization of titanium-based biomaterial surfaces: a flexible toolbox." Journal of the Royal Society Interface 7.suppl_1 (2009): S93-S105.
105. Lutz, R., et al. "Biofunctionalization of titanium implants with a biomimetic active peptide (P-15) promotes early osseointegration." Clinical Oral Implants Research 21(7), (2010): 726–734.
106. Anselme, K., and M. Bigerelle. "Topography effects of pure titanium substrates on human osteoblast long-term adhesion." Acta Biomaterialia 1(2), (2005): 211–222.
107. De Oliveira, Paulo Tambasco, and Antonio Nanci. "Nanotexturing of titanium-based surfaces upregulates expression of bone sialoprotein and osteopontin by cultured osteogenic cells." Biomaterials 25(3), (2004): 403–413.
108. Becker, D., et al. "Proliferation and differentiation of rat calvarial osteoblasts on type I collagen-coated titanium alloy." Journal of Biomedical Materials Research: An Official Journal of The Society for Biomaterials, The Japanese Society for Biomaterials, and The Australian Society for Biomaterials and the Korean Society for Biomaterials 59(3), (2002): 516–527.
109. Surmeneva, M. A., et al. "Nano-hydroxyapatite-coated metal-ceramic composite of iron-tricalcium phosphate: improving the surface wettability, adhesion and proliferation of mesenchymal stem cells in vitro." Colloids and Surfaces B: Biointerfaces 135 (2015): 386–393.
110. Smith, L. J., et al. "Increased osteoblast cell density on nanostructured PLGA-coated nanostructured titanium for orthopedic applications." International Journal of Nanomedicine 2(3), (2007): 493.
111. De Jonge, L. T., et al. "Organic–inorganic surface modifications for titanium implant surfaces." Pharmaceutical Research 25(10), (2008): 2357–2369.
112. Sun, H., et al. "Proliferation and osteoblastic differentiation of human bone marrow-derived stromal cells on akermanite-bioactive ceramics." Biomaterials 27(33), (2006): 5651–5657.
113. Ajita, J., S. Saravanan, and N. Selvamurugan. "Effect of size of bioactive glass nanoparticles on mesenchymal stem cell proliferation for dental and orthopedic applications." Materials Science and Engineering: C 53 (2015): 142–149.
114. Webster, T. J., et al. "Enhanced functions of osteoblasts on nanophase ceramics." Biomaterials 21(17), (2000): 1803–1810.

115. Valerio, P., et al. "The effect of ionic products from bioactive glass dissolution on osteoblast proliferation and collagen production." Biomaterials 25(15), (2004): 2941–2948.
116. Ho, M.-H., et al. "Improving effects of chitosan nanofiber scaffolds on osteoblast proliferation and maturation." International Journal of Nanomedicine 9 (2014): 4293.
117. Cui, H., and P. J. Sinko. "The role of crystallinity on differential attachment/ proliferation of osteoblasts and fibroblasts on poly (caprolactone-co-glycolide) polymeric surfaces." Frontiers of Materials Science 6(1), (2012): 47–59.
118. McBeath, R., et al. "Cell shape, cytoskeletal tension, and RhoA regulate stem cell lineage commitment." Developmental Cell 6(4), (2004): 483–495.
119. Schmidt, C., A. A. Ignatius, and L. E. Claes. "Proliferation and differentiation parameters of human osteoblasts on titanium and steel surfaces." Journal of Biomedical Materials Research: An Official Journal of The Society for Biomaterials and The Japanese Society for Biomaterials 54(2), (2001): 209–215.
120. Vandrovcová, M., and L. Bacakova. "Adhesion, growth and differentiation of osteoblasts on surface-modified materials developed for bone implants." Physiological Research 60(3), (2011): 403.
121. Boyan, B. D., et al. "Osteoblasts generate an osteogenic microenvironment when grown on surfaces with rough microtopographies." European Cells and Materials 6(24), (2003): 22–27.
122. Olivares-Navarrete, R., et al. "Direct and indirect effects of microstructured titanium substrates on the induction of mesenchymal stem cell differentiation towards the osteoblast lineage." Biomaterials 31(10), (2010): 2728–2735.
123. Ohgushi, H., et al. "Osteogenic differentiation of cultured marrow stromal stem cells on the surface of bioactive glass ceramics." Journal of Biomedical Materials Research: An Official Journal of The Society for Biomaterials and The Japanese Society for Biomaterials 32(3), (1996): 341–348.
124. Nordström, E., et al. "Osteogenic differentiation of cultured marrow stromal stem cells on surface of microporous hydroxyapatite based mica composite and macroporous synthetic hydroxyapatite." Bio-medical Materials and Engineering 9(1), (1999): 21–26.
125. Li, X., et al. "Osteogenic differentiation of human adipose-derived stem cells induced by osteoinductive calcium phosphate ceramics." Journal of Biomedical Materials Research Part B: Applied Biomaterials 97(1), (2011): 10–19.
126. Wang, S., et al. "Design of electrospun nanofibrous mats for osteogenic differentiation of mesenchymal stem cells." Nanomedicine: Nanotechnology, Biology and Medicine 14(7), (2018): 2505–2520.

CHAPTER 14

Characteristics and Functions of Mesenchymal Stem Cells Derived from Nonosteoarthritis and Osteoarthritis Conditions

SATAR JABBAR RAHI ALGRAITTEE[1,2] and RAJESH RAMASAMY[1,*]

[1]*Stem Cell & Immunity Research Group, Immunology Laboratory, Department of Pathology, Faculty of Medicine and Health Sciences, Universiti Putra Malaysia, 43400 Serdang, Selangor, Malaysia*

[2]*Department of Medical Microbiology, College of Medicine, University of Kerbala, 46001 Karbala, Iraq*

[*]*Corresponding author. E-mail: rajesh@upm.edu.my*

ABSTRACT

Mesenchymal stem cells (MSCs) possess in addition to multipotency, ease of isolation, rapid growth capacity, and ability to maintain differentiation potential under in vitro conditions, unique characteristics such as the secretion of essential cytokines and growth factors that bestows potent regenerative and immunosuppressive prowess. These characteristics facilitate the execution of MSCs' physiological function of tissue regeneration and cell pool replenishment, hence making them ideal for cell-based regenerative/immunological therapies. For this reason, the infusion of MSCs has over the recent years served as an effective technique for the treatment of degenerative diseases like osteoarthritis (OA). Identified as the 11th highest contributor to global disability, OA has attracted the attention of many researchers leading to the emergence of questions over its possible dual pathogenesis, that is, whether OA is inflammation-mediated or a condition resulting from an inherently defective stem cell or both. This chapter provides a paradigm shift in the approach toward answering the question by reviewing and scrutinizing the characteristics and functions of MSCs derived from OA cartilage in relation to those of MSCs derived from non-OA cartilage, thereby highlighting the ability of local inflammation hallmarked in OA to disrupt the characteristics and functions of MSCs in an OA cartilage.

14.1 INTRODUCTION

Tissue engineering and regenerative medicine endeavors to reconstruct damaged, diseased, and worn-out tissues by an assemblage of scaffold biomaterial with adequate amounts of clonogenic (stem) cells and a milieu of suitable bioactive molecules [1]. Undifferentiated multipotent adult stem cells are derivatives of embryonic stem cells during the development and are widely distributed throughout the body. These cells can multiply by cell division to carry out the function of replenishing dying cells and regenerating damaged tissues [2]. In living organisms, adult stem cells are geared with the responsibility of maintaining and repairing the tissue in which they reside and as a result are categorized based on the kind of tissue they repair or form, that is, hematopoietic stem cells, neural stem cells, epithelial stem cells, and mesenchymal stem cells (MSCs) [3].

Among the most widely investigated adult stem cells in tissue engineering, mesenchymal stem cells appeared to be the most promising stem cells that possibly fulfil all requirements [4]. Mesenchymal stem cells are generated from almost all body parts such as bone marrow [5], muscle [6] adipose tissue [7], synovium [8], cartilage [9], trabecular bone [10], dermis [11], umbilical cord [12], umbilical code blood [13], pericyte [14], periosteum [15], and peripheral blood [16]. Researchers have, over the recent years, isolated MSCs from different tissues and compared their characteristics vis-à-vis their differentiation potentiality [17], in vitro expansion [18], immunosuppressive prowess [19] as well as therapeutic prospects [20]. It remains unclear whether MSCs from a particular source are better than those from another, however, it can be inferred from these studies that tissue source determines to a large extent the physiological characteristics of their MSCs as previously highlighted by Urrutia et al. [17] and Xu et al. [21].

Additionally, MSCs secrete a spectrum of the essential bioactive molecules, including cytokines and growth factors that enable regenerative and immunosuppressive functions making MSCs as ideal for therapeutic use [22, 23]. New insights have revealed that MSCs are found as perivascular cells, that is, they are pericytes released at the injury site, where they exert immunomodulatory and trophic effects by secreting large amounts of bioactive factors [24]. These factors create a local microenvironment that directs the functionality of the MSCs in wound healing and tissue regeneration [25].

By virtue of possessing these unique characteristics, MSCs have been recognized and are utilized in cell-based therapies to treat conditions like asthma [26], radiation exposure [27], neurological disorders [28], metabolic disorders [29] in addition to immune-related disorders such as inflammatory arthritic conditions like rheumatoid arthritis [30] and osteoarthritis (OA) [31].

14.2 ISOLATION AND CHARACTERIZATION OF MESENCHYMAL STEM CELLS

The anatomical location of MSCs is defined by their functional expectations of self-renewability and multilineage differentiation potential (i.e., into lineages of mesenchymal origins like bone, fat, cartilage, and various other collagenous connective tissues) [32]. As a heterogeneous cell population tasked with the responsibility of repair of worn-out tissues, MSCs inhabit tissues of the mesenchymal origins in different niches. Moreover, MSCs have been described as tissue-resident cells cohabiting with tissue-specific cells in virtually all postnatal organs and tissues [33, 34]. Others have described MSCs as pericyte [35], migrant or circulatory (with no specific niche) [36], and extraosseous (cartilage derived MSCs) [37]. For these reasons, MSCs can be isolated from numerous tissues as highlighted earlier.

The bone marrow cavity is arguably the most abundant source of MSCs, and as a result, MSCs were classically isolated from the bone marrow tissues through manipulation of their ability to adhere to culture plastic and in vitro culture expansion, thereby being easily distinguishable from hematopoietic cells [38]. It is crucial to indicate that there is no standardization of isolation techniques for MSCs; however, the techniques employed by different researchers are greatly influenced by the source tissue. For example, fractionation using Ficoll density gradient centrifugation, followed by low-density plating methods, is mostly used for isolation of bone marrow MSCs and as a technique to improve cell purification [39]. Recent studies have also successfully isolated heterogeneous MSC populations from bone marrow, by utilizing their ability to express high levels of surface $\alpha 5\beta 1$ integrin, which promotes rapid adhesion to fibronectin-coated culture plates and provides a more effective selectivity from hematopoietic and endothelial cell types [40].

Isolation of MSCs from solid tissue sources (cartilages and umbilical cord) involves cleaning of tissue with buffers to remove blood cells, followed by cutting the tissue into small sizes (not smaller than approximately 1–2 mm^3, as excessive mincing can induce mechanical destruction of cells), then extraction of the cells either by direct culturing of explant or enzymatic digestion techniques [41].

In the explant culture method, the tissue fragments are cultured in a growth medium where their reduced size allows improved access to nutrients and easy diffusion of gases. As a result, MSCs migrate from the fragments and adhere to culture surfaces. The fragments can be removed after several days of culturing [42]. Additional steps are, however, employed in the enzymatic digestion protocol where the tissue fragments are incubated with an extracellular matrix (ECM) digesting enzyme solution. The dissociation of the ECM releases single cells or cellular fragments that are then collected and cultured in the growth medium. After a few days of culturing colonies of adherent MSCs are formed [43]. Therefore density gradient centrifugation, direct culturing of explant, and enzymatic digestion are the widely utilized techniques for isolation of MSCs from adipose tissues, Wharton's jelly, umbilical cord, umbilical cord blood, amniotic fluid, dental tissues, placenta, synovial fluid, skin [41], and cartilage [44].

Independent of tissue source, there are important factors influencing the successful isolation of MSCs. These factors include but not limited to:

1. Sample quality

The isolation of MSCs from aged and diseased tissues have recorded low cell yield, slow proliferation rate, and skewed differentiation potential, when compared to tissues from young and nondiseased donors [45, 46].

2. Isolation method

There is the unavailability of a comprehensive comparative study on the effects different isolation methods may have on the successful isolation of MSCs. Nevertheless, the harvest of cell populations with less heterogeneity, higher cell yield, superior viability, and higher proliferation rates has been reported to be isolated using explant method in comparison to enzymatic digestion [43]. This could be as a result of the presence of nondissociated ECM that provides a favorable environment for the migrating cells by protecting the cells from proteolytic attack as well as mechanical stress. Interestingly, the enzymatic digestion method can be modified to improve in terms of efficiency, cell yield, and cell proliferation rate by modification of the type and concentration of the enzyme used for dissociation as well as digestion incubation time [44].

3. Culturing technique

The use of appropriate culture medium and adequate supplementation with growth factors have been shown to improve significantly the cell yield and proliferation rate during the initial culturing and propagation stages, respectively. It is also possible to modify the culturing technique to improve cell yield. In our previous study, we were able to significantly improve the cell yield after enzymatic digestion, using a serial plating culture procedure where after the first adherent cells were observed, the supernatant was collected, augmented with fresh media, and recultured. The procedure was repeated for a total of two cycles [44].

Unlike isolation, efforts have been made to standardize the characterization of MSCs, yet there is no single biomarker available for the identification of human MSCs. This is because MSCs isolated from different sources may show varying levels in the expression of a particular cell surface marker and/or may vary to differentiate into particular cell types. However, regardless of tissue source or isolation method, all purified MSCs are expected to exhibit adherence to culture plastic, tri-lineage differentiation ability (i.e., chondrogenesis, adipogenesis, and osteogenesis), as well as the presence/absence of specific cell surface markers, have been the main criteria for the classification of MSCs [47].

14.3 MECHANISM OF IMMUNOSUPPRESSIVE AND IMMUNOMODULATORY EFFECTS OF MESENCHYMAL STEM CELLS

Apart from their tissue reparatory functions, MSCs have been found to exhibit cogent immunosuppressive abilities against nearly all immune cells, especially, T lymphocytes [48, 49]. It has been revealed, through many in vitro and in vivo students that MSCs execute regenerative functions partly by immunosuppression as they create a conducive microenvironment for the restorative process free from obstruction by immune surveillance [50]. The immunosuppressive property of MSCs has, over the last few decades, caught the attention of many researchers, leading to numerous studies exploring its possible therapeutic application in cancer immunotherapy [51], immune diseases [52], and metabolic disorders [29].

Investigations have revealed that MSC-mediated immunosuppression occurs through complex interactions with the target immune cells either by physical contact or via secretion of soluble factors, culminating into the prevention of immune cell activation or inhibiting their proliferation [48]. From another perspective, MSC-mediated immunosuppression may rely on activation of the immune cells where the resulting activation by-products trigger MSCs to become immunosuppressive. This is observable in MSCs' suppression of T cells as they naturally tolerate quiescence T cell survival but exert inhibition of proliferation when they are activated [53].

Studies involving co-culture of MSCs with T cells have revealed that they ensure that T cells remain in a naïve state by inducing the down-regulation of cell surface markers of T cell activation such as CD25 [54], CD28 [55], and CD69 [56], which are involved in antigenic activation of T cells by dendritic cells (DCs) and other antigen presenting cells [57].

The detailing of the intricate mechanism involved in T cell immunosuppression by MSCs requires understanding of the conditions that trigger MSCs' interaction with target immune cells. Since not only primed T cells can trigger MSC-mediated immunosuppression, some researchers suggest that inflammation is required as a prerequisite [58]. As reported by Djouad et al., MSC-mediated immunosuppression occurred in response to inflammation induced by the presence of TNF-α, IL-1α/β, and IFN-γ [59]. As a result of inflammation, MSCs produce many immunosuppressive factors, soluble factors, as well as different chemokines and adhesion molecules such as interleukin (IL)-10, IL-1 receptor antagonist (IL-1Ra), transforming growth factor (TGF)-β, CCR5 ligands, CXCR3 ligands, vascular cell adhesion molecule-1, and intercellular adhesion molecule-1; the accumulation of these factors creates a favorable microenvironment for an amplified immunosuppression [60].

Generally, MSC-mediated immunosuppression can occur through a direct mechanism involving physical contact between MSCs and T cells via receptor–ligand interactions or the binding of soluble factors to specific receptors. This physical interaction inhibits T cell proliferation, prevents activation of cytotoxic T cells, and alters differentiation of T helper cells [61, 62]. The MSCs can also modulate proliferation of T cells and evade immune response through indirect mechanism where paracrine interactions occur with other immune cells facilitated by soluble factors, growth inhibitors, chemoattractant, and degradative enzymes, resulting to an alteration of the metabolism as well as the functionality of immune cells like naïve T cells, regulatory T Cells (Tregs), and DCs [63].

14.4 OSTEOARTHRITIS

OA is a chronic inflammatory disease that ultimately results in the progressive degeneration of articular cartilage. The complex interplay of genetic, biochemical, metabolic, and biomechanical factors coupled with activation of the inflammatory response in the cartilage, synovium, and subchondral bone induce development of the disease condition [64, 65]. The most common areas affected by OA include the knees, spine, hips, and joints in the hands [64]. The OA is also associated with painful swelling and stiffness of the joints. This occurs as a result of the rubbing of the bones exposed by wear and tear of the cartilage tissue [65]. Globally, OA is the 11th contributor to global disability affecting nearly 80% of individuals above 60 years. It is being projected to be the 4th most significant impact on health by 2020, and hence considering death and disability. This is the fastest growing major health condition [66].

Remarkable progress in the understanding of OA was made in the 1990s. It was discovered that numerous soluble factors, like prostaglandins or cytokines, can increase the production and/or release of proteinases especially the matrix metalloproteinase (MMPs) by chondrocytes [67]. This served as the primary step for the formulation of the "inflammatory" theory of pathophysiology of OA. However, years later several studies were done considered synovitis as propulsion of OA and subsequently, the condition was recognized as a critical feature of the OA process [68].

The subchondral bone serves as a mechanical shock absorber in joints. However, OA can cause subchondral bone marrow lesions [69]. The osteoblasts in OA-associated defective subchondral bone secrete inflammatory mediators, such as IL-6, IL-8, and TGF-β1 as well as proinflammatory cytokines, like IL-1β, which further aggravates the inflammation leading to degeneration of the deep layer cartilage in OA [70]. Therefore, primarily, OA may be considered cartilage wear and tear disease, but it is in no small extent, a complex disease condition involving an interplay between inflammatory factors released by the cartilage, synovium, and bone as well as the presence of degradative agents in the synovial fluid [64]. To discuss in detail, the pathophysiology of OA can be grouped into two categories, namely, the trauma/injury associated OA and the age-associated OA.

14.4.1 TRAUMA/INJURY-ASSOCIATED OSTEOARTHRITIS

The superficial zone of articular cartilage provides the smooth low friction surface that resists the high tensile forces and ensures normal function and structure is retained by cartilage. Trauma to joints often occurs, and the superficial layer begins to show signs of damage. Damage to this layer exposes the deeper cartilage zones, propagating forces to the lower layers when loaded, triggering further degeneration [71]. If traumatic injury or imbalance in ECM catabolism and anabolism occurs, cartilage tissue lacks the ability to respond to damage, therefore limited in its reparative response [72]. Partial thickness chondral defects that are contained within the cartilaginous regions are incapable of spontaneous repair. Full-thickness or osteochondral defects <3 mm that extend down to the vascularized subchondral bone are able to repair, which could be as a result of the release of stem cells from the bone marrow to fill the defect [73]. However, when the osteochondral lesions are larger, that is, > 6 mm, they do not heal, leading to progressive degeneration as previously reported in a goat defect model [74]. The repair tissue formed in small osteochondral defects is fibrocartilaginous and does not integrate into the surrounding articular cartilage. The mechanical properties of fibrocartilage are inferior to hyaline cartilage and, therefore, prone to future injury [72]. Damage to articular cartilage ultimately leads to the debilitating condition of OA [75].

The discovery of mechano-signaling provides new insights in understanding the association between mechanical stress and cell signaling in OA. Mechanoreceptors such as integrins and ion channels can be found on the surface of chondrocytes and subchondral bone cells of the cartilage and subchondral bone, respectively [64]. These receptors convert abnormal mechanical stress usually in the form of shear stress, stretch, hydrostatic pressure, or compression, into activated intracellular signals [76]. Although the exact mechanism is still unclear, it is suggested that chondrocytes can experience changes in extracellular osmolarity, acute volume, and active volume regulation following constant exposure to mechanical stressors [77]. These result to the restructuring of F-actin located at the chondrocyte cytoskeleton and hence solute transport leading to the influx extracellular Ca^{2+} and the activation of intracellular signaling cascades [78]. The signals eventually lead to an increase in the expression of inflammatory soluble mediators such as cytokines, prostaglandins, and chemokines. The activation of inducible signaling pathways such as NF-kB and mitogen-activated protein kinase (MAPK) mediates the intracellular synthesis of inflammatory mediators in response to mechanical signals [64].

Recently, the Ca^{2+}-permeable, nonspecific cation channel transient receptor potential of the vanilloid 4 has been identified at the novel candidate associated with osmo-mechanotransduction in chondrocyte [79, 80]. An understanding of the intricate mechanism through which intracellular signals are produced from mechanical signals will hopefully provide useful insights into the development of targeted therapies capable of modifying the course of OA.

14.4.2 AGE-ASSOCIATED OSTEOARTHRITIS

Aging or senescence has an impact on cartilage functions and structure. Notably, the mechanical properties of the cartilage undergo a significant alteration that occurs coincide with morphological changes [81]. Although changes such as superficial fibrillation and softening observed on the surface but the evidence of extension into the deeper zones is not cohesive [82]. The quantity and quality of the chondrocytes appear to decline with age as such cells displayed a distorted mitotic index and reduced metabolism that further subject the articular cartilage to the susceptibility for damage [82]

Another age-related factor capable of affecting the stability of chondrocytes is the accumulation of oxidative stress [83]. Oxidative stress builds up when the generation of reactive oxygen species (ROS) exceeds the oxidants and free radical scavenging capacity of the antioxidant species [84] causing tissue damage and cell death [85]. ROS are endogenous by-products of metabolism that occurs in the mitochondria, and peroxisomes of chondrocytes by enzymes such as NADPH oxidase, lipoxygenases, and cytochrome P450. However, they can be released through several exogenous sources such as during inflammation by macrophages and neutrophils [84]. These highly oxidizing agents can readily attack chondrocyte biomolecules leading to degeneration of articular cartilage, or induce alteration of gene expression. For example, hydrogen peroxide causes apoptosis through upregulation of caspase-3 and induction of genomic instability [86]. Studies have also shown that ROS increases expression of procollagen IIA and chondroitin sulfate 3B3(-), which are early markers of chondrocyte hypertrophy and OA [87]. Macrophages stimulated by ROS in OA release inflammatory cytokines inducing ROS further release, which assists cartilage degeneration [88, 89].

14.5 IMMUNOPATHOLOGY OF OSTEOARTHRITIS

Inflammation is central to the pathogenesis and progression of OA. This is because the synovial fluid of an osteoarthritic knee joint is a mesh of inflammatory agents [90]. Several studies have reported the presence of many proinflammatory cytokines in the synovial fluid at different stages of OA progression [91]. The important cytokines found in the synovial fluid at the early onset of OA are IL-1α, IL-1β, and TNFα [92]. Immune cells accumulate at the synovium primarily in response to the presence of these inflammatory agents at the site of injury. The immune cells such as macrophages further aggravate the inflammatory by secreting more pro-inflammatory agents and proteinases [93]. As the disease progresses, the combined effect of these highly inflammatory agents in the synovial fluid leads to the degradation of the cartilage matrix and the pain associated with OA [92].

Understanding the relationship between continuous overloading of the articular cartilage and build-up of stress-related inflammatory signals provides useful information on the mechanism of inflammation-induced cartilage degeneration. The resting state of chondrocytes in articular cartilage is perturbed as a result of overloading stress; hence, triggers the activation of the NF-κB pathway that disrupts the chondrocyte resting state and induces the switch in chondrocyte metabolism from an anabolic state to a catabolic state [94]. As a result, stress-related pathways are activated, which in turn triggers the production and release of proinflammatory mediators that facilitate the disruption in the growth-arrested state of articular cartilage and breakdown of the ECM [95].

Continuous overloading and build-up of stress lead to repeated activation of NF-κB, which ultimately leads to the activation of regulatory transcription factors such as hypoxia inducing factor 2 alpha (HIF2α) and E74-like factor 3 (ELF3). An activated HIF2α induces the Indian hedgehog/runt-related transcription factor 2 (RUNX2) axis that is involved in the inhibition of chondrocyte proliferation and regulates endochondral ossification [96, 97]. The NF-kB-mediated activation of these transcription factors leads to the expression of MMP13 that facilitates the digestion of type II collagen, disrupts the extracellular matrix, and triggers calcification, osteophyte formation and apoptosis, and hence contributes to the onset and/or progression OA [98–100]. The signaling pathways implicated at different stages of OA progression include molecules such as growth arrest and DNA damage-inducible 45β, MAPK/extracellular signal-regulated kinases, Janus kinase, inhibitors of kappa B kinase alpha/beta (IKKα/β), p38, and transforming growth factor beta-activated kinase 1 (TAK1) as well as transcription factors like NF-κB, Ihh, HIF2α, ELF3, CCAAT-enhancer-binding proteins (C/EBPβ), activator protein 1 (AP-1),) and Runx2. Interestingly, nearly all of these regulatory factors involve the direct or indirect activation of transcription and activity of MMP-13 [95].

It has been previously hypothesized that the pro-inflammatory environment not only leads to the destruction of the joint but also impairs the intrinsic joint regeneration potential of the local MSCs. This hypothesis has been tested in rheumatoid arthritis by Jones et al. The results obtained showed that the synovial inflammation exerts an adverse effect on the function of the MSCs [101]. In OA, however, studies have implicated IL-1β and TNF-α in the epigenetic modification of chondrocytes. IL-1β and TNF-α downregulate cartilage regenerative genes such as [SRY-Box9 (Sox9), Collagen type II alpha 1 chain (Col2A1), and the gene encoding aggrecan (ACAN)], with upregulation of catabolic genes (like MMP13 and ADAMTS5), and inhibition of alkaline phosphates activity in chondrocytes leading to arrest of chondrogenesis and promotion of osteogenesis [102].

Also, IL-1β downregulates miRNA 140 [103], and miRNA 148a in chondrocytes. Studies have shown that these microRNAs are involved in cartilage regeneration and prevention of cartilage degeneration [104, 105]. Figure 14.1 is a schematic presentation of the role inflammation plays in the pathogenesis of OA through the interaction between immune cells, the synovial milieu, synovium, and cartilage of an osteoarthritic and a healthy knee joint. Many studies have suggested that the local inflammation may be the primary culprit in OA disease as it brings about the destruction of the cartilage matrix. However, investigations need to be done as to whether the local inflammation can also affect the potency of the human OA cartilage derived MSCs.

14.6 OSTEOARTHRITIS AND MESENCHYMAL STEM CELLS

The function of MSCs in OA is substantial, as the parent cells of osteoblasts and chondrocytes MSCs have the responsibility replenishing of the inflammation-mediated degradation of cartilage tissue by synthesizing the cartilage and bone extracellular matrix [5]. In light of this evidence, it is essential to assess the reparative ability of MSCs in OA as any alteration of this ability may lead to abnormal tissue development and homeostasis, thereby being a significant contributor to the pathogenesis of OA [106].

Although several investigations have established a defective or abnormal activity of MSCs over OA progression, there is very little understanding of the relationship between this abnormal activity with the progression of OA. Murphy et al. observed a lacking in differentiation potentials and proliferation when BM-MSCs isolated from hospice OA patients were compared with age-matched healthy individuals used as controls. Also, the group observed a reduced proliferative activity, modification of differentiation preferences (as the cells tended toward osteogenesis and adipogenesis at the expense of chondrogenesis) in the OA MSCs [107]. This phenomenon offers an explanation for the detection of osteophytes in the synovial fluid during the advanced stage of OA. Among many factors, age and epigenetic factors have been shown to affect the differentiation preference of OA MSCs [104]. Human periosteal MSCs obtained from donors with aged less than 30 years showed spontaneous chondrogenesis, whereas MSCs obtained from donors exceeding 30 years of

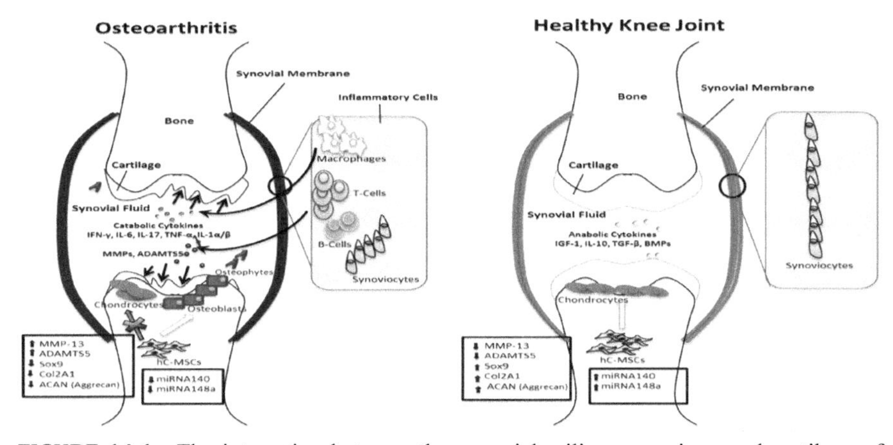

FIGURE 14.1 The interaction between the synovial milieu, synovium, and cartilage of osteoarthritic and healthy knee joint:
(A) In OA knee, the levels of proinflammatory cytokines (IL-1β, Interleukin-1 beta; TNF-α, Tumor Necrosis Factor 1; and IL-6, Interleukin-6) are elevated in OA. These cytokines contribute to the pathogenesis of OA through several mechanisms including Localization of immune cells (Macrophages, T-lymphocytes and B-lymphocytes) at the Synovium. Secretion of catabolic cytokines (IFN-γ, interferon gamma; IL-17, Interleukin-17 and IL-1α, Interleukin-1 alpha) as well as proteinases (like MMPs, matrix metalloproteinase; ADAMTS5, a disintegrin-like and metalloproteinase with thrombospondin motifs 5), leading to cartilage tissue degeneration. Downregulation of anabolic genes (like Sox9, SRY-Box9; Col2A1, collagen type II alpha 1 chain & ACAN, gene encoding Aggrecan), miRNA140 and miRNA148 with upregulation of catabolic genes (like MMP13 and ADAMTS5) in tissue-resident human cartilage MSC (hC-MSCs) and Chondrocytes leading to arrest of chondrogenesis and promotion of osteogenesis. (B) In healthy knee joints, anabolic cytokines (IGF-1, insulin-like growth factor 1; IL-10, Interleukin-10; TGF-β, transforming growth factor-beta and BMPs, bone morphogenetic proteins) are found in the synovial milieu, and hence the synovium is free of immune cells. Anabolic genes, miRNA140 and miRNA148, are upregulated while catabolic genes are downregulated, and tissue-resident C-MSCs readily differentiates towards chondrogenesis. Compiled using information sourced from [92, 104, 105].

age did not show spontaneous chondrogenesis in culture [15]. In vitro studies on cancellous bone MSCs, which were sorted based on the expression of CD271, revealed an age-related proliferation deficiency in OA-derived MSCs [101].

Studies have shown that chondrogenic genes like Sox9, ACAN, and Col2A1, which are typically expressed in healthy chondrocytes, are repressed in OA chondrocytes [104, 105] stimulated with IL-1β and observed consistent IL-1β-dependent repression of Aggrecan (ACAN) in the chondrocyte [104]. Also, through the microarray, a significant reduction in the expression of microRNAs such as miRNA140 and miRNA148a were revealed in OA chondrocytes [104, 105]. The studies transfected the microRNAs into chondrocytes and observed upregulation of the chondrogenic genes as well as suppression of IL-1β, MMP-13, and ADAMT5S. Whether this molecular variation observed in OA chondrocytes applies to MSCs in OA remains to be unraveled by future studies. Generally, in healthy joints, the synovial fluid compartment has a fewer number of MSCs compared to those in an OA patient, and the number of cells increased with the severity of OA progression [108]. MSCs separated from BM show less chondrogenic capability than synovial fluid MSC (CSF-MSCs) [109]. The normal function of MSCs, chondrocytes, synovial cells, and joint progenitor cells is affected by proinflammatory cytokines, chemokines, pathogenesis, and progression of OA.

14.7 STABILITY AND POTENTIAL OF NON-OA AND OA-DERIVED MESENCHYMAL STEM CELLS

Safety remains the utmost concern in cell therapy; therefore MSCs that can be used as "cell medicine" must possess the required phenotype, functional potential, remain untransformed with no experimental artifacts or microbiological contaminations [110]. Since MSCs are in general practice, isolated from different tissues and expanded under in vitro culture conditions, it is challenging to ignore complications that may arise from their utilization as therapeutic tools in vivo. There are concerns over the stability of MSCs during the transition from the laboratory to the bedside as it remains unclear whether the "culture-adapted" MSCs undergo an adaptive transformation during long-term passaging in vitro [111].

Despite these concerns, the utilization of MSCs derived from different sources in cell-based therapy and cell-based tissue engineering for the treatment of degenerative disease including OA has recorded tremendous success over the years.

The MSCs derived from bone marrow, adipose tissue, and cartilage of non-OA donors have displayed stability, healthy phenotype, functional potentials, and have shown the ability to effectively improve clinical results when injected into arthritic joints [112]. For MSCs from diseased tissues, however, great refrainment being observed in their utilization as cell medicines as several studies have reported alteration of inherent nature at the molecular level as well as skewed functional potentials in MSCs derived from diseased patients [113].

There is a lack of sufficient comparative studies exploring the stability and therapeutic potentials of cartilage-derived MSCs (C-MSCs) from non-OA donors in relation to those derived from OA patients. However, Xia et al. analyzed the proliferation activity, differentiation potential, and expression levels microRNAs (MiR-31–5p and miR-424–5p) of C-MSCs from degraded cartilage in comparison with C-MSCs from healthy cartilage. The study observed significant reduction in cell percentage, proliferation rate, and alteration of differentiation profile as well as dysregulation of miRNA expression in C-MSCs from the degraded cartilage compared to those from the healthy cartilage. They suggested that these alterations may be complicit in the OA C-MSCs dysfunction that contributes to imbalance in cartilage homeostasis and hence OA-related cartilage erosion [114]. Another study characterized OA C-MSCs at cellular and molecular levels and induced chondrogenic differentiation in the OA C-MSCs. They reported an alteration of gene regulation, elevation of hypertrophic OA phenotypes, and stimulation of the secretion of matrix metalloproteinase 13 (MMP-13) [115].

Whether OA C-MSCs can be used for OA therapy remains a tough question for future research. However, studies have identified cell clustering in response to high chondrocyte proliferation during the early stages of OA [115, 116]. Therefore it has been suggested that since the gene regulatory network in OA C-MSCs is altered, inactivation of these cells may be necessary prior to or during cell-based therapy for OA patients.

14.8 PATHOPHYSIOLOGY OF OSTEOARTHRITIC CARTILAGE-DERIVED MESENCHYMAL STEM CELLS

The isolation of MSCs from the cartilage tissue occurred concurrently with attempts to isolate the cells from a diseased condition, specifically, cartilage tissues of OA patients. Successful isolations of OA C-MSCs have been consistently recorded by many studies conducted over a period spanning 15 years [9, 115, 117, 118]. These studies have highlighted significant differences in the characteristics of OA C-MSCs when compared to cartilage MSCs from the physiological condition or MSCs from other sources, yet very little is known on the pathophysiology of OA C-MSCs. This can be attributed to the continuous absence of specific molecular markers defining and distinguishing these cells from other MSCs. Hence, recent research works on OA C-MSCs have attempted to address the following pressing questions:

1. Are the OA C-MSCs merely cartilage tissue-resident MSCs that survived the onset of the OA disease or are they an entirely different cell population coexisting in the cartilage?
2. Do the OA C-MSCs possess specific characteristics that support or hinder the pathogenesis and progression of OA?
3. Are these characteristics (if identified) part of their inherent genetic make-up or are they acquired as a result of epigenetic modification by the inflammatory joint microenvironment associated with the OA disease?

Although OA C-MSCs have been shown to adequately express the requisite MSCs' immunophenotypic markers [44], a study conducted by Jayasuriya et al revealed that OA C-MSCs, unlike BM-MSCs, do not express SSEA4. Also, they exhibited overexpression of CD54; a surface marker highly expressed by OA chondrocytes and lowly expressed by MSCs [115]. The same study showed that OA C-MSCs are prone to hypertrophy, release of MMP-13 via high mRNA expression of RUNX2, a transcription factor and positive regulator of chondrocyte epiphyseal phenotype in OA previously implicated in chondrocyte hypertrophy [119]; COL10A1, a hypertrophy marker in OA cartilage [120] and PRX1, a limb bud mesenchymal cell transcription factor associated with skeletogenesis [121, 122]. They also observed that C-MSCs secreted the cartilage tissue digesting metalloproteinase MMP-13 when induced to undergo chondrogenesis. From their findings, it was suggested that OA C-MSCs may be more inclined toward osteogenesis and hence contribute to OA pathogenesis and disease progression [115]. The reduced chondrogenic ability in OA C-MSCs has been previously associated with reduced expression of Aggrecan and type II collagen [107].

Although there is hardly any clear explanation why OA C-MSCs possess a high osteogenic tendency, this behavior is consistent with the significant bony changes, subchondral bone sclerosis, focal necrosis of subchondral bone, and osteophyte formation associated with the pathology of OA [123]. In physiological condition, osteogenic differentiation is initiated by committed preosteoblasts through the secretion of alkaline phosphatase. Subsequently, the cells change phenotypically into cells with a larger nucleus and Golgi as the endoplasmic reticulum becomes more extensive to facilitate the secretion of bone matrix proteins [124].

The differentiation of C-MSCs toward adipogenesis in OA has also been reported with the presence of adipose fat tissues around the cartilage and synovial area [125, 126]. This may not be the strong clinical manifestation of OA especially as some studies have reported decreased adipogenesis in OA hC-MSCs [127]. It is suggested that the altered differentiation pattern of OA hC-MSCs is a result of the inherent genetic makeup of the cartilage cells. Complex signaling cascades regulate adipogenesis via changes in the gene expression of peroxisome proliferator-activated receptor gamma (PPARγ), a transcription factor that interacts with members of the CCAAT/enhancer-binding proteins (C/EBP) family as well as many other adipocyte-specific genes to regulate adipogenesis [128, 129].

Generally, the unavailability of sufficient data on the biology of OA C-MSCs, especially their immunosuppressive prowess, leaves more questions than answers on the pathophysiological roles of OA C-MSCs in the OA disease. Nevertheless, it can be suggested that a better approach to preserving the cartilage integrity in OA may involve the suppression or deletion of the OA C-MSCs rather than activating them for tissue regeneration.

With the ultimate goal of opening new frontiers in the repair of damaged tissues in joint disorders, MSCs research has been given much attention. For instance, recent evidence regarding the presence of MSCs in the cartilage tissue as well as their ability to exert immunosuppression on the one hand and to differentiate into cartilage on the other have provided an ideal opportunity for therapeutic interventions in regenerative medicine. Under normal physiological conditions, MSCs contribute majorly in the repair and maintenance of joint tissues. However, for therapeutic purposes, a detailed explanation of the biology of MSCs in OA disease is fundamental as it will serve as a scientific basis for developing tailored interventions.

There should be a paradigm shift in the approach toward these interventions. Factors affecting the residual MSCs' function in the preparation of degradative tissues should be given much consideration. Hence, it is necessary for the first step in any therapeutic intervention using autologous C-MSC transplantation for inducing neochondrogenesis in OA, for including an effective suppression of local inflammation. This would prevent a futile cycle as the inflammatory milieu may continuously over time modify the infused MSCs and reduce the cells' immunosuppressive and chondrogenesis prowess.

14.9 CONCLUSION

This chapter has discussed the pathophysiology of OA, ascribing it to the accumulation of inflammation leading firstly to cartilage tissue breakdown and secondly to inhibition of the normal regenerative function of C-MSCs. However, scientific investigations should be done to identify whether an inherent genetic defect, modification by epigenetic factors, or the presence of fewer MSCs are responsible for the C-MSCs' failure to bring about cartilage tissue reparation as observed in OA. Also, the in vivo interplay between healthy immune cells and C-MSC from OA in an inflammatory milieu warrants further investigation. This will provide useful insights and spring up giant milestones in regenerative medicine and possibly lead to the development of a cure for OA.

KEYWORDS

- **mesenchymal stem cells**
- **osteoarthritis**
- **cartilage**
- **immunosuppression and inflammation**

REFERENCES

1. Mano, J., et al., *Natural origin biodegradable systems in tissue engineering and regenerative medicine: present status and some moving trends.* Journal of the Royal Society Interface, 2007. **4**(17): pp. 999–1030.
2. Valarmathi, M.T., et al., *The mechanical coupling of adult marrow stromal stem cells during cardiac regeneration assessed in a 2-D co-culture model.* Biomaterials, 2011. **32**(11): pp. 2834–2850.
3. Kalra, K. and P. Tomar, *Stem cell: basics, classification and applications.* American Journal of Phytomedicine and Clinical Therapeutics, 2014. **2**(7): pp. 919–930.
4. Tuan, R.S., G. Boland, and R. Tuli, *Adult mesenchymal stem cells and cell-based tissue engineering.* Arthritis Research and Therapy, 2002. **5**(1): p. 32.
5. Pittenger, M.F., et al., *Multilineage potential of adult human mesenchymal stem cells.* Science, 1999. **284**(5411): pp. 143–147.
6. Jankowski, R., B. Deasy, and J. Huard, *Muscle-derived stem cells.* Gene Therapy, 2002. **9**(10): pp. 642.
7. Zuk, P., *Adipose-derived stem cells in tissue regeneration: a review.* ISRN Stem Cells, 2013. **2013**.
8. De Bari, C., et al., *Multipotent mesenchymal stem cells from adult human synovial membrane.* Arthritis

& Rheumatism, 2001. **44**(8): pp. 1928–1942.
9. Alsalameh, S., et al., *Identification of mesenchymal progenitor cells in normal and osteoarthritic human articular cartilage.* Arthritis & Rheumatism, 2004. **50**(5): pp. 1522–1532.
10. Tuli, R., et al., *Characterization of multipotential mesenchymal progenitor cells derived from human trabecular bone.* Stem Cells, 2003. **21**(6): pp. 681–693.
11. Young, H.E., et al., *Human reserve pluripotent mesenchymal stem cells are present in the connective tissues of skeletal muscle and dermis derived from fetal, adult, and geriatric donors.* The Anatomical Record: An Official Publication of the American Association of Anatomists, 2001. **264**(1): pp. 51–62.
12. Lu, L.-L., et al., *Isolation and characterization of human umbilical cord mesenchymal stem cells with hematopoiesis-supportive function and other potentials.* Haematologica, 2006. **91**(8): pp. 1017–1026.
13. Lee, O.K., et al., *Isolation of multipotent mesenchymal stem cells from umbilical cord blood.* Blood, 2004. **103**(5): pp. 1669–1675.
14. Brighton, C.T., et al., *The pericyte as a possible osteoblast progenitor cell.* Clinical Orthopaedics and Related Research, 1992(275): pp. 287–299.
15. De Bari, C., F. Dell'Accio, and F.P. Luyten, *Human periosteum-derived cells maintain phenotypic stability and chondrogenic potential throughout expansion regardless of donor age.* Arthritis & Rheumatism: Official Journal of the American College of Rheumatology, 2001. **44**(1): pp. 85–95.
16. Zvaifler, N.J., et al., *Mesenchymal precursor cells in the blood of normal individuals.* Arthritis Research & Therapy, 2000. **2**(6): p. 477.
17. Urrutia, D.N., et al., *Comparative study of the neural differentiation capacity of mesenchymal stromal cells from different tissue sources: An approach for their use in neural regeneration therapies.* PLoS One, 2019. **14**(3): pp. e0213032.
18. Secunda, R., et al., *Isolation, expansion and characterisation of mesenchymal stem cells from human bone marrow, adipose tissue, umbilical cord blood and matrix: a comparative study.* Cytotechnology, 2015. **67**(5): pp. 793–807.
19. Karaöz, E., et al., *Comparative Analyses of immunosuppressive characteristics of bone-marrow, wharton's jelly, and adipose tissue-derived human mesenchymal stem cells.* Turkish Journal of Hematology, 2017. **34**(3): p. 213.
20. Tao, H., et al., *Proangiogenic features of mesenchymal stem cells and their therapeutic applications.* Stem Cells International, 2016. **2016**.
21. Xu, L., et al., *Tissue source determines the differentiation potentials of mesenchymal stem cells: a comparative study of human mesenchymal stem cells from bone marrow and adipose tissue.* Stem Cell Research & Therapy, 2017. **8**(1): p. 275.
22. Ramasamy, R., et al., *The immunosuppressive effects of human bone marrow-derived mesenchymal stem cells target T cell proliferation but not its effector function.* Cellular Immunology, 2008. **251**(2): pp. 131–136.
23. Le Blanc, K. and O. Ringden, *Immunomodulation by mesenchymal stem cells and clinical experience.* Journal of Internal Medicine, 2007. **262**(5): pp. 509–525.
24. Caplan, A., *Why are MSCs therapeutic? New data: new insight.* The Journal of Pathology: A Journal of the Pathological

Society of Great Britain and Ireland, 2009. **217**(2): pp. 318–324.
25. Weiss, A.R.R. and M.H. Dahlke, *Immunomodulation by Mesenchymal Stem Cells (MSCs): Mechanisms of Action of Living, Apoptotic, and Dead MSCs.* Frontiers in Immunology, 2019. **10**.
26. Kitoko, J., et al., *Therapeutic administration of bone marrow-derived mesenchymal stromal cells reduces airway inflammation without up-regulating Tregs in experimental asthma.* Clinical & Experimental Allergy, 2018. **48**(2): pp. 205–216.
27. Rodgers, K. and S.S. Jadhav, *The application of mesenchymal stem cells to treat thermal and radiation burns.* Advanced Drug Delivery Reviews, 2018. **123**: pp. 75–81.
28. Mukai, T., A. Tojo, and T. Nagamura-Inoue, *Mesenchymal stromal cells as a potential therapeutic for neurological disorders.* Regenerative Therapy, 2018. **9**: pp. 32–37.
29. Cantore, S., et al., *Recent advances in endocrine, metabolic and immune disorders: mesenchymal stem cells (MSCs) and engineered scaffolds.* Endocrine, Metabolic & Immune Disorders-Drug Targets (Formerly Current Drug Targets-Immune, Endocrine & Metabolic Disorders), 2018. **18**(5): pp. 466–469.
30. De Bari, C., *Are mesenchymal stem cells in rheumatoid arthritis the good or bad guys?* Arthritis Research & Therapy, 2015. **17**(1): p. 113.
31. Jo, C.H., et al., *Intra-articular injection of mesenchymal stem cells for the treatment of osteoarthritis of the knee: a proof-of-concept clinical trial.* Stem Cells, 2014. **32**(5): pp. 1254–1266.
32. Nombela-Arrieta, C., J. Ritz, and L.E. Silberstein, *The elusive nature and function of mesenchymal stem cells.* Nature Reviews Molecular cell biology, 2011. **12**(2): p. 126.
33. da Silva Meirelles, L., P.C. Chagastelles, and N.B. Nardi, *Mesenchymal stem cells reside in virtually all post-natal organs and tissues.* Journal of Cell Science, 2006. **119**(11): pp. 2204–2213.
34. Sandrasaigaran, P., et al., *Characterisation and immunosuppressive activity of human cartilage-derived mesenchymal stem cells.* Cytotechnology, 2018. **70**(3): pp. 1037–1050.
35. Caplan, A.I., *All MSCs are pericytes?* Cell stem cell, 2008. **3**(3): pp. 229–230.
36. Verfaillie, C.M. *Multipotent adult progenitor cells: an update.* in *Novartis Found Symposia*, 2005. Wiley Online Library.
37. McGonagle, D., T.G. Baboolal, and E. Jones, *Native joint-resident mesenchymal stem cells for cartilage repair in osteoarthritis.* Nature Reviews Rheumatology, 2017. **13**(12): p. 719.
38. Phinney, D.G., et al., *Plastic adherent stromal cells from the bone marrow of commonly used strains of inbred mice: variations in yield, growth, and differentiation.* Journal of Cellular Biochemistry, 1999. **72**(4): pp. 570–585.
39. Ferrin, I., et al., *Isolation, culture, and expansion of mesenchymal stem cells*, in *Stem Cell Banking*. 2017, Springer. pp. 177–190.
40. Yusop, N., et al., *Isolation and characterisation of mesenchymal stem cells from rat bone marrow and the endosteal niche: a comparative study.* Stem Cells International, 2018. **2018**.
41. Mushahary, D., et al., *Isolation, cultivation, and characterization of human mesenchymal stem cells.* Cytometry Part A, 2018. **93**(1): pp. 19–31.
42. Hendijani, F., *Explant culture: an advantageous method for isolation of mesenchymal stem cells from human*

tissues. Cell Proliferation, 2017. **50**(2): p. e12334.
43. Boey, K., et al., *Comparison of extraction methods and culture medium for umbilical cord lining-and Wharton's jelly-derived mesenchymal stromal cells.* Cytotherapy, 2019. **21**(5): p. S80.
44. Algraittee S.J.R, L.H., Boroojerdi M.H., Hosseinpour V., Hwa King, Ramasamy R. *Generation and characterization of human osteoarthritis cartilage-derived mesenchymal stem cells by modified sample processing and culture method.* Malaysian Journal of Medicine and Health Sciences, 2018. **14**(14): pp. 2–10.
45. Choudhery, M.S., et al., *Donor age negatively impacts adipose tissue-derived mesenchymal stem cell expansion and differentiation.* Journal of Translational Medicine, 2014. **12**(1): p. 8.
46. Traggiai, E., et al., *Bone marrow-derived mesenchymal stem cells induce both polyclonal expansion and differentiation of B cells isolated from healthy donors and systemic lupus erythematosus patients.* Stem Cells, 2008. **26**(2): pp. 562–569.
47. Dominici, M., et al., *Minimal criteria for defining multipotent mesenchymal stromal cells. The International Society for Cellular Therapy position statement.* Cytotherapy, 2006. **8**(4): pp. 315–317.
48. Uccelli, A., V. Pistoia, and L. Moretta, *Mesenchymal stem cells: a new strategy for immunosuppression?* Trends in Immunology, 2007. **28**(5): pp. 219–226.
49. de Castro, L.L., et al., *Current understanding of the immunosuppressive properties of mesenchymal stromal cells.* Journal of Molecular Medicine, 2019. **97**(5): pp. 605–618.
50. Qi, K., et al., *Tissue regeneration: the crosstalk between mesenchymal stem cells and immune response.* Cellular Immunology, 2018. **326**: pp. 86–93.
51. GOMES, J.P., et al., *Deepening a simple question: Can MSCs be used to treat cancer?* Anticancer Research, 2017. **37**(9): pp. 4747–4758.
52. Wang, L.-T., et al., *Human mesenchymal stem cells (MSCs) for treatment towards immune-and inflammation-mediated diseases: review of current clinical trials.* Journal of Biomedical Science, 2016. **23**(1): p. 76.
53. Benvenuto, F., et al., *Human mesenchymal stem cells promote survival of T cells in a quiescent state.* Stem Cells, 2007. **25**(7): pp. 1753–1760.
54. Yoo, H.S., et al., *Mesenchymal stromal cells inhibit CD25 expression via the mTOR pathway to potentiate T-cell suppression.* Cell Death & Disease, 2017. **8**(2): p. e2632.
55. Liu, M., et al., *Study of immunomodulatory function of exosomes derived from human umbilical cord mesenchymal stem cells.* Zhonghua yi xue za zhi, 2015. **95**(32): pp. 2630–2633.
56. Arutyunyan, I., et al., *Elimination of allogeneic multipotent stromal cells by host macrophages in different models of regeneration.* International Journal of Clinical and Experimental Pathology, 2015. **8**(5): p. 4469.
57. Wheat, W.H., et al., *Suppression of canine dendritic cell activation/maturation and inflammatory cytokine release by mesenchymal stem cells occurs through multiple distinct biochemical pathways.* Stem Cells and Development, 2017. **26**(4): pp. 249–262.
58. Mohammadpour, H., et al., *TNF-α modulates the immunosuppressive effects of MSCs on dendritic cells and T cells.* International Immunopharmacology, 2015. **28**(2): pp. 1009–1017.

59. Djouad, F., et al., *Reversal of the immunosuppressive properties of mesenchymal stem cells by tumor necrosis factor α in collagen-induced arthritis.* Arthritis & Rheumatism, 2005. **52**(5): pp. 1595–1603.
60. Gornostaeva, A., E. Andreeva, and L. Buravkova, *Factors governing the immunosuppressive effects of multipotent mesenchymal stromal cells in vitro.* Cytotechnology, 2016. **68**(4): pp. 565–577.
61. Li, Y., et al., *Cell–cell contact with proinflammatory macrophages enhances the immunotherapeutic effect of mesenchymal stem cells in two abortion models.* Cellular & Molecular Immunology, 2019: p. 1.
62. Espagnolle, N., et al., *CD54-mediated interaction with pro-inflammatory macrophages increases the immunosuppressive function of human mesenchymal stromal cells.* Stem Cell Reports, 2017. **8**(4): pp. 961–976.
63. Wu, M., et al., *Mesenchymal stem cells immunosuppressed IL-22 in patients with immune thrombocytopenia via soluble cellular factors.* Journal of Immunology Research, 2015. **2015**.
64. Berenbaum, F., *Osteoarthritis as an inflammatory disease (osteoarthritis is not osteoarthrosis!).* Osteoarthritis and Cartilage, 2013. **21**(1): pp. 16–21.
65. Mamidi, M.K., et al., *Mesenchymal stromal cells for cartilage repair in osteoarthritis.* Osteoarthritis and Cartilage, 2016. **24**(8): pp. 1307–1316.
66. Veronese, N., et al., *Adherence to a Mediterranean diet is associated with lower prevalence of osteoarthritis: data from the osteoarthritis initiative.* Clinical Nutrition, 2017. **36**(6): pp. 1609–1614.
67. Goldring, M.B., *The role of the chondrocyte in osteoarthritis.* Arthritis & Rheumatism: Official Journal of the American College of Rheumatology, 2000. **43**(9): pp. 1916–1926.
68. Van den Berg, W., *The role of cytokines and growth factors in cartilage destruction in osteoarthritis and rheumatoid arthritis.* Zeitschrift für Rheumatologie, 1999. **58**(3): pp. 136–141.
69. Hunter, D.J., et al., *Bone marrow lesions from osteoarthritis knees are characterized by sclerotic bone that is less well mineralized.* Arthritis Research & Therapy, 2009. **11**(1): p. R11.
70. Wen, C., W.W. Lu, and K.Y. Chiu, *Importance of subchondral bone in the pathogenesis and management of osteoarthritis from bench to bed.* Journal of Orthopaedic Translation, 2014. **2**(1): pp. 16–25.
71. Martin, J. and J. Buckwalter, *The role of chondrocyte–matrix interactions in maintaining and repairing articular cartilage.* Biorheology, 2000. **37**(1, 2): pp. 129–140.
72. Redman, S., S. Oldfield, and C. Archer, *Current strategies for articular cartilage repair.* European Cells & Materials, 2005. **9**(23–32): pp. 23–32.
73. Shapiro, F., S. Koide, and M.J. Glimcher, *Cell origin and differentiation in the repair of full-thickness defects of articular cartilage.* JBJS, 1993. **75**(4): pp. 532–553.
74. Jackson, D.W., et al., *Spontaneous repair of full-thickness defects of articular cartilage in a goat model: a preliminary study.* JBJS, 2001. **83**(1): pp. 53–53.
75. Hunziker, E.B., *The elusive path to cartilage regeneration.* Advanced Materials, 2009. **21**(32-33): pp. 3419–3424.
76. Guilak, F., *Biomechanical factors in osteoarthritis.* Best Practice & Research Clinical Rheumatology, 2011. **25**(6): pp. 815–823.
77. Issa, R. and T. Griffin, *Pathobiology of obesity and osteoarthritis: integrating biomechanics and inflammation.*

Pathobiology of Aging & Age-related Diseases, 2012. **2**(1): p. 17470.
78. Erickson, G., D. Northrup, and F. Guilak, *Hypo-osmotic stress induces calcium-dependent actin reorganization in articular chondrocytes.* Osteoarthritis and Cartilage, 2003. **11**(3): pp. 187–197.
79. Clark, A.L., et al., *Chondroprotective role of the osmotically sensitive ion channel transient receptor potential vanilloid 4: Age-and sex-dependent progression of osteoarthritis in Trpv4-deficient mice.* Arthritis & Rheumatism, 2010. **62**(10): pp. 2973–2983.
80. Phan, M.N., et al., *Functional characterization of TRPV4 as an osmotically sensitive ion channel in porcine articular chondrocytes.* Arthritis & Rheumatism: Official Journal of the American College of Rheumatology, 2009. **60**(10): pp. 3028–3037.
81. Martin, J.A. and J.A. Buckwalter, *Aging, articular cartilage chondrocyte senescence and osteoarthritis.* Biogerontology, 2002. **3**(5): pp. 257–264.
82. Horton, W., P. Bennion, and L. Yang, *Cellular, molecular, and matrix changes in cartilage during aging and osteoarthritis.* Journal of Musculoskeletal and Neuronal Interactions, 2006. **6**(4): p. 379.
83. LOESER, R.F., *The role of aging in the development of osteoarthritis.* Transactions of the American Clinical and Climatological Association, 2017. **128**: p. 44.
84. Finkel, T. and N.J. Holbrook, *Oxidants, oxidative stress and the biology of ageing.* Nature, 2000. **408**(6809): p. 239.
85. Wijeratne, S.S., S.L. Cuppett, and V. Schlegel, *Hydrogen peroxide induced oxidative stress damage and antioxidant enzyme response in Caco-2 human colon cells.* Journal of Agricultural and Food Chemistry, 2005. **53**(22): pp. 8768–8774.
86. Henrotin, Y., B. Kurz, and T. Aigner, *Oxygen and reactive oxygen species in cartilage degradation: friends or foes?* Osteoarthritis and Cartilage, 2005. **13**(8): pp. 643–654.
87. Rothenberg, A. and J. Elisseeff, *Bone and Cartilage*, in *Tissue Engineering for the Hand: Research Advances and Clinical Applications*. 2010, World Scientific Publishing Co. pp. 219–242.
88. Bonnet, C. and D. Walsh, *Osteoarthritis, angiogenesis and inflammation.* Rheumatology, 2005. **44**(1): pp. 7–16.
89. Mackie, E., et al., *Endochondral ossification: how cartilage is converted into bone in the developing skeleton.* The international journal of biochemistry & cell biology, 2008. **40**(1): pp. 46–62.
90. Gomez-Aristizabal, A., et al., *Stage-specific differences in secretory profile of mesenchymal stromal cells (MSCs) subjected to early-vs late-stage OA synovial fluid.* Osteoarthritis and Cartilage, 2017. **25**(5): pp. 737–741.
91. Kapoor, M., et al., *Role of proinflammatory cytokines in the pathophysiology of osteoarthritis.* Nature Reviews Rheumatology, 2011. **7**(1): p. 33.
92. Goldring, S.R. and M.B. Goldring, *The role of cytokines in cartilage matrix degeneration in osteoarthritis.* Clinical Orthopaedics and Related Research (1976–2007), 2004. **427**: pp. S27–S36.
93. Sandy, J.D., et al., *Human genome-wide expression analysis reorients the study of inflammatory mediators and biomechanics in osteoarthritis.* Osteoarthritis and Cartilage, 2015. **23**(11): pp. 1939–1945.
94. Stevens, A.L., et al., *Mechanical injury and cytokines cause loss of cartilage integrity and upregulate proteins associated with catabolism, immunity, inflammation, and repair.* Molecular

& Cellular Proteomics, 2009. **8**(7): pp. 1475–1489.
95. Goldring, M.B. and M. Otero, *Inflammation in osteoarthritis.* Current Opinion in Rheumatology, 2011. **23**(5): p. 471.
96. Laurie, L.E., et al., *The transcription factor Hand1 is involved in Runx2-Ihh-regulated endochondral ossification.* PLoS One, 2016. **11**(2): p. e0150263.
97. Rosen, C.J., J.E. Compston, and J.B. Lian, *ASBMR primer on the metabolic bone diseases and disorders of mineral metabolism.* 2009, John Wiley & Sons.
98. Husa, M., R. Liu-Bryan, and R. Terkeltaub, *Shifting HIFs in osteoarthritis.* Nature Medicine, 2010. **16**(6): p. 641.
99. Saito, T., et al., *Transcriptional regulation of endochondral ossification by HIF-2α during skeletal growth and osteoarthritis development.* Nature Medicine, 2010. **16**(6): p. 678.
100. Yang, S., et al., *Hypoxia-inducible factor-2α is a catabolic regulator of osteoarthritic cartilage destruction.* Nature Medicine, 2010. **16**(6): p. 687.
101. Jones, E., et al., *Large-scale extraction and characterization of CD271+ multipotential stromal cells from trabecular bone in health and osteoarthritis: Implications for bone regeneration strategies based on uncultured or minimally cultured multipotential stromal cells.* Arthritis & Rheumatism, 2010. **62**(7): pp. 1944–1954.
102. Joos, H., et al., *Interleukin-1 beta and tumor necrosis factor alpha inhibit migration activity of chondrogenic progenitor cells from non-fibrillated osteoarthritic cartilage.* Arthritis Research & Therapy, 2013. **15**(5): p. R119.
103. Barreto, A. and T.R. Braun, *A new treatment for knee osteoarthritis: Clinical evidence for the efficacy of Arthrokinex™ autologous conditioned serum.* Journal of Orthopaedics, 2017. **14**(1): pp. 4–9.
104. Miyaki, S., et al., *MicroRNA-140 is expressed in differentiated human articular chondrocytes and modulates interleukin-1 responses.* Arthritis & Rheumatism, 2009. **60**(9): pp. 2723–2730.
105. Vonk, L.A., et al., *Overexpression of hsa-miR-148a promotes cartilage production and inhibits cartilage degradation by osteoarthritic chondrocytes.* Osteoarthritis and Cartilage, 2014. **22**(1): pp. 145–153.
106. Tornero-Esteban, P., et al., *Study of the role of miRNA in mesenchymal stem cells isolated from osteoarthritis patients.* Revista Española de Cirugía Ortopédica y Traumatología (English Edition), 2014. **58**(3): pp. 138–143.
107. Murphy, J.M., et al., *Reduced chondrogenic and adipogenic activity of mesenchymal stem cells from patients with advanced osteoarthritis.* Arthritis & Rheumatism, 2002. **46**(3): pp. 704–713.
108. Sekiya, I., et al., *Human mesenchymal stem cells in synovial fluid increase in the knee with degenerated cartilage and osteoarthritis.* Journal of Orthopaedic Research, 2012. **30**(6): pp. 943–949.
109. Sakaguchi, Y., et al., *Comparison of human stem cells derived from various mesenchymal tissues: superiority of synovium as a cell source.* Arthritis & Rheumatism, 2005. **52**(8): pp. 2521–2529.
110. Volarevic, V., et al., *Ethical and safety issues of stem cell-based therapy.* International Journal of Medical Sciences, 2018. **15**(1): p. 36.
111. Chen, G., et al., *Monitoring the biology stability of human umbilical cord-derived mesenchymal stem cells during long-term culture in serum-free medium.* Cell and Tissue Banking, 2014. **15**(4): pp. 513–521.

112. Squillaro, T., G. Peluso, and U. Galderisi, *Clinical trials with mesenchymal stem cells: an update.* Cell Transplantation, 2016. **25**(5): pp. 829–848.
113. Uccelli, A., L. Moretta, and V. Pistoia, *Mesenchymal stem cells in health and disease.* Nature Reviews Immunology, 2008. **8**(9): p. 726.
114. Xia, Z., et al., *Altered function in cartilage derived mesenchymal stem cell leads to OA-related cartilage erosion.* American Journal of Translational Research, 2016. **8**(2): p. 433.
115. Jayasuriya, C.T., et al., *Molecular characterization of mesenchymal stem cells in human osteoarthritis cartilage reveals contribution to the OA phenotype.* Scientific Reports, 2018. **8**(1): p. 7044.
116. Akkiraju, H. and A. Nohe, *Role of chondrocytes in cartilage formation, progression of osteoarthritis and cartilage regeneration.* Journal of Developmental Biology, 2015. **3**(4): pp. 177–192.
117. Grogan, S.P., et al., *Mesenchymal progenitor cell markers in human articular cartilage: normal distribution and changes in osteoarthritis.* Arthritis Research & Therapy, 2009. **11**(3): p. R85.
118. Hattori, S., C. Oxford, and A.H. Reddi, *Identification of superficial zone articular chondrocyte stem/progenitor cells.* Biochemical and Biophysical Research Communications, 2007. **358**(1): pp. 99–103.
119. Yoshida, C.A., et al., *Runx2 and Runx3 are essential for chondrocyte maturation, and Runx2 regulates limb growth through induction of Indian hedgehog.* Genes & Development, 2004. **18**(8): pp. 952–963.
120. Van der Kraan, P. and W. Van den Berg, *Chondrocyte hypertrophy and osteoarthritis: role in initiation and progression of cartilage degeneration?* Osteoarthritis and Cartilage, 2012. **20**(3): pp. 223–232.
121. Giffin, J.L., D. Gaitor, and T.A. Franz-Odendaal, *The forgotten skeletogenic condensations: a comparison of early skeletal development amongst vertebrates.* Journal of Developmental Biology, 2019. **7**(1): p. 4.
122. ten Berge, D., et al., *Prx1 and Prx2 in skeletogenesis: roles in the craniofacial region, inner ear and limbs.* Development, 1998. **125**(19): pp. 3831–3842.
123. Pritzker, K.P., et al., *Osteoarthritis cartilage histopathology: grading and staging.* Osteoarthritis and Cartilage, 2006. **14**(1): pp. 13–29.
124. Olsen, B.R., A.M. Reginato, and W. Wang, *Bone development.* Annual Review of Cell and Developmental Biology, 2000. **16**(1): pp. 191–220.
125. Siggelkow, H., *Does osteoblast to adipocyte differentiation play a role in osteoarthritis?* Current Rheumatology Reviews, 2008. **4**(3): pp. 202–205.
126. Madsen, O., et al., *Body composition and muscle strength in women scheduled for a knee or hip replacement. A comparative study of two groups of osteoarthritic women.* Clinical Rheumatology, 1997. **16**(1): pp. 39–44.
127. Campbell, T.M., et al., *Mesenchymal stem cell alterations in bone marrow lesions in patients with hip osteoarthritis.* Arthritis & Rheumatology, 2016. **68**(7): pp. 1648–1659.
128. Farmer, S., *Regulation of PPARγ activity during adipogenesis.* International Journal of Obesity, 2005. **29**(S1): p. S13.
129. Rastegar, F., et al., *Mesenchymal stem cells: molecular characteristics and clinical applications.* World Journal of Stem Cells, 2010. **2**(4): p. 67.

CHAPTER 15

Bioinspired Nanofibers: State of Art in Tissue Regeneration

POORNIMA BALAN, JANANI INDRAKUMAR, ANBUTHIRUSELVAN SOLAIMUTHU, PADMAJA MURALI, AISHWARYA SATISH, and PURNA SAI KORRAPATI,*

Biological Materials Laboratory, CSIR-Central Leather Research Institute, Adyar, Chennai, Tamil Nadu 600020, India

**Corresponding author.*
E-mail: purnasaik.clri@gmail.com, purnasai@clri.res.in

ABSTRACT

Engineering new products for tissue regeneration received a profound growth in the recent years due to the wide availability of the synthesized molecules and polymers. Polymeric nanofibers offer excellent interaction of the material with the cells, chemical versatility, and mechanical strength making them ideal molecules for tissue engineering. Integrating polymers along with bioactives yielded quantitative and qualitative treatment for various ailments. The electrospun nanofibers have been intensively explored due to their architecture that mimics native extracellular matrix. Surface modification of the nanofibers, incorporation of nanoparticles for timely delivery of drugs, delayed decomposition of the nanofibers, and biocompatibility rendered fabrication of polymeric nanofibers an efficient method for the delivery of drugs for tissue engineering. Biomaterial-based scaffolds serve as a critical platform to overcome the deleterious sequelae in various organs. This chapter seeks the conceptual understanding of synthetic and natural polymeric blends and their parameters for differentiation and regeneration for specific tissue. We then highlight current developments that utilize electrospun nanofibers to manipulate biological processes significant to skin, nerve, bone, and cardiac tissue regeneration. This provides a valuable insight on favorable approaches and the potential applications of the biomaterial scaffolds for tissue regeneration.

15.1 INTRODUCTION

Tissue repair involves highly synchronized signaling mechanism between the external environment and internal cell milieu. The biomaterial scaffolds designed are aimed at initiating cell signaling molecules required for the repair and hence mimic the cellular environment [1]. Despite the availability of several commercial substitutes, the problems of high cost, and subnormal microstructure, inconsistent engraftment has coerced the continuous evolution of tissue-engineered substitutes for time-dependent release of signaling biomolecules, differentiation factors and protein domains facilitating cell migration, adhesion, proliferation, and differentiation [2, 3]. Therefore, there is a dire demand for a robust, more conducive combinatorial therapeutic regimen to tackle the dysfunctional process at multiple cellular levels. Scientists have drawn inspiration from various natural and complex chemical composition and physical properties of native tissue architecture in various animals that led to a paradigmatic emergence of nanoscience for biomimetic design [4]. These findings accentuate the need for development and implementation of novel nanotechnology-driven therapies.

The advent of newer technologies such as microfabrication, nanotechnology, and three-dimensional (3D) printing have opened newer and complex therapeutic approaches for structural and functional integrity of the damaged tissues and provides scope for major breakthroughs in the field of medicine [5].

Nanotherapeutic approaches involve material fabrication in a nanoscale dimension ranging between 1 and 100 nm, and thus possess the advantages of versatility and tunability of the physicochemical properties (e.g., hydrophobicity, charge, size). Electrospinning is one such promising and popular fabrication method applied for the development of functionalized nanomaterials with a potential for a wide range of applications ranging from tissue engineering to biosensors [6]. The idea of nanofibers actually arose when researchers got inspired by the thin slender webs spun by spiders and the cotton candy which arises with molten sugar. The similar principle was used in nanofibers which used hydrostatic and centrifugal force to spin thin interconnected dry fibers made up of synthetic and natural polymers. Later, step-by-step nanofibers were developed for different applications based on various natural sources. The matrix on the surface of Hercules beetle can reversibly change its color from green to black in response to humidity changes in the environment. This structural organization of fibers was exploited to produce nanofibers that can sense humidity in the environment [7]. Similarly, many plant systems like pine cone have a specific structural arrangement that plays a major role in water retention and absorption [8]. However, the choice of the polymers for the fabrication of electrospun nanofibers ranges from a variety of polymers, which may include synthetic organic polymers, biopolymers, and blends of organic and biopolymers so as to obtain the desirable and essential applicative properties. In medicine, tissue regeneration and repair of a damaged tissue is considered as a branch of translational research in tissue engineering and molecular biology. This essentially deals with either stimulating body's own repair mechanisms or implants that can replace the lost or damaged tissues. The tissue engineering for regeneration is again highly complex owing to the differences in the structure and function of different tissues. Each tissue is characterized by a specific developmental origin, genotype, and phenotype. The repair and regeneration properties are also highly tissue specific and thereby a thorough understanding of the anatomy and physiology alone would pave a way for designing the nanofiber-based therapeutic regime. The nanofibers that are developed for regeneration should therefore be drafted carefully according to the therapeutic need [9]. The fabrication of nanofibers requires a system that mimics the natural tissue morphology for effective regeneration. The methodology adopted for one tissue may not be ideal for other tissues, and hence a thorough understanding of the repair mechanisms alone would facilitate the success of the regenerative medicine [10, 11]. Furthermore, smart functionalization of materials involving targeted site specificity, drug delivery, cellular hosting, and interactivity simulating the biomorphic organization would pave a way for transduction pathways and hence exhibit enhanced functional efficiency to overcome the lacunae in the therapeutic regimen. The previous chapter deals with the methods and choices of smart nanofibers and their complete characterization to understand the suitability for functional requirements.

This chapter therefore essentially deals with the review of the essential structure and physiology of the specific tissue followed by the ideal scaffold requisites for regeneration of different tissues especially with reference to skin, nerve, bone, and cardiac cartilage.

15.2 SKIN ANATOMY

The skin is the largest organ system of the body, and it plays fundamental role in protecting the body against bacterial infection, mechanical forces, fluid imbalance, and thermal dysregulation. Skin (Figure 15.1) is composed of three layers: epidermis, dermis, and hypodermis, including the hair follicles and sebaceous glands [12]. As a protective layer in the body, skin is constantly being exposed to potential injury from the external environmental factors, and thus wound healing is an essential process for survival in all higher organisms. A wound is defined as an injury to the cellular and anatomical structure of the tissue. The wound healing process ensures the complete reestablishment of the lost tissue integrity and functional restoration of the damaged tissues. Although the process of wound is more or less similar in all animals, the mechanism and capacity of complete restoration varies dramatically between various species. However, some including lower vertebrates such as fish (Zebrafish) and amphibians retain the ability to perfectly regenerate the skin [13]. It is known that after full-thickness excisional wounds in *Xenopus* froglets and axolotols, the entire skin, including secretory appendages, regenerates [13]. Throughout this process, even the pigmentation pattern of the skin is found to be fully re-established. In contrast, the epidermis is regenerated alone in adult human skin and the dermis is healed by repair mechanism leading to formation of scarring.

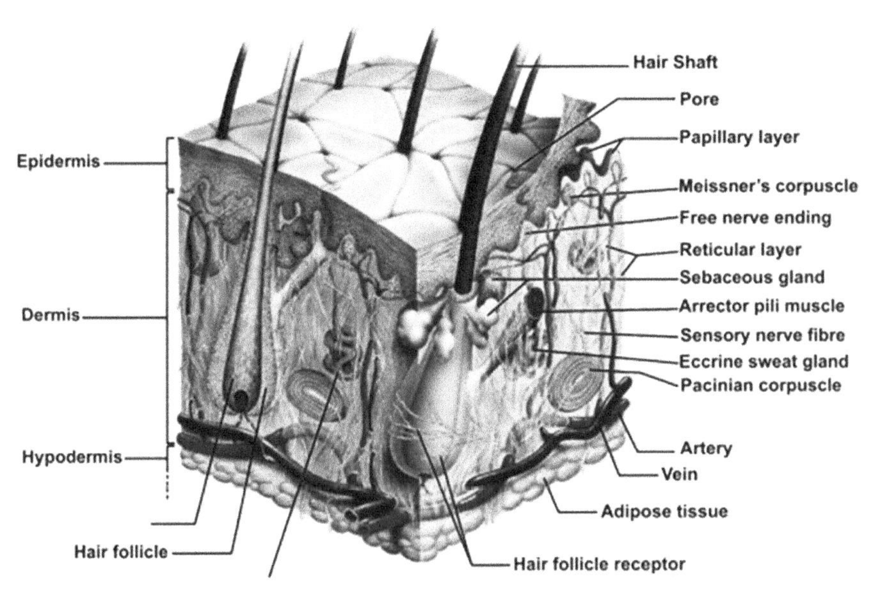

FIGURE 15.1 Schematic representation of the skin layers and appendage.
Source: Reprinted from with permission from Ref. [14]. © 2012 Elsevier.

15.2.1 WOUND HEALING STAGES

Wound healing is a cascade of complex biological processes that are characteristically described as the succession of four overlapping phases: hemostasis, inflammation, proliferation, and tissue remodeling [15] (Figure 15.2). All these processes require the coordinated effect of cells like fibroblasts, keratinocytes, endothelial cells, platelets, and immune cells to initiate and culminate the tissue restoration process. In the initial hemostasis phase, platelet activation occurs leading to clot formation resulting in cytokine secretion such as platelet-derived growth factor (PDGF), transforming growth factor-β (TGF-β), and vascular endothelial growth factor (VEGF) [16]. The onset of inflammation is marked by the infiltration of immune cells. Neutrophil is the first cell to invade the injured site. Its function is to cleanse the wound site and to remove the necrotic tissue, nonfunctional host cells through a process called wound debridement. The inflammatory phase deals with cleansing the wound from foreign cells and molecules before rebuilding the skin [17, 18]. The proliferative phase aids in repairing the wound site by restoring the damaged dermal and epidermal layer. The secreted growth factors (GFs) help in migration of fibroblasts and keratinocytes that are crucial for wound healing. The last stage of wound healing is remodeling phase that is essential for functional restoration and normal appearance of the healed tissue. The

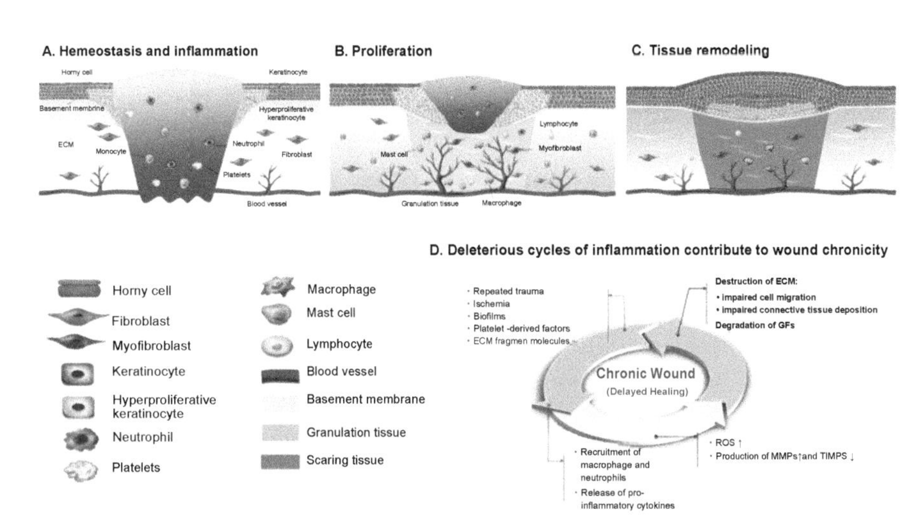

FIGURE 15.2 Different phases of acute wound healing

Source: Reprinted from with permission from Ref. [20]. © 2018 Elsevier.

maturation phase helps in tentatively regularize epidermal thickness, composition of extracellular matrix (ECM), cellular content, and blood vessel count as close as possible to unwounded skin [19].

There are generally two major classes of wounds: chronic and acute. Acute wounds are mostly superficial that involves damages to both the epidermis and superficial dermis or full thickness in which the underlying subcutaneous layer is compromised. Common examples of acute wounds include abrasions, thermal wounds, and surgical incisions, which are often associated with complication of infection. Generally, acute wound is a well-orchestrated process that follows the normal healing stages and heals within 3 weeks of time. However, chronic wound does not follow the regular order of healing and are prolonged up to 3 months from the time of injury. Often, chronic wound results in prolonged inflammatory phase with excessive secretion of proinflammatory cytokines, reactive oxygen species, and proteases. Chronic wounds include diabetic foot ulcer, pressure ulcer, venous ulcers, arterial ulcers, and ulcers that arise due to complications of other diseases (result of neurodegenerative processes like Pick's disease) [21]. Many of the issues related to chronic wound healing focus on the deleterious effects of various disease processes on the mechanisms of biochemical signaling, ECM deposition, and cell migration.

15.2.2 CURRENT APPROACHES FOR WOUND MANAGEMENT

The current standard of care for wound consists of cleaning, dressing, and in some cases debridement of the wound. The ultimate aim of wound treatment is to achieve accelerated wound closure with restored native skin function that improves the aesthetic appearance of the wound. The existing wound healing approaches are based on autografts, allografts, and cultured epithelial autografts, and wound dressings based on biocompatible and biodegradable polymers [22]. Despite the abundance of allografts and xenografts, their application is limited due to the risk of complications, including pain, graft rejection, and disease transmission. To circumvent the immune rejection, allografts, cultured autologous keratinocytes, and split thickness skin grafts have been utilized as an alternative [23, 24]. Hitherto, their application results in poor skin graft matching, wound contraction, and defects in skin tissue remodeling. Therefore, there is an existing clinical urge for tissue engineering approaches to develop an alternative therapeutic modality.

Materials developed for therapeutic applications in skin wound healing includes hydrogels, decellularized porcine dermal matrix, and freeze-dried or gas-foaming formed scaffold. These materials, however, lack capability to recapitulate the architecture of ECM of skin. Recently, electrospinning, an enabling nanotechnology, has attracted a lot of attention in wound healing, because this technology can produce biomimetic nanofibrous materials from a wide variety of natural and synthetic polymers with biologically relevant features.

15.2.2.1 EXTRACELLULAR MATRIX

ECM is a 3D network of macromolecules that provides structural and biochemical support to the surrounding cells. It is composed of polysaccharides and natural polymers such as collagen, elastin, and fibrinogen. Cell–ECM interaction plays a major role during the process of wound healing, formation of clot, granulation tissue, and tissue remodeling. Other functions including cell migration, adhesion, growth, differentiation, and apoptosis are all controlled by the interaction between the cell and ECM through transmission of signals [25]. The multifarious communication between the cell and ECM takes place through membrane-bound integrin and nonintegrin receptors, which is a complex and dynamic process, playing critical role during development, wound healing, and environmental maintenance [26, 27].

15.2.3 ELECTROSPUN NANOFIBERS FOR SKIN TISSUE REGENERATION

The fabricated tissue-engineered scaffold in the nanoscale range could offer a safe and promising alternative to skin wound management. Electrospun nanofibers have been extensively investigated for applications in skin tissue engineering. Nanofibers offer several advantages as skin substitutes. They mimic the native ECM of skin, protect wound from fluid loss, enable exudate removal from the wound site, allow oxygen permeation, inhibit exogenous microorganism invasion, and improve the aesthetic appearance of the healed tissue. A broad range of polymers have been electrospun and evaluated for their ability to support skin regeneration [28, 29]. Nanofibrous scaffolds made of natural polymers not only mimic the native ECM but also have the architectural resemblances. Natural polymers have been widely used in biomedical applications due to their biocompatibility, biodegradability, biological characteristics, and similarity to biological macromolecules. However, due to their poor mechanical stability and easy degradability they are often blended with synthetic polymers, which have higher mechanical strength and lower degradation rate.

Collagen (Col) is the main structural protein in the ECM constituted in skin and other connective tissues. It is present in different forms in different tissues and plays a pivotal role in maintaining the biological and structural integrity of the ECM architecture. Apart from its widespread availability and function, it has numerous properties such as high abundancy, low antigenicity, low inflammatory and cytotoxic responses, high water affinity, good cell compatibility, and biodegradability that makes it an attractive biomaterial for tissue engineering applications. Studies have demonstrated the efficacy of collagen nanofibers in normal human keratinocyte adhesion, proliferation, and early-stage wound healing. Powell et al. compared the wound healing potential of freeze-dried and electrospun collagen nanofibers and found that both scaffolds supported the formation of keratinized epidermal layer and stratified dermal layer in vitro when cultured with human epidermal keratinocytes and dermal fibroblasts at the air–liquid interface. A layer of the basement membrane was detected at the dermal–epidermal junction. Grafting of these cultured skin substitutes in vivo showed that electrospun collagen nanofibers accelerated wound contraction rate. Collagen nanofibers are cross-linked to reduce its water solubility, to improve the resistance to enzymatic degradation, and to enhance the mechanical strength. Cross-linking can be done using chemical (e.g. glutaraldehyde, genipin, carbodiimides), enzymatic (e.g. transglutaminase, tyrosinase, laccase) or physical method (e.g. UV radiation, gamma radiation, dehydrothermal treatment). Bonvallet et al. [30] had fabricated Col/PCL (polycaprolactone) nanofibers with different ratios and mechanically introduced pores into the fibers that enabled fibroblast infiltration thereby filled the pores with fibroblast-secreted ECM molecules, including fibronectin and Col I. The scaffold served as a substrate for keratinocyte proliferation and stratification. When implanted into full-thickness skin wounds, the microporous scaffolds promote faster and better healing than the scaffolds lacking micropores.

Gelatin is obtained through partial hydrolysis of collagen in which the triple-helical structure of collagen is turned into single-strand molecules. Electrospun gelatin/chitosan nanofiber mat was fabricated with high porosity and with tensile strength of 1.1 MPa. The fabricated scaffold displayed better compatibility with human dermal fibroblast (HDF), and scanning electron microscope (SEM) analysis showed well spread and stretched spindle-shaped HDF cells upon growing on highly porous gelatin/chitosan nanofibrous scaffold. The scaffold displayed intact fiber morphology when maintained in culture medium for 7 days [31]. Sardou et al. [32] electrospun a core–shell nanofibrous scaffold using PCL/gelatin. Lawsone (2-hydroxy-1,4-naphthoquinone) at different concentrations ((0.5%, 1%, 1.5%) was used to encapsulate into the nanofibrous scaffold and was evaluated for its potential in skin regenerative application. Core–shell structure was confirmed through transmission electron microscopy (TEM) analysis and release studies showed prolonged release of the drug up to 20 days. In vitro bioactivity of fibers on human gingiva fibroblast cells was evaluated by the 3-(4,5-dimethylthiazol-2-yl)-2,5-diphenyltetrazolium bromide (MTT) assay. TGF β-1 and COL1 gene expression were found to be upregulated on treatment with the scaffold. Furthermore, the scaffold promoted faster re-epithelialization on rat excision wound studies.

Other natural polymers including silk, fibrinogen, chitosan, and elastin have been electrospun either as such or blended with synthetic polymers to improve its mechanical and physiochemical properties for application in tissue engineering approaches. Due to antimicrobial property of the chitosan and biodegradability, it is blended with synthetic polymer such as PCL to fabricate core–shell structure nanofibers for delivering dual bioactive agents. The scaffold exhibited biocompatibility with keratinocyte cells and blood cells. In vitro migration assay exhibited faster keratinocytes migration, thus explicating its re-epithelialization capacity. Rat excision wound model results confirmed the efficiency of the scaffold in regeneration of injured tissues within 15 days (Figure 15.3) compared with the control rat groups [33].

Synthetic biopolymers pose advantage over natural polymers for being able to be tailored to specific functions and properties by tuning the chemical composition, crystallinity, molecular weight, and copolymerization. FDA-approved polymers such as polylactic acid (PLA), polyglycolic acid (PGA), PCL, and their copolymers have been studied extensively for tissue engineering applications. Hitherto, there is a lacuna in cell recognition signals that can be overcome by amalgamating with natural polymers. Shoba et al. [143] fabricated a PCL/PVA (polyvinyl alcohol) gelatin (GE) core–shell nanofiber composite encapsulated with enzyme bromelain and salvianolic acid. In the study, PCL was used as core and PVA GE was

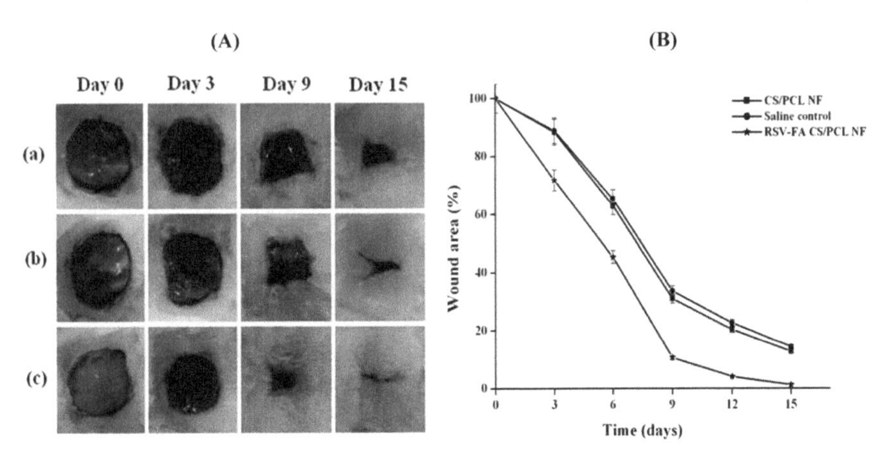

FIGURE 15.3 In vivo wound healing evaluation. (A) Representative photographic images of full thickness excision wounds treated with (a) saline, (b) chitosan (CS)/PCL nanofiber (NF), and (c) resveratrol-ferulic acid (RSV-FA) CS/PCL NF scaffolds at day 0, 3, 9, 15. (B) Graphical representation of quantitative wound closure percentage.

Source: Reprinted from with permission from Ref. [33]. © 2017 Elsevier.

used for shell. The blend of natural and synthetic polymers improved keratinocyte and endothelial cell adhesion and proliferation. Neovascularization and vessel density were observed in the chick chorioallantoic membrane assay, which attributes to the angiogenic potential of the scaffold. Through in vivo rat studies the potential regenerative ability of the scaffold was confirmed. In yet another study, endogenous molecule such as triiodothyronine hormone plays a critical role during tissue regeneration after injury has been studied for its wound healing potential. The hormone delivering scaffold showed prolonged sustained release improving migration of cells required during re-epithelialization process and lowered the systemic toxicity of the hormone in in vivo studies [34].

Exogenous GF administration for tissue repair and damage has been investigated for skin tissue regeneration. Among the copious types of GFs available, VEGF, PDGF, fibroblast growth factor (FGF), and hepatocyte GF could be used for instigating molecular events during wound healing cascade through activating molecules required for regeneration of damaged tissue [35]. Recently, application of electrospun nanofibers to deliver GFs has gained wider research attention, since GF releasing scaffold has shown to promote cellular and biological activities that are required to facilitate wound healing and tissue regeneration. Epidermal growth factor (EGF) is a potent stimulator of keratinocyte proliferation and migration. It has been reported that EGF can promote cell mitosis and chemotaxis to accelerate formation of granulation tissue and the epidermis. However, sustained and localized administration of EGF is required for its effectiveness due to the short half-life and rapid dissolution. Wang et al. [36] developed a novel biodegradable nanofibrous scaffold by emulsion electrospinning. PCL and hydroxyapatite (HA) were blended in the oil phase to form shell, and EGF was dissolved in deionized water to form the core. Burst release was observed in case of both EGF and HA. In vitro experiments suggested that EGF and HA released from nanofibrous scaffolds promoted cell infiltration, upregulated Col and TGF-b1 gene expression, and increased the ratio of Col III to Col I. Application of the scaffold in in vivo model demonstrated accelerated epidermal regeneration in the early phases of wound healing. Treatment with PCL/HA/EGF scaffold displayed thicker epidermal layer and an organized dermal layer. A novel coaxial electrospinning method was developed to fabricate poly (lactic-co-glycolic acid) (PLGA) nanofibers loaded with basic fibroblast growth factor (bFGF). This novel biofunctionalized construct was capable of sustained release of GFs and also enhanced adhesion and proliferation of bone marrow-derived mesenchymal stem cells (BMMSCs) [37].

Similarly, Norouzi et al. [38] fabricated PLGA with a core–shell structure encapsulating EGF through the emulsion electrospinning technique. The fabricated scaffold showed sustained release of EGF till 1 week and also found to be biocompatible with human fibroblast cells. In yet another approach, emulsion electrospinning was employed to encapsulate bFGF into ultrafine poly (ethylene oxide-co-lactic acid) (PELA) fibers with a novel core–sheath structure to promote skin regeneration. PELA scaffolds loaded with bFGF revealed significantly higher wound-healing rate with complete re-epithelialization and skin appendage regeneration in a diabetic rat model when compared to control [39]. In their study, Xie et al. [40] fabricated dual GF releasing nanofiber system in which PDGF-BB encapsulated PLGA nanoparticles were dispersed in VEGF-loaded chitosan/polyethylene oxide (PEO) nanofiber composite. The release studies showed faster release of VEGF and sustained release of PDGF-BB, which attributes to the promotion of angiogenesis and increased proliferation of HDFs in vitro. Despite the benefits of GF, the prepared scaffold also displayed antibacterial activity due to the presence of chitosan. The sustained release of GF accelerated wound contraction rate and re-epithelialization (Figure 15.4), more vascularization, faster collagen deposition, and earlier remodeling relative to the open wound control and commercial Hydrofera Blue wound dressing.

Currently, BMMSCs have been used as the main donor cells in tissue engineering approaches. Stem cell scaffolds with nanogeometry can be explored as a possibility to develop novel tissue-engineered scaffolds for terminal differentiation of stem cells, which can be used as artificial skin. Several reports have demonstrated that stem cells can

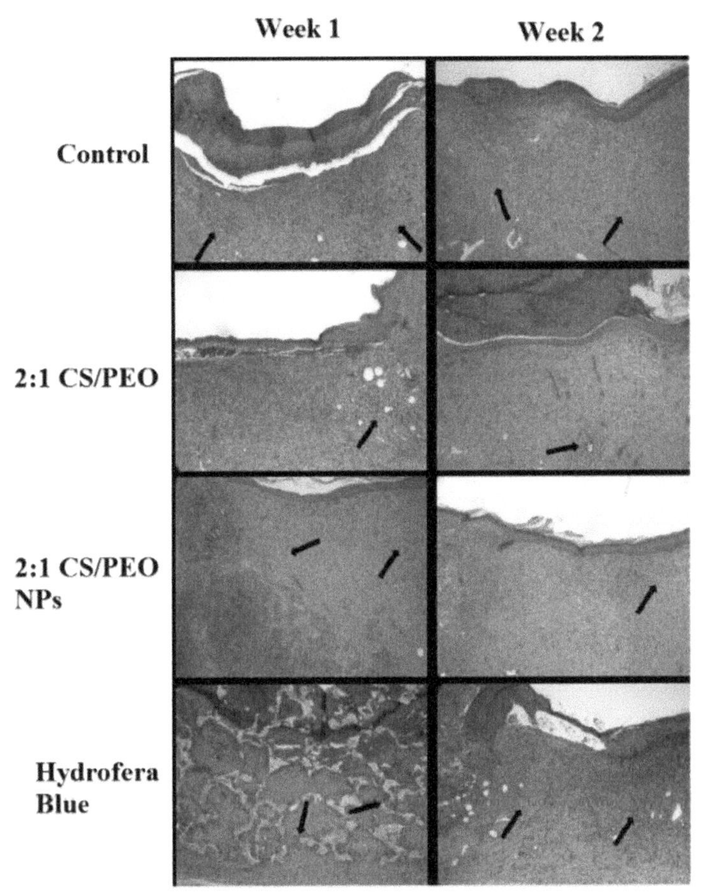

FIGURE 15.4 Histological evaluation of wounds treated by CS/PEO-NP meshes after week 1 and 2. The arrow mark indicates the inflammatory cells.

Source: Reprinted from with permission from Ref. [40]. © 2013 Elsevier.

react to their microenvironment and parade apposite response as well as interaction with nanofibers that can amplify the expression of proteins like vimentin, F-actin, γ-tubulin, and α-tubulin [42]. It is well known that upon skin injury, MSCs migrate from bone marrow to the injured site and contribute to the skin repair and regeneration. Ma et al. [43] fabricated a Col/PLGA-blended nanofibrous scaffold modified by conjugation with CD29 antibody for attachment and growth of MSCs. Combined application of nanofibrous scaffold along with BMMSCs improved full thickness skin wound healing with enhanced collagen deposition and intact epithelium formation. In another study, defined artificial microenvironments were synthesized by electrospinning blends of sericin/hyaluronan/chondroitin sulfate/cationic gelatin to mimic the microstructure and composition of the natural ECM of skin. The scaffold was co-cultured with human mesenchymal stem cells (hMSCs)-keratinocytes and fibroblasts. After 5 days of contact co-culture, results revealed that electrospun scaffold containing sericin promoted epithelial differentiation of hMSCs (Figure 15.5) and expression of protein markers such as keratin 14, ΔNp63α, and pan-cytokeratin [41].

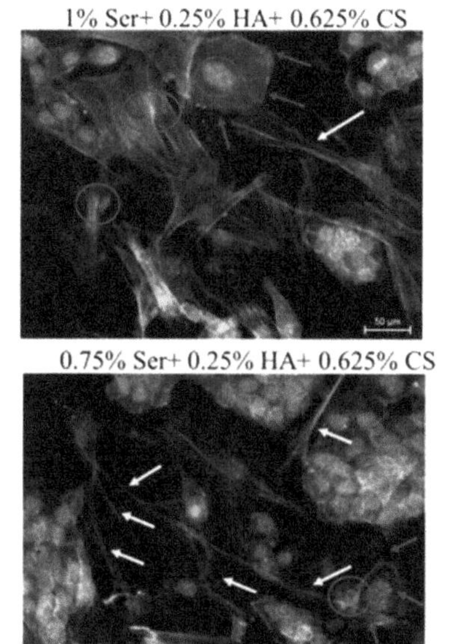

FIGURE 15.5 A human mesenchymal stem cell subpopulation attained keratinocyte like round morphology (red arrow), while other became elongated and took fibroblast like morphology (yellow arrow), independent of scaffold type. Multinucleated fused cells (red circle) were also observed during co-cultivation.

Source: Reprinted from with permission from Ref. [41]. © 2016 Elsevier.

15.3 NERVE TISSUE REGENERATION

When the injury is intense (caused by accidents, transections, lacerations, intense fractures, etc.), the impact could penetrate deeper and disrupt the internal tissues like nerves, muscles, and bones. Nerve injuries require special attention as nerve regeneration is challenging and spontaneous regeneration seldom occurs. Hence, it requires specialized dressings to bridge the gap between nerves (Figure 15.6).

15.3.1 NERVOUS SYSTEM

The nervous system is instrumental in propagating impulses and cellular communication necessary for carrying out the normal day to day activities. Brain and spinal cord make up the central nervous system (CNS) and the nerves that extend outwards from the CNS to the tissues make up the peripheral nervous system (PNS) that enables the conduction of impulses to and from the CNS [44].

Neuron is the structural unit of the nervous system and is responsible for the signal transduction. The structure of a neuron comprises of,

- *Cell body* or *soma* which comprises of all the cell organelles.
- *Dendrites* which are small extensions arising from the cell body and are involved in receiving impulses from other cells and transferring these signals to the cell body; and
- *Axon* is a nerve fiber that is the extension of neuron from the cell body. It is the conducting region responsible for nerve impulse transmission from soma to another nerve or to a muscle cell/effector.

The axons have an additional phospholipid covering called myelin sheath which is produced by Schwann cells in PNS and oligodendrocytes in the CNS. It provides an insulating layer to enhance the rate of signal transduction through

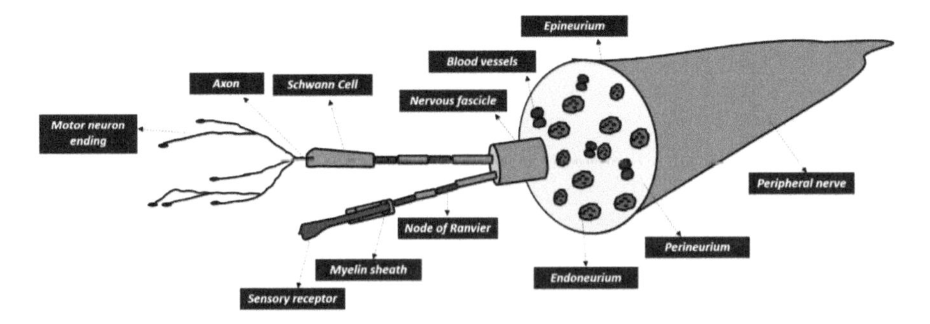

FIGURE 15.6 Pictorial representation of the structure of nerve (Adapted from [45]).

the axons with an intermittent gap called the node of Ranvier. A series of depolarization and repolarization sequence caused by sodium-potassium ion exchanges that lead to generation of impulses in nodes of Ranvier [44]. The axon transmits signal to the neighboring neuron by the release of neurotransmitters at the synapse between the current neuron's axon and the dendrite of the neighbouring neuron or the effector muscle. The signal transmission process is unidirectional [46].

15.3.1.1 NERVE INJURY

Injuries to the peripheral nerves could be caused by mechanical, chemical, thermal, or ischemic damages due to traumatic accidents or degenerative disorders [47]. The transection of nerves results in functional loss of the corresponding effector organ. This is of critical concern as the regeneration of nerve is a grueling task. According to the severity of injury, nerve injuries were classified by Seddon [148] as *Neurapraxia:* the axon does not degenerate but myelin sheath could be affected. In this case, recovery occurs spontaneously and completely and there is mostly no Wallerian degeneration; *Axonotmesis:* axons are involved resulting in loss of axonal continuity; however, the endoneurium and perineurium are intact, thus providing guidance for regeneration and recovery; *Neurotmesis:* the third degree and most severe form of nerve injury according to the Seddon classification, wherein there is a near complete interruption of the nerve and the functional recovery from this kind of injury is rare [48].

Based on the impact of injury Sunderland classified nerve injuries into five degrees [49]. The first and second degrees are similar to Seddon's classification; however, the third-degree injury involves an interruption in the endoneurium and the recovery from this type of injury is variable as the axons may lose directionality to regenerate. The fourth degree of injury involves a complete breach of the nerve bundles; yet, the continuity of the nerve is preserved and may be accompanied by neuroma. Fifth-degree injury results in complete loss of nerve trunk leading to a total paralysis of the severed nerve function. A surgical intervention is mandatory and unavoidable for grade 4 and 5 of nerve injury.

15.3.1.2 NERVE REGENERATION

The regeneration of the nerves ensues based on the severity of the injury as discussed earlier. Regeneration (Figure 15.7) is initiated by a series of complex process that starts by an initial degeneration distally from the site of injury by a process termed as Wallerian degeneration. It involves the degeneration of axons, cytoskeleton, and myelin from the node of Ranvier proximal to the injury site and continues distally throughout the nerve causing denervation. Axonal fragmentation, myelin clearance, and Wallerian degeneration occur much slower in the CNS compared to PNS (7–14 days in PNS, while in CNS it varies from months to years). This difference could be due to the presence of Schwann cells in the PNS, as they respond to nerve injury rapidly, serve to clear up myelin debris by phagocytosis, and also degrade its own myelin. The slow clearance of myelin-associated debris by the oligodendrocytes could attribute to the failure of CNS axons to regenerate [50]. In PNS, Schwann cells clear the debris initially and release chemokines like leukemia inhibitory factor (LIF) and TNFα that recruits the macrophages to continue the clearance by the process of phagocytosis. During this process, neurotrophic chemokines like IL-6, nerve growth factor (NGF), brain-derived neurotrophic factor (BDNF), glial cell-derived neurotrophic factor

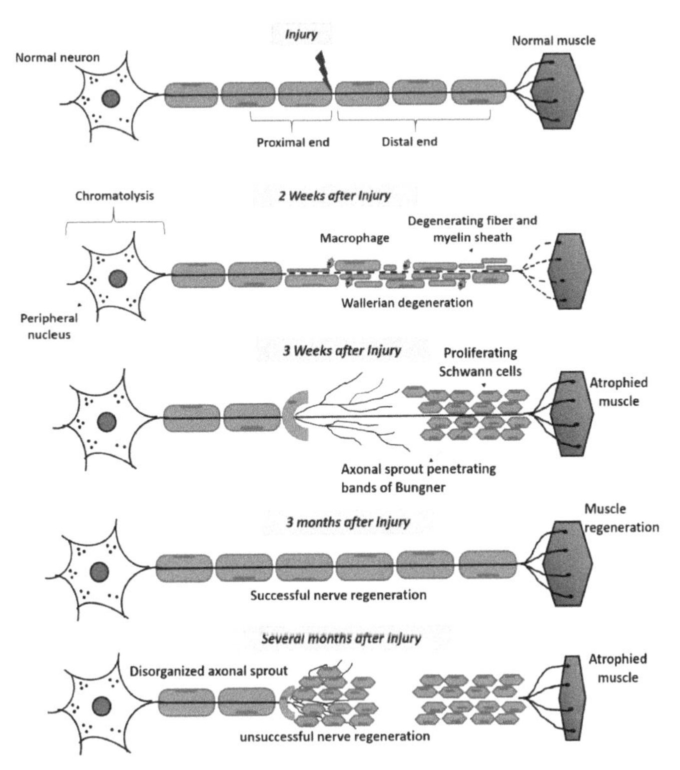

FIGURE 15.7 Nerve degeneration and regeneration responses following nerve injury (Adapted from [52]).

(GDNF), LIF, IGF, and FGF are also released, which instigates cell proliferation. The Schwann cells in contact with the injured axons begin proliferation and form a structure called "bands of Bunger." This forms a pillar for the sprouting axons to regenerate from the proximal end of the damaged nerve up to the distal end, along the bands of Bunger. In addition, Schwann cells also release laminin, an ECM constituent of the basal laminar tube that enhances axonal adhesion and outgrowth. Paradoxically, in the CNS, the oligodendrocytes require the presence of axons and if they are not in contact with the axons, they undergo apoptosis or go into resting phase. Hence, they are inefficient in clearing the myelin debris compared to Schwann cells. Similarly, the CNS microglia involved in clearing debris is slow compared to macrophages in PNS [50, 51].

The regeneration of the nervous system is fraught with difficulties like loss of directionality and uncontrolled branching of the regenerating axons, which leads to loss of nerve function [48]. Also in the case of neurotmesis, when there is a large gap (>2 cm), spontaneous regeneration does not occur and requires surgical intervention [53]. It should also be taken into account that when sufficient time (approximately 8 weeks after injury) has passed without contact of the axons to their target tissue, the chance of functional recovery of the nerve is less. If the time has crossed 6 months to 1 year, the probability of a successful regeneration is very low [54].

15.3.2 THERAPEUTIC INTERVENTION

Hence, a nerve guide is required to patch the nerve gap so as to direct the reinnervation process. Reanastamosis of the damaged nerve is the preferred treatment when the nerve gap is small. In the case of larger nerve gaps, autografts are the gold standard to bridge the gap. However, harvesting autografts causes the creation of a secondary surgery site (leading to functional loss of the donor site) which causes additional trauma to the patient and a chance of infection in the site. Also, the availability of donor nerve of the same diameter as that of the damaged nerve is limited [55]. Thus, to alleviate mismatch in the donor nerve and to decrease patient trauma, a biomaterial could be designed to be used as a nerve conduit [54]. Furthermore, the microenvironment of the injured nerve in CNS does not permit regeneration and this is not attributable to the incongruity of the neuron [56]. Thus, CNS regeneration could be prompted by providing a suitable microenvironment in the form of nerve graft material with bioactives incorporated.

15.3.2.1 CHARACTERISTICS OF NERVE CONDUITS [53, 57, 58]

The conduit should,

- Bridge across the nerve gap.
- Provide an optimal environment for the axons sprouting from the regenerating nerve end.
- Provide topographical and biochemical cues favorable for initiating regeneration.
- Possess biocompatibility and be hollow to allow growth of the regenerating axons.
- Remain intact and possess sufficient mechanical strength to allow axonal growth, withstand the stress of mechanical compression, and thereby evade guide collapse.
- Demonstrate slow and sustained biodegradation with minimal swelling and immune-reactivity.
- Possess permeability to allow exchange of nutrients and oxygen to the microenvironment.

It is imperative to construct a biomaterial that provides a biomimetic environment that is conducive to cell adhesion, proliferation, and differentiation and permits host tissue integration [59]. The niche for neural tissue engineering should be engineered with biochemical cues (like peptides, hormones, GFs, ECM to provide both haptotactic and chemotactic cues), physical cues (topographical features to provide directionality for cell growth and differentiation), and mechanical properties (strength and ductility to serve as a nerve guide material). These features need to be combined to fabricate a biomaterial that is favorable for nerve tissue regeneration.

Some of the nerve graft-based biomaterials fabricated are sheets, hydrogels, silicone tubes, patterned biomaterials prepared by solvent casting, imprinting or photolithography, and nanofibers [52]. The emergence of biomaterials fabricated from a choice of several natural and synthetic polymers has an advantage that it can be tailored to mimic the mechanical strength, topography, and other essential attributes of the nerve microenvironment. As nerve tissue is comprised of highly polarized cells that are oriented in a specific direction to transduce signals, the nanotopography of the supporting matrix is of utmost importance. Electrospun nanofiber matrices have good mechanical properties and act as a promising scaffolding material which could facilitate the release of bioactives in a controlled/sustained fashion [60], in addition to providing a topographically favorable microenvironment for the nerve tissue by engineering them in the form of aligned nanofibers. Aligned nanofibers possess structural resemblance to the microenvironment of nerve tissue and are thus germane to the field of nerve regeneration [61].

15.3.3 ELECTROSPUN NANOFIBERS FOR NERVE REGENERATION

Electrospinning is a versatile technique used for fabrication of scaffolds for application in tissue engineering and regenerative medicine. The morphology of these nanofibers has great similarity to the ECM, which is crucial for cell interaction and integration [62]. An added advantage of these nanofibers is that it can be tailored for the conjugation/encapsulation of various biomolecules and facilitate its release. These biomolecules can be chosen based on the application and can be moieties such as proteins, hormones, drugs, nucleic acids, peptides, and nanoparticles that can initiate and guide cell adhesion, proliferation, and differentiation to a particular cell lineage. Furthermore, the construction of aligned nanofibers through electrospinning process has enhanced the field of neural tissue engineering as the alignment of the matrices plays an essential topographical cue for enhancing directional growth, outgrowth, and neural cell differentiation [60]. Aligned nanofibers have thus gained impetus in the recent years as a material of vibrant potential to initiate nerve regeneration [63, 64]. The presence of a favorable structural support in the form of electrospun fibers on the surface of nerve guidance channel walls would increase the total surface area that is available for cell adhesion and interaction and also initiate contact guidance to cell in growth. Various in vivo tests on rats with a nerve gap have suggested that the structure of nanofibers in itself (both randomly oriented and aligned) has a substantial role for initiation of nerve regeneration due to the increased surface area available for cell attachment [65]. Yu et al. [149] used electrospinning technique to fabricate a collagen/poy(ε-caprolactone) nanofibrous scaffold, which showed compatibility by promoting Schwann cell adhesion, elongation, and proliferation. It was later grafted into an 8-mm sciatic nerve gap in adult rats as the in vivo testing model. These nanofibrous graft conduits were successfully integrated and supported nerve regeneration. In addition, the study deduced that these grafts achieved similar electrophysiological and muscle reinnervation results as compared to autografts. At 4 months postoperation, it was observed that the implanted Col/PCL nerve conduits facilitated more axons regenerating through the conduit lumen with a simultaneous degradation rate matching the nerve regeneration rate. This study confirms the positive influence of providing a nanofibrous environment (fabricated with an appropriate selection of polymers) for nerve regeneration.

Nanofibers for nerve regeneration applications can be fabricated from both natural and synthetic polymers. Synthetic polymers have superior mechanical resistance, good mechanical properties in physiological fluids, limited swelling, and longer degradation time. On the contrary, natural polymers have low mechanical resistance, faster swelling and degradation rates, but possess admirable biomimetic properties. In the case of nerve guide fabrication, it is preferred that the conduit can remain unaltered for a longer period of time in contact with the biological fluids and possess appreciable mechanical strength, as regeneration of a damaged nerve takes a longer time compared to other tissues. Hence, synthetic polymers are preferred for conduit fabrication. However, due to its low biomimetic properties, natural polymers are blended at desired ratios or designed as core–shell matrices to provide the much needed biomimetic properties to ease cell integration [66].

The influence of directionality and orientation provided by the nanofibrous scaffold in nerve guidance and regeneration has been accentuated in many studies by comparing the role of random and aligned nanofibers over neural cell growth and differentiation. The neurite outgrowth extended along the direction of parallel aligned nanofibers, whereas in the randomly oriented nanofibers, the cells were also randomly distributed. Highly aligned nanofibers supported the outgrowth of aligned neural cells compared to intermediate and random alignment, which ascertained the role of topographical cue (in the form of aligned nanofibers) [61, 67, 68]. Cooper et al. [69] fabricated random and aligned chitosan–polycaprolactone (chitosan–PCL) fibrous scaffold and investigated the influence of alignment over nerve cell organization and function. Schwann cells cultured over the aligned chitosan–PCL fibers oriented along the fiber alignment direction and the cells exhibited a bipolar morphology with an enhanced β-tubulin gene expression, whereas those on the randomly oriented fibers and films of the same material depicted a multipolar morphology (Figure 15.8) [69].

Similarly, Liu et al. [151] evaluated the potential of electrospun collagen nanofibers for spinal cord injury treatment. The study deciphered that the aligned fibers resulted in elongated astrocytes and directed the orientation of neurite outgrowth from dorsal root ganglia (DRGs) along the axis of nanofiber orientation. In contrast, the neurites emanated radially over randomly oriented collagen fibers. Through in vivo rat studies, it was observed

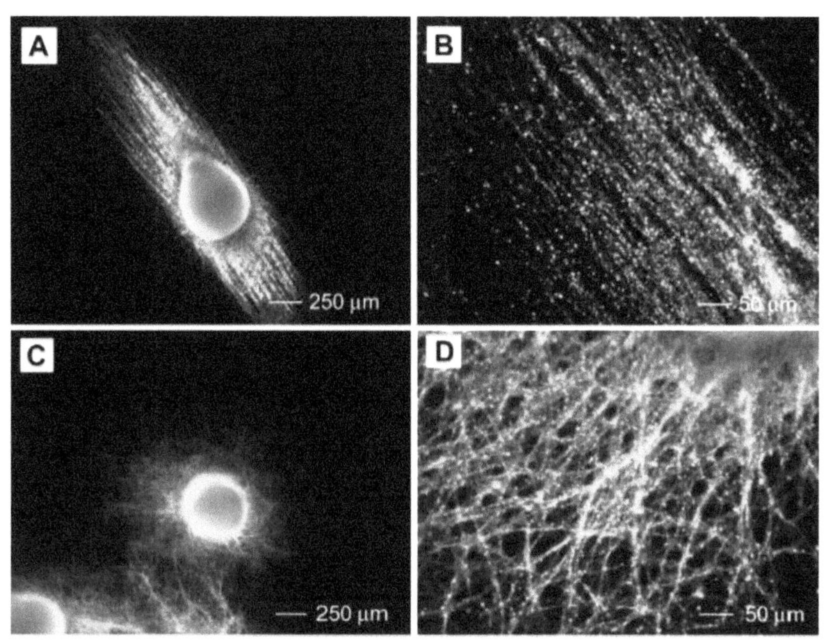

FIGURE 15.8 Immunohistochemistry performed on embryoid bodies after 14 days of culture on aligned PCL nanofibers for mature cell markers, including (A, B) β-tubulin (for neurons), random PCL nanofibers for mature cell markers (C, D) β-tubulin (for neurons).

Source: Reprinted from with permission from Ref. [67]. © 2009 Elsevier.

that postimplantation, extensive cellular penetration into the constructs was observed regardless of nanofiber orientation. Neverthe-less, after a period of 30 days, the scaffolds with aligned fibers had demonstrated better structural intactness, which further supports the potential of aligned nanofibers for neural applications [62]. Xie et al. [70] also observed the highly directional outgrowth of neurites when grown over a parallel array of aligned nanofibers compared to the radial outgrowth over random nanofibers. They also witnessed the expression of both random and aligned outgrowth of neurites from DRG when cultured in the border between regions of aligned and random nanofibers (Figure 15.9). In addition, the influence of different layers of the nanofibers over the orientation of neurite outgrowth was perceived by culturing the cells over a double-layered scaffold where the nanofibers in each layer were aligned along a different direction. This study highlights the influence of nanofibers in different layers of a scaffold over the directionality of the neurites, rather than the topmost layer only. The study by Xie et al. [70] hence divulged the valuable information concerning the apt design of nanofiber scaffolds for neuroregenerative applications, and also the effects of topological cues on neurite outgrowth and axonal regeneration [70].

Apart from the topographical cues provided by the nanofibers, the application of electrically conductive biomaterial scaffolds has great potential in neural tissue regeneration. As the nervous system is highly influenced by the electrical stimuli, the application of electrical stimulation through the conductive polymers, which acts as a scaffold for the growth of neural cells, takes advantage of the electrical properties that are inherent within the cells of the nervous system to enhance its differentiation and axonal outgrowth. Hence, electrical stimulation provides an environment that promotes better nerve regeneration and functional recovery of the nerve. Polymers such as polyaniline (PANI), poly(3,4-ethylenedioxythiophene) (PEDOT), and poly(pyrrole) (PPy) are some of the conducting polymers used for fabricating nanofibers for nerve tissue engineering. The application of an electrical stimulus

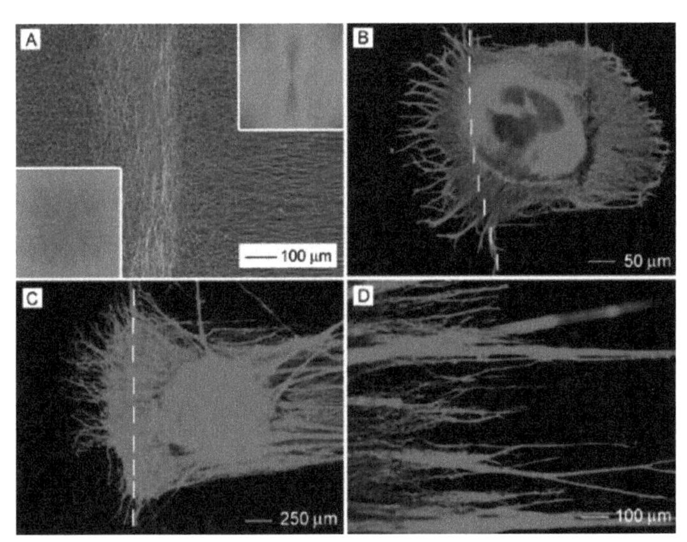

FIGURE 15.9 (A) SEM image of random-to-aligned fiber mat. Inset in pink indicates the FFT patterns that the fibers were aligned on one side and randomly oriented on the other side. (B) Typical morphology of DRG cultured at the border between random and aligned PCL nanofibers (no coating). (C) Typical morphology of DRG cultured at the border between random and aligned PCL fibers coated with laminin. The dashed line indicates the borderline between aligned (right side) and randomly oriented (left side) fibers. (D) An enlarged view of (C) indicating the alignment.

Source: Reprinted from with permission from Ref. [70]. © 2009 Elsevier.

to the conducting polymer creates molecular distortion due to the presence of dopants, thereby enabling the dopants to move throughout the structure and allowing charge to flow through [71].

A conducting mat of aligned nanofibrous matrix fabricated by electrospinning poly-L-lactic acid (PLLA), and further coating it with PPy was attempted. The aligned nanofibers were found to act as topological cue and enabled the directional growth of PC12 cells over the scaffold along the fiber length. In addition, the electrical stimulation of the PC12 cells cultured on aligned fibers along the direction of alignment further improved the outgrowth and extended the distribution of neurites. Hence, the supplementary role played by providing directional electrical stimulation through conducting polymeric nanofibrous scaffold for neural regeneration can be comprehended [72]. In another study, polylysine was coated on the PPy–PLLA fibers and over which NGF was coated by 1-ethyl-3-(3-dimethylaminopropyl) carbodiimide (EDC) chemistry. The presence of NGF stimulated the neurite outgrowth and axonal elongation. The stimulation of the scaffold with 100 mV/cm constant voltage further increased the neurite-bearing cells and neurite length. The synergistic effect of aligned nanofiber topography, conjugated NGF, and electrostimulation mediated through the fabricated nanofibrous scaffold confirms the role of physical, chemical, and topographical cues over nerve tissue engineering. In yet another study, regenerated spider silk protein was blended with the PLLA as the shell, and lysine-doped PPy blended with NGF as the core polymer was used to electrospin aligned nanofibers using a rotating mandrel. The electrospun composite nanofibers were found to have relatively stable conductivity and were effective at bridging a 2-cm sciatic nerve gap in adult rat within 10 months [73]. The role played by chemical cues in addition to the topological cues and electrical stimulation in instigating neural regeneration could be hence appreciated from the study.

Gu et al. [74] had fabricated conductive nanofibers through electrospinning using natural chitin and PANI-blended solutions. The combination of conductive PANI and chitin along with the nanofibrous structure was found to have superior potential for nerve tissue engineering application [74]. Zamani et al. [75] attempted the synthesis of a 3D core–shell nanofibrous matrix of two types: one with a PCL core and PANI-PLGA as sheath, and the other as PCL/PANI blend fibers in core and PLGA/PANI blend electrospun fibers in sheath, respectively. The study involving in vivo experiments concluded that the 3D nanofibers with core–shell structure were effective for nerve tissue regeneration [75]. In a study by Sirivisoot et al. [76], electrically conductive collagen gel scaffolds were prepared by synthesizing PANI and PEDOT nanofibers and adding these fibers to collagen solution. This was performed to incorporate biological groups that are essential for supporting and enhancing cell growth and differentiation. It was deciphered from this study that the presence of PANI and PEDOT in the 3D collagen scaffold promoted neurite outgrowth compared to the collagen only scaffold. Further, the stimulation of the scaffold with electrical signals would result in even higher cell density and differentiation [76].

The impact of biomolecules as the chemical cues in nerve regeneration was ascertained by several studies, wherein one or more biomolecules are released in a sustained/controlled manner or even as staged release of two or more molecules to affect regeneration and differentiation. These biomolecules, though playing a major role in initiating cell adhesion, proliferation, and/or differentiation, are usually chemically unstable and have short half-life in vivo. The required dosage of these biomolecules to affect the necessary biological activity is also usually less (in the range nanograms or even lesser). If provided in excess, it could sometimes lead to toxicity or other undesirable side effects. Hence, the biomaterials and nanomaterials that provide a favorable milieu for tissue regeneration and also possess the propensity to act as a carrier material can be a boon to act as a delivery system for these biomolecules. It can prompt the sustained/controlled release of the biomolecule in the range that can be tailored by the proper choice of the material used for fabricating the biomaterial and also its design [29]. A variety of biomolecules including retinoic acid (RA), BDNF, GDNF, FGF1, triiodothyronine (T_3), NGF, amongst others [62, 77] were encapsulated into the electrospun nanofibers to stage a sustained release or a staged release of one biomolecule initially, followed by the other biomolecule. Low et al. [78] used a copolymer of ε-caprolactone and ethyl ethylene phosphate to form a nanofibrous scaffold, which was encapsulated with RA and BDNF. The nanofibers enabled the sustained release of the biomolecules and it was observed that even at lower concentrations, the nanofibers affected comparative differentiation of mouse neural progenitor cells as that of bolus delivery of a higher concentration of these molecules [78]. Similarly, Satish and Korrapati [79] fabricated random and aligned core–shell nanofibers with PCL as shell and gelatin as core. RA was encapsulated within zein nanoparticles and was further restricted within the core of the nanofibers, while the shell contained T_3. A staged release of T_3 followed by RA was achieved, wherein initially T_3 affected N2A (progenitor neural cells) proliferation and later RA, which along with T_3 bolstered the differentiation to neural specific cell type. This was evidenced by the higher expression of neuron-specific gene and protein markers like MAP2 and β-tubulin and was especially significant in the aligned nanofiber group.

As the nervous system does not support robust regeneration of damaged nerves, the involvement of specific cell adhesion peptides, which possess an inherent ability to enhance the adhesion of neural cells and facilitate reinnervation, is advantageous. The presence of these peptides would improve the biomimetic nature of the scaffold [80, 81]. However, extra precaution should be taken during conjugation procedures to retain the biological activity of the peptide sequence. In a study by Hamsici et al. [82], laminin-derived IKVAV epitope was conjugated over electrospun cyclodextrin nanofibers to fabricate implantable scaffolds for peripheral nerve regeneration. The peptides promote attachment and differentiation of PC-12 cells and the aligned nanofibers provide a favorable scaffolding environment to enhance viability and adhesion of cells. These composite nanofibers supported a higher neurite outgrowth and also a higher expression of mature neuron-specific markers like β III-tubulin and synaptophysin. In another study, surface decoration of peptides derived from laminin (CDPGYIGSR, CQAASIKVAV) using a heterobifunctional cross linker was attempted over aligned nanofibers to manifest a biomaterial that is favorable for neural cell adhesion, proliferation, and differentiation. Aligned nanofibers were fabricated using polyvinyl cinnamate, a photoresponsive polymer over which peptides were surface conjugated by exposure to UV rays. T_3 was incorporated within the nanofibers. It was observed that adhesion and cell–scaffold interaction were greatly enhanced in the group containing the peptide moiety. Gene and protein expression studies illustrated the synergistic efficacy of peptides and triiodothyronine to enable cell differentiation to neural phenotype. The efficacy of the nanofibers over reinnervation was also investigated in zebrafishes with physical posterior lateral line injury. The peptide decorated and T_3 incorporated scaffolds showed a faster twitch response to stimuli compared to other groups indicating the faster functional recovery of the damaged nerve [83].

Cells preseeded into nanofibrous scaffolds have also been used as nerve guide conduits. In a study by Beigi et al. [84], PCL/gelatin nanofibers were fabricated, rolled around a copper wire, and fixed by medical grade adhesive to obtain a tubular-shaped biograft. This was used to bridge a 10-mm sciatic nerve gap in in vivo rat models. In addition, stem cells from human exfoliated deciduous tooth (SHED) were seeded in the nanofibrous nerve guide and transplanted to the site of nerve injury. The vascularization of the injured site and the functional recovery of the nerve were found to be higher in the group treated with SHED nanofibrous nerve guide [84]. Hence, this study accentuates that the nanofibrous structure providing a favorable environment for cell growth, proliferation, and differentiation.

In an attempt to increase the mechanical strength of the nerve guide composed of electrospun nanofibers, Kim et al. [85] fabricated a scaffold using PLGA and PU individually, by a modified electrospinning method using cellophane tapes and copper wires as demonstrated in Figure 15.10. In this method, the inner surface of the nerve guide is fabricated with highly aligned electrospun nanofibers and is thus able to enhance the adhesion and proliferation of neural cells. The central portion of the nerve guide has a double coating of random nanofibers over the aligned, hence providing an additional mechanical strength to the prospective nerve guide material. As the aligned nanofibers (present at the two edges) are transparent and random fibers (as a coating over

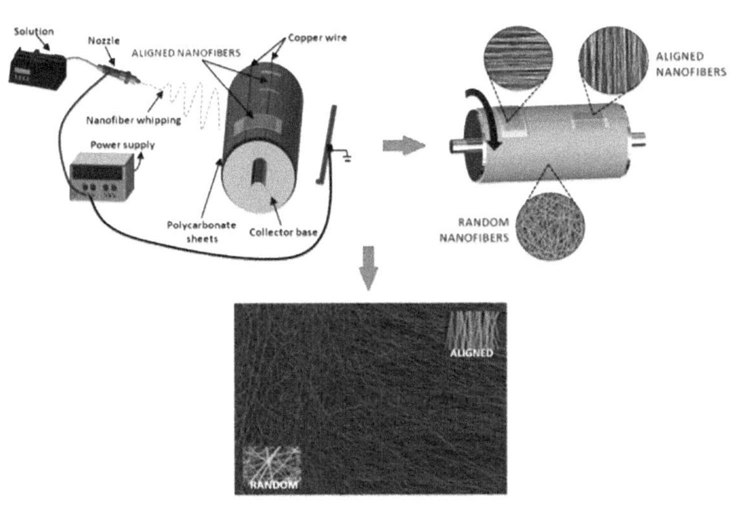

FIGURE 15.10 Fabrication for the controlled design of aligned and randomly oriented nanofibers, a schematic illustration and a SEM image demonstrating border of aligned and random nanofibrous mat. Reprinted with permission from [85] https://creativecommons.org/licenses/by/3.0/).

aligned nanofiber in central portion) are opaque to the naked eye, the grafting of the nerve guides can be performed with ease during suturing.

15.4 BONE TISSUE ENGINEERING

Bones are specialized hard tissues that make up the skeletal system of our body; they support and provide rigidity to human body. This tissue serves as a storage organ for numerous micronutrients and also houses the bone marrow. There are different types of cells that make up the bone; they are osteoblasts, osteocytes, and osteoclasts. It is estimated that a human body contains about 206 that are grouped into different types dependent on their functional structure. Based on the structure bone it is divided into three different parts such as the cortical bone (the outer rigid layer of the bone), bone tissue (inner layer of the bone), and the medullary rays that contain the bone marrow. The bone majorly contains two parts: one is the diaphysis and the epiphysis. The diaphysis is a hollow structure that contains the medullary rays filled with yellow bone marrow internally. The external surface of the diaphysis is made up of an osseous tissue that is made up of collagen and calcium phosphate crystals; the collagen provides a supportive matrix for the calcium crystals to attach and develop into full grown tissue. In contrast, the epiphysis is made up of the spongy bone tissue that contains the red bone marrow.

Bone is a metabolically active tissue involved in mineral homeostasis that regulates the concentration of electrolytes in the blood. It is a distinct tissue that has a superlative property of regeneration without development of a scar. Bones are calcium structures that have a high regenerative capacity, so that it can remodel itself at normal fracture [86]. All these characters put together make it an ultimate smart material that provides structural shape and support to our body. Major changes and alterations in this structure due to injury or disease can amend one's equilibrium and quality of life.

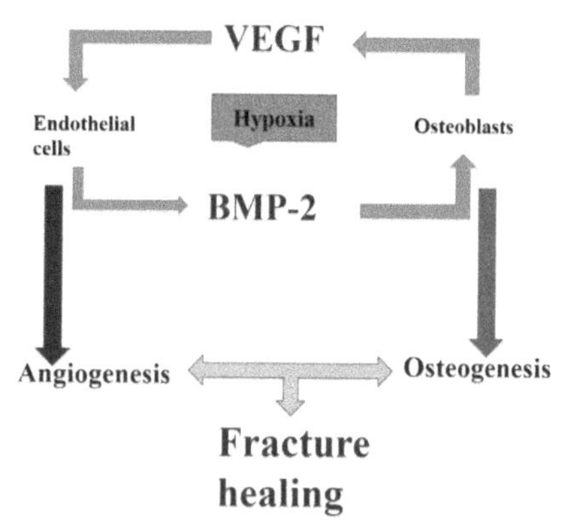

FIGURE 15.11 Steps involved in bone tissue regeneration.

The major events that take place in the verge of bone tissue engineering is the growth of bones (osteogenesis) and the development of blood vessels (angiogenesis) that nourishes the budding osteoblast to mature and develop into a fully formed bone tissue (Figure 15.11). The major proteins that guide osteogenesis and angiogenesis are the BMP-2 (bone morphogenic protein) and the VEGF proteins, respectively.

The major cell types that play a function in bone formation and resorption at the remodeling site are the osteoblasts and osteoclasts which are regulated by variety of hormones and GFs. Any imbalance between the osteoblastic bone formation and osteoclastic resorption would result in bone defects such as osteoporosis which is due to the unrestrained production of regulators [87].

A number of pathological conditions like congenital defects, nonunion fractures, tumor, and traumatic injuries necessitate orchestrated regeneration of bone tissue (Figure 15.12). Autologous bone grafts, allografts, and bone graft substitutes are the universal approaches commonly used in clinical practice to augment bone repair and regeneration [88]. Despite its histocompatibility, nonimmunogenicity, mechanical strength, and integrity, its use is hampered by the requirement of multiple surgery, associated donor site morbidity, chronic pain, and infection [89].

There are four major phases that are involved in bone repair and remodeling.

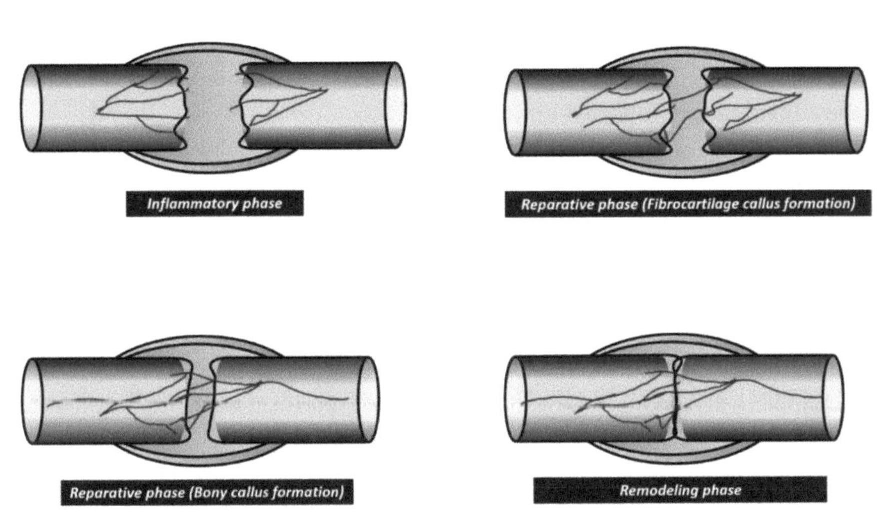

FIGURE 15.12 Schematic representation of fracture repair.

15.4.1 INFLAMMATORY PHASE

The impairment in the bones usually causes a mutilation in the vascular endothelial tissue, which causes formation of hematoma, activation of complement, and clotting cascades. Clotting is caused by the accumulation of polymorphonuclear leukocytes (PMNs) and immune cells. They also show an accelerated expression of various genes such as TGF-β, FGF-1, PDGF, BMP-2, osteonectin, IL-1, and IL-6.

15.4.2 REPARATIVE PHASE (FIBROCARTILAGE CALLUS FORMATION)

Revascularization accompanied by the formation of fibrous tissue and cartilage are the beginning of the reparative phase. This can be visualized by the presence of osteoclasts, fibroblasts, and endothelial cells along with the expression TGF-β, FGF-1, PDGF, BMP-2, osteonectin, IL-1 IL-6, and Col II, III, IV, V, VI, and IX.

15.4.3 REPARATIVE PHASE (BONY CALLUS FORMATION)

The new bone tissue is formed as a result of endochondral and intramembranous ossification of the impaired bone.

15.4.4 REMODELING PHASE

Replacement of the bony callus by the lamellar bone takes place. The excess callus formed is absorbed internally and the lamellar bone that was fractured will be substituted by the normal bone restoring the functional architecture of the bone.

These phases and the genes involved in the regulation of healing process are clearly mentioned in the diagram. Any alteration or genetic deficiency of the proteins involved in these phases leads to bone deformity and requires assistance.

Bone tissue engineering holds a great promise in providing an alternative solution for the functional restoration of impaired bone. This would initiate the recruitment of osteoprogenitor cells followed by their migration, proliferation, differentiation, and matrix formation of bone [90]. Biomaterial scaffolds with osteoinductive and osteoconductive properties that mimic natural ECM have gained prominence in recent years [91]. The ideal scaffolds could be achieved through the delivery of biomolecules or GFs in order to direct the cells to the desired lineages upon receiving morphogenic signals.

An ideal scaffold or a graft which is designed for the engineering of the bone tissue should have the following features.

15.4.4.1 BIOCOMPATIBILITY

All the graft or scaffolds that are to be used should always be biocompatible, so that it does not elucidate any immune response

15.4.4.2 POROSITY

Bone is the hard tissue of the body which is required for a number of metabolic process; hence, good porosity and interconnectivity are required for exchange of gasses and metabolites within the developing tissue and its matrix [92].

15.4.4.3 PORE SIZE

Porosity refers to the volume of pores present in a matrix; similarly, the pore size of a scaffold is very important in developing an ideal biomedical scaffold. The communication of signals in between the ECM and neovascularization of cells depends upon the pore size of the scaffold. An approximate size of about 200–900 μm is best for development of a bone regeneration scaffold [93].

15.4.4.4 SURFACE PROPERTIES

Surface properties, both topographical and chemical, play a crucial responsibility in the fabrication *of* a bone implant. The topographical properties play a major role in osteoinductive properties: the scaffold fabricates a pathway for the migration of the osteogenic cells to the wounded area thus paving way in the development of new tissue. It is said that rough topography would be better suitable than a smooth topology for integration of the scaffold to the defective bone tissue. There are a variety of nanofibers, and among them the most commonly synthesized nanofibers are the random nanofibers.

15.4.4.5 OSTEOINDUCTIVITY

Osteoinductivity is a programmed process that initiates the differentiation of osteogenic progenitor cells to a developed bone tissue. The large number and variety of cells involved in this process harbors the presence of a compound which would assist in the differentiation in addition to the biodegradable scaffolds (Figure 15.13).

15.4.5 MECHANICAL PROPERTIES

Bone is the vital hard tissue, which provides shape and structure to the body. This hard tissue of the body is under continuous hydrostatic pressure and stress exerted the body. The mechanical properties of the scaffold are ideal for maintaining space for the cell infiltration, growth, and differentiation.

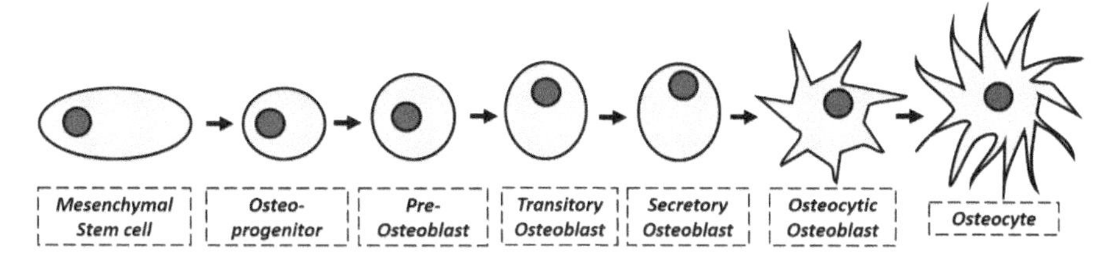

FIGURE 15.13 Stages of stem cell differentiation into osteocytes.

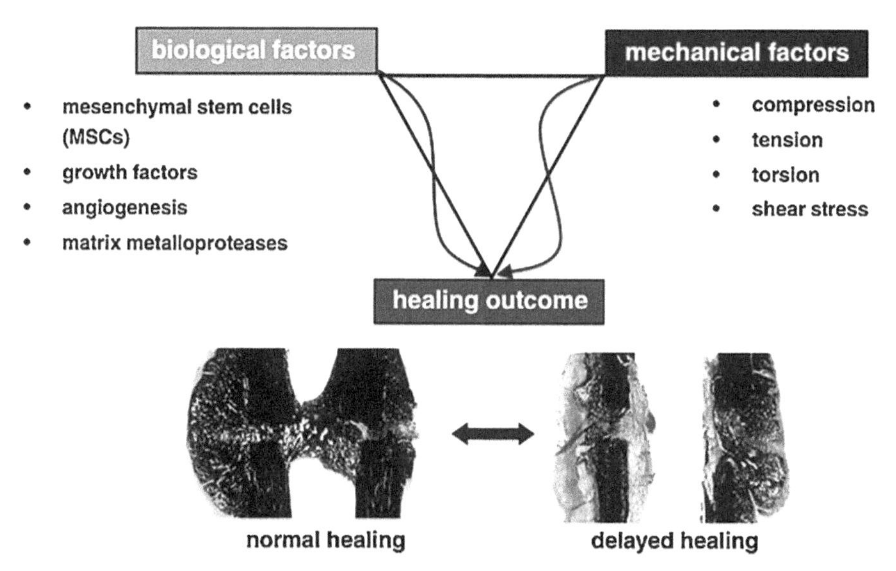

FIGURE 15.14: Factors involved in bone repair and regeneration.

Source: Reprinted from with permission from Ref. [94]. © 2015 Elsevier.

Since bone is a complex network of several constituents, its healing requires a synchronization of biological factors such as GFs (FGF, BMP-2), VEGF, matrix metalloproteases (MMPs), and mechanical factors such as compression, torsion, and sheer stress (Figure 15.14). In certain conditions such as improper calcium in the body or in the case of ailments, bone healing fails to occur. Bone is the hard tissue of the body that comprises a number of complexities in its structure, thereby requiring a long time to heal when compared with other tissues, and hence to accelerate the healing, bio materials have been fabricated to fasten up the reaction. There are numerous treatment methodologies that use bone grafts in assisted healing. There are almost a million of bones and nearly 1 million bone deformation cases that necessitate bone grafts to assist in healing. The current treatment methodologies involve the usage of autologous grafts, allografts, or metal ceramics [95].

Autologous grafts have only been the best suitable items for bone tissue engineering. The samples would be excised from the patient's own body and are placed at the site of injury. This has an added advantage of using the injured person's own osteoconductive and osteoinductive factors for healing the bruised tissue [1, 33, 34]. Nevertheless, it cannot be applied throughout the therapeutic regime due to the limited availability of the tissues to be excised and used for replacement [92].

Allografts are grafts that are used by excising from other individual and replace them within the patient, though if this is successful there is a major shortcoming. This system has high chances of immune system rejection due to the incompatibility of the histocompatibility complex between the host and the recipient.

Metal ceramic: The other major contribution to grafts is provided by metals and ceramics. Metals provide an excellent mechanical support and strength to the system, and on the flip side they are unable to amalgamate with the other tissues of the body and may cause infection or formation of tumors around the foreign substance [96].

Ceramic grafts: Ceramics are very fragile and brittle substances that cannot withstand sheer stress and tension, causing them to take a back seat in the field of bone tissue engineering.

The use of polymer composites as an alternative bone substitute materialized in the 1980s. The role of polymers and reinforcing fillers with improved mechanical, thermal, and biological properties has been immensely explored in the fabrication of novel polymeric metal composites [97]. Conventional materials such as carbon black, glass, etc. are commonly used in the polymeric matrix to increase the mechanical performance of the composites. However, fillers are effective only in the weight percentage of 20%–50% to the polymeric matrix. On the other side, high loading of traditional fillers affects the weight increment and polymer workability.

Nanofillers comprising of nanoparticles are capable of forming intercalating bonds with polymeric chains without hindering the material workability. Depending upon the particles size, shape, and huge specific surface area, the polymer matrix properties like electrical and thermal conductivity, polymer phase behavior, flame retardancy, density, mechanical properties (stiffness, young's modulus, wear, fatigue), and physical properties (magnetic, optic and electrical) can be modified [98].

Nanotechnology has been an emerging field that has revolutionized therapeutics for the last few decades. The field of nanotechnology involves several forms of nanomaterials that have been exploited in the area of bone tissue engineering; some of the examples of materials used are copper, zirconia [99], silver [100], titanium [101], strontium [102], etc. All the above mentioned nanoparticles have decreased cell proliferation and increased differentiation of osteoblast cells. Even though nanoparticles have a number of advantages in the area of bone tissue engineering, there is always a problem in formulating the exact concentration of the nanoparticles at the site of injury. Nanoparticles also have a property of crossing the blood–brain barrier, thereby causing toxicity to the entire system [103]. In addition, they also cannot be easily expelled away from the body.

To overcome all these disadvantages caused by nanoparticles, research started focusing of developing nanofibers for therapeutic applications. Nanofibers are nanomatrices that can encapsulate the therapeutic moiety and release them in a sustained fashion. In addition, they have a unique capability to mimic cushion effect provided by the ECM of the tissues [103]. Their extremely small size, easy modification, etc. make them ideal for tissue engineering constructs.

Among all the nanotechnological interventions, nanofibers serve a major role in the development of bone tissue engineering. Electrospun nanofibers occupy an ideal position in tissue engineering due to their versatile properties such as mimicking the ECM, slow and sustained release of the encapsulated moiety, tunable mechanical and degradation capabilities, biodegradability, and easy

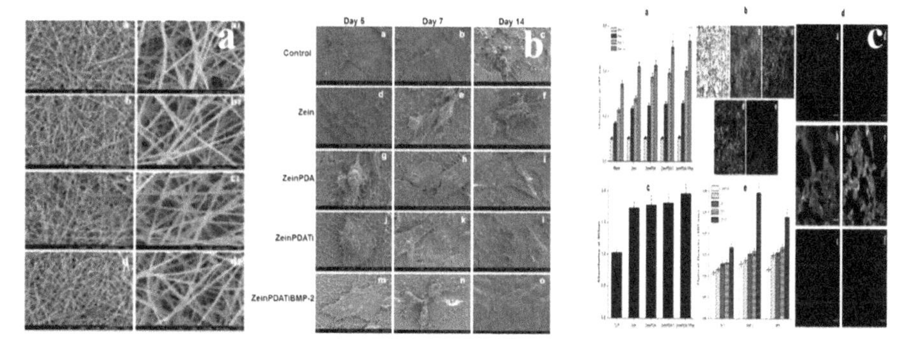

FIGURE 15.15 Application of ZeinPDA nanofibrous scaffold impregnated with BMP-2 peptide-conjugated TiO2 nanoparticle for bone tissue engineering.

Source: Reprinted from with permission from Ref. [101]. © 2017 Wiley.

to transport. Figure 15.15 clearly illustrates the fabrication procedure of nanofibrous scaffold and their subsequent testing in vitro. Immunocytochemistry and the alizarin red results visibly indicate that the nanofibrous scaffolds show better cell differentiation and shape than that of the normal tissue culture plate. After initial standardization in lab scale, the scaffolds are then taken for in vivo animal trials [104].

Structurally bone is made up of two important phases: the inorganic phase that majorly comprises HA and the organic phase (contains of proteins such as glycoproteins and collagen). Based on the structure of bone, scientists have fabricated nanofiber matrices comprising of bone constituent materials, such as HA and collagen. Since Collagen fibers do not have sufficient strength and mechanical properties, collagen has been made has been made as a composite with several natural and synthetic polymers in order to tailor the needs of an ideal bone graft. Among all polymeric materials, a group of poly(α-hydroxyl acid), such as PGA, PLA, PCL, and their copolymers, have been the most broadly explored nanofiber systems for the regeneration of tissues, including bone [95]. PCL, a synthetic polymer, was the first polymer that was experimented to be a degradable nanofiber matrix for the bone regeneration [105]. However, the use of synthetic polymers reduced cell adhesion and condensed migration of the cells on the fabricated scaffolds; to overcome this, synthetic polymers were amalgamated with natural polymers. The most widely used polymer after collagen in bone regeneration studies is gelatin. Gelatin was blended with PCL to increase the cell penetration property of the scaffold [106]. On the same ground, PLA nanofibers were blended in different ratios with gelatin to increase the viability of osteoblasts. Incorporation of HA also increases the porosity and mechanical strength of the grafting nanofiber. To further enhance the strength and thermal properties of the nanofibers, they have been incorporated with several metallic nanoparticles such as TiO_2 [104], magnetic fibers made up of Fe_3O_4 [107], silver [89], etc.

Bioactive glass is a substance that is widely used in the field of bone tissue engineering. This increases the bioactivity and differentiation of bone progenitor cells. To further increase the adhesion, fibronectin, a natural cell adhesive, was layered in the surface of the bioactive glass nanofiber. In addition to mechanical properties, nanofiber scaffolds are fabricated by encapsulating GFs and molecules that are required for regeneration. Huang et al. [108] reported that PLGA scaffolds containing a combination of plasmids encoding DNA for BMP-4,

VEGF, and human bone marrow stromal cells promoted greater bone formation when implanted into the subcutaneous tissue of SCID mice than those containing a single factor or a combination of two factors [108]. The nanofibers were further modified to deliver the plasmid DNA that codes for BMP-2 proteins to further ensure the stability of the required protein. BMP-2 was encapsulated directly within the blending polymer of silk and polyethylene oxide to show enhanced MSC differentiation into the osteogenic linage and calcification [106]. For gene delivery within the nanofibrous matrix, DNA was first encapsulated within a block copolymer polylactide–poly (ethylene glycol), which was further electrospun in concert with the PLGA solution. The results showed that the nanofibrous matrix delivered DNA that was capable of cellular transfection and encoding protein β-galactosidase. Recent study on specific targeting for bone tissue has been reported by Nie et al. [150], where they used the PLGA/HA composite nanofibers to deliver BMP-2 plasmid DNA [150].

In addition to all these properties and characteristics of the encapsulated drug, the form of nanofiber is very important in order to achieve bone regeneration. There are different forms of nanofibers such as aligned and random nanofibers. There are different forms of nanofibers that are used for different applications in therapeutics. Amongst them aligned nanofibers were found to be successful in bone regeneration. Aligned fibers show better cell proliferation and cell migration of MSCs that are ideal for bone tissue engineering. In addition, aligned nanofibers have a better mechanical strength that further supports bone differentiation and morphology [109]. Figure 15.16 shows the aligned arrangement of PLGA nanofibers incorporated with HA

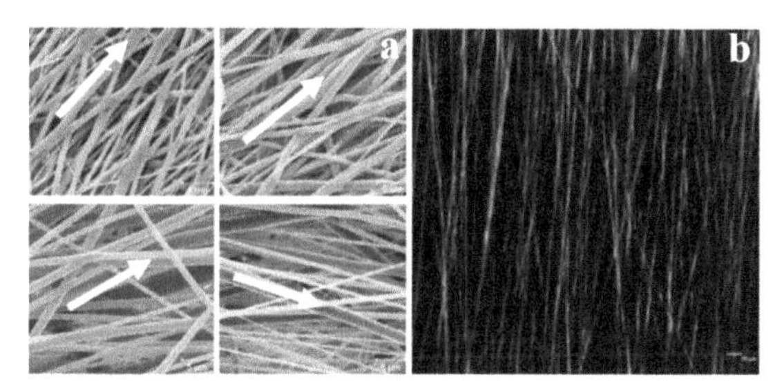

FIGURE 15.16 (a) SEM images of aligned PLGA nanofibers incorporated with HA (hydroxyapatite nanoparticles). (b) Confocal laser image showing the aligned dispersion of HA nanoparticles.

Source: Reprinted from with permission from Ref. [110]. © 2009 Elsevier.

nanoparticles fabricated and tested for bone tissue regeneration.

Therefore, a suitable scaffold can be tailored based on the specific bone ailment, supplementing the required proteins that support in healing.

15.5 CARTILAGE TISSUE REGENERATION

Cartilage is a specialized connective tissue that is present in between the bone surfaces. It is an important friction-resistant coating that is present between our movable joints. The cartilage contains a number of chondrocytes, synovial fluid, and highly interconnected ECM (proteoglycans and collagen). There are two important types of cartilages, namely the hyaline cartilage and the articular cartilage.

Hyaline cartilage is located in soft tissue areas such as the ears, nose and trachea of the human body. On the other hand, the articular cartilage is found in between the bones and joint of the human body. The articulate cartilage is difficult to regenerate since it has a very low mitotic activity. Even if there is proliferation of cells that forms the tissue, they have very low mechanical strength and they are not functionally active [111].

Since the articular cartilage is the one that is subjected to maximum degeneration that cannot be repaired by the body itself, more concentration is given to development of therapeutic moieties to overcome this damage. The articular cartilage has three important zones as follows.

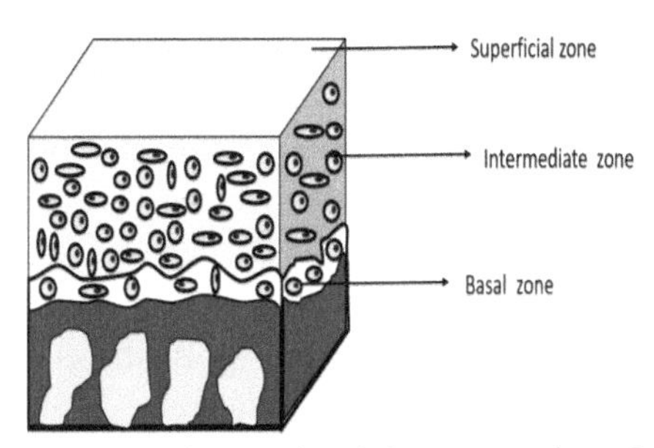

FIGURE 15.17 Pictorial representation of different zones of the articular cartilage.

15.5.1 ZONES OF ARTICULAR CARTILAGE

15.5.1.1 SUPERFICIAL ZONE (TANGENTIAL ZONE)

This is the zone that contains maximum collagen and minimal amount of proteoglycans. The chondrocytes that are present in this region are flattened in shape. This is the region where the progenitor cells that can differentiate into mature chondrocytes are spotted (Figure 15.17).

15.5.1.2 INTERMEDIATE ZONE

This zone consists of collagen layer that are crosslinked in random organization with round chondrocytes. This layer contains high amount of proteoglycans compared to the superficial zone.

15.5.1.3 DEEP LAYER (BASAL LAYER)

This layer consists of collagen and round chondrocytes that are arranged in a column. This consists of highest amount of proteoglycans. The major GFs that are present in the cartilage are PDGF, TGF-β, b-FGF, and IGF-1 (insulin growth factor).

15.5.2 CARTILAGE DAMAGE AND REPAIR

The articular cartilage contains cells that are of very low mitotic activity, and hence the repair and regeneration is a major challenge. There are two types of damages made to the articular cartilage, which are categorized based on the depth of the lesion.

15.5.2.1 CHONDRAL

A superficial damage that is healed by the chondrocytes that have very less mitotic activity is called as a chondral or partial lesion. This healing does not depend on the vascular supply of blood from the host.

15.5.2.2 OSTEOCHONDRAL

Osteochondral or the full thickness lesions are complete degeneration of the cartilage. This regeneration depends on the blood supply or vasculature that is provided by the body's host blood supply. In these cases, the bone marrow provides the MSCs and vasculature for repair and regeneration [112].

Though the body's regenerative mechanism can manage this injury at the initial stage, later stages need therapeutic attention. Ineffective treatment methodologies or negligence leads to the formation of fibrocartilage, which paves way for the beginning of early degenerative osteoarthritis. A few of the current treatment methodologies and their drawbacks are given in Table 15.1. There are a number of current treatment methods that are practiced to regenerate the lost cartilage.

Although there are number of treatment methods that are being followed, cartilage regeneration has still been a challenging task due to the shortcomings in the current methods. An ideal alternate to aid in regeneration of cartilage tissue is the use of biomaterials that provide a suitable microenvironment for the growth and regeneration of the defective tissue [116]. In the present decade, nanofibrous matrices are utilized in therapeutics for their unique topographical structures that mimic the natural ECM [117].

Nanofibers are nanosized therapeutic modalities made up of natural and synthetic polymers. The choice

TABLE 15.1 Different Treatment Modalities Involved in Cartilage Regeneration

Treatment Method	Procedure	Shortcoming
Micro fracture	Drilling of small holes in the defected area to make it accessible to the bone marrow cells to repair the damage [113].	The chondral lesions are not fully healed by this process
Mosaicplasty/ osteochondral grafting	Drafting the cartilage plug from a healthy area to the area of defect [114]	Limited availability of grafts and possibility of bacterial and viral contamination
Autologous chondrocyte implantation	The cartilage cells are harvested from healthy tissue and grown in vitro. These cells are then implanted to the defective region [115]	Time-consuming process

amongst them is made according to the requirement. Synthetic polymers are mechanically stable and thereby give strength to the scaffold, whereas natural polymers support cell infiltration and facilitate easy biodegradability. A wide number of studies have been attempted in this area using nanofibrous scaffolds to aid in complete regeneration of the cartilage. Since collagen is the major protein that is present in the cartilages, different scaffolds with collagen were fabricated in order to mimic the tissue microenvironment. Mechanical stability and functional degradability were major focus areas in fabrication of nanofibers for cartilage regeneration. Li et al. [151] fabricated PCL nanofibers and tested their response to assist differentiation of MSCs and fetal bovine chondrocyte into mature bone tissue, and PCL showed good mechanical properties to support cartilage regeneration [151, 118]. PCL being a hydrophobic polymer restricted cell infiltration to a certain extent to overcome such difficulties. PVA, a hydrophilic nanofiber, was spun into nanofibers and tested for regeneration of hyaline cartilage [119]. In comparison with synthetic polymers, natural polymers had better cell adhesion and migration properties; PCL was blended with PLLA and cationized with gelatin to enhance the biocompatibility of the scaffold with chondrocytes [120]. PCl nanofibers were also blended with chitosan and were placed at the site of injury followed by TGF-β injected separately to aid regeneration [121]. In addition to the scaffolds, stem cells and GFs responsible for regeneration were encapsulated with nanofibers to enhance the activity of the developed scaffold. PCL was blended with sodium hyaluronate and was tested on the MSCs from bone marrow. The presence of sodium hyaluronate along with PCL support increased the expression of chondrocyte markers and swiftly initiated differentiation of chondrocyte end. The differentiation of chondrocytes was investigated with the sustained release of TGF-β from PCL nanofibrous scaffolds and was found to be successful in achieving expression of cartilage markers higher than that of microfibrous scaffold [111]. Figure 15.18 gives a bird's eye view of formulating an ideal therapeutic scaffold for the functional restoration and re-development of cartilage tissue.

Similar to bone regeneration, morphology and the topography of the nanofibers play a major role in cartilage regeneration. To test this, Shafiee et al. [117] constructed random and aligned nanofibers spun with PLLA/PCL hybrids. This experiment demonstrated the fact that aligned fibers showed better cell attachment and patterning of chondrocytes when compared to the random nanofibers [111]. Thereby, fabricating of scaffolds with the knowledge of various bioactives involved in cartilage regeneration at different stages of damage understanding the topography of the tissue microenvironment would open new therapeutic avenues in the areas of cartilage regeneration and repair.

15.6 CARDIAC TISSUE REGENERATION

15.6.1 CARDIOVASCULAR SYSTEM

Cardiovascular system is composed of heart, blood vessels, and blood.

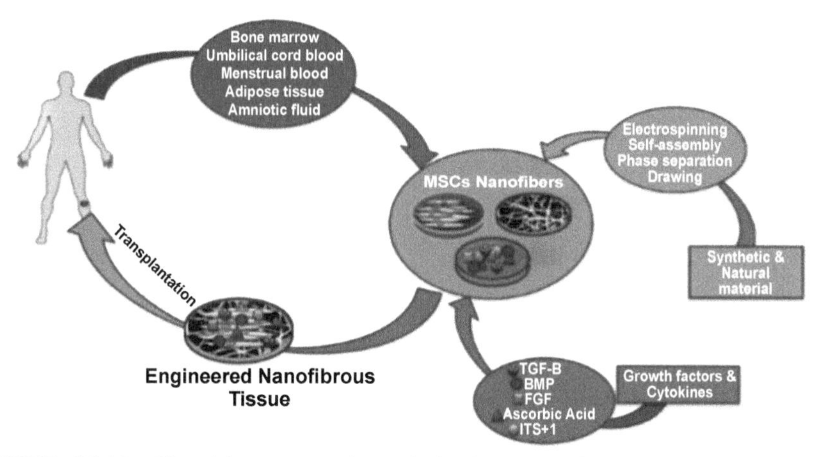

FIGURE 15.18 Pictorial representation of development of therapeutic nano fibrous scaffolds for cartilage tissue regeneration. (Reprinted with permission from [111] https://creativecommons.org/licenses/by/4.0/)

The main function of the system is to circulate blood that carries oxygen and essential nutrients to various parts of the body which is the prime requisite for survival in higher organisms.

15.6.2 THE MYOCARDIUM

The heart is a hollow muscular organ with a conical geometry, located in the thoracic cavity between the lungs. Its size is approximately that of a clenched fist (about 13–15 cm in length, 9–10 cm in breadth, 6 cm in thickness), while its weight is about 250–300 g in adults [122]. The heart wall consists of three distinct layers.

- ***The epicardium***: It is a tissue layer that covers the hearts external surface and is mainly composed of connective tissue and fat.
- ***The myocardium***: This is a thick layer of cardiac muscle that constitutes the muscular mass of the heart and is responsible for the contraction and relaxation of ventricles and atria. This layer is almost completely composed of cardiomyocytes (CMs).
- ***The endocardium***: This is a smooth membrane of endothelial cells that upholsters the inner cavities of the heart, that is, ventricles and atria, as well as heart valves.

The myocardium is sandwiched between the inner endocardium and the outer epicardium. It is a hybrid between skeletal and smooth muscle tissues, showing properties of both of them. Like the skeletal muscle, the myocardium ensures a fast and powerful action, showing the characteristic striations, and consists of distinct cells from each other. On the other hand, like the smooth muscle, the myocardium is an involuntary muscle and contracts as a single fiber in response to an electrical stimulus. The simultaneous presence of properties of both skeletal and smooth muscles allows the heart to reach better performance in terms of pump function.

CMs are the muscle cells that constitute the myocardium of both atria and ventricles. They are different from multinucleated skeletal cells with only one nucleus and show striations like myoblast. CMs show a roughly cylindrical shape in their central portion and divide into prolongations at their ends, which are responsible for cell–cell connections through cellular bridges, called intercalated discs. At regular intervals along the disc, gap junctions can be identified that allow sodium, potassium, and calcium diffusion from cell to cell, making polarization and depolarization easier and faster. CM cells are responsible for generating contractile forces in the intact heart. When there is a demand for increased contractile force, CMs undergoes enlargement (hypertrophy) and its inability to meets these requisites results in insufficient cardiac functioning and failure. This is one of the most common causes of death in the Western world [123].

In a normal adult heart, CMs constitute about 25% of the total number of myocardial cells. The remaining 75% consists of non-CM cell types, among which cardiac fibroblasts represent the majority. Cardiac fibroblasts are specifically positioned inside the myocardium in order to contribute to the structural, biochemical, mechanical, and electrical properties of the normal heart. In detail, they play a pivotal role in the synthesis of ECM constituents and the secretion of cytokines and GFs [124].

15.6.2.1 MYOCARDIAL INFARCTION

Heart diseases remain one of the leading causes of mortality in industrialized nations. Damage to heart muscle, acute or chronic, has long been considered a tipping point for individual health outlook and progression to heart failure. Globally, 29.6% of altogether deaths occurred are found to be caused majorly due to cardiovascular diseases according to a study by Global Burden of Disease study [125]. Among which, more than 7 million causes are due to ischemic cardiomyopathies that mainly lead to acute myocardial infarction (MI) and chronic heart dysfunction.

MI, also known as heart attack, defines the acute coronary syndrome resulting from the obstruction of one or more branches of the coronary arteries caused by a clot or a span [126] (Figure 15.19). MI is the most common cause of cardiac injury and results in acute loss of a large number of myocardial cells. Because the heart has negligible regenerative capacity, CM death triggers a reparative response that ultimately results in formation of a scar and is associated with dilative remodeling of the ventricle [127].

After MI, a complex healing cascade is elicited that is divided into three main phases: necrosis, fibrosis, and remodeling (Figure 15.20). The necrotic phase commences immediately after occurrence of MI and lasts up to 1 week. Necrosis mainly affects ECM proteins and myocytes and is characterized by an increase in tissue stiffness as a consequence of titin and collagen degradation and a reduction in ventricular wall thickness with loss in contractility due to the expansion of the infarct. CM death also induces inflammatory response that increases the secretion of cytokines and chemokines which recruit neutrophils and macrophages to clear the infarct

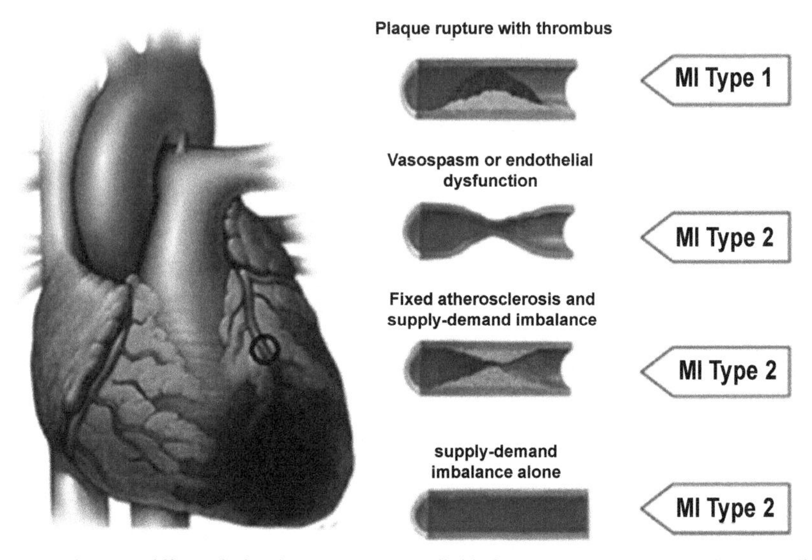

FIGURE 15.19 Differentiation between myocardial infarction (MI) types 1 and 2 according to the condition of the coronary arteries.

Source: Reprinted from with permission from Ref. [128]. © 2021 Wolters Kluwer Health.

area of necrotic myocytes and ECM debris [127]. The following phase, fibrosis, lasts about 3 weeks and is characterized by the deposition of a collagen matrix that will originate a scar, providing a mechanical support to the damaged tissue and resulting in a further increase in infarct stiffness. At the same time, in the border zone, the activated fibroblasts proliferate and differentiate toward myofibroblasts, while endothelial cells organize themselves to form new microvascular system providing oxygen and nutrients to the damaged area.

The early remodeling phase begins about 4–5 days post-MI in mice, granulation tissue rich in ECM proteins begins to form as myofibroblasts and endothelial cells proliferate, and macrophages continue to phagocytoze the remaining necrotic myocardium [129]. As healing continues, myofibroblasts deposit collagen that will be organized into the infarct scar, and the granulation tissue begins to be reabsorbed. By about 21 days post-MI, granulation tissue absorption is complete and a collagenous scar remains. Under the influence of continued inflammation, expression of MMPs, apoptosis, and mechanical factors, ventricular remodeling continues for months to years after the initial injury in both the infarct and noninfarct regions and often leads to heart failure.

15.6.3 CURRENT TREATMENT STRATEGIES

Conventional treatment methodologies for MI include thrombolytic therapy, coronary artery bypass graft surgery, drug therapy, and percutaneous coronary intervention. Thrombolytic therapy is currently the prime approach for the treatment of MI. In addition, drug therapies that include angiotensin receptor blockers, aldosterone receptor antagonists, angiotensin-converting enzyme inhibitors, β-receptor blockers, and so on [130]. Principles of thrombolytic therapy in treating acute MI are restoration of the perfusion as soon as possible and recanalization of the occluded coronary artery [131]. Similarly, coronary artery bypass grafting is an effective surgical treatment of coronary heart disease and myocardial ischemia, which can effectively relieve symptoms.

Cardiac transplant is the gold standard treatment for cardiac replacement therapy. However, limited availability of donors and risk of rejection makes this approach challenging and unsuitable for many patients [132]. Although traditional therapies to some extent can help in reliving the symptoms related to myocardial ischemia, these treatments cannot restore the actual functionality of the myocardial tissue.

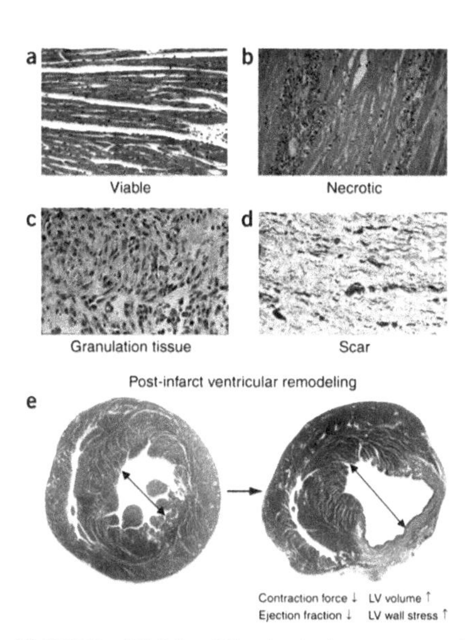

FIGURE 15.20 Histological stages of the infarcted myocardial tissue during the healing cascade (a– d). Changes in ventricle architecture, geometry, and function after an myocardial infarction (e).

Source: Reprinted from with permission from Ref. [133]. © 2005 Springer.

For these reasons, alternative cardiac replacement therapies have been investigated. Cell transplantation is an area of growing interest in clinical cardiology, as a potential means of treating patients after acute MI. The application of cellular cardiomyoplasty for the treatment of MI involves tissue revascularization, isolation of autologous stem cells from the patients, and administration through repeated cardiac catheterization or intramyocardial injection. The major drawback related to the application of cellular therapy is stem cell retention and engraftment after intramyocardial implantation [134]. Other unresolved issues include cell delivery method and route, cell distribution, time transplantation, cell type, cell number, and viability. Solutions to the unmet challenges in cardiovascular regenerative medicine may encompass the application of engineered materials to enhance cardiac differentiation and replicate the 3D structure of the heart following MI. 3D scaffolds could provide a suitable environment for cell homing, proliferation, and differentiation as well as structural and mechanical support to the repairing tissue.

Numerous tissue-engineered scaffolds have been studied as a cardiac patch for myocardial repair in recent studies. Similarly, gels and transplantable materials are applied as injectable materials that polymerizes after implementation into the target site has been studied. In yet another study, Shimizu et al. [135] developed a scaffold-free cell sheets achieved by culturing cells on temperature-sensitive plates which forms a sheet upon cooling. Despite all the studies, these materials limit clinical acceptance owing to the drawbacks, including relatively lower concentrations of CM cell survival, inadequate vascularization, and practical difficulty. Consequently, alternative nanotechnology-based biomaterial scaffold/patches are been studied for application in cardiac tissue engineering.

15.6.3.1 CHARACTERISTIC REQUIREMENT OF MYOCARDIAL PATCH

The designed scaffolds should possess the following features [136, 137].

- Must be biocompatible and biodegradable.
- Induce no or minimal inflammatory response.
- Support cell adhesion, migration and proliferation.
- Possess suitable mechanical properties to that of native cardiac tissue.
- Appropriate biomimetic structure that mimics the microenvironment of the native cardiac tissue
- An apposite conductivity of the scaffold to allow dissemination of electrical stimulation.

15.6.4 ELECTROSPUN SCAFFOLDS FOR MYOCARDIAL REGENERATION

The various requirements for functional cardiac tissue scaffolds have resulted in the development of a plethora of different fabrication methods and material utilization. Over the last few decades, electrospun nanofibers fabricated from natural, synthetic, and composite polymers loaded with drugs, bioactives, GFs, and stem cells have been tremendously studied as cardiac regenerative patch. Clinically approved electrospun collagen scaffolds have been used as support materials for cardiomyoblast culture. Among different type of collagen samples, it was found that atelocollagen produced better electrospun fibers than acid and basic fibrous collagen. The cardiomyoblasts showed t3D morphology when grown on the scaffold which was further ascertained by immunostaining for actin. In vivo study showed good compatibility of the scaffold up to 2 weeks after implantation [138]. To improve the mechanical strength and to overcome the drawbacks of electrospinning natural polymers, it is often blended with synthetic polymers as composite. In a particular study, Liu et al. [139] fabricated composite PCL electrospun nanofibrous sheets, containing elastin and collagen. The sheets mimicked the cardiac microenvironment and promoted cell survival and proliferation. Furthermore, bone marrow cell seeded sheets was used as an implant in MI mice model which showed improved cardiac function and after 4 weeks of transplantation, reduced infarction area and restricted left ventricular (LV) remodeling. Similarly, Venugopal et al. [140] proposed a blend of PCL and Col I and III. The tensile modulus of the structures was 18 MPa with a tensile strength of 7.79 MPa, appropriate for a blood vessel conduit. Furthermore, there was a significant increase in the proliferation of coronary artery smooth muscle cells when seeded on to PCL/Col scaffolds.

Bhaarathy et al. [141] have incorporated two different components in the initial polymer to improve different functionalities. Their study involved the fabrication of nanofiber by blending copolymer poly (L-lactic acid)-co-poly (ε-caprolactone) (PLACL), silk fibroin (SF), and aloe vera (AV) through electrospinning process to create a composite scaffold. The study showed that the myocardial microenvironment was influenced positively in terms of cellular adhesion, morphology, proliferation, and protein expression for neonatal rat CMs (Figure 15.21). The nanofibers of PLACL/SF/AV showed elasticity and mechanical properties similar to those of the myocardium, also giving a positive environment in terms of contractility. The presence of AV also adds up to the functionality of the scaffold and its potential use as a cardiac patch, since it induces anti-allergic and anti-inflammatory effects to the electrospun scaffold.

In yet another approach, injectable poly (glycerol sebacate) (PGS) short fibers were developed to lessen the trauma associated with surgical invasion and allow targeted transfer at otherwise unapproachable sites of the infarcted myocardium. In the current approach, core–shell nanofibers were fabricated initially with PGS as core material and PLLA as shell material. The PLLA was then removed by treatment of the PGS/PLLA core–shell fibers with DCM: hexane (2:1) to obtain PGS short fibers. The scaffolds were characterized by SEM, FTIR, contact angle, and cell–scaffold interactions using CMs. The in vitro studies showed that the cardiac marker proteins actinin, troponin, myosin heavy chain, and connexin 43 were expressed more on short PGS fibers compared to PLLA nanofibers. Thus, use of an apropos biomaterial like short PGS fibers which are capable of providing mechanical strength as well as the biological cues necessary for functional tissue formation can be an alternative to currently employed like transplantation and open heart surgery [142].

One of the methods to improve the functionality of the nanofibrous scaffold is through incorporation of bioactive materials and electrospinning them into composite fibrous mats. Through this way, the efficiency of cardiac cell attachment, viability, biocompatibility, and differentiation can be improved. Shoba et al. [143] fabricated a PCl/GE core–shell nanofibrous scaffold loaded with dual bioactives (salvianolic acid B and magnesium L-ascorbic acid 2 phosphate). The functionalization of polymeric PCL/GE core–shell nanofibrous scaffold with dual bioactives enhanced the physico-chemical and biomechanical properties of the scaffolds and resulted in a 3D topography that mimicked the natural cardiac like ECM. The designed scaffold provided myotube inductive cues in a slow and sustained fashion to the cell surface receptor resulting in increased expression of cardiac makers. In addition, in ovo chorioallantoic membrane assay corroborated the efficacy of the scaffold in inducing angiogenesis required for cardiac regeneration after MI.

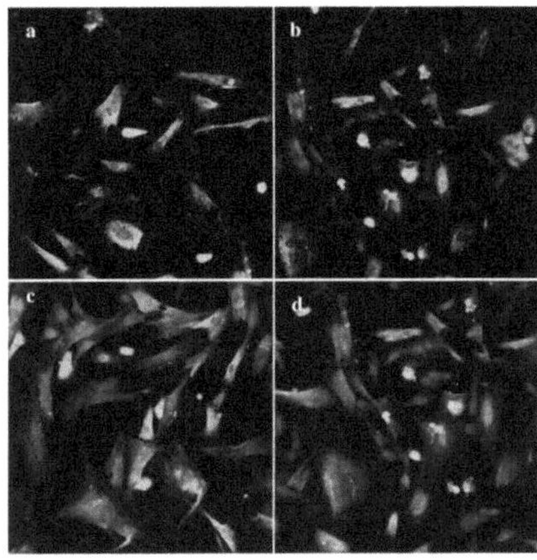

FIGURE 15.21 Immunocytochemical analysis for the expression of cardiomyocyte marker protein myosin at 60× magnification on day 9 (a) TCP, (b) PLACL, (c) PLACL/SF, and (d) PLACL/SF/AV nanofibers.

Source: Reprinted from with permission from Ref. [141]. © 2014 Elsevier.

Recent studies exhibit the potential application of stem cell-based approaches in regeneration and restoration of infarcted heart after MI and blood vessel formation [144]. Kang et al. [145] investigated the umbilical-cord-blood-derived MSC (UCB-MSC)-seeded fibronectin (FN)-immobilized PCL nanofibrous scaffold efficiency in cardiac regeneration and found that it improved cardiac function and inhibited LV remodeling (Figure 15.22) in a rat MI model.

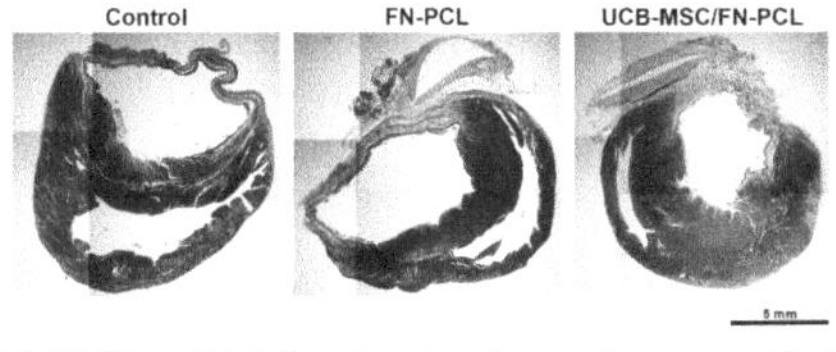

FIGURE 15.22 Evaluation of myocardial infarction size, left ventricular fibrosis, and scar thickness and heart sections stained with Masson's trichrome show fibrosis and wall thinning in the infarcted area. Fibrotic areas are colored in blue and viable myocardium in red.

Source: Reprinted from with permission from Ref. [145]. © 2014 Elsevier.

Fabrication of tissue-engineered constructs mimicking the structural properties of the myocardium is more prospective in promoting integration of CMs and electrical-mechanical coupling [147]. Khan et al. [146] developed human-inducible pluripotent stem cell-derived cardiomyocytes (hiPSC-CMs) cultured on a highly aligned PLGA nanofibrous cardiac patch. The prepared aligned nanofiber scaffold demonstrated potential benefits, such as facilitated the tissue construct to develop anisotropic mechanical properties that mimic native myocardium and also helped in restoring CM morphology and improved mechanical function. Robust expression of actinin and connexin-43 was seen in hiPSC-CMs seeded on cardiac patch. Moreover, the results exhibited significantly higher beating rates when seeded with hiPSC-CM cells on aligned nanofibers as compared to those seeded on the flat tissue culture plates. The obtained results thus advocate that for proper cardiac function, appropriate alignment of CMs with neighboring cells may render efficient coupling required for precise electrical signal transduction and synchronous cell contractions.

15.7 CONCLUSION

Nature has efficient multifunctional structures with optimized functional integration which has been the source of inspiration. Understanding the structural and functional complexities of the natural biological materials led to the rapid growth of biomimetic and bioinspired research.

Bioinspired materials like adhesives of mussels, Gecko feet, lotus waxy coating, structural coloring of insect and bird feathers., with unique nanostructural and functional organization, have evoked the designing and fabrication of self-healing gels, nanofiber composites, wound dressings, nano implants, nano conduits, etc., based on the targeted site and functional requirement. The molecular organization attributing to the mechanical strength, physical characteristics, and biological functions is achieved by unique combination of one or more materials either synthetic or natural, improvised to replace and restore the lost structural and functional integrity.

Bioinspired regeneration is not only restricted to the tissue-based repair and reorganization but also has an enormous progression in the bioactive delivery to the targeted site with optimum pharmacological activity. The key to different regeneration strategies lies in the novel approaches for biodevices with smart, multiple, and hierarchical functionalities simulating the biomimetic composition. The challenges that limit the expansion of nanofiber applications are the pore size and scaffold thickness, which are now being addressed by surface modifications, bioactive incorporation, material functionalization, and nanoparticle conjugation. The biomimetic systems being dynamic, the precise regulations with reference to interaction with the biomacromolecules foresee the success of the nanomaterials.

KEYWORDS

- **biomaterials**
- **tissue engineering**
- **regeneration**
- **nanotechnology**

REFERENCES

1. Miyoshi, H.; Suzuki, K.; Ju, J.; Ko, J. S.; Adachi, T.; Yamagata, Y. A Perturbation Analysis to Understand the Mechanism How Migrating Cells Sense and Respond to a Topography in the Extracellular Environment. *Anal. Sci.* **2016**, *32* (11), 1207–1212. https://doi.org/10.2116/analsci.32.1207.
2. Lee, E. J.; Kasper, F. K.; Mikos, A. G. Biomaterials for Tissue Engineering. *Ann Biomed Eng* **2014**, *2* (1), 323–337. https://doi.org/10.1007/s10439-013-0859-6.Biomaterials.
3. Robert Lanza Robert Langer Joseph Vacanti. *Principles of Tissue Engineering*, 3rd Editio.; Vacanti, R. L. R. L. J., Ed.; Academic Press, 2007.
4. Patterson, J.; Martino, M. M.; Hubbell, J. A. Biomimetic Materials in Tissue Engineering. *Mater. Today* **2010**, *13* (1–2), 14–22. https://doi.org/10.1016/S1369-7021(10)70013-4.
5. Ramos, A. P.; Cruz, M. A. E.; Tovani, C. B.; Ciancaglini, P. Biomedical Applications of Nanotechnology. *Biophys. Rev.* **2017**, *9* (2), 79–89. https://doi.org/10.1007/s12551-016-0246-2.
6. Lannutti, J.; Reneker, D.; Ma, T.; Tomasko, D.; Farson, D. Electrospinning for Tissue Engineering Scaffolds. *Mater. Sci. Eng. C* **2007**, *27* (3), 504–509. https://doi.org/10.1016/j.msec.2006.05.019.
7. Symone L. M. Alexander, Lindsay E. Matolyak, L. T. J. K. Intelligent

Nanofiber Composites : Dynamic Communication between Materials and Their Environment. *Mater. Eng. Macromol.* **2017**, *201700133*, 1–18. https://doi.org/10.1002/mame.201700133.
8. Reyssat E, Hygromorphs L. M. From Pine Cones to Biomimetic Bilayers. *J. R. Soc. Interface* **2009**, *6*, 951–957.
9. Sharanya Sankar, Sharma C. S., Subha N., Rath, S. R. Electrospun Nanofibers to Mimic Natural Hierarchical Structure of Tissues: Application in Musculoskeletal Regeneration. *J Tissue Eng Regen Med* **2018**, *12* (1). https://doi.org/10.1002/term.2335.
10. Babitha S, Rachita L, Karthikeyan K, Shoba E, Janani I, Poornima B, P. S. K. Electrospun Protein Nanofibers in Healthcare: A Review. *Int. J. Pharm.* **2017**, *523* (1), 52–90. https://doi.org/10.1016/j.ijpharm.2017.03.013.
11. Jenkins, T. L.; Little, D. Synthetic Scaffolds for Musculoskeletal Tissue Engineering: Cellular Responses to Fiber Parameters. *npj Regen. Med.* **2019**, *4* (1), 1–14. https://doi.org/10.1038/s41536-019-0076-5.
12. Rittié, L. Cellular Mechanisms of Skin Repair in Humans and Other Mammals. *J. Cell Commun. Signal.* **2016**. https://doi.org/10.1007/s12079-016-0330-1.
13. Hitoshi Yokoyama, Tamae Maruoka, Akio Aruga, Takanori Amano, Shiro Ohgo, T. S. and K. T. Prx-1 Expression in Xenopus Laevis Scarless Skin-Wound Healing and Its Resemblance to Epimorphic Regeneration. *J. Invest. Dermatol.* **2011**, *131* (12), 2477–2485. https://doi.org/10.1038/jid.2011.223.
14. Alexander A, Dwivedi S, Ajazuddin, Tapan K. Giri Swarnlata Saraf, Shailendra Saraf, D. K. T. b. Approaches for Breaking the Barriers of Drug Permeation through Transdermal Drug Delivery. *J. Control. Release* **2012**, *164* (1), 26–40. https://doi.org/10.1016/j.jconrel.2012.09.017.
15. Guo, S.; Dipietro, L. A. Factors Affecting Wound Healing. *J Dent Res* **2010**, *89* (3), 219–229. https://doi.org/10.1177/0022034509359125.
16. Gitay-Goren, H.; Cohen, T.; Tessler, S.; Sokerl, S.; Gengrinovitch, S.; Rockwell, P.; Klagsbrun, M.; Levi, B. Z.; Neufeld, G. Selective Binding of VEGF121 to One of the Three Vascular Endothelial Growth Factor Receptors of Vascular Endothelial Cells. *J. Biol. Chem.* **1996**, *271* (10), 5519–5523. https://doi.org/10.1074/jbc.271.10.5519.
17. David M. Mosser, J. P. E. Exploring the Full Spectrum of Macrophage Activation. *Nat. Rev. Immunol.* **2008**, *8* (12), 958–969. https://doi.org/10.1038/nri2448.
18. Antonio C.L. Campos, Anne K. Groth, A. B. B. Assessment and Nutritional Aspects of Wound Healing. *Curr Opin Clin Nutr Metab Care* **2008**, *11* (3), 281–288.
19. Hani Sinno, S. P. D. Complements and the Wound Healing Cascade : An Updated Review. *Plast. Surg. Int.* **2013**, *2013*.
20. Hye Sung Kima, Xiaoyan Sun, Jung-Hwan Lee, Hae-Won Kim, Xiaobing Fub, K. W. L. Advanced Drug Delivery Systems and Artificial Skin Grafts for Skin Wound Healing. *Adv. Drug Deliv. Rev.* **2018**, #pagerange#. https://doi.org/10.1016/j.addr.2018.12.014.
21. Mustoe, T. A.; O'Shaughnessy, K.; Kloeters, O. Chronic Wound Pathogenesis and Current Treatment Strategies: A Unifying Hypothesis. *Plast. Reconstr. Surg.* **2006**, *117* (7 SUPPL.), 35–41. https://doi.org/10.1097/01.prs.0000225431.63010.1b.
22. Han, G.; Ceilley, R. Chronic Wound Healing: A Review of Current Management and Treatments. *Adv.*

Ther. **2017**, *34* (3), 599–610. https://doi.org/10.1007/s12325-017-0478-y.
23. Hickerson, WL; Compton, C; Fletchall, S. L. Cultured Epidermal Autografts and Allodermis Combination for Permanent Burn Wound Coverage. *Burns* **1994**, *20*.
24. Dreifke, M. B.; Jayasuriya, A. A.; Jayasuriya, A. C. Current Wound Healing Procedures and Potential Care. *Mater. Sci. Eng. C* **2015**, *48*, 651–662. https://doi.org/10.1016/j.msec.2014.12.068.
25. Caroline Bonnans, Jonathan Chou, and Z. W. Remodelling the Extracellular Matrix in Development and Disease. *Nat. Rev. Mol. Cell. Biol.* **2014**, *15* (12), 786–801. https://doi.org/10.1038/nrm3904.Remodelling.
26. Chiquet, J. E. An Overview of Extracellular Matrix Structure and Function. *Biology of Extracellular Matrix*, 2011, pp. 1–39.
27. Scott A. Sell Patricia S. Wolfe, Koyal Garg, Jennifer M. McCool, Isaac A. Rodriguez, G. L. B. The Use of Natural Polymers in Tissue Engineering: A Focus on Electrospun Extracellular Matrix Analogues. *Polymers (Basel).* **2010**, *2*, 522–553. https://doi.org/10.3390/polym2040522.
28. Veleirinho, B.; Coelho, D. S.; Dias, P. F.; Maraschin, M.; Ribeiro-do-Valle, R. M.; Lopes-da-Silva, J. A. Nanofibrous Poly(3-Hydroxybutyrate-Co-3-Hydroxyvalerate)/Chitosan Scaffolds for Skin Regeneration. *Int. J. Biol. Macromol.* **2012**, *51* (4), 343–350. https://doi.org/10.1016/j.ijbiomac.2012.05.023.
29. Mohammad Norouzi Samaneh Moghadasi Boroujeni Noushin Omidvarkordshouli, M. S. Advances in Skin Regeneration : Application of Electrospun Scaffolds. *Mater. Views* **2015**, 1–20. https://doi.org/10.1002/adhm.201500001.
30. Bonvallet, P. P.; Schultz, M. J.; Mitchell, E. H.; Bain, J. L.; Culpepper, B. K.; Thomas, S. J.; Bellis, S. L. Microporous Dermal-Mimetic Electrospun Scaffolds Pre-Seeded with Fibroblasts Promote Tissue Regeneration in Full-Thickness Skin Wounds. *PLoS One* **2015**, *10* (3), 1–17. https://doi.org/10.1371/journal.pone.0122359.
31. Rajabi, M. P. · M. Z. · S. Tailoring the Gelatin/Chitosan Electrospun Scaffold for Application in Skin Tissue Engineering : An in Vitro Study. *Prog. Biomater.* **2018**, *7* (3), 207–218. https://doi.org/10.1007/s40204-018-0094-1.
32. Mahboobeh Adeli-Sardou, Mohammad Mehdi Yaghoobi, Masoud Torkzadeh-Mahani, M. D. *Controlled Release of Lawsone from Polycaprolactone/Gelatin Electrospun Nano Fibers for Skin Tissue Regeneration*; Elsevier BV, 2018. https://doi.org/10.1016/j.ijbiomac.2018.11.237.
33. Poornima, B.; Korrapati, P. S. Fabrication of Chitosan-Polycaprolactone Composite Nanofibrous Scaffold for Simultaneous Delivery of Ferulic Acid and Resveratrol. *Carbohydr. Polym.* **2017**, *157*, 1741–1749. https://doi.org/10.1016/j.carbpol.2016.11.056.
34. Satish, A.; Korrapati, P. S. Fabrication of Triiodothyronine Incorporated Nanofibrous Biomaterial:Its Implications on Wound Healing. *RSC Adv.* **2015**, *5*, 83773–83780. https://doi.org/10.1039/C5RA14142G.
35. Park, U.; Kim, K. Multiple Growth Factor Delivery for Skin Tissue Engineering Applications. *Biotechnol. Bioprocess Eng.* **2017**, *22* (6), 659–670. https://doi.org/10.1007/s12257-017-0436-1.
36. Wang Z, Qian Y, Li L, Pan L, Njunge LW, Dong L, Y. L. Evaluation of Emulsion Electrospun Polycaprolactone/Hyaluronan/Epidermal Growth Factor Nanofibrous Scaffolds for Wound

HealingNo Title. *J. Biomater. Appl.* **2016**, *30* (6), 686–698. https://doi.org/10.1177/0885328215586907.

37. Sahoo, S.; Ang, L. T.; Goh, J. C. H.; Toh, S. L. Growth Factor Delivery through Electrospun Nanofibers in Scaffolds for Tissue Engineering Applications. *J. Biomed. Mater. Res. Part A* **2010**, *93* (4), 1539–1550. https://doi.org/10.1002/jbm.a.32645.
38. Mohammad Norouzi, Iman Shabani, Fatemeh Atyabi, M. S.; Young. EGF-Loaded Nanofibrous Scaffold for Skin Tissue Engineering Applications. *Fibers Polym.* **2015**, *16* (4), 782–787. https://doi.org/10.1007/s12221-015-0782-6.
39. Yang, Y.; Xia, T.; Zhi, W.; Wei, L.; Weng, J.; Zhang, C.; Li, X. Promotion of Skin Regeneration in Diabetic Rats by Electrospun Core-Sheath Fibers Loaded with Basic Fibroblast Growth Factor. *Biomaterials* **2011**, *32* (18), 4243–4254. https://doi.org/10.1016/j.biomaterials.2011.02.042.
40. Zhiwei Xie Christian B. Paras Hong Weng Primana Punnakitikashem, Lee-Chun Su, Khanh Vu, Liping Tang, Jian Yang, K. T. N. *Dual Growth Factor Releasing Multi-Functional Nanofibers for Wound Healing Department of Bioengineering Materials Research Institute The Huck Institutes of the Life*; Acta Materialia Inc., 2013. https://doi.org/10.1016/j.actbio.2013.07.030.
41. Sirsendu Bhowmick, Dieter Scharnweber, D. V. K. Co-Cultivation of Keratinocyte-Human Mesenchymal Stem Cell (HMSC) on Sericin Loaded Electrospun Nanofibrous Composite Scaffold (Cationic Gelatin /Hyaluronan /Chondroitin Sulfate) Stimulates Epithelial Differentiation in HMSCs: In Vitro Study. *Biomaterials* **2016**. https://doi.org/10.1016/j.biomaterials.2016.02.034.
42. Cohen, D. M.; Estes, B. T.; Chen, C. S.; Guilak, F.; Liedtke, W.; Gimble, J. M. Control of Stem Cell Fate by Physical Interactions with the Extracellular Matrix. *Cell Stem Cell* **2009**, *5* (1), 17–26. https://doi.org/10.1016/j.stem.2009.06.016.
43. Kun Ma, Susan Liao, Liumin He, Jia Lu, Seeram Ramakrishna, C. K. C. Effects of Nanofiber/Stem Cell Composite on Wound Healing in Acute Full-Thickness Skin Wounds. *Tissue Eng. Part A* **2011**, *17*. https://doi.org/10.1089/ten.tea.2010.0373.
44. Sugden, F. D. A Novel Dual Modeling Method for Characterizing Human Nerve Fiber Activation, California Polytechnic State University, 2014.
45. Rui Damásio Alvites, Ana Rita Caseiro Santos, A. S. P. V. and A. C. P. de C. O. M. *Olfactory Mucosa Mesenchymal Stem Combination Therapies Nerve Injury New Combination to Regenerative Therapies after Peripheral Nerve Injury*; InTech, 2017. https://doi.org/10.5772/intechopen.68174.
46. Malmivuo, J.; Universit, T.; View, E.; View, B.; Malmivuo, J. Bioelectromagnetism. **2017**, No. November.
47. Arslantunali, D.; Dursun, T.; Yucel, D.; Hasirci, N.; Hasirci, V. Peripheral Nerve Conduits : Technology Update. *Med. Devices Evid. Res.* **2014**, *7*, 405–424.
48. Alvites, R. D.; Santos, A. R. C.; Varejão, A. S. P.; Maurício, A. C. P. de C. O. Olfactory Mucosa Mesenchymal Stem Cells and Biomaterials: A New Combination to Regenerative Therapies after Peripheral Nerve Injury. In *Mesenchymal Stem Cells— Isolation, Characterization and Applications*; Pham, P. Van, Ed.; InTech: Rijeka, 2017; pp 77–102. https://doi.org/10.5772/intechopen.68174.

49. Sunderland, S. A Classification of Peripheral Nerve Injuries Producing Loss of Function. *Brain* **1951**, *74* (4), 491–516.
50. Vargas, M. E.; Barres, B. A. Why Is Wallerian Degeneration in the CNS So Slow? **2007**. https://doi.org/10.1146/annurev.neuro.30.051606.094354.
51. Strittmatter, E. A. H. and S. M. Axon Regeneration in the Peripheral and Central Nervous Systems. *Results Probl. Cell Differ.* **2009**, *48*, 339–351. https://doi.org/10.1007/400.
52. Arslantunali, D, Dursun, T, D Yucel, N Hasirci, v H. Peripheral Nerve Conduits : Technology Update. *Med. Devices Evid. Res.* **2014**, *7*, 405–424.
53. Chiono, V.; Tonda-Turo, C. Trends in the Design of Nerve Guidance Channels in Peripheral Nerve Tissue Engineering. *Prog. Neurobiol.* **2015**, *131*, 87–104. https://doi.org/10.1016/j.pneurobio.2015.06.001.
54. Neal, R. A.; Tholpady, S.S; Foley, P. L.; Swami, N.; Ogle, R. C; Botchwey, E. A. Alignment and Composition of Laminin–Polycaprolactone Nanofiber Blends Enhance Peripheral Nerve Regeneration. *J. Biomed. Mater. Res. A* **2013**, *100* (2), 406–423. https://doi.org/10.1002/jbm.a.33204.Alignment.
55. Li, A.; Hokugo, A.; Yalom, A.; Berns, E. J.; Stephanopoulos, N.; Mcclendon, M. T.; Segovia, L. A.; Spigelman, I.; Stupp, S. I.; Jarrahy, R. Biomaterials A Bioengineered Peripheral Nerve Construct Using Aligned Peptide Amphiphile Nano Fi Bers. *Biomaterials* **2014**, *35* (31), 8780–8790. https://doi.org/10.1016/j.biomaterials.2014.06.049.
56. Horner, P. J.; Gage, F. H. Regenerating the Damaged Central Nervous System. *Nature* **2000**, *407* (6807), 963–970. https://doi.org/10.1038/35039559.
57. Godard C. W. de Ruiter; Malessy, Martijn J. A.; YaszeMski, M. J.; ,Aanthony, J. W.; and Spinner R.J. Designing Ideal Conduits for Peripheral Nerve Repair. **2009**, *26* (February), 1–9. https://doi.org/10.3171/FOC.2009.26.2.E5.
58. Chiono, V.; Tonda.-Turo, C. Trends in the Design of Nerve Guidance Channels in Peripheral Nerve Tissue Engineering. *Prog. Neurobiol.* **2015**, 1–18. https://doi.org/10.1016/j.pneurobio.2015.06.001.
59. Sai, P.; Karthikeyan, K.; Satish, A.; Raghavan, V.; Reddy, J.; Ramakrishna, S. Recent Advancements in Nanotechnological Strategies in Selection Design and Delivery of Biomolecules for Skin Regeneration. *Mater. Sci. Eng. C* **2016**, *67*, 747–765. https://doi.org/10.1016/j.msec.2016.05.074.
60. Xie, J.; MacEwan, M. M. R.; Schwartz, A. G. A.; Xia, Y. Electrospun Nanofibers for Neural Tissue Engineering. *Nanoscale* **2010**, *2* (1), 35–44. https://doi.org/10.1039/b9nr00243j.
61. Cooper, A.; Bhattarai, N.; Zhang, M. Fabrication and Cellular Compatibility of Aligned Chitosan-PCL Fibers for Nerve Tissue Regeneration. *Carbohydr. Polym.* **2011**, *85* (1), 149–156. https://doi.org/10.1016/j.carbpol.2011.02.008.
62. Ting Liu, John D. Houle, Jinye Xu, Barbara P. Chan, S. Y. C. Nanofibrous Collagen Nerve Conduits for Spinal Cord Repair. *Tissue Eng. Part A* **2012**, *18*. https://doi.org/10.1089/ten.tea.2011.0430.
63. Cho, Y. Il; Choi, J. S.; Jeong, S. Y.; Yoo, H. S. Nerve Growth Factor (NGF)-Conjugated Electrospun Nanostructures with Topographical Cues for Neuronal Differentiation of Mesenchymal Stem Cells. *Acta Biomater.* **2010**, *6* (12), 4725–4733. https://doi.org/10.1016/j.actbio.2010.06.019.
64. Hamsici, S.; Cinar, G.; Celebioglu, A.; Uyar, T.; Tekinay, A. B.; Guler, M.

O. Bioactive Peptide Functionalized Aligned Cyclodextrin Nanofibers for Neurite Outgrowth. *J. Mater. Chem. B* **2017**, *5*, 517–524. https://doi.org/10.1039/C6TB02441F.
65. Panseri, S.; Cunha, C.; Lowery, J.; Del Carro, U.; Taraballi, F.; Amadio, S.; Vescovi, A.; Gelain, F. Electrospun Micro- and Nanofiber Tubes for Functional Nervous Regeneration in Sciatic Nerve Transections. *BMC Biotechnol.* **2008**, *8* (February), 39. https://doi.org/10.1186/1472-6750-8-39.
66. Satish, A.; Korrapati, P. S. Tailored Release of Triiodothyronine and Retinoic Acid from a Spatio-Temporally Fabricated Nanofiber Composite Instigating Neuronal Differentiation. *Nanoscale* **2017**, *9*, 14565–14580. https://doi.org/10.1039/c7nr05918c.
67. Xie, J.; Willerth, S. M.; Li, X.; Macewan, M. R.; Rader, A.; Sakiyama-elbert, S. E.; Xia, Y. Biomaterials The Differentiation of Embryonic Stem Cells Seeded on Electrospun Nanofibers into Neural Lineages. *Biomaterials* **2009**, *30* (3), 354–362. https://doi.org/10.1016/j.biomaterials.2008.09.046.
68. Satish, A.; Korrapati, P. S. Tailored Release of Triiodothyronine and Retinoic Acid from a Spatio-Temporally Fabricated Nanofiber Composite Instigating Neuronal Differentiation. *Nanoscale* **2017**, *9* (38), 14565–14580. https://doi.org/10.1039/c7nr05918c.
69. Ashleigh Cooper, Narayan Bhattarai1, M. Z. Fabrication and Cellular Compatibility of Aligned Chitosan PCL Fibers for Nerve Tissue Regeneration. *Carbohydr. Polym.* **2011**, *85* (1), 149–156. https://doi.org/10.1016/j.carbpol.2011.02.008.
70. Jingwei Xie, Matthew R. MacEwan, Xiaoran Li, Shelly E. Sakiyama-Elbert, Y. X. Neurite Outgrowth on Nanofiber Scaffolds with Different Orders Structures and Surfaces Properties. *ACS Nano* **2009**, *3* (5), 1151–1159.
71. Richard Balint, Nigel J.Cassidy, S. H. C. Conductive Polymers: Towards a Smart Biomaterial for Tissue Engineering. *Acta Biomater.* **2014**, *10* (6), 2341–2353.
72. Xu, H.; Holzwarth, J. M.; Yan, Y.; Xu, P.; Zheng, H.; Yin, Y.; Li, S.; Ma, P. X. Conductive PPY/PDLLA Conduit for Peripheral Nerve Regeneration. *Biomaterials* **2014**, *35* (1), 225–235. https://doi.org/10.1016/j.biomaterials.2013.10.002.
73. Zhang, H.; Wang, K.; Xing, Y.; Yu, Q. Lysine-Doped Polypyrrole/Spider Silk Protein/Poly(l-Lactic) Acid Containing Nerve Growth Factor Composite Fibers for Neural Application. *Mater. Sci. Eng. C* **2015**, *56*, 564–573. https://doi.org/10.1016/j.msec.2015.06.024.
74. Gu, B. K.; Kim, M. S.; Kang, C. M.; Kim, J.-I.; Park, S. J.; Kim, C.-H. Fabrication of Conductive Polymer-Based Nanofiber Scaffolds for Tissue Engineering Applications. *J. Nanosci. Nanotechnol.* **2014**, *14* (10), 7621–7626. https://doi.org/10.1166/jnn.2014.9575.
75. Zamani, F.; Amani-tehran, M.; Zaminy, A.; Shokrgozar, M. Conductive 3D Structure Nanofibrous Scaffolds for Spinal Cord Regeneration. **2017**, *18* (10), 1874–1881. https://doi.org/10.1007/s12221-017-7349-7.
76. Sirivisoot, S.; Pareta, R.; Harrison, B. S. Protocol and Cell Responses in Threedimensional Conductive Collagen Gel Scaffolds with Conductive Polymer Nanofibres for Tissue Regeneration. *Interface Focus* **2014**, *4* (1). https://doi.org/10.1098/rsfs.2013.0050.
77. Low, W. C.; Rujitanaroj, P.-O.; Wang, F.; Wang, J.; Chew, S. Y.

Nanofiber-Mediated Release of Retinoic Acid and Brain-Derived Neurotrophic Factor for Enhanced Neuronal Differentiation of Neural Progenitor Cells. *Drug Deliv. Transl. Res.* **2015**, *5* (2), 89–100. https://doi.org/10.1007/s13346-013-0131-5.

78. Wei Ching Low; Pim-On Rujitanaroj; Feng Wang; Jun Wang; Sing Yian Chew. Nanofiber-Mediated Release of Retinoic Acid and Brain-Derived Neurotrophic Factor for Enhanced Neuronal Differentiation of Neural Progenitor Cells. *Drug Deliv. Transl. Res.* **2013**. https://doi.org/10.1007/s13346-013-0131-5.
79. Aishwarya, S.; Korrapati, P. S. Tailored Release of Triiodothyronine and Retinoic Acid from a Spatio-Temporally Fabricated Nanofiber Composite Instigating Neuronal Differentiation. *Nanoscale* **2017**, *9*, 14565–14580. https://doi.org/10.1039/c7nr05918c.
80. Massia, S. P.; Rao, S. S.; Hubbell, J. A. Covalently Immobilized Laminin Peptide Tyr-Ile-Gly-Ser-Arg (YIGSR) Supports Cell Spreading and Co-Localization of the 67-Kilodalton Laminin Receptor with ??-Actinin and Vinculin. *J. Biol. Chem.* **1993**, *268* (11), 8053–8059.
81. Kam, L.; Shain, W.; Turner, J. N.; Bizios, R. Selective Adhesion of Astrocytes to Surfaces Modified with Immobilized Peptides. *Biomaterials*. **2002**, *23*, 511–515.
82. Seren Hamsici, Goksu Cinar, Asli Celebioglu, Tamer Uyar, Ayse B. Tekinay, M. O. G. Bioactive Peptide Functionalized Aligned Cyclodextrin Nanofibers for Neurite Outgrowth. *J. Mater. Chem. B* **2016**. https://doi.org/10.1039/C6TB02441F.
83. Satish, A.; Korrapati, P. S. Strategic Design of Peptide-decorated Aligned Nanofibers Impregnated with Triiodothyronine for Neural Regeneration. *J. Tissue Eng. Regen. Med.* **2019**. https://doi.org/10.1002/term.2822.
84. Mohammad-Hossein Beigi, Laleh Ghasemi-Mobarakeh, Molamma P. Prabhakaran, Khadijeh Karbalaie, Hamid Azadeh, Seeram Ramakrishna, Hossein Baharvand, M.-H. N.-E. In Vivo Integration of Poly (e -Caprolactone)/Gelatin Nanofibrous Nerve Guide Seeded with Teeth Derived Stem Cells for Peripheral Nerve Regeneration. *Soc. Biomater.* **2014**, 1–14. https://doi.org/10.1002/jbm.a.35119.
85. Jeong In Kim1,Tae In Hwang, Ludwig Erik Aguilar, Chan Hee Park, C. S. K. A Controlled Design of Aligned and Random Nanofibers for 3D Bi-Functionalized Nerve Conduits Fabricated via a Novel Electrospinning Set-Up. *Nat. Sci. Reports* **2016**, *6* (March), 1–12. https://doi.org/10.1038/srep23761.
86. Dimitriou, R.; Jones, E.; McGonagle, D.; Giannoudis, P. V. Bone Regeneration: Current Concepts and Future Directions. *BMC Med.* **2011**, *9* (1), 66. https://doi.org/10.1186/1741-7015-9-66.
87. Jimi, E.; Hirata, S.; Osawa, K.; Terashita, M.; Kitamura, C.; Fukushima, H. The Current and Future Therapies of Bone Regeneration to Repair Bone Defects. *Int. J. Dent.* **2012**, *2012*, 1–7. https://doi.org/10.1155/2012/148261.
88. J. E. Mumford; Simpson, A. H. R. W. Management of Bone Defects: A Review of Available Techniques. *Iowa Orthop. J.* **1992**, *12* (July), 42–49.
89. Oryan, A.; Alidadi, S.; Moshiri, A. Current Concerns Regarding Healing of Bone Defects. *Hard Tissue* **2014**, *2* (2), 1–12. https://doi.org/10.13172/2050-2303-2-2-374.

90. Bedilu A. Allo, Daniel O. Costa S. Jeffrey Dixon, Kibret Mequanint, A. S. R. Bioactive and Biodegradable Nanocomposites and Hybrid Biomaterials for Bone Regeneration. *J. Funct. Biomater.* **2012**, *3* (2), 432–463. https://doi.org/10.3390/jfb3020432.
91. Matassi, F.; Nistri, L.; Paez, D. C.; Innocenti, M. New Biomaterials for Bone Regeneration. *Clin. cases Miner. bone Metab.* **2011**, *8* (1), 21–24.
92. Freed, L. E.; Vunjak-Novakovic, G. Culture of Organized Cell Communities. *Adv. Drug Deliv. Rev.* **1998**, *33* (1–2), 15–30. https://doi.org/10.1016/S0169-409X(98)00017-9.
93. Boccaccini, A. R.; Notingher, I.; Maquet, V.; Jérôme, R. Bioresorbable and Bioactive Composite Materials Based on Polylactide Foams Filled with and Coated by Bioglass® Particles for Tissue Engineering Applications. *J. Mater. Sci. Mater. Med.* **2003**, *14* (5), 443–450. https://doi.org/10.1023/A:1023266902662.
94. Qifei Wang, Jianhua Yan, Junlin Yang, B. L. Nanomaterials Promise Better Bone Repair. *Biochem. Pharmacol.* **2016**. https://doi.org/10.1016/j.mattod.2015.12.003.
95. Karen J.L. Burg, Scott Porter, J. F. K. Biomaterial Developments for Bone Tissue Engineering. *Biomaterials* **2000**, *21*, 2347–2359. https://doi.org/10.5663/aps.v1i1.10138.
96. Rose, F. R. A. J.; Oreffo, R. O. C. Bone Tissue Engineering: Hope vs Hype. *Biochem. Biophys. Res. Commun.* **2002**, *292* (1), 1–7. https://doi.org/10.1006/bbrc.2002.6519.
97. Bonfield, W.; Grynpas, M. D.; Tully, A. E.; Bowman, J.; Abram, J. Hydroxyapatite Reinforced Polyethylene—A Mechanically Compatible Implant Material for Bone Replacement. *Biomaterials* **1981**, *2* (3), 185–186. https://doi.org/10.1016/0142-9612(81)90050-8.
98. Hanemann, T.; Szabó, D. V. *Polymer-Nanoparticle Composites: From Synthesis to Modern Applications*; 2010; Vol. 3. https://doi.org/10.3390/ma3063468.
99. Zhu, Y.; Zhu, R.; Ma, J.; Weng, Z.; Wang, Y.; Shi, X.; Li, Y.; Yan, X.; Dong, Z.; Xu, J.; et al. In Vitro Cell Proliferation Evaluation of Porous Nano-Zirconia Scaffolds with Different Porosity for Bone Tissue Engineering. *Biomed. Mater.* **2015**, *10* (5), 55009. https://doi.org/10.1088/1748-6041/10/5/055009.
100. Nirmala, R.; Kang, H. S.; Park, H. M.; Navamathavan, R.; Jeong, I. S.; Kim, H. Y. Silver-Loaded Biomimetic Hydroxyapatite Grafted Poly(ε-Caprolactone) Composite Nanofibers: A Cytotoxicity Study. *J. Biomed. Nanotechnol.* **2012**, *8* (1), 125–132. https://doi.org/10.1166/jbn.2012.1359.
101. Korrapati, P. S.; Venugopal, J. R.; Poddar, K.; Annamalai, M.; Ramakrishna, S.; Saha, S.; Dykas, M. M.; Venkatesan, T.; Babitha, S. Fabrication of a Biomimetic ZeinPDA Nanofibrous Scaffold Impregnated with BMP-2 Peptide Conjugated TiO 2 Nanoparticle for Bone Tissue Engineering *J. Tissue Eng. Regen. Med.* **2017**, No. 65. https://doi.org/10.1002/term.2563.
102. Xie, H.; Gu, Z.; He, Y.; Xu, J.; Xu, C.; Li, L.; Ye, Q. Microenvironment Construction of Strontium-Calcium-Based Biomaterials for Bone Tissue Regeneration: The Equilibrium Effect of Calcium to Strontium. *J. Mater. Chem. B* **2018**, *6* (15), 2332–2339. https://doi.org/10.1039/c8tb00306h.
103. Jong, W. H. De. Ijn-0302-133.Pdf. **2008**, *3* (2), 133–149. https://doi.org/10.2147/IJN.S596.
104. Babitha S, Meenakshi Annamalai, Michal Marcin Dykas, Surajit Saha,

K. P.; Jayarama Reddy Venugopal, Seeram Ramakrishna, Thirumalai Venkatesan, P. S. K. Fabrication of a Biomimetic ZeinPDA Nanofibrous Scaffold Impregnated with BMP-2 Peptide Conjugated TiO 2 Nanoparticle for Bone Tissue Engineering. *J. Tissue Eng. Regen. Med.* **2017**, No. 65. https://doi.org/10.1002/term.2563.

105. Yoshimotoa, H.; Shina, Y.M.; Teraia, H.; Vacanti, J. P. A Biodegradable Nanofiber Scaffold by Electrospinning and Its Potential for Bone Tissue Engineering. *Biomaterials* **2003**, *24* (12), 2077–2082. https://doi.org/10.1016/S0142-9612(02)00635-X.
106. Li, C.; Vepari, C.; Jin, H. J.; Kim, H. J.; Kaplan, D. L. Electrospun Silk-BMP-2 Scaffolds for Bone Tissue Engineering. *Biomaterials* **2006**, *27* (16), 3115–3124. https://doi.org/10.1016/j.biomaterials.2006.01.022.
107. Lin, Y.; Wei, Y.; Song, Y.; Zhang, X.; Wang, X.; Hu, X.; Deng, X.; Han, B. Magnetic Biodegradable Fe 3 O 4 / CS/PVA Nanofibrous Membranes for Bone Regeneration *Biomed. Mater.* **2011**, *6* (5), 055008. https://doi.org/10.1088/1748-6041/6/5/055008.
108. Yen-Chen Huang, Darnell Kaigler, Kevin G Rice, Paul H Krebsbach, D. J. M. Combined Angiogenic and Osteogenic Factor Delivery Enhances Bone Marrow Stromal Cell-Driven Bone Regeneration. *J. Bone Miner. Res.* **2005**, *20* (5), 848–857. https://doi.org/10.1359/JBMR.041226.
109. Doustgani, A.; Vasheghani-Farahani, E.; Soleimani, M. Aligned and Random Nanofibrous Nanocomposite Scaffolds for Bone Tissue Engineering Nanofibrous Scaffolds for Bone Tissue Engineering. *Nanomed. J.* **2013**, *1* (1), 20–27.
110. Moncy, V. Jose; Thomas, V.; Johnson, K. T.; Dean, R.; Nyairo, E. Aligned PLGA/HA Nanofibrous Nanocomposite Scaffolds for Bone Tissue Engineering. *Acta Biomater.* **2009**, *5* (1), 305–315. https://doi.org/10.1016/j.actbio.2008.07.019.
111. Kazemnejad, S.; Khanmohammadi, M.; Baheiraei, N.; Arasteh, S. Current State of Cartilage Tissue Engineering Using Nanofibrous Scaffolds and Stem Cells. *Avicenna J. Med. Biotechnol.* **2017**, *9* (2), 50–65.
112. Redman S. N., O. S. F. and A. C. W. Current Strategies for Articular Cartilage Repair. *Rev. Cartil. Repair Strateg.* **2005**, *44*, 23–32. https://doi.org/10.22203/eCM.v009a04.
113. Steadman, J. R.; Rodkey, W. G.; Rodrigo, J. J. Microfracture: Surgical Technique and Rehabilitation to Treat Chondral Defects. *Clin. Orthop. Relat. Res.* **2001**, *391* (Suppl), S362–S369. https://doi.org/10.1097/00003086-200110001-00033.
114. Hangody, L.; Vásárhelyi, G.; Hangody, L. R.; Sükösd, Z.; Tibay, G.; Bartha, L.; Bodó, G. Autologous Osteochondral Grafting-Technique and Long-Term Results. *Injury* **2008**, *39* (1 SUPPL.), 32–39. https://doi.org/10.1016/j.injury.2008.01.041.
115. Fu, F.; Erggelet, C.; Browne, J. E.; Mandelbaum, B.; Moseley, J. B.; Micheli, L. J.; Zurakowski, D. Autologous Chondrocyte Implantation of the Knee: Multicenter Experience and Minimum 3-Year Follow-Up. *Clin. J. Sport Med.* **2003**, *11* (4), 223–228. https://doi.org/10.1097/00042752-200110000-00003.
116. Eva Filová; Michala Rampichová; Andrej Litvinec; Milan Drzík; Andrea Míchková; Matej Buzgo; Eva Kost'áková; Lenka Martinová; Dusan Usvald; Eva Prosecká Jirí Uhlíkh; Jan Motlíkg; Ludek Vajnerh; E. A. A Cell-Free Nanofiber Composite Scaffold Regenerated Osteochondral Defects in Miniature Pigs. *Int. J. Pharm.*

2013, *447*, 139–149. https://doi.org/10.1016/j.ijpharm.2013.02.056.
117. Shafiee, A.; Soleimani, M.; Chamheidari, G. A.; Seyedjafari, E.; Dodel, M.; Atashi, A.; Gheisari, Y. Electrospun Nanofiber-Based Regeneration of Cartilage Enhanced by Mesenchymal Stem Cells. *J. Biomed. Mater. Res. Part A* **2011**, *99 A* (3), 467–478. https://doi.org/10.1002/jbm.a.33206.
118. Fong, C.-Y.; Gauthaman, K.; Ramakrishna, S.; Bongso, A.; Subramanian, A.; Biswas, A.; Venugopal, J. Human Umbilical Cord Wharton's Jelly Stem Cells Undergo Enhanced Chondrogenic Differentiation When Grown on Nanofibrous Scaffolds and in a Sequential Two-Stage Culture Medium Environment. *Stem Cell. Rev. Rep.* **2011**, *8* (1), 195–209. https://doi.org/10.1007/s12015-011-9289-8.
119. Coburn, J. M.; Gibson, M.; Monagle, S; Patterson, Z.; Elisseeff, J. H. Bioinspired Nano Fi Bers Support Chondrogenesis for Articular Cartilage Repair. *PNAS* **2012**, *109* (25). https://doi.org/10.1073/pnas.1121605109/-/DCSupplemental.www.pnas.org/cgi/doi/10.1073/pnas.1121605109.
120. Chen, J. P.; Su, C. H. Surface Modification of Electrospun PLLA Nanofibers by Plasma Treatment and Cationized Gelatin Immobilization for Cartilage Tissue Engineering. *Acta Biomater.* **2011**, *7* (1), 234–243. https://doi.org/10.1016/j.actbio.2010.08.015.
121. Casper, M. E.; Fitzsimmons, J. S.; Stone, J. J.; Meza, A. O.; Huang, Y.; Ruesink, T. J.; O'Driscoll, S. W.; Reinholz, G. G. Tissue Engineering of Cartilage Using Poly-E{open}-Caprolactone Nanofiber Scaffolds Seeded in Vivo with Periosteal Cells. *Osteoarthr. Cartil.* **2010**, *18* (7), 981–991. https://doi.org/10.1016/j.joca.2010.04.009.
122. Mannan, S.; Khalil, M; Rahman M.; A. Mohammad, S. A. Measurement of Different External Dimensions of the Heart in Adult Bangladeshi Cadaver. *Mymensingh Med. J* **2009**, *18* (2), 175–178.
123. Woodcock, E. A.; Matkovich, S. J. Cardiomyocytes Structure, Function and Associated Pathologies. *Int. J. Biochem. Cell Biol.* **2005**, *37* (9), 1746–1751. https://doi.org/10.1016/j.biocel.2005.04.011.
124. Camelliti, P.; McCulloch, Andrew D.; Kohl, P. Microstructured Cocultures of Cardiac Myocytes and Fibroblasts: A Two-Dimensional In Vitro Model of Cardiac Tissue. *Microsc. Microanal.* **2005**, *11* (3).
125. Lozano, R.; Naghavi, M.; Foreman, K.; Lim, S.; Shibuya, K.; Aboyans, V.; Abraham, J.; Adair, T.; Aggarwal, R.; Ahn, S. Y.; et al. Global and Regional Mortality from 235 Causes of Death for 20 Age Groups in 1990 and 2010: A Systematic Analysis for the Global Burden of Disease Study 2010. *Lancet* **2012**, *380* (9859), 2095–2128. https://doi.org/10.1016/S0140-6736(12)61728-0.
126. Jeffrey, W. Holme, Thomas, K. Borg, J. W. C. Structure And Mechanics of Healing Myocardial Infarcts. *Annu. Rev. Biomed. Eng* **2005**, *7*, 223–253. https://doi.org/10.1146/annurev.bioeng.7.060804.100453.
127. Marcin Dobaczewski; Carlos Gonzalez-Quesada; N. G. F. Journal of Molecular and Cellular Cardiology The Extracellular Matrix as a Modulator of the in Fl Ammatory and Reparative Response Following Myocardial Infarction. *J. Mol. Cell. Cardiol.* **2010**, *48* (3), 504–511. https://doi.org/10.1016/j.yjmcc.2009.07.015.

128. Kristian Thygesen; Joseph S. Alpert; Allan S. Jaffe; Maarten L. Simoons; Bernard R. Chaitman; H. D. W. Third Universal Definition of Myocardial Infarction. *JACC* **2012**, *60* (16), 1581–1598. https://doi.org/10.1016/j.jacc.2012.08.001.
129. Jitka Ismail Virag, C. E. M. Myofibroblast and Endothelial Cell Proliferation during Murine Myocardial Infarct Repair. *Am. J. Pathol.* **2003**, *163* (6), 2433–2440. https://doi.org/10.1016/S0002-9440(10)63598-5.
130. Er, F.; Dahlem, K. M.; Nia, A. M.; Erdmann, E.; Waltenberger, J.; Hellmich, M.; Kuhr, K.; Le, M. T.; Herrfurth, T.; Taghiyev, Z.; et al. Randomized Control of Sympathetic Drive With Continuous Intravenous Esmolol in Patients With Acute ST-Segment Elevation Myocardial Infarction: The BEtA-Blocker Therapy in Acute Myocardial Infarction (BEAT-AMI) Trial. *JACC Cardiovasc. Interv.* **2016**, *9* (3), 231–240. https://doi.org/10.1016/j.jcin.2015.10.035.
131. Bates, E. R.; Nallamothu, B. K. Commentary: The Role of Percutaneous Coronary Intervention in ST-Segment–Elevation Myocardial Infarction. *Circulation* **2008**, *118* (5), 567–573. https://doi.org/10.1161/circulationaha.108.788620.
132. Ramakrishna, H.; Pajaro, O. Heart Transplantation in the Era of Continuous Flow Ventricular Assist Devices and the Total Artificial Heart: Will New Technologies Surpass the Gold Standard? *Ann. Card. Anaesth.* **2011**, *14* (3), 174. https://doi.org/10.4103/0971-9784.83981.
133. Michael A Laflamme, C. E. M. Regenerating the Heart. *Nat. Biotechnol.* **2005**, *23* (7), 845–856. https://doi.org/10.1038/nbt1117.
134. Chachques, J. C.; Acar, C.; Herreros, J.; Trainini, J. C.; Prosper, F.; D'Attellis, N.; Fabiani, J. N.; Carpentier, A. F. Cellular Cardiomyoplasty: Clinical Application. *Ann. Thorac. Surg.* **2004**, *77* (3), 1121–1130. https://doi.org/10.1016/j.athoracsur.2003.09.081.
135. Shimizu, T.; Sekine, H.; Yang, J.; Isoi, Y.; Yamato, M.; Kikuchi, A.; Kobayashi, E.; Okano, T. Polysurgery of Cell Sheet Grafts Overcomes Diffusion Limits to Produce Thick, Vascularized Myocardial Tissues. *FASEB J.* **2006**, *20* (6), 708–710. https://doi.org/10.1096/fj.05-4715fje.
136. Boffito, M.; Sartori, S.; Ciardelli, G. Polymeric Scaffolds for Cardiac Tissue Engineering: Requirements and Fabrication Technologies. *Polym. Int.* **2014**, *63* (1), 2–11. https://doi.org/10.1002/pi.4608.
137. Ma, P. X. Biomimetic Materials for Tissue Engineering. *Adv. Drug Deliv. Rev.* **2008**, *60*, 184–198. https://doi.org/10.1016/j.addr.2007.08.041.
138. Kitsara, M.; Agbulut, O.; Kontziampasis, D.; Chen, Y.; Menasché, P. Fibers for Hearts: A Critical Review on Electrospinning for Cardiac Tissue Engineering. *Acta Biomater.* **2017**, *48*, 20–40. https://doi.org/10.1016/j.actbio.2016.11.014.
139. Yang Liu; Yachen Xu; Zhenhua Wang Dezhong Wen; Wentian Zhang; Sebastian Schmull; H.; Li; Yao Chen; S. X. Electrospun Nanofibrous Sheets of Collagen /Elastin /Polycaprolactone Improve Cardiac Repair after Myocardial Infarction. *Am J Transl Res* **2016**, *8* (4), 1678–1694.
140. Venugopal, J.; Zhang, Y. Z.; Ramakrishna, S. Fabrication of Modified and Functionalized Polycaprolactone Nanofibre Scaffolds for Vascular Tissue Engineering. *Nanotechnology* **2005**, *16* (10), 2138–2142. https://doi.org/10.1088/0957-4484/16/10/028.
141. Bhaarathy V, Venugopal J, Gandhimathi C, Ponpandian N, Mangalaraj D, R.

Biologically Improved Nanofibrous Scaffolds for Cardiac Tissue Engineering. *Mater. Sci. Eng. C Mater. Biol. Appl.* **2014**, *44*, 268–277.

142. Ravichandran, R.; Venugopal, J. R.; Sundarrajan, S.; Mukherjee, S.; Sridhar, R.; Ramakrishna, S. Minimally Invasive Injectable Short Nanofibers of Poly(Glycerol Sebacate) for Cardiac Tissue Engineering. *Nanotechnology* **2012**, *23* (38). https://doi.org/10.1088/0957-4484/23/38/385102.
143. Ekambaram Shoba; Rachita Lakra; Manikantan Syamala Kiran; P. S. K. Strategic Design of Cardiac Mimetic Core-Shell Nano Fi Brous Sca Ff Old Impregnated with Salvianolic Acid B and Magnesium L-Ascorbic Acid 2 Phosphate for Myoblast Differentiation. *Mater. Sci. Eng. C* **2018**, *90*, 131–147. https://doi.org/10.1016/j.msec.2018.04.056.
144. Donald Orlic; Jan Kajstura; Stefano Chimenti; Igor Jakoniuk; Stacie M. Anderson; Baosheng Li; James Pickel; Ronald McKay; Bernardo Nadal-Ginard; David M. Bodine; Annarosa Leri; P. A. Bone Marrow Cells Regenerate Infarcted Myocardium. *Lett. Nat.* **2001**, *410* (April), 701–705.
145. Byung-Jae Kang; Hwan Kim; Seul Ki Lee; Joohyun Kim; Yiming Shen; Sunyoung Jung Kyung-Sun Kang; Sung Gap Im; So Yeong Lee; Mincheol Choi; Nathaniel S. Hwang; J.-Y. C. Umbilical-Cord-Blood-Derived Mesenchymal Stem Cells Seeded onto Fibronectin-Immobilized Polycaprolactone Nanofiber Improve Cardiac Function. *ACTA Biomater.* **2014**, No. April, 1–11. https://doi.org/10.1016/j.actbio.2014.03.013.
146. Bhaarathy, V.; Venugopal, J.; Gandhimathi, C.; Ponpandian, N; Mangalaraj, D.; Ramakrishna, S. Biologically Improved Nanofibrous Scaffolds for Cardiac Tissue Engineering. *Mater. Sci. Eng. C* **2014**, *44*, 268–277. https://doi.org/10.1016/j.msec.2014.08.018.
147. Weining Bian, Christopher P Jackman, N. B. Controlling the Structural and Functional Anisotropy of Engineered Cardiac Tissues. *Biofabrication* **2014**, *6*. https://doi.org/10.1088/1758-5082/6/2/024109.
148. Seddon H. J. Three Types of Nerve Injury. *Brain* 1943, 66,4.
149. Wenwen Yu; Wen Zhao; Chao Zhu; Xiuli Zhang; Dongxia Ye; Wenjie Zhang; Yong Zhou; Xinquan Jiang; Zhiyuan Zhang. Sciatic Nerve Regeneration In Rats By A Promising Electrospun Collagen/Poly(ε-caprolactone) Nerve Conduit With Tailored Degradation Rate. *BMC Neuroscience*. **2011**, *12*, 68. https://doi.org/10.1186/1471-2202-12-68.
150. Hemin Nie; Mei-Ling Ho; Chih-Kuang Wang; Chi-Hwa Wang; Yin-Chih Fu. BMP-2 Plasmid Loaded PLGA/HAp Composite Scaffolds for Treatment of Bone Defects in Nude Mice. *Biomaterials*. **2009**,*30* (5), 892–901. doi: 10.1016/j.biomaterials.2008.10.029.
151. Wan-Ju Li; Richard Tuli; Chukwuka Okafor; Assia Derfoul; Keith G. Danielson; David J. Hall; Rocky S. Tuan. A Three-Dimensional Nanofibrous Scaffold For Cartilage Tissue Engineering Using Human Mesenchymal Stem Cells. *Biomaterials*. **2005**, *26*, 599–609. doi:10.1016/j.biomaterials.2004.03.005.

CHAPTER 16

Phyto-Nano Bioengineered Scaffolds: A Promise to Tissue Engineering Research

GAYATHRI SUNDAR[1,2], JOSNA JOSEPH[3], REBU SUNDAR[1], ANNIE JOHN[1], and ANNIE ABRAHAM[1,*]

[1]*Department of Biochemistry, University of Kerala, Thiruvananthapuram 695581, India*

[2]*Department of Biotechnology, CEPCI Laboratory & Research Institute, Kollam, India*

[3]*Advanced Centre for Tissue Engineering, Department of Biochemistry, University of Kerala, Thiruvananthapuram 695581, India*

[*]*Corresponding author. E-mail: annieab2001@gmail.com*

ABSTRACT

Nanostructures are ideal matrices for drug delivery and tissue regeneration; hence, designing such constructs with specific biochemical, mechanical, and electrical properties enhances their efficacy. Phytochemicals used in traditional medicine are treasure houses, which have a wide range of biochemical and functional activities that are yet to be explored and if tagged to innovative technologies like nanotechnology and electrospinning could generate a new array of tissue-engineered products. These green synthesized phyto-nanocomposite incorporated scaffolds (PNCS), combining traditional knowledge with cutting edge nanotechnology, are a “bench to bedside” concept with translational potential. In view of these, the specific chapter is focused on the combinatorial concept of tissue engineering, nanotechnology, and phytochemistry as an innovative approach into the field of regenerative medicine. Thus, the chapter voyages through traditional medicinal plants with proven tissue regenerative capabilities, isolation methods of the select phytochemicals/plant extracts, synthesis of phyto-nanoparticles and composites, their possible cellular interaction pathways, construction, characterization, and evaluating cytocompatibility of PNCS.

16.1 INTRODUCTION

Tissue engineering concepts and techniques aid in laying foundation for complete organ regeneration. Natural and synthetic biocompatible biomimetic scaffolds are employed to create complete normal tissues, rebuild, retain, or recover damaged tissues and whole organs. Recent research has highlighted the promising potential of nanotechnology and phytomedicine in regenerative medicine [1]. Materialization of nanoscaled particles assures few unique physical properties like size, stability, solubility, and target specificity to the drug moieties. In addition, nanoparticles can enhance the solubility and stability of phytochemical compounds and absorption capacity, prevent early degradation, and extend their circulation time in the body [2, 3].

Plants are enriched with numerous bioactive compounds that are responsible for the prevention and treatment of numerous diseases [4]. This chapter is focused on the combinatorial concept of tissue engineering, nanotechnology, and phytochemistry as an innovative approach into the field of regenerative medicine. Hence, the discussion addresses the traditional medicinal plants with proven tissue regenerative capabilities, isolation methods of the select phytochemicals/ plant extracts, synthesis of phyto-nanoparticles and composites, their possible cellular interaction pathways, construction, characterization, and evaluating cytocompatibility of phyto-nanocomposite incorporated scaffolds (PNCS).

16.1.1 TRADITIONAL AYURVEDIC PLANTS AND THEIR TISSUE INDUCTIVE CAPACITIES

Tribal communities have prodigious knowledge about the rich medicinal heritage of various traditional plants. Ethnobotanical knowledge of tribal people unravels myriads of treatment in several communities of the developing world. Many ayurvedic herbal plants have a great role in the treatment of different health conditions like burns, wounds, tumors, infectious, and inflammatory diseases. Phytochemicals are organic compounds produced by plants that have no direct functions in their growth and development known as plant secondary products/ metabolites. Three major classes of secondary metabolites are reported: terpenes, phenolics, and nitrogen-containing compounds. The primary function of these compounds includes protection against herbivores and other pathogens, and they act as attractants (color, odor, and taste) for insects and animals for pollination purpose. Plant kingdom produces different forms of secondary metabolites that mostly provide its own protection against competitors and predators. In addition, these phytochemicals are used widely as herbal remedies for many diseases by native medical practitioners. Furthermore, the dietary intake of these natural plant products is crucial for human health and these phytochemicals are present in vegetables, fruits, roots, flowers, legumes, nuts, seeds, and spices. They possess antioxidant, antidiabetic, antiinflammatory, anticancer, and antimicrobial properties.

Certain phytochemical constituents isolated from plant extract influence osteogenesis, cell regeneration, and other cellular developments, as listed in Tables 16.1 and 16.2. Plant-based drugs have high commercial importance because of their cost-effectiveness, easy availability, nontoxicity, and extractable properties [5].

TABLE 16.1 Plants/Phytochemicals with Osteogenic and Neural Regenerative Potential

Osteogenic Regeneration	Neural Regeneration	Regeneration in Other Cell Lineages
Cissus quadrangularis [6]	*Butea superba* [19]	*Panax ginseng* [26]
Butea monosperma [7]	*Ginkgo biloba* [20]	*Rehmannia glutinosa* [27]
Withania somnifera [8]	*Phyllanthus emblica* [21]	*Salvia miltiorrhiza* [28]
Rhizoma drynariae [9]	*Scutellaria baicalensis* [22]	*Geum japonicum* [29]
Herba epimedii [10]	*Epimedium* flavonoids [23]	*Rosa laevigata* [30]
Morinda citrifolia leaf [11]	*Acanthopanax* [24, 25]	*Liquorice* [31]
Acemannan [12]	*Angelica* *Rhodiola*	*Cichorium intybus* leaf [32]
Osthole (coumarin extracted from *Fructus cnidii*) [13]	*Ganoderma* spore *Polygala* *Gardenia*	*Sinomenium acutum* [33]
	Astragaloside,	*Mangifera indica* [34]
Hesperetin (flavonoid found in citrus fruit) [14]	Ginsenoside Rg1 *Panax*	Peonidin (blackberry) [35]
Zanthoxylum schinifolium [15]	*notoginseng* saponins	Corilagin *(Terminalia chebula)* [36]
Centella asiatica (Asiaticoside) [16]		*Boswellia serrata* [37]
Icariin (flavonoid found in *Epimedium* species) [17]		
Thymbra spicata var. *intricata* [18]		

TABLE 16.2 Plants/Phytochemicals With Wound Healing Potential

Wound healing plants	
Quercus infectoria [38]	*Calendula officinalis* [64]
Aloe barbadensis [39, 40]	*Argemone mexicana* [65]
Bryophyllum pinnatum [41]	*Terminalia chebula* [66]
Achyranthes aspera [42]	*Entada africana* [67]
Acorus calamus [43]	*Azadirachta indica* [68]
Aegle marmelos [40, 44]	*Cyperus rotundus* [69]
Centella asiatica [45, 46]	*Napoleona imperialis* [70]
Schinus terebinthifolius [47]	*Buddleja* [71]
Buchanania lanzan [48]	*Butea monosperma* [72]
Lantana camara [49]	*Thespesia populnea* [73]
Hydnocarpus weighuana [50]	*Saba florida* [74]
Cuminum cyminum [51]	*Ficus religiosa* [75, 76]
Eucalyptus globulus [52]	*Peperomia pellucida* [77]
Scrophularia nodosa [53]	*Opuntia ficus-indica* [78]
Achillea millefolium [54]	*Cydonia oblonga* [79]
	Cydonia vulgaris
Tridax procumbens [55, 56]	*Pelargonium reniforme* [80, 81]
	Pelargonium radala
Ginkgo biloba [57]	*Ruta graveolens* [82]
	Ruta chalepensis
Cissus quandrangularis [58]	*Tephrosia purpurea* [83]
Indigofera enneaphylla [59]	*Thuhar Euphorbia neriifolia* [84]
Jasminum auriculatum [60]	*Hypericum mysorense* [61]
Nelumbo nucifera [61]	*Helianthus annuus* [85]
Arnica montana [62]	*Ocimum tenuiflorum* [86]
	Ocimum sanctum
Glycyrrhiza glabra [63]	*Curcuma longa* [87]
	Solanum xanthocarpum [88]

16.1.2 ISOLATION OF PHYTOCHEMICAL OF CHOICE

Plant-derived phytochemicals play an essential role in the treatment of various diseases. Multiple factors influence the availability of bioactive phytochemicals from plant sources and bioactive compounds contained in each part of the plant is variable. Based on the type of collection and isolation process, they exist as either pure compounds or plant extracts. The yield may vary with the extraction processes. Phytochemicals can be extracted from different plant parts like leaves, bark, stem, flowers, roots, fruits, etc. Every plant species have physical and chemical defense mechanisms to protect their plant body by producing abundant chemical constituents and active components. Physical protection by thorns, spines, prickles, and other structures helps to defend them from surrounding attacks. Diverse combinations of secondary metabolites provide chemical protection, mainly classified as terpenes, phenolics, and nitrogen-containing compounds, formulated from mevalonic acid, simple sugars, and nitrogen-containing compounds, respectively. All these class of compounds help to devise new drug formulations with varied pharmacological activities. Phytochemical extraction and evaluation are the most important screening criteria for the investigation of new phytochemical components to be used in the drug designing process. Several modern techniques are available for evaluation and quantification of efficacy and purity of the isolated molecules [89].

After phytochemical extraction, plant materials should be evaluated for obtaining qualitative information and quantitative data about their yield, composition (gas chromatography-mass spectrometry [MS]), antioxidant (total antioxidant content, reducing power assay) and antimicrobial properties (agar well diffusion assay), and physico-chemical characterization techniques like Fourier-transform infrared (FTIR) spectroscopy, nuclear magnetic resonance (NMR) spectroscopy and MS chromatographic techniques, etc. Based on their size, shape, and charge, phytochemicals could be separated by different chromatographic techniques [90] as listed below:

1. Adsorption chromatography
2. Partition chromatography
3. Ion-exchange chromatography
4. Affinity chromatography
5. Size-exclusion chromatography
6. Paper chromatography
7. Thin layer chromatography
8. Column chromatography
9. Gas chromatography
10. High-performance liquid chromatography
11. High-performance thin layer chromatography
12. Optimum performance laminar chromatography.

Upon further evaluation, phytochemicals are used as mitogens/growth factors in different therapies either as purified forms or suitable extracted forms. Recent studies demonstrate the successful usage of plant-based growth factors in tissue engineering applications. Phyto-components are derived from *Cissus quadrangularis* [91, 92]. *Butea monosperma,* and *Mimosa tenuiflora cortex* [93]. Safflower seed extracts [94]. which promote the osteogenic differentiation of cells, cellular differentiation, and development, are exploited for tissue engineering applications. The cost-effectiveness of these plant-derived trophins/mitogens may explore wide possibilities among drug manufacturing fields.

16.1.3 PHYTO-NANOCOMPOSITE CONSTRUCTION

Nanoparticles are defined as particles that exist below 100 nm dimension. They possess many physical and chemical properties and are stable, mostly biocompatible, cost-effective, easily diffusible, absorbable, and has high retention capacity. Small size of the nanoparticle helps in easy migration to target sites and circulation through the bloodstream. Hence, the novel concept of phyto-nanoconstruct has tremendous applications in drug delivery. Phyto-nanoparticles are synthesized by different methods like nanoprecipitation, bioreduction methods, or green synthesis method. Recent studies conducted in our lab by Soumya et al. and Kumar et al. have resulted in the development of allicin functionalized locust bean gum nanoparticle and naringenin nanoparticle system with polyvinylpyrrolidone, respectively, by the simple nanoprecipitation technique [95, 96]. Several metals like silver, zinc, and gold could initiate bioreduction in the synthesis of plant-based phyto-nanoparticles. Almost all plants contain plenty of flavonoids, alkaloids, terpenes, and several other classes of compounds, which could be utilized in the reduction of metals into ions and the formation of phyto-nanoparticles. Usage of medicinal plants in synthesis of nanoparticles by the bioreduction method using *Dioscorea bulbifera* tuber extracts [97]. pomegranate fruit seeds [98]. *Citrus sinensis* peel extract [99]. *Iresine herbstii* leaf aqueous extract,[100]. leaf extract of *Elephantopus scaber* [101]. *Morinda citrifoliaroot* extract [102]. and *Curcuma longa* tuber powder [103] has been reported.

16.1.4 CELL–PHYTO NANOCOMPOSITE INTERACTION

Nanoparticles easily interact with cells due to its nano size and stability. They can easily diffuse through blood and enter into the target cell. Specific nature of nanoparticles is prolonged degradation and stability in bloodstream. Many studies have described about the cellular interactions with nanoparticles and its role in cellular pathways. Easily diffusible nature of nanoparticles enhances their cellular interaction, and they could act as a ligand to many receptors. Nanoparticles can trigger various cellular activations like apoptosis, inflammation, proliferation, angiogenesis, differentiation, migration, and oxidative stress. In addition to receptor activation, nanoparticles can enter into the cell via the mechanism of phagocytosis, macropinocytosis, clathrin-mediated endocytosis, caveolin-mediated endocytosis, and nonclathrin- and noncaveolin-mediated endocytosis [104]. Through these mechanisms, nanoparticles are engulfed by cell membrane and further trigger the defined cellular activities.

Size, shape, and surface chemistry of nanoparticle enhance their interaction with many cell membrane receptors like EGF (epidermal growth factor) receptor and integrin receptors. Specific ligand-–receptor interaction activates downstream signaling cascades with numerous second messengers and carrier proteins, leading to the activation of kinase cascades like rat sarcoma (RAS) kinase/(mitogen-activated protein) kinase (MAP) kinase pathway and PI3 kinase pathway. Epidermal growth factor receptor (EGFR) stimulation leads to a parallel process of apoptosis and proliferation through the activation of c-Jun N-terminal kinase (JNK) and protein kinase B (AKT) proteins. These downstream signaling cascades activate some transcription factors (NfκB, IκB,

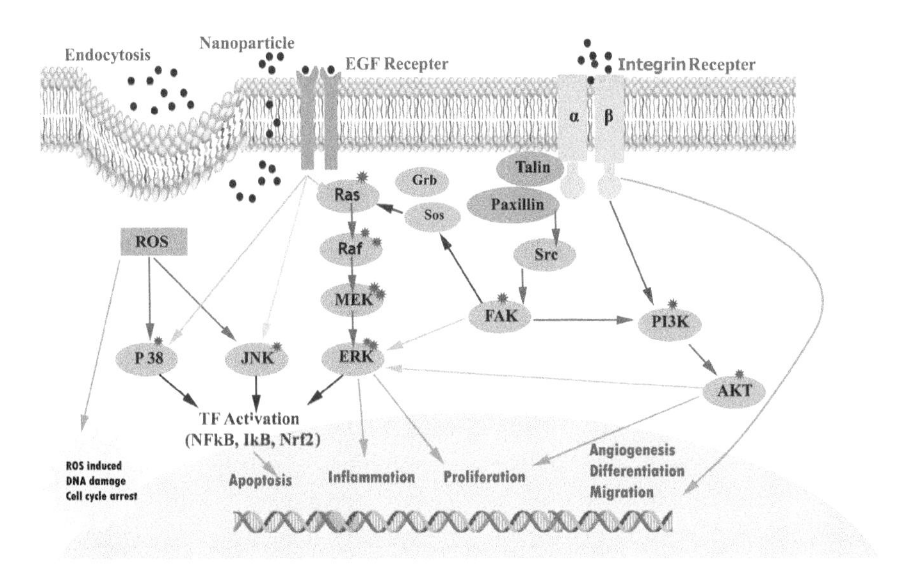

FIGURE 16.1 Nanoparticle (NP)-mediated receptor signaling.

Nrf2) responsible for the gene expression of antioxidant enzymes (HO1) in the nucleus due to ROS activation in stress signaling pathways (Figure16.1).

16.1.5 TYPES OF MATERIALS AS CARRIER SCAFFOLDS

Scaffolds/biomaterials, the key components of tissue engineering, play a crucial role in tissue regeneration as they mimic extracellular matrix. A better scaffold should act as a platform for cell growth and development, support cell attachment, and further tissue development. Scaffolds, cells, and signaling molecules together called as the tissue engineering triad is the main pillar of tissue regeneration. The scaffolds of choice should meet some criteria like biodegradability, nontoxicity, matching mechanical strength, porosity, etc. Highly porous scaffold with adequate pore size permits migration of cells into the scaffold. The scaffolds also serve as delivery vehicles for the transport of exogenous cells, growth factors, and drugs [105]. Different types of biomaterials used to devise different scaffolds are synthetic and natural polymers, ceramics, metals, hydrogels, etc. Metals are widely used as load-bearing implants mainly for hip and knee replacement. Because of the similarity to the bone composition, ceramic materials are often used as orthopedic and dental implants.

The major class of biomaterials—polymers—has a broad variety of application in the field of regenerative medicine. This large group comprises both natural (collagen, chitosan, cellulose, etc.) and synthetic polymers (polycaprolactone [PCL], polylactic-*co*-glycolic acid [PLGA], etc.). Hydrogels are one of the subclasses of polymers, and they are special because of their ability to swell in water but without complete dissolution. Due to their high water content, they have immense application in soft tissue regeneration. Different types of scaffolds accessible in the field of tissue engineering include porous scaffolds, fibrous scaffolds, hydrogel scaffolds, microsphere scaffolds, polymer–bioceramic composite scaffolds, and acellular scaffolds [106]. which introduce a new platform to either new growing cells or damaged cells to recover their original phase [107].

16.1.5.1 NANO FIBERS/FIBROUS SCAFFOLDS

These scaffolds can be synthesized mainly by three methods: electrospinning, self-assembly, and phase separation. Among them, the widely accepted technique and the most promising one is electrospinning [108]. Electrospinning is an efficient technique for the preparation of nanofibers from polymeric biomaterials. With the help of electrospinning, nanofibers with optimum thickness, porosity, and composition, favorable for cell attachment and cell integration, could be manufactured. Electrostatic repulsion of polymeric solution is the principle used for making polymeric fibers. As the name indicates, it is a method of using electric field for the spinning of nanofibers, the electric potential applied to the polymer solution imparts a charge to the solution and by increasing the potential the solution starts to eject from the tip of the syringe which hold the polymeric solution along with the formation of a structure called taylor cone. The charged jet of solution forms scattered nanofibers that can be collected on a suitable collector. Natural polymers, such as collagen, chitosan, silk fibrion, and hyaluronic acid, and synthetic polymers, such as polylactic acid, and PCL, PLGA, are widely used for synthesizing electrospun sheets.

16.1.5.2 HYDROGELS

They are three-dimensional structures made of hydrophilic polymers, which can absorb a huge amount of water and swell readily without dissolving the polymer. Because of its soft, rubbery appearance, these scaffolds are widely used for soft tissue regeneration and as delivery scaffolds as they can encapsulate both desirable cell types and bioactive molecules, such as growth factors and phytochemicals. Natural, synthetic, and semisynthetic hydrogels are the classifications of hydrogels on the basis of their origin. Proteins, polysaccharides, and DNA are some of the natural polysaccharides, whereas polyethene glycol (PEG) and polypropylene fumarate (PPF)-PEG are some of the synthetic biodegradable hydrogels. The common methods of hydrogel preparations are physical cross-linking, chemical cross-linking, free radical polymerization, and irradiation cross-linking. Among them, physical cross-linking and chemical cross-linking are widely accepted methods [109].

16.1.5.3 MICROSPHERES

These are polymeric scaffolds mainly used for drug delivery applications. They have great potential in bone and cartilage tissue engineering because of their well-known mechanical strength, porosity, cell, and drug loading efficiency and controllable drug release.

16.1.5.4 ACELLULAR OR DECELLULARIZED SCAFFOLDS

They are one of the emerging scaffolds for regenerative medicine.

Decellularized whole organs/tissues can act as scaffolds, and they can serve as platforms for organ tissue engineering. However, the current applications are limited to simple organs, such as blood vessels and cardiac valves, and to tissue levels, which may eventually replace all other scaffolds for functional organ regeneration.

16.1.5.5 BIOCOMPOSITE SCAFFOLDS

Biocomposites are blends or combinations of different materials for generating a better outcome. When different materials like polymers, ceramics, metals, etc. are combined together, it will give a synergistic effect to enhance the tissue regeneration potential.

16.1.6 FABRICATION OF PNCS

The combination of nanotechnology and phytochemistry offers a wide area of therapeutics in regenerative medicine. The incorporation of phyto-nanoparticles into suitable carrier scaffolds has great significance in the development of tissue engineered constructs with tissue/organ-specific tailored functions. Many studies revealed that phyto-nanoparticles incorporated biocomposite scaffolds can be fabricated by different approaches [110, 111] as listed below:

- Nanofiber self-assembly
  - Electrospinning
  - Self-assembly
  - Phase separation/ emulsification/freeze-drying
- Solvent casting and particulate leaching
- Gas foaming
- Thermally induced phase separation
- Laser-assisted bioprinting (LaBP)
- 3D formulations
- Rapid-prototyping techniques nanofiber self-assembly. For the formulation of nanofibres, there are different mechanisms: electrospinning, self-assembly, and phase separation.
  - *Electrospinning*: It is the widely used technique to synthesize nanofibers in different nanometer ranges developed by applying a high voltage electric field, as discussed in detail in Section 16.6.1.
  - *Self-assembly*: It describes the self-association of molecules in the mixture to help the formation of nanofibers without any external forces.
  - *Phase separation*: In this method, phase separation or creation of an emulsion by homogenization of an organic solvent from a water mixture is subsequently followed by freeze-drying

process to eliminate the water content.

- Solvent casting and particulate leaching: Here, particular solvents called porogen (sodium chloride, sodium citrate, etc.) are mixed into the polymer solution to synthesize sponge or foam-like structures. After the casting of solution, the porogen causes leaching out and helps to create pores in the scaffold. The porosity can be controlled by changing the concentration of salt crystals.
- Gas foaming: In this method, gas is used as a porogen agent. Here, solid polymers are exposed to high pressure, helping in release of gas (carbon dioxide) from the polymer, leading to thermodynamic instability of the solid polymer system, and causing the formation of pores in the scaffolds.
- Thermally induced phase separation: Here, the phase separation of solvents occurs through a thermal-induced process. In this process, the polymer is first dissolved in a solvent at a high temperature and separation is induced by lowering the solution temperature.
- LaBP: This technique is done by using laser beams. Here, powder materials (wax, synthetic polymers, etc.) are mixed with polymer samples and the laser beams selectively scan the powder material. The interaction of laser beam and powder materials causes the formation of a new layer on the sample by elevating the temperature of powder and scaffold material.
- 3D formulations: In this method, the shape of scaffold formulated by layering process by adhesive bonding using powder as a base material.
- Rapid-prototyping techniques: This method is based on the advanced development of computer science and manufacturing industry to produce complex products with the help of computer-aided designs.

16.1.7 CHARACTERIZATION OF THE PNCS

A scaffold synthesized through different methods has to be well characterized before implanting into a living organism for eliminating the chance of rejection. The characterization studies provide information about the attributes of the scaffold, its chemical composition, size and shape of the nanoparticles, structural information on an atomic level, functional groups, and microstructure of the material, mechanical strength, porosity, crystallinity, and wettability. Physico-chemical and biological characterization of phyto-nano scaffolds are done by different techniques like FTIR spectroscopy, scanning electron microscopy (SEM) and energy dispersive X-ray analyzer (EDX) analysis, TEM, X-ray powder diffraction (XRD), and NMR analysis.

Through FTIR analysis, active biomolecules present in the scaffold are identified, so the incorporation of nanoparticles can be confirmed by FTIR analysis. Internal microstructure analysis is important for the better understanding of physical and chemical properties of the material. SEM and TEM images are useful for the micro/nanostructure analysis as well as for morphological analysis, that is, material size, shape, and the spatial distribution of different particles. XRD technique reveals finer details about the crystallographic orientation of nanostructures, crystallite size, etc. NMR spectroscopy is widely used to prove and confirm the incorporation of drug moieties in the PNCS. For the better optimization of scaffold architecture, mechanical strength analysis, porosity measurement, and wettability measurements are crucial. Compressive and elastic/tensile mechanical testing could provide details about the strength of a scaffold. In addition, in vitro and in vivo characterization of the material is necessary for the clinical application. Cytocompatability is required for understanding the cellular interaction with material via cell proliferation and cell attachment on the material [112].

16.1.7.1 ROLE OF SURFACE TOPOGRAPHY AND FUNCTIONAL/CHEMICAL GROUPS OF THE PNCS

Surface topography is the prime factor controlling the cellular attachment and migration on scaffolds. Topological features of scaffolds help in initial protein adsorption, which leads to further protein-receptor receptor binding. It enhances the signaling cascades of cell attachment, spreading, migration, and functions. Micro and nanostructural analysis of the material imparts fine details about the structure and size of pores, size and shape of incorporated nanoparticles, etc. In addition to these morphological analyses, they provide information about the physical and chemical properties of material. The surface topology of the scaffold seeded with cells is required for assessing the cell attachment, proliferation, migration, and differentiation on the material (Figure 16.2).

Techniques for surface topological characterizations are as follows:

1. Wettability-contact angle methods
2. Electron spectroscopy for chemical analysis (a.k.a. X-ray photoelectron spectroscopy)

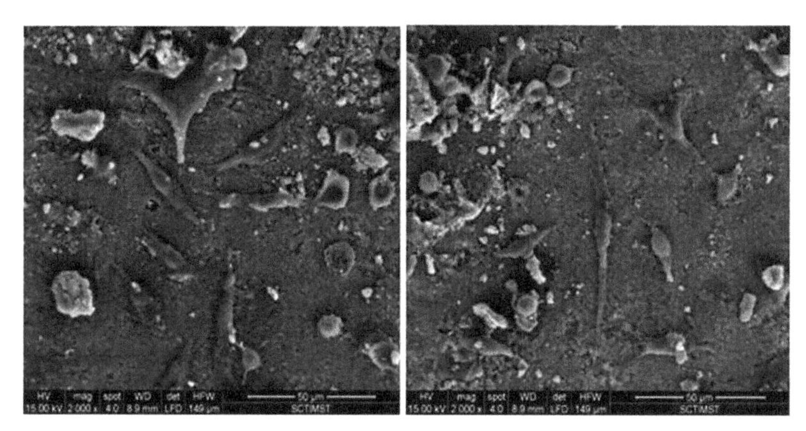

FIGURE 16.2 Scanning electron microscopy images of the surface topology of cell adhesion (L929 fibroblasts) on curcumin spiroborate ester incorporated ceramic scaffolds.

3. Secondary ion mass spectrometry
4. Infrared spectroscopy
5. Scanning electron microscopy
6. Scanning tunneling microscopy
7. Atomic force microscopy

16.1.7.2 CYTOTOXICITY AND CYTOCOMPATIBILITY OF PNCS

The cytotoxicity and cytocompatibility studies of the PNCSs may be evaluated on cell lines, primary cells, and stem cells by in vitro studies. This biological characterization of the scaffold is mandatory for further testing on animal models (ISO 10993). Cell viability and proliferation on the material could be checked by measuring the mitochondrial dehydrogenase activity by (3-(4,5-dimethylthiazol-2-yl)-2,5-diphenyl tetrazolium bromide) (MTT), cytosolic lactate dehydrogenase (LDH) activity by LDH, neutral red assays, live dead cell staining (calcein AM–ethidium homodimer/acridium orange–propidium iodide staining), and (4′,6-diamidino-2-phenylindole) (DAPI)/Hoechst stain (nuclear stain for detecting cell proliferation). The number of viable cells present can be detected by MTT assay and LDH assay. Alamar blue assay is another method based on the detection of metabolic activity.

16.2 FUTURE PERSPECTIVES

PNCS is a novel concept in the field of tissue engineering. Several indigenous plants and phytochemicals possess tissue regenerative capability with proven therapeutic efficacy in traditional medicine. Therefore, a combinatorial approach of merging traditional medical knowledge with cutting edge technologies like nanotechnology has a tremendous impact in tissue engineering applications. Future perspective of the study is to fabricate different forms of plant-based compounds with suitable scaffolds for the application of skin, bone, and tendon regeneration. These green synthesized PNCS, combining traditional knowledge with cutting edge nanotechnology, are a "Make in India" concept toward "bench to bedside."

16.3 CONCLUSION

Usage of traditional ayurvedic medicine in congruence with tissue engineering and nanotechnology for the construction of better biomaterials is a novel arena in regenerative medicine. Nano-sized phytochemicals have higher stability and have faster interaction with the cells, and thereby their action in combination with scaffold would yield better effectiveness. Innumerable number of scaffolds such as nanofibers, hydrogels, microspheres etc. is able to carry the phyto-nanoconstruct and incorporation of phyto-nanocomposites into the scaffolds can induce enhanced cell viability, proliferation, and differentiation which might be at par with the usage of growth factors. Thus, these cost-effective phyto-nanocomposite loaded scaffolds could aid tissue regeneration and have treatment potential injury associated and degenerative diseases.

ACKNOWLEDGMENT

The first and third authors are grateful to University Grants Commission for their Junior Research Fellowship (UGC-JRF) and University of Kerala for providing infrastructure facility for carrying out this work. Authors also acknowledge UGC Emeritus Fellowship to Dr Annie John, KSCSTE-YIPB Project, Government of Kerala to Dr Josna Joseph and UGC-BSR Faculty Fellowship to Dr Annie Abraham, corresponding author.

KEYWORDS

- **phyto-nano constructs**
- **biocomposites**
- **phytochemicals**
- **nanotechnology**
- **tissue engineering scaffolds**

REFERENCES

1. Conte, R.; Luca, I. D.; Luise, A. D.; Petillo, O.; Calarco, A.; Peluso, G. New therapeutic potentials of nanosized phytomedicine. *J Nanosci Nanotechnol.* 2016, *16*, 8176–8187.
2. Wang, S.; Su, R.; Nie, S.; Sun, M.; Zhang, J.; Wu, D.; Moustaid-Moussa, N. Application of nanotechnology in improving bioavailability and bioactivity of diet-derived

phytochemicals. *J Nutr Biochem.* 2014, *25*, 363–376.
3. Martínez Ballesta, M.; Gil Izquierdo, Á.; García Viguera, C.; Domínguez Perles, R. Nanoparticles and controlled delivery for bioactive compounds: Outlining challenges for new "Smart-foods" for health. *Foods,* 2018, *7*, 72.
4. Teodoro, A. J. Bioactive compounds of food: Their role in the prevention and treatment of diseases. *Oxid Med Cell Longev.* 2019, *4,* https://doi.org/10.1155/2019/3765986 (accessed March 11, 2019).
5. Koçak, E.; Pazir, F. Effect of extraction methods on bioactive compounds of plant origin. *Turk J Agric Food Sci Technol.* 2018, *6*, 663–675.
6. Soumya, S.; Sajesh, K. M.; Jayakumar, R.; Nair S. V.; Chennazhi, K. P. Development of a phytochemical scaffold for bone tissue engineering using cissus quadrangularis extract. *Carbohydr Polym.* 2012, *87*,1787–1795.
7. Maurya, R.; Yadav, D. K.; Singh, G.; Bhargavan, B.; Narayana Murthy, P. S.; Sahai, M.; et al. Osteogenic activity of constituents from Butea monosperma. *Bioorg Med Chem Lett.* 2009, *19*, 610–613.
8. Ichikawa, H.; Takada, Y.; Shishodia, S.; Jayaprakasam, B.; Nair, M. G.; Aggarwal, B. B. Withanolides potentiate apoptosis, inhibit invasion, and abolish osteoclastogenesis through suppression of nuclear factor-kappaB (NF-kappaB) activation and NF-kappaB-regulated gene expression. *Mol Cancer Ther.* 2006, *5*, 1434–1445.
9. Zhang, P.; Dai, K. R.; Yan, S. G.; Yan, W. Q.; Zhang, C.; Chen, D. Q.; et al. Effects of naringin on the proliferation and osteogenic differentiation of human bone mesenchymal stem cell. *Eur J Pharmacol.* 2009, *607*, 1–5.
10. Zhang, J.; Li G.; Chan, C.; Meng, C.; Lin, M. C.; Chen, Y.; et al. Flavonoids of Herba Epimedii regulate osteogenesis of human mesenchymal stem cells through BMP and Wnt/beta-catenin signaling pathway. *Mol Cell Endocrinol.* 2010, *314*, 70–74.
11. Boonanantanasarn, K.; Janebodin, K.; Suppakpatana, P.; Arayapisit, T.; Rodsutthi, J.; Chunhabundit, P.; et al. Morinda citrifolia leaves enhance osteogenic differentiation and mineralization of human periodontal ligament cells. *Dent Mater J.* 2014, *33*, 149–149.
12. Chantarawaratit, P.; Sangvanich, P.; Banlunara, W.; Soontornvipart, K.; Thunyakitpisal, P. Acemannan sponges stimulate alveolar bone, cementum and periodontal ligament regeneration in a canine class II furcation defect model. *J Periodontal Res.* 2014, *49*, 164–178.
13. Gao, L. N.; An, Y.; Lei, M.; Li, B.; Yang, H.; Lu, H.; et al. The effect of the coumarin-like derivative osthole on the osteogenic properties of human periodontal ligament and jaw bone marrow mesenchymal stem cell sheets. *Biomaterials.* 2013, *34*, 9937–9951.
14. Kim, S. Y.; Lee, J. Y.; Park, Y. D.; Kang, K. L.; Lee, J. C; Heo, J. S. Hesperetin alleviates the inhibitory effects of high glucose on the osteoblastic differentiation of periodontal ligament stem cells. *PLoS One.* 2013, *8*, 67504.
15. Kim, S. Y.; An, S. Y.; Lee, J. S.; Heo, J. S. Zanthoxylum schinifolium enhances the osteogenic potential of periodontal ligament stem cells. *In Vitro Cell Dev Biol Anim.* 2015, *51*, 165–173.
16. Nowwarote, N.; Osathanon, T.; Jitjaturunt, P.; Manopattanasoontorn, S.; Pavasant, P. Asiaticoside induces type I collagen synthesis and osteogenic differentiation in human periodontal ligament cells. *Phytother Res.* 2013, *27*, 457–462.

17. Pei, Z.; Zhang, F.; Niu, Z.; Shi, S. Effect of icariin on cell proliferation and the expression of bone resorption/formation-related markers in human periodontal ligament cells. *Mol Med Rep*. 2013, *8*, 1499–1504.
18. Mendi, A.; Yağcı, B. G.; Kızıloğlu, M.; Saraç, N.; Uğur, A.; Yılmaz, D.; et al. Thymbraspicata var. intricata induces mesenchymal stem cell proliferation and osteogenic differentiation. *Braz Arch Biol Technol*. 2017, *60*, 1678–4324.
19. Arai, M. A.; Koryudzu, K.; Koyano, T.; Kowithayakorn, T.; Ishibashi, M. Naturally occurring Ngn2 promoter activators from Butea superba. *Mol Biosyst.* 2013, *9*, 2489–2497.
20. Wang, J.; Chen, W.; Wang, Y. A Ginkgo biloba extract promotes proliferation of endogenous neural stem cells in vascular dementia rats. *Neural Regen Res.* 2013, *8*, 1655.
21. Swain, U.; Sindhu, K. K.; Boda, U.; Pothani, S.; Giridharan, N. V.; Raghunath, M.; et al. Studies on the molecular correlates of genomic stability in rat brain cells following Amalakirasayana therapy. *Mech Ageing Dev.* 2012, *133*, 112–117.
22. Li, M.; Choi, S. T.; Tsang, K. S.; Shaw, P. C.; Lau, K. F. DNA microarray expression analysis of baicalin-induced differentiation of C17.2 neural stem cells. *Chembiochem Eur J Chem Biol.* 2012, *13*, 1286–1290.
23. Yao, R.; Zhang, L.; Li, X.; Li, L. Effects of Epimedium flavonoids on proliferation and differentiation of neural stem cells in vitro. *Neurol Res.* 2010, *32*, 736–742.
24. Si, Y.; Li, Q.; Xie, C.; Niu, X.; Xia, X.; Yu, C. Chinese herbs and their active ingredients for activating xue (blood) promote the proliferation and differentiation of neural stem cells and mesenchymal stem cells. *Chin Med.* 2014, *9*, 13.
25. Lin, P. C.; Chang, L. F.; Liu, P. Y.; Lin, S. Z.; Wu, W. C.; Chen, W. S.; et al. Botanical drugs and stem cells. *Cell Transplant.* 2011, *20*, 71–83.
26. Sasaki, T.; Oh, K. B.; Matsuoka, H.; Saito, M. Effect of Panax ginseng components on the differentiation of mouse embryonic stem cells into cardiac-like cells. *Yakugaku Zasshi.* 2008, *128*, 461–467.
27. Wang, Y. B.; Liu, Y. F.; Lu, X. T.; Yan, F. F.; Wang, B.; Bai, W. W.; Zhao, Y. X. Rehmannia glutinosa extract activates endothelial progenitor cells in a rat model of myocardial infarction through a SDF-1 α/CXCR4 cascade. *PLoS One.* 2013, *8*, 54303.
28. Li, K.; Li, S.; Zhang, Y.; Wang, X. The effects of dan-shen root on cardiomyogenic differentiation of human placenta-derived mesenchymal stem cells. *Biochem Biophys Res Commun.* 2011, *415*, 147–151.
29. Lin, X.; Peng, P.; Cheng, L.; Chen, S.; Li, K.; Li, Z. Y.; et al. Natural compound induced cardiogenic differentiation of endogenous MSCs for repair of infarcted heart. *Differentiation* 2012, *83*, 1–9.
30. Zhou, Z.; Li, D.; Zhou, H.; Lin, X.; Li, C.; Tang, M.; et al. Plants and their bioactive compounds with the potential to enhance mechanisms of inherited cardiac regeneration. *Planta Med.* 2015, *81*, 637–647.
31. Su, W. T.; Chen, X. W. Stem cells from human exfoliated deciduous teeth differentiate into functional hepatocyte-like cells by herbal medicine. *Biomed Mater Eng.* 2014, *24*, 2243–2247.
32. Ebrahiminia, M.; Esmaeili, F.; Kharazian, N.; Houshmand, F.; Ebrahimi, M. Effect of flavonoids of extract of cichorium intybus L. leaf on induction of P19 stem-cells

differentiation to insulin-producing cells. *J Mazandaran Univ Med Sci.* 2014, *23*, 93–100.
33. Mark, W.; Schneeberger, S.; Seiler, R.; Stroka, D. M.; Amberger, A.; Offner, F.; et al. Sinomenine blocks tissue remodeling in a rat model of chronic cardiac allograft rejection. *Transplantation.* 2003, *75*, 940–945.
34. García, D.; Leiro, J.; Delgado, R.; Sanmartín, M. L.; Ubeira, F. M. Mangifera indica L. extract (Vimang) and mangiferin modulate mouse humoral immune responses. *Phytother Res.* 2003, *17*, 1182–1187.
35. Wang, L. S.; Stoner, G. D. Anthocyanins and their role in cancer prevention. *Cancer Lett.* 2008, *269*, 281–290.
36. Chang, C. L.; Lin, C. S. Phytochemical composition, antioxidant activity, and neuroprotective effect of Terminalia chebula Retzius extracts. *Evid Based Complement Alternat Med.* 2012, *2012*, 125247.
37. Singh, S.; Khajuria, A.; Taneja, S. C.; Johri, R. K.; Singh, J.; Qazi, G. N. Boswellic acids: A leukotriene inhibitor also effective through topical application in inflammatory disorders. *Phytomedicine Int J Phytother Phytopharm.* 2008, *15*, 400–407.
38. Jalalpure, S. S.; Agrawal, N.; Patil, M. B.; Chimkode, R.; Tripathi, A. Antimicrobial and wound healing activities of leaves of *Alternanthera sessilis* Linn. *Int J Green Pharm.* 2008, *2*, 3.
39. Chithra, P.; Sajithlal, G. B.; Chandrakasan, G. Influence of Aloe vera on collagen turnover in healing of dermal wounds in rats. *Indian J Exp Biol.* 1998, *36*, 896–901.
40. Udupa, S. L.; Udupa, A. L.; Kulkarni, D. R. Studies on the anti-inflammatory and wound healing properties of Moringa oleifera and Aegle marmelos. *Fitoterapia.* 1994, *65*, 119–123.
41. Khan, M.; Patil, P. A.; Shobha, J. C. Influence of Bryophyllum pinnatum (Lim.) leaf extract on wound healing in albino rats. *J Nat Remedies.* 2004, *4*, 41–46.
42. Ghosh, P. K.; Gupta, V. B.; Rathore, M. S.; Hussain, I. Wound-healing potential of aqueous and ethanolic extracts of apamarga leaves. *Int J Green Pharm.* 2011, *5*, 12.
43. Jain, N.; Jain, R.; Jain, A.; Jain, D. K.; Chandel, H. S. Evaluation of wound-healing activity of Acorus calamus Linn. *Nat Prod Res.* 2010, *24*, 534–541.
44. Jaswanth, A.; Loganathan, V.; Manimaran, S. Wound healing activity of *Aegle marmelos*. *Indian J Pharm Sci.* 2001, *63*, 41.
45. Parameshwaraiah, S.; Shivakumar, H. G. Evaluation of topical formulations of aqueous extract of *Centella asiatica* on open wounds in rats. *Indian J Exp Biol.* 1998, *36*, 569–572.
46. Shetty, B. S.; Udupa, S. L.; Udupa, A. L.; Somayaji, S. N. Effect of *Centella asiatica* L (Umbelliferae) on normal and dexamethasone-suppressed wound healing in Wistar albino rats. *Int J Low Extrem Wounds*. 2006, *5*, 137–143.
47. Lucena, P. L. H. D.; Ribas Filho, J. M.; Mazza, M.; Czeczko, N. G.; Dietz, U. A.; Correa Neto, M. A.; et al. Evaluation of the aroreira (*Schinus terebinthifolius* Raddi) in the healing process of surgical incision in the bladder of rats. *Acta Cir Bras.* 2006, *21*, 46–51.
48. Chitra, V.; Dharani, P. P.; Pavan, K. K.; Alla, N. R. Wound healing activity of alcoholic extract of *Buchanania lanzan* in Albino rats. *Int J ChemTech Res.* 2009, *1*, 1026–1031.
49. Nayak, B. S.; Raju, S. S.; Eversley, M.; Ramsubhag, A. Evaluation of wound healing activity of *Lantana camara* L.—a preclinical study. *Phytother Res.* 2009, *23*, 241–245.

50. Oommen, S. T.; Rao, M.; Raju, C. V. Effect of oil of hydnocarpus on wound healing. *Int J Lepr Mycobact Dis.* 1999, *67*, 154–158.
51. Patil, D. N.; Kulkarni, A. R.; Shahapurkar, A. A.; Hatappakki, B. C. Natural cumin seeds for wound healing activity in albino rats. *Int J Biol Chem.* 2009, *3*, 148–152.
52. Hukkeri, V. I.; Karadi, R. V.; Akki, K. S.; Savadi, R. V.; Jaiprakash, B.; Kuppast, I. J.; et al. Wound healing property of *Eucalyptus globulus* L. leaf extract. *Indian Drugs* 2002, *39*, 481–483.
53. Stevenson, P. C.; Simmonds, M. S. J.; Sampson, J.; Houghton, P. J.; Grice, P. Wound healing activity of acylated iridoid glycosides from *Scrophularia nodosa*. *Phytother Res.* 2002, *16*, 33–35.
54. Nirmala, S.; Karthiyayini, T. Wound healing activity on the leaves of *Achillea millefolium* L. by excision, incision, and dead space model on adult Wistar albino rats. *Int Res J Pharm* 2011, *2*, 240–245.
55. Diwan, P. V.; Tilloo, L. D.; Kulkarni, D. R. Steroid depressed wound healing and Tridax procumbens. *Indian J Physiol Pharmacol.* 1983, *27*, 32–36.
56. Raina, R.; Prawez, S.; Verma, P. K.; Pankaj, N. K. Medicinal plants and their role in wound healing. *Online Vet. J.* 2008, *3*, 21.
57. Bairy, K. L.; Rao, C. M. Wound healing profiles of Ginkgo biloba. *J Nat Remedies.* 2001, *1*, 25–27.
58. Inngjerdingen, K.; Nergård, C. S.; Diallo, D.; Mounkoro, P. P.; Paulsen, B. S. An ethnopharmacological survey of plants used for wound healing in Dogonland, Mali, West Africa. *J Ethnopharmacol.* 2004, *92*, 233–244.
59. Hemalatha, S.; Subramanian, N.; Ravich, V.; Chinnaswamy, K. Wound healing activity of *Indigofera enneaphylla* Linn. *Indian J Pharm Sci.* 2001, *63*, 331.
60. Deshpande, S. M.; Upadhyay, R. R. Chemical studies of *Jasminum auriculatum* (Vahl) leaves—II. *Cell Mol Life Sci.* 1968, *24*, 421.
61. Mukherjee, P. K.; Verpoorte, R.; Suresh, B. Evaluation of in-vivo wound healing activity of *Hypericum patulum* (family: hypericaceae) leaf extract on different wound model in rats. *J Ethnopharmacol.* 2000, *70*, 315–321.
62. Karow, J. H.; Abt, H. P.; Fröhling, M.; Ackermann, H. Efficacy of Arnica montana D4 for healing of wounds after Hallux valgus surgery compared to diclofenac. *J Altern Complement Med.* 2008, *14*, 17–25.
63. Kishore, G. S.; Kumar, B. S.; Ramachandran, S.; Saravanan, M.; Sridhar, S. K. Antioxidant and wound healing properties of Glycyrrhiza glabra root extract. *Indian Drugs.* 2001, *38*, 355–357.
64. Preethi, K. C.; Kuttan, R. Wound healing activity of flower extract of *Calendula offlcinalis*. *J Basic Clin Physiol Pharmacol.* 2009, *20*, 73–80.
65. Dash, G. K.; Murthy, P. N. Evaluation of Argemone mexicana Linn. leaves for wound healing activity. *J Nat Prod Plant Resour.* 2011, *1*, 46–56.
66. Choudhary, G. P. Wound healing activity of the ethanol extract of Terminalia bellirica Roxb. fruits. *Nat. Prod. Radiance*. 2008, *1*, 19–21.
67. Diallo, D.; Paulsen, B. S.; Liljebäck, T. H.; Michaelsen, T. E. Polysaccharides from the roots of Entada Africana Guill. et Perr., Mimosaceae, with complement fixing activity. *J Ethnopharmacol.* 2001, *74*, 159–171.
68. Nagesh, H. N.; Basavanna, P. L.; Kishore, M. S. Evaluation of wound healing activity of ethanolic extract of Azadirachta Indica leaves on incision and excision wound models in Wistar

albino rats. *Int J Basic Clin Pharmacol.* 2017, *4*, 1178–1182.
69. Puratchikody, A.; Devi, C. N.; Nagalakshmi, G. Wound healing activity of *Cyperus rotundus* Linn. *Indian J Pharm Sci.* 2006, *68*, 97.
70. Esimone, C. O.; Ibezim, E. C.; Chah, K. F. The wound healing effect of herbal ointments formulated with Napoleona imperialis. *J Pharm Allied Sci.* 2006, *3*, 294–299.
71. Mensah, A. Y.; Sampson, J.; Houghton, P. J.; Hylands, P. J.; Westbrook, J.; Dunn, M.; et al. Effects of Buddleja globosa leaf and its constituents relevant to wound healing. *J Ethnopharmacol.* 2001, *77*, 219–226.
72. Sumitra, M.; Manikandan, P.; Suguna, L. Efficacy of Butea monosperma on dermal wound healing in rats. *Int J Biochem Cell Biol.* 2005, *37*, 566–573.
73. Nagappa, A. N.; Cheriyan, B. Wound healing activity of the aqueous extract of Thespesia populnea fruit. *Fitoterapia.* 2001, *72*, 503–506.
74. James, O.; Victoria, I. A. Excision and incision wound healing potential of Saba florida (Benth) leaf extract in Rattus novergicus. *Int J Pharm Biol Res.* 2010, *1*, 101–107.
75. Roy, K.; Shivakumar, H.; Sarkar, S. Wound healing potential of leaf extracts of Ficus religiosa on Wistar albino strain rats. *Int J Pharm Tech Res.* 2009, 1, 506–508.
76. Garg, V. K.; Paliwal, S. K. Wound-healing activity of ethanolic and aqueous extracts of Ficus benghalensis. *J Adv Pharm Technol Res.* 2011, *2*, 110–114.
77. Villegas, L. F.; Marçalo, A.; Martin, J.; Fernández, I. D.; Maldonado, H.; Vaisberg, A. J.; et al. (+)-epi-Alpha-bisabolol [correction of bisbolol] is the wound-healing principle of Peperomia galioides: investigation of the in vivo wound-healing activity of related terpenoids. *J Nat Prod.* 2001, *64*, 1357–1359.
78. Park, E. H.; Chun, M. J. Wound healing activity of Opuntia ficus-indica. *Fitoterapia.* 2001, *72*, 165–167.
79. Hemmati, A. A.; Kalantari, H.; Jalali, A.; Rezai, S.; Zadeh, H. H. Healing effect of quince seed mucilage on T-2 toxin-induced dermal toxicity in rabbit. *Exp Toxicol Pathol.* 2012, *64*, 181–186.
80. Kayser, O.; Kolodziej, H. Antibacterial activity of extracts and constituents of Pelargonium sidoides and Pelargonium reniforme. *Planta Med.* 1997, *63*, 508–510.
81. Pepeljnjak, S.; Kalodera, Z.; Zovko, M. Antimicrobial activity of flavonoids from Pelargonium radula (Cav.) L'Hérit. *Acta Pharm.* 2005, *55*, 431–435.
82. Ivanova, A.; Mikhova, B.; Najdenski, H.; Tsvetkova, I.; Kostova, I. Antimicrobial and cytotoxic activity of Ruta graveolens. *Fitoterapia.* 2005, *76*, 344–347.
83. Lodhi, S.; Pawar, R. S.; Jain, A. P.; Singhai, A. K. Wound healing potential of Tephrosia purpurea (Linn.) Pers. in rats. *J Ethnopharmacol.* 2006, *108*, 204–210.
84. Rasik, A. M.; Shukla, A.; Patnaik, G. K.; Dhawan, B. N.; Kulshrestha, D. K.; Srivastava, S. Wound healing activity of latex of *Euphorbia neriifolia* Linn. *Indian J Pharmacol.* 1996, *28*, 107.
85. Subashini, R.; Rakshitha, S. U. Phytochemical screening, antimicrobial activity and in vitro antioxidant investigation of methanolic extract of seeds from *Helianthus annuus* L. *Chem Sci Rev Lett.* 2012, *1*, 30–34.
86. Udupa, S. L.; Shetty, S.; Udupa, A. L.; Somayaji, S. N. Effect of *Ocimum sanctum* Linn. on normal and dexamethasone suppressed wound healing. *Indian J Exp Biol.* 2006, *44*, 49–54.

87. Mehra, K. S.; Mikuni, I.; Gupta, U.; Gode, K. D. *Curcuma longa* (Linn) drops in corneal wound healing. *Tokai J Exp Clin Med.* 1984, *9*, 27–31.
88. Dewangan, H.; Bais, M.; Jaiswal, V.; Verma, V. K. Potential wound healing activity of the ethanolic extract of *Solanum xanthocarpum* schrad and wendl leaves. *Pak J Pharm Sci.* 2012, *25*, 189–194.
89. Swami Handa, S.; Khanuja, S. S.; Longo, G.; Rakesh, D. D; United Nations Industrial Development Organization. Extraction technologies for medicinal and aromatic plants. International Centre for Science and High Technology, Trieste, Italy. 2008, 29.
90. Ingle, K. P.; Deshmukh, A. G.; Padole, D. A.; Dudhare, M. S.; Moharil, M. P.; Khelurkar, V. C. Phytochemicals: Extraction methods, identification and detection of bioactive compounds from plant extracts. *J Pharmacogn Phytochem*. 2017, *6*, 32–36.
91. Raghavan, R. N.; Somanathan, N.; Sastry, T. P. Evaluation of phytochemical-incorporated porous polymeric sponges for bone tissue engineering: a novel perspective. *Proc Inst Mech Eng [H]*. 2013, *227*, 859–865.
92. Jain, A.; Dixit, J.; Prakash, D. Modulatory effects of Cissus quadrangularis on periodontal regeneration by bovine-derived hydroxyapatite in intrabony defects: exploratory clinical trial. *J Int Acad Periodontol.* 2008, *10*, 59–65.
93. Martel Estrada, S. A.; Rodríguez Espinoza, B.; Santos Rodríguez, E.; Jiménez Vega, F.; García Casillas, P. E.; Martínez Pérez, C. A.; et al. Biocompatibility of chitosan/Mimosa tenuiflora scaffolds for tissue engineering. *J Alloys Compd.* 2015, *643*, 119–123.
94. Kim, H. Y.; Kim, C. S.; Jhon, G. J.; Moon, I. S.; Choi, S. H.; Cho, K. S.; et al. The effect of safflower seed extract on periodontal healing of 1-wall intrabony defects in beagle dogs. *J Periodontol.* 2002, *73*, 1457–1466.
95. Soumya, R. S.; Sherin, S.; Raghu, K. G.; Abraham, A. Allicin functionalized locust bean gum nanoparticles for improved therapeutic efficacy: An in silico, in vitro and in vivo approach. *Int J Biol Macromol.* 2018, *109*, 740–747.
96. Kumar, R. P.; Abraham, A. PVP-coated naringenin nanoparticles for biomedical applications—In vivo toxicological evaluations. *Chem Biol Interact.* 2016, *257*, 110–118.
97. Ghosh, S.; Patil, S.; Ahire, M.; Kitture, R.; Kale, S.; Pardesi, K.; et al. Synthesis of silver nanoparticles using Dioscorea bulbifera tuber extract and evaluation of its synergistic potential in combination with antimicrobial agents. *Int J Nanomed.* 2012, *7*, 483–496.
98. Chauhan, S.; Upadhyay, M. K.; Rishi, N.; Rishi, S. Phytofabrication of silver nanoparticles using pomegranate fruit seeds. *Int J Nanomater Biostructures.* 2011, *1*, 17–21.
99. Kaviya, S.; Santhanalakshmi, J.; Viswanathan, B.; Muthumary, J.; Srinivasan, K. Biosynthesis of silver nanoparticles using citrus sinensis peel extract and its antibacterial activity. *Spectrochim Acta A Mol Biomol Spectrosc*. 2011, *79*, 594–598.
100. Dipankar, C.; Murugan, S. The green synthesis, characterization and evaluation of the biological activities of silver nanoparticles synthesized from Iresine herbstii leaf aqueous extracts. *Colloids Surf B Biointerfaces.* 2012, *98*, 112–119.
101. Francis, S.; Joseph, S.; Koshy, E. P.; Mathew, B. Microwave assisted green synthesis of silver nanoparticles using leaf extract of elephantopus scaber

and its environmental and biological applications. *Artif Cells Nanomed Biotechnol.* 2018, *46*, 795–804.
102. Suman, T. Y.; Rajasree, S. R.; Kanchana, A.; Elizabeth, S. B. Biosynthesis, characterization and cytotoxic effect of plant mediated silver nanoparticles using *Morinda citrifolia* root extract. *Colloids Surf B Biointerfaces.* 2013, *106*, 74–78.
103. Shameli, K.; Ahmad, M. B.; Zamanian, A.; Sangpour, P.; Shabanzadeh, P.; Abdollahi, Y.; et al. Green biosynthesis of silver nanoparticles using Curcuma longa tuber powder. *Int J Nano med.* 2012, *7*, 5603.
104. Rauch, J.; Kolch, W.; Laurent, S.; Mahmoudi, M. Big signals from small particles: Regulation of cell signaling pathways by nanoparticles. *Chem Rev.* 2013, *113*, 3391–3406.
105. Hollinger, J. O.; Einhorn, T. A.; Doll B. Sfeir, C. *Bone Tissue Engineering.* CRC Press, 2004.
106. Dhandayuthapani, B.; Yoshida, Y.; Maekawa, T.; Kumar, D. S. Polymeric scaffolds in tissue engineering application: A review. *Int J Polym Sci.* 2011, *2011*, 19.
107. Temenoff, J. S.; Mikos, A. G. *Biomaterials: The Intersection of Biology and Materials Science.* Pearson/Prentice Hall Upper Saddle River, NJ, USA, 2008.
108. Vasita, R.; Katti, D. S. Nanofibers and their applications in tissue engineering. *Int Nanomed.* 2006, *1*, 15.
109. El Sherbiny, I. M.; Yacoub, M. H. Hydrogel scaffolds for tissue engineering: Progress and challenges. *Glob CardiolSciPract.* 2013, *3*, 38.
110. Patel, H.; Bonde, M. Srinivasan G. Biodegradable polymer scaffold for tissue engineering. *Trends Biomater Artif Organs.* 2011, *25*, 20–29.
111. Weigel, T.; Schinkel, G.; Lendlein, A. Design and preparation of polymeric scaffolds for tissue engineering. *Expert Rev Med Devices.* 2006, *3*, 835–851.
112. Bandyopadhyay, A.; Bose, S. *Characterization of Biomaterials.* Newnes; 2013.

CHAPTER 17

Effect of Surface Nanostructuring on Gene Expression for Protein Synthesis, Osteoblast Cells Recruitment and Size of Focal Adhesion: A Review

NANCY RAJ[1], RAHUL AGRAWAL[2], K. CHATTOPADHYAY[3], R. BANSAL[1], and V. SINGH[3,*]

[1]*Faculty of Dental Science, Banaras Hindu University, Varanasi, Uttar Pradesh, India*

[2]*Department of Metallurgical Engineering and Materials Science, IIT Bombay, Mumbai, Maharashtra, India*

[3]*Department of Metallurgical Engineering, IIT (BHU), Varanasi, Uttar Pradesh, India*

[*]*Corresponding author. E-mail: drvakilsingh@gmail.com*

ABSTRACT

Dental implants provide oral rehabilitation by mimicking bone topography, faster bone formation, and rapid bone healing to enable early loading protocol with better stability. The dental implant surface affects the mechanical stability of the implant affecting cell adhesion and osteogenic cell response. Different techniques have shown biological advantages through increase in surface wettability, surface roughness, and altered surface topography. These have benefitted osseointegration in both healthy as well as a compromised bone through its underlying cellular mechanisms. The effects of nanostructuring are discussed in this chapter in terms of gene expression, biochemical stimuli, biocompatibility, corrosion resistance, cell metabolism, and its differentiation.

17.1 INTRODUCTION

Metals, polymers, ceramics, and composites are a major class of materials used as a biomaterial. Metallic biomaterials comprise 70%–80% of all types of implant biomaterials and have longest history among the various classes of biomaterials [1–3]. Stainless steel was first successfully used as implant material in mid-1920s. Co–Cr alloys, titanium (Ti)-alloys, Mg-alloys, Nb-alloys, and so forth are among various metallic biomaterials with acceptable biocompatibility. Metals are generally used as stem in dental implants in partially and completely edentulous patients as bone plates, screws and pins, knee joint, and hip joint [4, 5].

Osseointegration is the process of formation of a bony interface between the implant surface and bone. The implant with surface irregularities and active surface for the migration of osteoblast cells and connective tissues is known as an osseointegrated implant. Surface characteristics of implant affect the process of osseointegration through which the molecular and cellular behaviors are affected. Osseointegration is the foremost consideration for the success of bone implants and commonly illustrates the response of bone tissue to Ti [6–8].

The process of osseointegration is similar to that occurring in the healing of bone in a fractured or damaged bone. The stages involved in fracture healing are granulation, tissue formation, callus formation, and remodeling of fractured bone into original bone contour. Most of the implants fail due to poor adhesion with bone (osseointegration), stress shielding, or aseptic loosening caused due to mismatch of elastic modulus, inflammation caused by wear debris, and toxicity induced by release of metallic ions such as those of Al, V, and Ni, and so forth [9].

For better osseointegration, it is essential that a physically effective long-term relation between the implant surface and peri-implant bone should be established through modifications of the surface of the implant to reduce the failure rate of the implant and exert a vital role in bone healing. Surface plays a crucial role in biological interactions as it is the only part of biomaterial in contact with the body environment. Also, surfaces seldom have composition and morphology similar to the bulk [10].

Implantation of an orthopedic and dental device results in exposure of the surface of the biomaterial to biological fluids and the surface interacts with the host ions and cells. Likewise, alteration in surface prior to the implant placement affects these responses and thus the process of osseointegration [11]. Surface modifications of implants are thus needed to enhance protein adsorption or protein desorption, over the implant depending on the application [12–15].

Whenever the surface of the biomaterial is altered, not only its biological properties are changed but also corrosion properties are rehabilitated. A highly corrosive environment including blood, water, sodium, chlorine, plasma, proteins, amino acids, saliva, and so forth surround the implant. Human body fluid consists of various anions such as chloride, phosphate, bicarbonates, cations those of sodium, potassium, calcium, magnesium, and so forth, low and high weight biopolymers, and dissolved oxygen [16].

17.2 SURFACE MODIFICATION OF BIO-IMPLANTS USED FOR NANOSTRUCTURING

In the application of implant, emphasis is there on the manufacturing of functional materials, tailored especially to perform an intended biological function. For most of the cases, instead of altering the entire material, modification of the surface has proven more economical and biocompatible. The surface treatments such as acid etching, sand blasting, surface coating, alkali-heat treatment, plasma treatment, and ion implantation techniques of Ti and its alloys have been reported for enhancement of the osseointegration. The important parameters of the various techniques of surface modification are listed in Table 17.1.

Many of the metals and alloys form a bioinert protective oxide film which is strongly adherent and chemically impermeable and thus prevents a reaction between the metal and the surrounding tissues. Passive surface oxide film prevents corrosion but in a constant aqueous environment under cyclic loading, the passive oxide layer gets impaired. Surface modification improves the biological performances through stabilization of the surface reactivity and electrical protection of implants [17, 18]. Surface oxide layer reacts with water and hydroxyl group readily forms protective layer within 30 ms which has high corrosion resistance.

Different techniques have been used to alter surface topography such as adding particles on the surface and forming surface heaps in contrast to techniques where particles are removed from the surface and pits or ditches are created on the surface.

Additive Processes

1. Hydroxyapatite (HA), Ti, and other calcium phosphate coatings by plasma-spraying.
2. Ion deposition.

Subtractive Processes

1. Blasting
2. Etching

TABLE 17.1 Different Techniques of Surface Modification of Implants and Their Characteristic Features

S. No.	Surface Treatment	Advantages	Disadvantages
1.	Sandblasting, large grit and acid etching (SLA) [28, 45–48]	• Increase in surface roughness and high potential of improving the healing of implant bone [45]. • Combination of both macro-roughness (~5–20 μm in diameter) and micro-pits (~0.5–3 μm in diameter) by sandblasting with large-grit (25–50 mm) followed by acid etching [42]. • SLA surface were found to have a 50%–60% mean value of bone implant contact after 6 weeks [42]. • Two-step treatment (acid-alkali) showed optimized morphology (1–2 μm) high shear strength and good bioactivity resulted in good and early osseointegration [28].	• Gene expression for protein synthesis at nanoscale roughness revealed that only human mesenchymal stem cells (hMSCs) can be sensitive.
2.	Sandblasting [28]	• Increases both surface roughness as well as osseointegration.	• Lacks information on properties like bone implant contact, values of removal torque, response of tissues, and biocompatibility [28] • Surface of the SLA treated surface was rough and irregular after sandblasting; water contact angle was >117° (hydrophobic).

TABLE 17.1 *(Continued)*

S. No.	Surface Treatment	Advantages	Disadvantages
3	Plasma spray coating (calcium phosphate, ZrO_2 HA, TiO_2) [19, 23, 28, 29]	• Shows higher surface roughness compared to acid etched and grit blasted surface • Implant surfaces are porous and promote bone contact • Zirconia also possesses good potential for dental implants whereby it increases microhardness and roughness [5.7 ± 0.2 μm) and mechanical properties when coated onto Ti alloy.	HA coating has less bone contact compared to other surface modifications. • Bond strength of HA coating weakens after long time due to degradation of cohesive bonding which results in poor long-term adherence of the coating to the substrate material [19–23] • Nonuniform thickness in the deposited layer, and variation in crystallinity and composition of the coating. • Titanium plasma sprayed surface had only a 30%–40% mean value of BIC after a period of 6 weeks [28, 29]
4	Grit blasting [31, 33–35]	• Microhardness is very high (425–600 μm) [31] • Alumina grit blast produces surface roughness of 4.15 ± 0.26 μm [34] • Ti blast shows highest surface roughness of 8.55 ± 0.78 μm	• Zirconia blasted titanium exhibited bacterial adhesion [33, 35]. • Good for surface roughness but not for osseointegration.
5	Acid etching $HF/HNO_3/H_2SO_4$ [38]	• Enhanced the process of osseointegration increasing cell adhesion and bone formation [38]. • Micro rough surface (1–10 μm) allows bone ingrowth.	• Suitability of these acids in etching was not determined which is required for bone implant contact and torque removal.
6	Dual acid etching (DAE) [25, 28, 39]	• Rapid osseointegration because of increase in bone implant contact [25, 28] • Greater resistance to reverse torque removal. • Roughness from 0.44 to 3.51 μm [39]	• The treatment of acid etching is heavily dependent on selection of the acid and the process.

17.2.1 PLASMA SPRAY COATING

Plasma spray technique is the spraying of melted materials on implant surface; it usually gives thick layer of depositions, such as HA and Ti and has a good combination of biocompatibility and mechanical properties [19]. The surface area of implants is substantially increased by plasma spray through an increase in their surface roughness [20, 21]. An excellent bioactivity of the composite coating of HA/YSZ/Ti-6Al-4V is achieved by a metastable calcium phosphate solution. These coatings induce bone-like apatite nucleation and growth on the implant surface [22]. Fouda et al. [23] reported that rapid heating occurred in HA-coated implants in comparison to uncoated implants. HA coatings were found to enhance cell proliferation [24]. However, according to Liu et al. [25] and Yang et al. [26] an immersion of the HA-coated implant in simulated body fluid (SBF) for a long time leads to weakening of the HA coating due to the degradation of cohesive bonding in the coating. Knabe et al. [27] observed that plasma-sprayed Ti surface showed larger surface roughness compared to acid-etched and grit-blasted surface; and in an in-vitro test, relatively less bone contact was observed with HA coating processes compared with other surface modifications. Addition of yttria-stabilized zirconia (YSZ) was found to cause a significant improvement in mechanical properties of HA coating [28]. It was observed earlier that reinforced mini HA-coating with zirconia performed better in bond strength and dissolution behavior of the Ti implants. During the period of 4 weeks, following the SBF immersion, there was ~27.7% reduction in tensile strength of the HA/YSZ/Ti-6Al-4V composite coating whereas, it was ~78.8% in the case of pure HA coating. Alkaline-modified plasma-sprayed implants showed more bone growth and less clinical healing times [29]. The homogenous distribution of YSZ particles in the coating matrix increased the strength of the matrix [30] by increase in bonding within the composite and improving the mechanical properties.

17.2.2 GRIT BLASTING

The process of blasting ceramic, silica, HA, alumina, or TiO_2 particles onto the implant surface has been termed as grit blasting. An acid etching always follows grit blasting to remove the residual blasting particles. Microhardness of the zirconia blasted Ti surface was found to be higher than that of a controlled polished Ti surface [31, 32]. Al-Radha et al. evaluated bacterial adhesion on a rough surface and found greater adhesion on ZrO_2 blasted Ti than on the smooth surface [33]. Aparicio et al. used

425–600 μm alumina particles in a similar case to achieve high surface roughness [34]. In-vivo studies by Bacchelli et al. revealed that CP-Ti showed highest surface roughness in microns, followed by ZrO_2 sand blasting with improved osteogenesis. However, this technique improved only the surface roughness but not the osseointegration [35].

17.2.3 ACID ETCHING

The process of acid etching cleans the metallic surface and also increases the roughness [36]. Strong acids like hydrofluoric (HF), nitric (HNO_3), and sulfuric (H_2SO_4) or combinations of these are commonly used [37]. Implant surfaces etched with acid had increased cell adhesion and bone formation because the homogeneous roughening enhances bonding strength. Alla et al. reported that nanotopography due to acid etching allows more bone ingrowth [38]. Disadvantage of this technique is the dependence of the rate of etching both on the type and concentration of the acid used, however the suitability of these acids in etching was not assessed to evaluate the contact of bone and implant and also the removal torque [39].

17.2.3.1 THE PROCESS OF DUAL ACID ETCHING

Like acid etching, process of dual acid etching provides microsurface roughness and enhances osseointegration [40, 41]. Yang et al. examined the effect of surface roughness, inserting 15 implants into tibias of rabbits, a histomorphometric analysis revealed higher reverse torque for removal after 2, 4, and 8 weeks due to better osseointegration through peri-implant bone formation [25, 28]. Juodzbalys et al. observed that an acid-etched surface exhibited similar topography as large-grit sand-blasted and acid-etched surface [40]. An appreciable surface roughness of 1–10 μm micropits was obtained from etching with H_2SO_4 followed with HCl in comparison to a poor surface microtexture achieved from HCl and followed by H_2SO_4. Also porosities of size ranging from 0.5 to 2 μm formed from the use of these acids. With an increase in the torque rotation force, bone-to-implant contact also increased and led to greater osseointegration [3, 9].

17.2.4 BLAST BY SAND, LARGE-GRIT, FOLLOWED BY ACID-ETCHING

This technique of blasting with large-grit and sand particles followed by acid etching was applied sequentially to obtain macro-roughness and micro-pits for increasing the surface roughness and also osseointegration in Ti implants [42–44]. Cho and Jung reported that surface processed by this technique possessed both macro-roughness (~5–20 μm in diameter) and micro-pits (~0.5–3 μm in diameter) which increases surface area [44]. Taborelli et al. confirmed that the water contact angle of the sand blast, large grit, and acid etching (SLA) surface was about 117°±2.7, while Buser et al. measured the dynamic contact angle (DCA) of the SLA surface and found that the SLA surface was hydrophobic (DCA=138.3°±4.2). In-vivo studies were carried out on six adult dogs by Xue et al. [45] which showed that alkaline treatment after blasting exhibited high shear strength, improved early bone growth, and osseointegration. In a recent study on a two-step chemical treatment (acid-alkali), it was observed that optimized morphology, higher shear strength, and better bioactivity promoted osseointegration during the early stage of the implantation. He et al. [28] also observed that the implants treated with blasting followed by double acid etching of HCl and H_2SO_4 enhance osseointegration during the healing stage, indicating significant improvement in the bioactivities. Also, the biological evaluation by He et al. [28] and Kim et al. [42] revealed that human osteoblasts grow profusely on the treated surfaces which provide a larger area for attachment of cells and their proliferation. Initially, surface morphology after sand-blasting became rough and irregular, but following the acid etching treatment the surface became more uniform and small micro-pits (1–2 μm in diameter) formed. The surface of titanium oxide formed after sand-blasting shows low surface energy due to the absorption of carbonates and hydrocarbons from air. During acid etching, the layer of titanium oxide gets dissolved and small native hydrogen ions diffuse into the subsurface of the implant and enrich the implant surface with hydrogen and precipitate as titanium hydride (TiH). X-ray photoelectron spectroscopy (XPS) analysis indicated that SLA surface had a 44.2±1.9 at% of oxygen, 18.4±1.6 at% Ti, and 37.3±3.4 at% carbon [46].

Currently, the processes of surface roughening by grit blasting, acid etching, and SLA with coatings of CaP and HA are commonly used in practice. Both the methods have their advantages and limitations as discussed in this chapter. Improvement of the bone-implant interface and higher resistance against failure has been reported by acid-etched surface [47]. Sandblasted surface with large-grit (25–50 mm) followed by acid etching were found to have 50%–60% mean value of bone-implant contact (BIC) in comparison with Ti plasma-sprayed surface which exhibited only 30%–40% mean value of bone BIC after 6 weeks [48]. The BIC value is very important in the long-term success of dental implants. Several investigations have shown that rough implants surface have better bone apposition and BIC than smooth surface implants [23]. Surface roughness also stimulates cell migration and their proliferation which in turn promotes BIC [47]. Surfaces treated by different techniques such as sandblasted, large-grit, acid-etched (SLA), and coated were chemically different but their physical properties were same those were conducted to assess BIC as a measure of osseointegration. It was also reported that fatigue properties of Ti-6Al-4V alloy improved after SLA-treatment [49].

17.3 SURFACE NANOCRYSTALLIZATION (SNC) TECHNIQUES

In respect of the different additive and subtractive surface modification techniques presented above, nanostructuring of surface offers great potential. Any modification technique which can induce nanostructuring is preferable as the human cells are in the nanometric range [50]. Lu et al. have classified such SNC techniques into three classes:

1. SNC through deposition of various coatings of nanocrytalline materials over the substrate by physical vapor deposition chemical vapor deposition, and plasma spraying, and so forth.
2. Surface self nanocrystallization by transforming surface layer of materials into nanocrystalline states while keeping overall composition and phases unchanged.
3. By producing a layer transformable to nanostructure range (though the layer may not be nanocrystalline) and then using chemical, thermal, or metallurgical process to produce nanocrystalline layer with different chemical composition or different phase.

17.3.1 TYPES OF SELF-NANOCRYSTALLIZATION SURFACE

1. Mechanically induced self-nanocrystallization of surface involves grain refinement to nanometric regime through mechanical treatment. Surface mechanical attrition treatment (SMAT) is an example of this process.
2. Thermally induced self-nanocrystallization surface is achieved via phase transformations such as melting and solidification. In such cases, nucleation rate is very high and growth rate is small.

Tao et al. [51] proposed that nanocrystals at the surface can be obtained by severe plastic deformation techniques like SMAT, in which there are repeated impacts of the surface by high-speed balls driven by generator, air blasts, and so forth. Such techniques can be of two types depending on the mechanism of acceleration of balls.

In type-I SMAT, balls are accelerated via collision with a vibrating chamber driven by an electric motor. In type-II process, ultrasonic generators are used. Vibration frequency ranges from 50 Hz in type-I to 20 kHz in type-II. Ball size can range between 1 and 10 mm diameters and can attain velocity up to 20 m/sec. Figure 17.1 shows a schematic of SMAT type-II, ultrasonic shot peening (USSP) unit. Craters like features are formed over the treated surface due to repeated impacts (Figure 17.2).

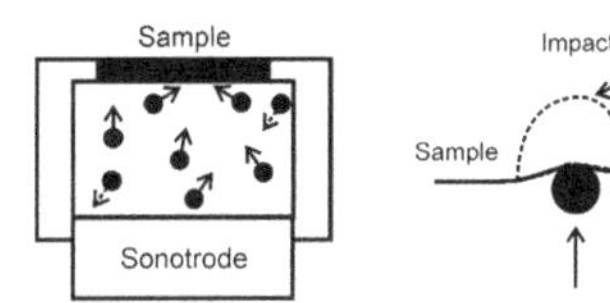

FIGURE 17.1 (a) Schematic of USSP setup, (b) schematic showing formation of plastic zone due to impact of shots. *Adapted from V. Singh et al., Trans. IIM., 69(2) (2016) pp. 295—301.

In SMAT, strain and strain rates vary along with the depth of the specimen and a gradient microstructure is obtained. This modified layer consists of a nanostructured layer at the top and up to 15-μ depth, below this it is submicron-sized regime, backed by micro-sized strained regime. Matrix or the unaffected region is as deep as 100 μm from the top surface [52]. The mechanism of nanocrystallization by SMAT for different metals and alloys has been well studied [53].

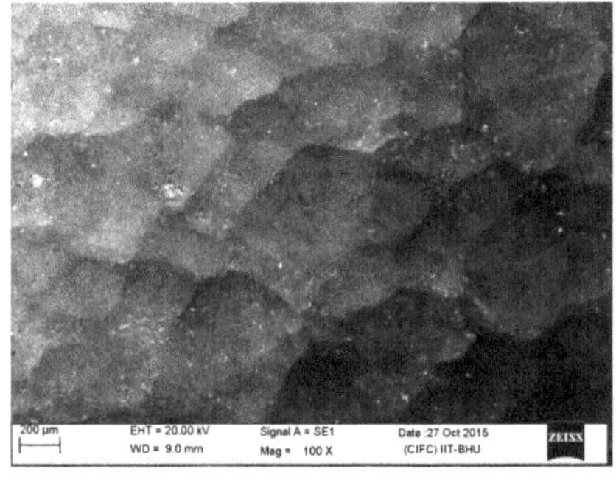

FIGURE 17.2 Surface morphology of a USSP treated CP-Ti sample, showing craters formed due to impact of shots.

The process of SMAT involves severe plastic deformation at the surface of the material. At the surface, there is a high strain and strain rate [54–56]. Surfaces of SMATed samples are characterized by the presence of craters due to the impact of balls during the treatment. Due to this initially, there is increase in surface roughness. As more and more region is covered by craters of impact, with an increase in the duration of shot peening, roughness saturates and at this instant, the height of peak to valley of craters becomes constant [57]. Consequently, the surface undergoes a large amount of plastic deformation and a gradient microstructure with equiaxed nanograins at the top, followed by equiaxed ultrafine grains and a deformed matrix is created [51].

SMAT additionally induces residual stresses in the component. These residual stresses are compressive in nature and the magnitude increases with depth to a peak value and thereafter decreases. On heat treatment, the residual stresses are relieved due to which corrosion and biological properties are increased. Zhu et al. studied the effect of different processing parameters such as the duration of peening, diameter of shot, amplitude of sonotrode, and distance of peening of the USSP process on the hardness and microstructure evolution of CP Ti and correlated the hardness with USSP duration [58].

The topmost layer in the USSP treated sample has ultrafine and/or nanograins as shown in Figure 17.3, below this, there is a region of refined grains, progressively increasing with depth but smaller than that of the matrix, associated with a large amount of plastic deformation. Below this, there is a substrate of undeformed grains. Figure 17.4 shows a cross-section of the USSPed CP-Ti sample. Due to large plastic deformation at the surface, extensive twinning occurs, evident by sets of parallel lines like features. The density of such features decreases on moving away from the treated surface. The thickness of different regions are shown in Figure 17.3 varies with different parameters. The thickness of the topmost ultrafine/nanograins increases with peening duration. However, on longer duration of peening, the thickness of the grain refined layer reduces as shown in Figure 17.5. The microstrain also follows a similar trend with peening duration.

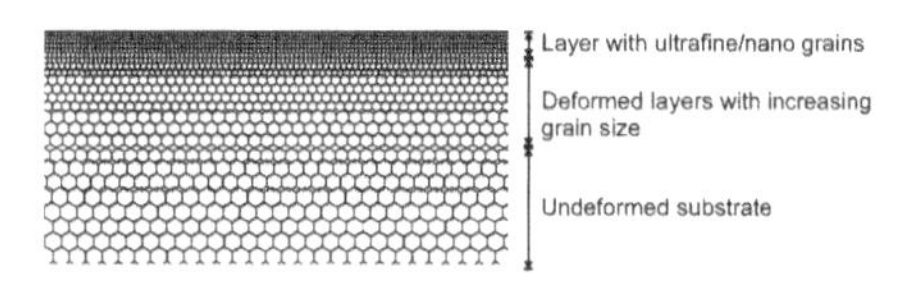

FIGURE 17.3 Schematic showing different layers formed by USSP treatment.

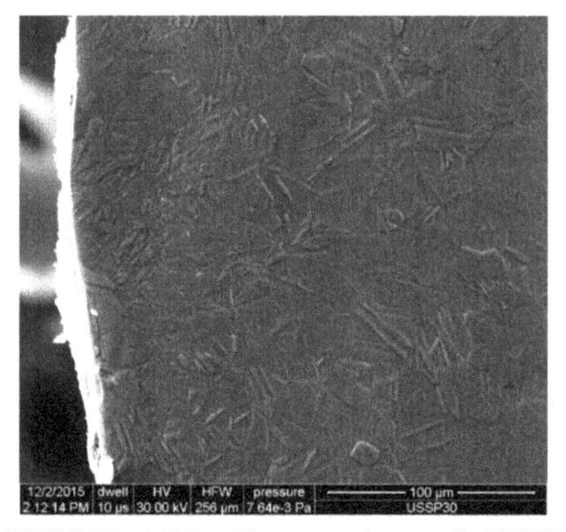

FIGURE 17.4 Cross-section of USSP treated CP-Ti sample showing extensive twinning near the treated surface.

Surface hardness increases rapidly with a small increase in peening duration, in the beginning of shot peening, however it saturates after a long duration of peening. Concurrently, the rate of surface hardening also saturates after a long peening time. Initially, the treated surface undergoes roughening but tends to smoothen on a longer duration of peening.

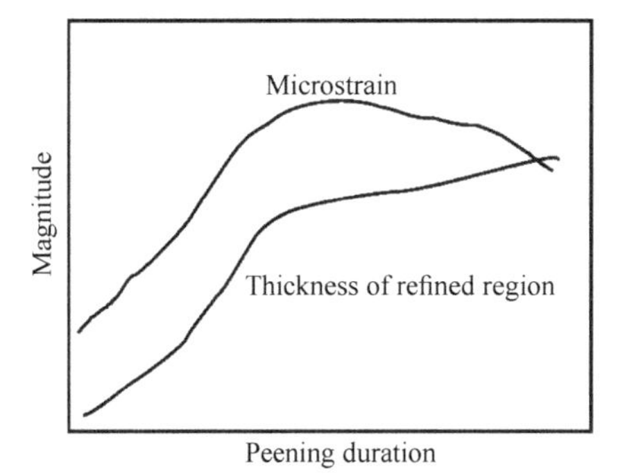

FIGURE 17.5 Variation in microstrain and thickness of refined layer as a function of peening duration.

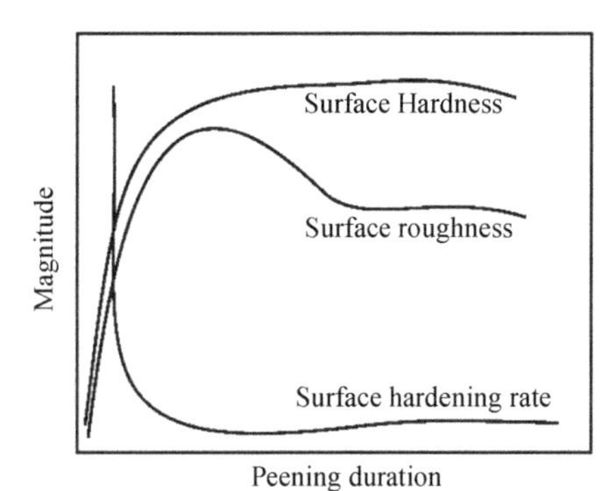

FIGURE 17.6 Variation in surface hardness, hardening rate at surface, and roughness with peening duration.

The other important peening parameter is the diameter of shot. For a given peening duration, the surface hardness increases with the shot diameter (Figure 17.7). With an increase in shot diameter, the surface roughness increases but very large diameter shots yield a smoother surface. An important effect of shot peening is the introduction of residual stresses. A typical residual stress variation from the top most surface to interior is shown in Figure 17.8. The residual stress is an important parameter in controlling mechanical properties and corrosion resistance of the material.

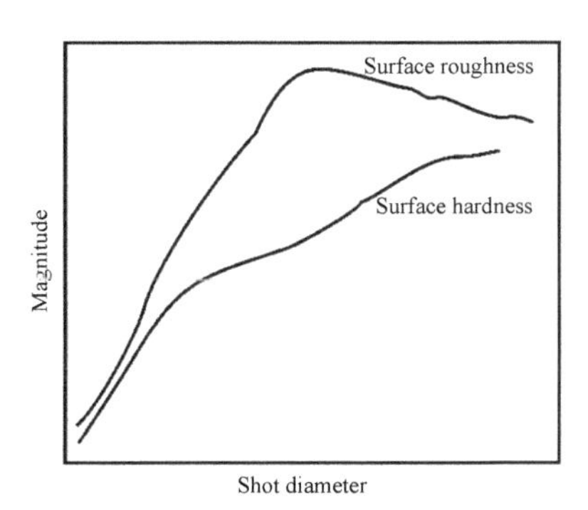

FIGURE 17.7 Variation of surface roughness and surface hardness with shot diameter (all other parameters being same).

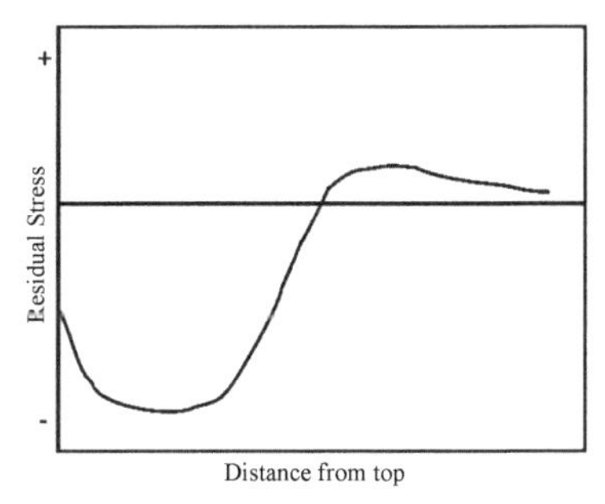

FIGURE 17.8 Residual stress distribution as a function of distance from the top surface for USSP-treated samples.

Liu et al. demonstrated that SMAT is an innovative technique of producing nanocrystalline (NC) layers of several tens of micrometer thickness in the surface region of metallic materials. The microstructural observations showed the formation of (1) dislocation tangles; (2) dislocation bands; and (3) dynamic recrystallization of the dislocation bands until the formation of nano-crystalized Ti [50]. The severe plastic deformation leads to grain refinement (surface nanostructuring) [12] and produces more number of grain boundaries. SMAT induces compressive residual stress, imparts passivation, cell proliferation, and increases the hardness [46, 59–61].

In-vitro and in-vivo testing was carried by Guo et al. for the ultrasonic shot-peened Ti surface [62]. In comparison of the initial coarse-grained microstructure of the surface, following USSP, surface with nanograins showed enhanced osseointegration on evaluation by push-out test histological observations and fluorescent labeling. The enhanced cell growth over the SMATed surface can be from surface nanostructuring, optimum surface roughness, and presence of points of high positive potential on the surface. In a review of suitable severe plastic deformation processes for obtaining nanocrystalline metallic substrate for biomedical application, it was found that at grain boundaries of such metals numerous nanoscale groove-like structures are present. Such topographies are beneficial in improving cell adhesion [63]. Nanostructures on the surface of the implant can have the adsorbed proteins in certain configurations for favorable bioactivity. These retained proteins help in cellular adhesion and differentiation [64, 65].

A porous structure over the implant is generally preferred because of a larger contact area. Also, the cells are capable to migrate inside such pores. This interpenetration helps in better osseointegration [66]. The oxidized film formed over the treated specimen may provide such a porous structure. There are large numbers of sites for nucleation of oxide films because of defects generated from SMAT. The cumulative oxide film formed on such surfaces would be porous, giving a beneficial effect to cell behavior [67].

It is very well established that cell adhesion and viability over the implant surface is related to protein adsorption over these surfaces. Cells are indirectly controlled by surface adsorbed proteins and not by direct signaling from the nanoscale surface [64]. The treated samples have shown points of higher positive potential on the surface. Such sites could act as preferential adhesion points for proteins. These adhered proteins can further help in increasing cell viability. Non-negative potential on the surface enhances cell spreading as compared with negative potential [68]. Negative potential of 300 mV (with respect to Ag/AgCl) on surface of Ti has been shown to cause cell death [69]. It was shown that a properly modeled cell potential could modulate protein production along with modulating gene expression and thus cell proliferation [70]. Several studies have confirmed the beneficial effect of positive potential on the implant surface [71–73].

The role of shot peening as a process of surface modification of bio-metallic Ti implant has been demonstrated by Singh [74]. The cell proliferation was improved by 10% for the shot peened sample and was enhanced by 50% in case of annealed shot-peened samples. Formation of nanograins was considered responsible for improved cell viability. However, the role of residual stresses was found to be detrimental for cell viability.

It has been shown that a net positive charge develops on SMATed surface of 316L stainless steel [9]. This positive charge helps in cellular adhesion and prevents denaturing of adsorbed proteins such as fibronectin. Corrosion fatigue resistance of the treated samples was also increased by 50% in a saline medium.

USSP induces SNC even when the duration of treatment is as low as 120 seconds [9, 58, 62, 75]. Grain refinement occurs from the combined process of twinning and slip. The beneficial effect of USSP on cell proliferation is attributed to SNC of the surface along with an increase of positive potential at the treated surface. While nanoscale structure at the surface provides dimensionally proportionate bonding sites to cells, sites of positive potential act as preferential adhesion points to proteins.

17.3.2 BRIEF DESCRIPTION OF FOCAL ADHESION, CELL PROLIFERATION, MATURATION AND PROTEIN SYNTHESIS

17.3.2.1 FOCAL ADHESION COMPLEXES

The focal adhesions comprise of large size molecular complexes that link transmembrane receptors; like integrins, to the actin cytoskeleton and mediate signals modulating cell attachment, migration, proliferation, differentiation, and gene expression. These complexes are heterogeneous and have dynamic structures that are apparent targets of regulatory signals that control the function of focal adhesions [76, 77].

After implant placement, there is a huge accumulation of platelet containing growth factors of high level including platelet-derived growth factors, transforming growth factors β1 and β2, vascular endothelial growth factors, platelet derived endothelial growth factors, interleukin 1 and 2, basic fibroblast growth factor, platelet-activating factor 4. The cascade of reaction involves the immediate binding of secreted growth factors to the transmembrane receptors which are present on the external surface of cell membranes at the site of implantation. This activates an endogenous internal signal protein, which further initiates the expression of a normal gene sequence of cell such as formation of matrix, cellular proliferation, production of osteoid, and collagen synthesis [76].

17.3.2.2 CELL ADHESION VIA INTEGRIN

Bone marrow mesenchymal stem cells can be recruited from the neighboring bone marrow or the peripheral circulation and involve in the process of osseointegration, following implantation. Adhesion of cells takes place through the binding of integrins, the receptors that adhere specifically to matrix proteins such as fibronectin and different collagens. Interactions between integrin receptors and extracellular matrix proteins not only provide cells with a physical link to the extracellular matrix and also act as a channel for propagating signals from the extracellular environment [67]. Moreover, surface nanostructure promotes cell adhesion [78].

17.3.2.3 CELL METABOLISM AND PROLIFERATION ACTIVITY

The proliferation rate of cells is calculated by MTT cell metabolic activity assay. The metabolic condition of individual cells and the numbers of cells both promote the total cell metabolic activity [79, 80].

17.3.2.4 OSTEOGENIC DIFFERENTIATION ACTIVITY IN BONE MARROW MSC

Expression of osteogenic differentiation-related genes, including the early m-RNAs markers *Runx-2* and *ALP*, the medium-stage marker osteopontins OPN, and late-stage marker osteocalcin *OCN* are important *during differentiation activities* [81, 82].

17.3.2.5 SEQUENCE LEADING TO OSSEOINTEGRATION

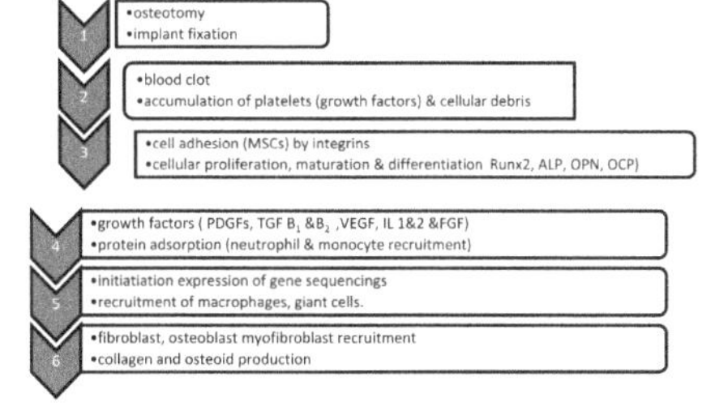

17.4 EFFECT OF NANOSTRUCTURED TOPOGRAPHY ON OSSEOINTEGRATION

Webster et al. characterized and evaluated osteoblast adhesion on nanophase materials such as Ti, Ti-6Al-4V, Cr–CoMo alloys. It was observed that osteoblastic adhesion was preferentially at boundaries of surface particles for both nano- and conventional metals and that increased osteoblastic adhesion occurred on nanotopography because more particle boundaries were present in nanophase topography than in conventional topography [83].

The idea of creating nanotopography of material surfaces with decreased surface dimensions is to mimic the roughness of extracellular matrices in bone [84]. The treatment of Ti and Ti-6Al-4V alloy with H_2SO_4 and H_2O_2 resulted in the formation of nanophase surfaces and enhanced osteoblast osteopontin and bone sialoprotein synthesis and also improved the function of osteoblasts [85].

There is a much higher number of atoms at the surface of nanophase materials compared with the bulk, greater area of increased surface defects such as edge, corner sites and particle boundaries, and also larger proportion of electron delocalization [86]. Nanotopography created by various methods like additives or substractive ultimately stimulates osteogenic differentiation of human mesenchymal stem cells. Focal adhesion was quantified in S-phase of the cells, bone-related transcription factors RUNX2 and osteocalcin were noted [87].

Metabolic and morphologic differences have been shown between plasma Ti and nano-patterned Ti in relation to focal adhesion, transcription factor phospho-RUNX2, cell metabolism, cytoskeletal description of early and decision-making stage of stem cells. MSCs on 15 nm topography were enriched in metabolism that should facilitate gene and protein modulation required for differentiation of MSCs and also increased the level of lipoate phenotype on 15 nm nanotopography. Sphingosine upregulation reduces osteolytic bone loss by the promoting migration of osteoclastic precursors away from the implant site [88]. Labeling of adhesion in S-phase shows reduction in the number of adhesion but size of adhesion is retained reactive to planar control. This study showed that the formation of super mature adhesion (>5 μm in length) is important for osteogenesis. The 15 nm high nanotopography showed enhancement of super mature adhesion than those of 55 nm and 90 nm.

Cellular mechanical properties may serve as a biological marker of cell phenotype (cellular transformation). Stiffness moduli of osteoblast is greatest > chondrocytes > adipocytes > neurons. This study showed that the changes in mechanical properties of hMSCs caused by nanotopography differ as they depend on the stiffness of the substrate. Nanotopography was found to induce hMSCs to differentiate into neuronal-like phenotype with a significant increase in expression of neuronal markers [89].

Stiffness of the substrate in addition to biochemical stimuli plays a role in the differentiation of hMSCs toward selective lineages. Osteoblastic differentiation occurred on the stiffest material like Ti grade V, myogenic on intermediate stiffness, neurogenic on softest material [42]. Formation of nanonetwork surface oxide layer of TiO_2 created significant improvement in in-vitro and in-vivo hMSCs growth [12].

Bicinchoninic acid assay was done for three different Ti films; they exhibited the same surface chemistry but different nanotopography. There was very little influence of surface roughness on albumin or fibrinogen adsorption, osteoblast proliferation, and cell viability among different groups [46]. It was shown that dental implants are affected by their composition and surface roughness [59]. They described different methods used for increasing surface roughness or applying osteoconductive coatings like Ti plasma spray, gritblasting, acid etching, anodization, calcium phosphate coating. They found that osseointegration of dental implants was enhanced by these modifications to enable their immediate loading.

The hMSCs growth and bone specific mRNAs were increased on the surface with superimposed nanoscale features compared to machined and acid-etched [90]. On the 14th day, OSX and mRNA levels were increased by 2, 3.5, 4, and 3 fold for anatase, rutile, alumina, and zirconia, respectively. This improvement in osseointegration response provided a faster and more reliable contact of bone to the implant. PCR array showed improvement in osseointegration response on the implants coated with Al_2O_3 as compared to others. Further, the expression of cartilage oligomeric matrix protein was greater on nanoscale Al_2O_3 coated Ti. Collagen types I and II, BMP1 or pro-collagen C proteinase, BMP 2, 4, and 5 are also upregulated. Transforming growth factor β1, vascular endothelial growth factor A increased at 14th and third day, respectively. Also, an increased value of ALP and osteocalcin mRNAs was reported [91].

The hMSCs and progenitors are sensitive to a small change in topography height on the polymer surface. The cellular response to Ti topographic features of varying heights between 15 and100 nm was studied. Nanopit structured polymer surfaces and increase in pit depth from 14 45 nm reduced cell spreading and adhesion [92]. Reduced cell interaction with increased topography height was also seen from looking at endothelium fibroblast where low features (<20 nm) increased and higher features reduced cell spreading, adhesion, and cytoskeletal formation [17, 18]. Observation of increased cell spreading, formation of larger adhesions, and improved cytoskeleton organization have been reported as early-stage indicators of increased osteospecific cell spreading [18] and the increase in hMSCs spreading has been linked to increased osteospecific differentiation [16, 93, 94]. Formation of larger than numerous adhesions is important in osteospecific differentiation. The focal adhesion is reduced only when the lateral spacing between integrins are greater than 73 nm [92].

Nanotopography changes hMSCs from their shape to differentiation. Zyxin protein expression plays a key role in hMSCs response to nanotopography [95]. Zyxin is recruited to focal adhesion upon their maturation [96]. Zyxin dissociates from focal adhesion upon dissipation of force. Mechanical forces are known to influence the structure of focal adhesions; cellular contractility is essential for the development of focal adhesion [97] and under the force, focal adhesion grows larger [98]. The hMSCs on 350 nm topography displayed mature focal adhesion of smaller size and exhibited a higher migratory speed and caused a decrease in zyxin protein expression.

Formation of TiO_2 nanomesh layer on the anodized surface of Ti increased its corrosion resistance and reduced the corrosion rate and passive current in SBF [99]. The nanomesh also caused better cell adhesion and spreading was achieved by the nanomesh layer; also there was faster cell proliferation and initial differentiation. Thus, the biocompatibility of the Ti surface was improved by the nanomesh layer. Crystallized form of oxide on the surface of CP-Ti has been found to be highly effective in spreading and proliferation of cells [100–102]. Surface chemistry of Ti film prepared by electron beam deposition was the same but exhibited different topography at the nanometer scale (2–21 nm) with different grain sizes. Albumin and fibrinogen adsorption showed no statistical difference between the different groups of Ti and there was no significant effect of surface roughness on osteoblast proliferation and cell viability [46].

Nanotopography causes greater osteoblastic gene expression (Runx2, osterix, alkaline phosphatase bone sialoprotein) and cell adherence in HF-treated nano-surface compared to microscale surface [80]. It indicates rapid osteoblastic differentiation. The surface modification promotes a high level of IGF2, BMP2, and BMP6 expression of adherent hMSCs for prolonged periods of time. Nanoscale roughness modifications have demonstrated increased bone formation and torque removal value [103]. A gap model of osseointegration was used in the canine mandible to demonstrate that implants with HF-modified nanoscale surface ossteointegrate in the early phase of healing following the installation of implant. High survival of implants in effectiveness trial involving HF-modified TiO_2 grit-blasted surface was observed [104].

Surface topography of implants is an important factor which affects the behavior of both protein as well as cells on the implant [19]. It is generally accepted that protein typically responds to surface structural features (pits, pillar, steps) of about 1–10 nm, whereas cells could be sensitive to structural features on the scale of 15–10 μm. Femtometer laser produces nanostructure of the size down to 20 nm on multiple parallel grooved surface with a period of submicron level (1–15 μm).

Stem cells start differentiating into mature tissue cells on exposure to intrinsic properties of the extracellular matrix, matrix structure, elasticity, and composition. These parameters modulate the forces that a cell can exert upon its matrix. Mechanosensitive pathways convert these biophysical cues into biochemical signal so ECM parameters are extremely dynamic and suggest that they play a morphogenetic role in guiding differentiation and arrangement of cells [105].

A new method for deposition of a thin layer of HA on a biomaterial surface is pulsed laser ablation. One set of samples was subjected to annealing treatment at 575 °C to increase the crystallinity of the deposited films. Primary human osteoblasts were seeded onto the surface of the material and cytoskeletal actin organization was examined. The annealed surfaces supported greater cell attachment and more defined cytoskeletal actin organization. Cell activity was observed to be appreciably higher on the annealed sample, also the annealed samples showed higher alkaline phosphatase activity than the unannealed ones. The effect of nanostructured surface morphology on osseointegration and cell interaction process is summarized in Table 17.2.

Ramires et al. studied biocompatibility of composite coatings of titania (TiO)/HA at different ratios achieved by the sol-gel process, studying the behavior of human MG63 osteoblast-like cells. Cytotoxicity and cytocompatibility tests were performed to evaluate biocompatibility. Cytotoxicity test was

TABLE 17.2 Role of Nanoscale Surface Topography on Cell Response

Contributors	Techniques	Remarks
Liu et al. [50]	Surface mechanical attrition treatment (SMAT) is an innovative process which can produce nanocrystalline layer	Grain refinement, induces residual compressive stresses, imparts passivation, and increase in hardness.
Webster et al. [83]	Ti, Ti-6Al-4V, Cr-Co-Mo alloys treated with H_2SO_4 and H_2O_2 create nanophase surfaces.	Osteoblastic adhesion was more at boundaries of nanosize particles than on the conventional metals because of more grain boundaries.
Siegel et al. [86, 87]	Higher number of atoms are there on nanophase materials at the surface compared to the bulk	Nanotopography stimulates osteogenic differentiation of human mesenchymal stem cells
McNamara et al. [88]	Metabolic and morphologic differences between plasma Ti and nano-patterned Ti (between 15, 55, and 90 nm].	15 nm nanotopography: large super mature focal adhesion >5 μ. 55 nm and 90 nm: smaller focal adhesion. Larger focal adhesion is required for good osseointegration.
L. Le. Guehennec et al (17]	Ti plasma spray, grit blasting, acid etching, anodization, calcium phosphate coating	Increased surface roughness enhances osseointegration for immediate loading.
Gustavo Mendonca et al. (29]	Titanium implants coated with anatase (An), rutile, alumina (Al_2O_3) and zirconia (ZrO_2) particles.	Increased value of ALP, osteocalcinmrnas and the expression of cartilage oligomeric matrix protein (COMP] was greater on nanoscale Al_2O_3 coated titanium.
Sjö ström et al [92, 17]	The cellular response to titanium topographic features of varying height between 15 and 100 nm was studied.	<20 nm pits depths and heights, increases cell spreading, adhesion and cytoskeletal formation and (>20 nm had low cellular responses].
Karina Kulangara et al. [96, 51]	Topography, focal adhesion	350 nm topography displayed mature focal adhesion of smaller size and low zyxin expression.
Wei-en Yang [99]	Nanomesh layer of TiO_2 on anodized surface of titanium.	Nano-mesh layer caused better cell adhesion and spreading as well as faster cell proliferation and initial differentiation

TABLE 17.2 *(Continued)*

Contributors	Techniques	Remarks
Singh [74]	Created nano structure (14–20 nm) at the surface of CP-Ti by ultrasonic shot peening	Significant enhancement of almost 50% in cell proliferation in the shot peened and annealed condition.
Yuwq and Daraio et al. [100, 101]	Crystallized form of surface oxide on CP-Ti	Enhanced cell spreading and proliferation.
Cai et al. [46]	Ti film prepared by electron beam deposition surface chemistry	2–21nm grain refinement, no statistical difference between the different groups of titanium on the basis of albumin-fibrinogen adsorption, osteoblastic cell viability and cell proliferation.
Berglundh et al. [104]	Used a gap model of osseointegration in canine mandible	TiO_2 grit blasted and HF treated nano scale surface implants promote the process of ossteointegration in early phase of healing following implant installation.
Vorobyev et al. [19]	Femtometer laser produces nanostructure with a size down to 20 nm, multiple parallel grooved surfaces with a period on submicron level (1–15 μm).	Protein typically responds to surface structural features (pits, pillar, steps] about 1–10 nm, while cells can be sensitive to structural features on the scale 15–10 μm
Balls et al. [105]	Pulsed laser ablation is a novel process of deposition of thin layer of hydroxyapatite (HA) on surface of biomaterial.	Cell activity was significantly higher on the annealed sample. Annealed samples also showed higher alkaline phosphatase activity than the samples not annealed.
Ramires et al. [107, 109]	Applied titania/hydroxyapatite (TiO_2/HA) composite coatings to study the behavior of human MG63 osteoblast-like cells	Bioactive TiO_2/HA coating deposited by sol-gel process causes calcium and phosphate precipitation onto the surface and stimulates the expression of alkaline phosphatase activity, collagen and osteocalcin production
Kokubo et al. [122]	In-vivo study	Good bone bonding ability of treated titanium implants after alkali treatment and heat treatment and it is essential in order to induce bioactivity.

performed with neutral red (NR), MTT, and kenacid blue assays, to evaluate the influence of the material extracts on lysosomes, mitochondria, and cell proliferation, respectively. The MG63 cell was cultured onto TiO/HA substrate to evaluate cell, cell morphology, cell proliferation, alkaline phosphatase activity, collagen, and osteocalcin production of MG63 cells, it was found that these materials had no toxic effects. Cytotoxicity test results explained osteoblastic cell growth, proliferation, and morphology being comparable on all the tested materials and detected by the total protein contents (kenacid blue) assay. Even more concentrated extract had some damaging effects on mitochondria and lysosomes and was found via MTT and NR assay, respectively. However, collagen production of osteoblasts and alkaline-phosphatase-specific activity were significantly higher on the coatings of TiO/HA than on the uncoated Ti and polystyrene of culture plate and were influenced by the chemical composition of the coatings [106]. A comparable cell morphology was observed by SEM analysis on all tested samples except for TiO/HA 0.5 substrate where round cells without ruled membranes were observed. The osteoblastic cells grown on TiO/HA (1:1) coating showed a significant increase in osteocalcin production. This calcium-binding polypeptide represents the peculiar marker of the final differentiation of osteoblasts and plays a fundamental role in bone remodeling. The bioactive coatings of TiO/HA deposited by the process of sol-gel cause precipitation of calcium and phosphate onto the surface and stimulate the expression of alkaline phosphatase activity, osteocalcin, and collagen production. The TiO/HA (1:1) coatings seem to stimulate more differentiation markers of osteoblastic phenotype. The hydroxyl groups, such as Ti/OH, detected on the coating's surface could promote the interactions with bone cells by providing the sites for nucleation of calcium and phosphate [107, 108]. The decrease in calcium concentration in the medium containing TiO/HA coatings, the appearance of a fine structure having a needle-like morphology on the surface of the coatings containing calcium and phosphorous, as detected by EDS analysis, was ascribed to precipitation of this ion. Similar needle-type crystallites with P and Ca have been reported earlier [109, 110].

To consider the bioactivity ability of Ti-6Al-7Nb alloy after alkali treatment and heat treatment in sand-blasted and mirror-polished surfaces were selected as these two represent opposite mechanical state [86]. The morphology and structural analyses were performed during microporous layer formation by alkali treatment and heat treatment, both on mirror polished and sandblasted surfaces of Ti-6Al-7Nb. The morphology of mirror-polished samples (MPA600, MA600) was different due to passivation, but in the case of sand-blasted samples (BA600, BPA600) it was alike in both the cases from the structural point of view since substrate roughness causes a higher degree of crystalline fraction than treated surfaces [111–113].

It is evident that the degree of oxidation during heat treatment is related to their surface roughness and mechanical treatment. Cold-working and blasting increase the degree of oxidation at a fixed temperature [114]. Different chemical (acid) treatments were proposed in literature, using H_2O_2 [115], HCl, and H_2SO_4 [116] as etching solutions but etching was most commonly done with concentrated NaOH solution and subsequent heat treatments at 500–800 °C [117]. Formation of sodium titanate layer on the treated samples showed a bioactive behavior during soaking in vitro in SBF by precipitation of an apatite layer. The layer of apatite formed on the treated Ti is much more strongly bonded to the substrates than that precipitated on the traditional bioactive materials (Bioglass 45S5-type, glass-ceramic apatite-wollastonite and sintered dense HA) [118]. There is a relation between the graded interface structure of the apatite and of titanate layer to the Ti metal substrates [119, 120].

The proposed mechanism, in order to explain the bioactivity of the treated samples, involves the transformation of the alkali titanate layer into TiO_2 hydrogel, via release of alkali ions into SBF and ion exchange with H_3O+ ions from the solution (121]. Kokubo et al. registered good bone-bonding ability of treated Ti implants by in vivo tests. Their results evidenced also that both alkali and heat treatment are essential in order to induce bioactivity in vivo [122, 123].

17.5 CONCLUSION

The surface nanostructuring influences protein adsorption, platelets adhesion, hemostasis, inflammation, and osteogenic cell response. The use of Ti and its alloys in biomedical field is now established due to their lower Young's modulus, superior biocompatibility, and enhanced corrosion resistance compared to the conventional stainless steels and cobalt-based alloys.

Mechanisms of the different tissue responses to different surface topographies have been presented based on in-vitro studies with a due uncertainty in the interpretation of courses. It promotes molecular response, cellular interaction, bone regeneration, mesenchymal stem cell involvement, and cell to cell communication at bone–implant interface. Nanostructured surface promotes rapid bone formation.

KEYWORDS

- **nanostructuring**
- **surface mechanical attrition treatment**
- **osseointegration**
- **grain boundaries**
- **biochemical stimuli**

REFERENCES

1. Saini, M.; Singh, Y.; Arora, P.; Arora, V.; Jain, K. Implant biomaterials: A comprehensive review. *World Journal of Clinical Cases*. 3(1), 2015, 52–57.
2. Liu, Y.; Li, H.; Zhang, B.T. Nanostructured ceramic coating biomaterials. *Advanced Nanomaterials and Coatings by Thermal Spray*. 2019, 291–311.
3. Choi, A.H.; Akyol, S.; Bendavid, A.; Ben-Nissan, B. Nanobioceramic thin films: Surface modifications and cellular responses on titanium implants. *Titanium in Medical and Dental Applications*. 2018, 147–173.
4. de Oliveira, P.T.; Nanci, A. Nanotexturing of titanium-based surfaces upregulates expression of bone sialoprotein and osteopontin by cultured osteogenic cells. *Biomaterials*. 25, 2004, 403–413.
5. Ralls, A.; Kumar, P.; Misra, M.; Menezes, P.L. Material design and surface engineering for bio-implants. *Advances in Surface Engineering*. 72, 2020, 684–694.
6. Breding, K.; Jimbo, R.; Hayashi, M.; Xue, Y.; Mustafa, K.; Andersson, M. The effect of hydroxyapatite nanocrystals on osseointegration of titanium implants: An in vivo rabbit study. *International Journal of Dentistry (Nano in Implant Dentistry Special Issue)*. 2014, 2014, 1–9.
7. Guglielmotti, M.B.; Olmedo, D.G.; Cabrini, R.L. Research on implants and osseointegration. *Periodontology. 2000*, 79, 2019, 178–189.
8. Simunek, A.; Vokurkova, J.; Kopecka, D.; Celko, M.; Mounajjed, R.; Krulichova, I.; Skrabkova, Z. Evaluation of stability of titanium and hydroxyapatite-coated osseointegrated dental implants: A pilot study. *Clinical Oral Implants Research*. 13(1), 2002, 75–79.
9. Bahl, S.; Shreyas, P.; Trishul, M.A.; Suwas, S.; Chatterjee, K. Enhancing the mechanical and biological performance of a metallic biomaterial for orthopedic applications through changes in the surface oxide layer by nanocrystalline surface modification. *Nanoscale*. 7 2015, 7704–16.
10. Agnes, A.J.; Jaganathan, S.K.; Supriyanto, E.; Manikandan, A. Surface modification of titanium and its alloys for the enhancement of osseointegration in orthopaedics. *Current Science*. 111(6), 2016, 1003–1015.
11. Elkin, B.S.; Azeloglu, E.U.; Costa, K.D.; Morrison 3rd, B. Mechanical heterogeneity of the rat hippocampus measured by atomic force microscope indentation. *Journal of Neurotrauma*. 24(5), 2007, 812–22.
12. Chianga, C.Y. Formation of TiO_2 nano-network on titanium surface increases the human cell growth. *Dental Materials*. 25, 2009, 1022–1029.
13. Allan, C.; Ker, A.; Smith, C.A.; Tsimbouri, P.M.; Borsoi, J.; O'Neill,

S.; Gadegaard, N.; Dalby, M.J.; Meek, R.M.D. Osteoblast response to disordered nanotopography. *Journal of Tissue Engineering*. 9, 2018, 1–7.
14. Sub, J.; Yao, Q.T.; Zhang, Y.H.; Du, X.D.; Wu, Y.C.; Tong, W.P. Simultaneously improving surface mechanical properties and in vitro biocompatibility of pure titanium via surface mechanical attrition treatment combined with low-temperature plasma nitriding. *Surface and Coatings Technology*. 309, 2017, 382–389.
15. Marenzi, G.; Impero, F.; Scherillo, F.; Sammartino, J.C.; Squillace, A.; Spagnuolo, G. Effect of Different Surface Treatments on Titanium Dental Implant Micro-Morphology. *Materials*. 12, 2019, 733.
16. Dalby, M.J.; McCloy, D.; Robertson, M.; Wilkinson, C.D.W.; Oreffo, R.O.C. Osteoprogenitor response to defined topographies with nanoscale depths. *Biomaterials*. 27(8), 2006, 1306–1315.
17. Sjostrom, T.; Dalby, M.J.; Hart, A.; Tare, R.; Oreffo, R.O.; Su, B. Fabrication of pillar-like titania nanostructures on titanium and their interactions with human skeletal stem cells. *Acta Biomaterialia*. 5, 2009, 1433–1441.
18. Biggs, M.J.P.; Richards, R.G.; McFarlane, S.; Wilkinson, C.D.W.; Oreffo, R.O.C.; Dalby, M.J. Adhesion formation of primary human osteoblasts and the functional response of mesenchymal stem cells to 330 nm deep microgrooves. *Journal of the Royal Society Interface*. 5(27), 2008, 1231–1242.
19. Vorobyev, A.Y.; Guo, C. Femtosecond laser structuring of titanium implants. *Applied Surface Science*. 253, 2007, 7272–7280.
20. Wang, Y.M.; Jiao, T.; Ma, E. Dynamic Processes for Nanostructure development in Cu after severe cryogenic rolling deformation. *Materials Transactions*. 44, 2003, 1926–1934.
21. Judd, K.G.; Sharma, M.M.; Eden, T.J. Multifunctional bioceramic composite coatings deposited by cold spray. *Key Engineering Materials*. 813, 2019, 228–233.
22. Gu, Y.W.; Khor, K.A.; Pan, D.; Cheang, P. Activity of plasma sprayed yttria stabilized zirconia reinforced hydroxyapatite/Ti6Al-4V composite coatings in simulated body fluid. *Biomaterials*. 25(16), 2004, 3177–3185.
23. Fouda, M.F.A.; Nemat, A.; Gawish, A.; Baiuomy, A. R. Does the coating of titanium implants by hydroxyapatite affect the elaboration of free radicals. An experimental study. *Australian Journal of Basic and Applied Sciences*. 3, 2009, 1122–1129.
24. Xie, Y.; Liu, X.; Zheng, X.; Ding, C.; Chu, P.K. Improved stability of plasma-sprayed dicalcium silicate/zirconia composite coating. *Thin Solid Films*. 515(3), 2006, 1214–1218.
25. Pariente, I.F.; Guagliano, M. About the role of residual stresses and surface work hardening on fatigue ΔK_{th} of a nitrided and shot peened low-alloy steel. *Surface and Coating Technology*. 202, 2008, 3072.
26. Yang, G.L.; He, F.M.; Yang, X.F.; Wang, X.X.; Zhao, S.F. Bone responses to titanium implants surface-roughened by sandblasted and double etched treatments in a rabbit model. *Oral Surgery, Oral Medicine, Oral Pathology, Oral Radiology, and Endodontology*. 106(4), 2008, 516–524.
27. Knabe, C.; Klar, F.; Fitzner, R.; Radlanski, R.J.; Gross, U. In vitro investigation of titanium and hydroxyapatite dental implant surfaces using a rat bone marrow stromal cell

culture system. *Biomaterials*. 23(15), 2002, 3235–3245.
28. Chang, E.; Chang, W.J.; Wang, B.C.; Yang, C.Y. Plasma spraying of zirconia-reinforced hydroxyapatite composite coatings on titanium: Part I Phase, microstructure and bonding strength. *Journal of Materials Science: Materials in Medicine.* 8(4), 1997, 193–200.
29. He, F.M.; Yang, G.L.; Li, Y.N.; Wang, X.X.; Zhao, S.F. Early bone response to sandblasted, dual acid-etched and H_2O_2/HCl treated titanium implants: An experimental study in the rabbit. *International Journal of Oral & Maxillofacial Surgery*. 38(6), 2009, 677–681.
30. Fu, L.; AikKhor, K.; Lim, J.P. The evaluation of powder processing on microstructure and mechanical properties of hydroxyapatite (HA)/ yttria stabilized zirconia (YSZ) composite coatings. *Surface and Coatings Technology*. 140(3), 2001, 263–268.
31. Ortega, E.V.; Jos, A.; Camean, A.M.; Mourelo, J.P.; Egea, J.J.S. In vitro evaluation of cytotoxicity and genotoxicity of a commercial titanium alloy for dental implantology. *Mutation Research—Genetic Toxicology and Environmental Mutagenesis*. 702(1), 2010, 17–23.
32. He, W.; Yin, X.; Xie, L.; Liu, Z.; Li, J.; Jou, S.; Chen, J. Enhancing osseointegration of titanium implants through large-grit sandblasting combined with micro-arc oxidation surface modification. *Journal of Materials Science: Materials in Medicine*. 30, 2019, 73.
33. Al-Radha, A.S.D.; Dymock, D.; Younes, C.; O'Sullivan, D. Surface properties of titanium and zirconia dental implant materials and their effect on bacterial adhesion. *Journal of Dentistry.* 40(2), 2012, 146–153.
34. Aparicio, C.; Padros, A.; Gil, F.J. In vivo evaluation of micro-rough and bioactive titanium dental implants using histometry and pull-out tests. *Journal of the Mechanical Behavior of Biomedical Materials*. 4(8), 2011, 1672–1682.
35. Ban, S.; Iwaya, Y.; Kono, H.; Sato, H. Surface modification of titanium by etching in concentrated sulfuric acid. *Dental Materials*. 22(12), 2006, 1115–1120.
36. Liu, X.; Poon, R.W.Y.; Kwok, S.C.H.; Chu, P. K.; Ding, C. Plasma surface modification of titanium for hard tissue replacements. *Surface and Coatings Technology.* 186(1–2), 2004, 227–233.
37. Guo, C.Y.; Matinlinna, J.P.; Tang, A.T.H. Effects of surface charges on dental implants: past, present, and future. *International Journal of Biomaterials (Dental Implants Special Issue)*. 2012, 2012, 1–5.
38. Cho, S.A.; Park, K.T. The removal torque of titanium screw inserted in rabbit tibia treated by dual acid etching. *Biomaterials.* 24(20), 2003, 3611–3617.
39. Iwaya, Y.; Machigashira, M.; Kanbara, K. Surface properties and biocompatibility of acid-etched titanium. *Dental Materials Journal.* 27(3), 2008, 415–421.
40. Juodzbalys, G.; Sapragoniene, M.; Wennerberg, A. New acid etched titanium dental implant surface. *Stomatologija—Baltic Dental and Maxillofacial Journal.* 5, 2003, 101–105.
41. Chou, B.Y.; Chang, E. Interface investigation of plasma sprayed hydroxyapatite coating on titanium alloy with ZrO2 intermediate layer as bond coat. *Scripta Materialia.* 45(4), 2001, 487–493.

42. Kim, H.; Choi, S.H.; Ryu, J.J.; Koh, S.Y.; Park, J.H.; Lee, I.S. The biocompatibility of SLA-treated titanium implants. *Biomedical Materials*. 3(2), 2008, 25011.
43. Massaro, C.; Rotolo, P.; De Riccardis, F. Comparative investigation of the surface properties of commercial titanium dental implants. Part I: Chemical composition. *Journal of Materials Science: Materials in Medicine*. 13(6), 2002, 535–548.
44. Cho, S.A.; Jung, S.K. A removal torque of the laser-treated titanium implants in rabbit tibia. *Biomaterials*. 24(26), 2003, 4859–4863.
45. Xue, W.; Liu, X.; Zheng, X.; Ding, C. In vivo evaluation of plasma-sprayed titanium coating after alkali modification. *Biomaterials*. 26(16), 2005, 3029–3037.
46. Cai, K.; Bossert, J.; Jandt, K.D. Does the nanometre scale topography of titanium influence protein adsorption and cell proliferation? *Colloids and Surfaces B: Biointerfaces*. 49, 2006, 136–144.
47. Jaeggi, C.; Kern, P.; Michler, J.; Zehnder, T.; Siegenthaler, H.; Anodic thin films on titanium used as masks for surface micropatterning of biomedical devices. *Surface and Coatings Technology*. 200, 2005, 1913–1919.
48. Ong, J.L.; Chan, D.C.N. Hydroxyapatite and their use as coatings in dental implants: a review. *Critical Reviews in Biomedical Engineering*. 28(5–6), 2000, 667–707.
49. Medvedev, A.E.; Ng, H.P.; Lapovok, R.; Estrin, Y.; Lowe, T.C.; Anumalasetty, V.N. Effect of bulk microstructure of commercially pure titanium on surface characteristics and fatigue properties after surface modification by sand blasting and acid etching. *Journal of the Mechanical Behavior of Biomedical Materials*. 57, 2016, 55–68.
50. Engler, A.J.; Sen, S.; Sweeney, H.L.; Discher, D.E. Matrix elasticity directs stem cell lineage specification. *Cell*. 126(4), 2006, 677–89.
51. Balaban, N.Q.; Schwarz, U.S.; Riveline, D.; Goichberg, P.; Tzur, G.; Sabanay, I.; Mahalu, D.; Safran, S.; Bershadsky, A.; Addadi, L.; Geiger, B. Force and focal adhesion assembly: A close relationship studied using elastic micropatterned substrates. *Nature Cell Biology*. 3(5), 2001, 466–72.
52. Tao, N.R.; Lu, J.; Lu, K. Surface nanocrystallization by surface mechanical attrition treatment. *Materials Science Forum*. 579, 2008, 91–107.
53. Lu, K. Lu, J. Nanostructured surface layer on metallic materials induced by surface mechanical attrition treatment. *Materials Science and Engineering A*. 375–377, 2004, 38–45.
54. Chichkov, B.N.; Momma, C.; Nolte, S.; Von Alvensleben, F.; Tunnermann, A. Femtosecond, picosecond and nanosecond laser ablation of solids. *Applied Physics A*. 63, 1996, 109.
55. Wu, X.; Tao, N.; Hong, Y.; Liu, G.; Xu, B.; Lu, J.; Lu, K. Strain-induced grain refinement of cobalt during surface mechanical attrition treatment. *Acta Mater*. 53, 2005, 681–691.
56. Wu, X.; Tao, N.; Hong, Y.; Xu, B.; Lu, J.; Lu, K.; Microstructure and evolution of mechanically-induced ultrafine grain in surface layer of AL-alloy subjected to USSP. *Acta Mater*. 50, 2002, 2075–2084.
57. Chauvy, P.F.; Hoffman, P.; Landolt, D. Applications of laser lithography on oxide film to titanium micromachining. *Applied Surface Science*. 208/209, 2003, 165.
58. Zhu, L.; Guan, Y.; Wang, Y.; Xie, Z.; Lin, J. Influence of process parameters of ultrasonic shot peening on surface nanocrystallization and hardness

of pure titanium. *The International Journal of Advanced Manufacturing Technology.* 89, 2017, 1451–1468.
59. Le Gu′ehennec, L.; Soueidan, A.; Layrolle, P.; Amouriq, Y. Surface treatments of titanium dental implants for rapid osseointegration. *Dental Materials*. 23, 2007, 844–854.
60. Tevlek, A.; Aydin, H.M.; Maleki, E.; Varol, R.; Unal, O. Effects of severe plastic deformation on pre-osteoblast cell behavior and proliferation on AISI 304 and Ti-6Al-4V metallic substrates. *Surface and Coatings Technology.* 366, 2019, 204–213.
61. Huang, R.; Zhang L.; Huang L.; Zhu, J. Enhanced in-vitro osteoblastic functions on β-type titanium alloy using surface mechanical attrition treatment. *Materials Science and Engineering C.* 97, 2019, 688–697.
62. Guo, Y.; Hu, B.; Tang, C.; Wu, Y.; Sun, P.; Zhang, X.; Jia, Y. Increased osteoblast function in vitro and in vivo through surface nanostructuring by ultrasonic shot peening. *International Journal of Nanomedicine.* 10, 2015, 4593–4603.
63. Bagherifard, S.; Ghelichi, R.; Khademhosseini, A.; Guagliano, M. Cell response to nanocrystallized metallic substrates obtained through severe plastic deformation. *ACS Applied Materials and Interfaces.* 6, 2014, 7963–7985.
64. Lord, M.S.; Foss, M.; Besenbacher, F. Influence of nanoscale surface topography on protein adsorption and cellular response. *Nano Today.* 5, 2010, 66–78.
65. Lai, M.; Cai, K.; Hu, Y.; Yang, X.; Liu, Q. Regulation of the behaviors of mesenchymal stem cells by surface nanostructured titanium. *Colloids Surface B. Biointerfaces.* 97, 2012, 211–220.
66. Lavos-Valereto, I.C.; Wolynec, S.; Ramires, I.; Guastaldi, A.C.; Costa, I. Electrochemical impedance spectroscopy characterization of passive film formed on implant Ti 6Al 7Nb alloy in Hank's solution. *Journal of Materials Science: MaterIals in Medicine.* 15, 2004, 55–59.
67. Pan, J.; Thierry, D.; Leygraf, C. Electrochemical impedance spectroscopy study of the passive oxide film on titanium for implant application. *Electrochimica Acta.* 41, 1996, 1143–1153.
68. Gilbert, J.L. Electrochemical behavior of metals in the biological milieu. *Comprehensive Biomaterials.* 2011, 21–48. doi:10.1016/b978–0-08–055294–1.00013–1.
69. Ehrensberger, M.T.; Sivan, S.; Gilbert, J.L. Titanium is not 'the most biocompatible metal' under cathodic potential: The relationship between voltage and MC3T3 preosteoblast behavior on electrically polarized CP-Ti surfaces. *Journal of Biomedical Materials Research Part A.* 93, 2010, 1500–1509.
70. Aizawa, M.; Koyama, S.; Kimura, K.; Haruyama, T.; Yanagida, Y.; Kobatake, E. Electrically stimulated modulation of cellular function in proliferation, differentiation, and gene expression. *Electochemistry.* 67, 1999 118–125.
71. Gilbert, J.L.; Zarka, L.; Chang, E.; Thomas, C.H. The reduction half cell in biomaterials corrosion: Oxygen diffusion profiles near and cell response to polarized titanium surfaces. *Journal of Biomedical Materials Research.* 42, 1998, 321–330.
72. Zhou, T.; Braunhut, S.J.; Medeiros, D.; Marx, K.A. Potential dependent endothelial cell adhesion, growth and cytoskeletal rearrangements. *MRS Proceedings.* 489, 1997, 211.

73. Tominaga, M.; Kumagai, E.; Harada, S. Effect of electrical stimulation on HIV-1-infected HeLa cells cultured on an electrode surface. *Applied Microbiology and Biotechnology.* 61, 2003, 447–450.
74. Singh, V. Effect of surface nanocrystallization on osseointegration of CP-titanium. *Patent No. 315419*, 2019, Intellectual Property India.
75. Zou, D.; Han, W.; You, S.; Ye, D.; Wang, L.; Wang, S.; Zhao, J.; Zhang, W.; Jiang, X.; Zhang, X.; Huang, Y. In vitro study of enhanced osteogenesis induced by HIF-1α-transduced bone marrow stem cells. *Cell Proliferation.* 44(3), 2011, 234–243.
76. Rikitake, Y.; Takai, Y. Chapter three directional cell migration: regulation by small G proteins, nectin-like molecule-5, and afadin, *International Review of Cell and Molecular Biology.* 287, 2011, 97–143.
77. Hynes, R.O. Integrins: a family of cell surface receptors. *Cell.* 48(4), 1987, 549–554.
78. Kheradmandfard, M.; Kashani-Bozorg, S.F.; Lee, J.S.; Kim, C.L.; Hansaki, A.Z.; Pyun, Y.S.; Cho, S.W.; Amanov, A.; Kim, D.E. Significant improvement in cell adhesion and wear resistance of biomedical β-type titanium alloy through ultrasonic nanocrystal surface modification. *Journal of Alloys and Compounds.* 762, 2018, 941–949.
79. Stevens, M.M.; George, J.H. Exploring and engineering the cell surface interface. *Science.* 310(5751), 2005, 1135–1138.
80. Peng, L.; Eltgroth, M.L.; LaTempa, T.J.; Grimes, C.A.; T.A.Desai, C.A. The effect of TiO_2 nanotubes on endothelial function and smooth muscle proliferation. *Biomaterials.* 30(7), 2009, 1268–1272.
81. Salasznyk, R.M.; Klees, R.F.; Boskey A.; Plopper, G.E. Activation of FAK is necessary for the osteogenic differentiation of human mesenchymal stem cells on laminin-5. *Journal of Cell Biochemistry.* 100(2), 2007, 499–514.
82. Salasznyk, R.M.; Klees, R.F.; Williams, W.A.; Boskey A.; Plopper, G.E. Focal adhesion kinase signaling pathways regulate the osteogenic differentiation of human mesenchymal stem cells. *Experimental Cell Research.* 313(1), 2007, 22–37.
83. Webster, T.J.; Ejio, J.U. Increased osteoblast adhesion on nanophase metals: Ti, Ti6Al4V, and CoCrMo. *Biomaterials.* 25, 2004, 4731–4739.
84. Niinomi, M. Recent metallic materials for biomedical applications, *Metall Urgical and Materials Transactions A.* 33, 2002, 477–486.
85. Kawaguchi, H.; McKee, M.D.; H.Okamoto, MD,; Nanci, A. Immunocytochemical andlectin-gold characterization of the interface between alveolar bone and implanted hydroxyapatite in the rat. *Cells Mater.* 3, 1993, 337–350.
86. Siegel, R.W. Creating nanophase materials. *Scientific American.* 275, 1996, 74–79.
87. Klabunde, K.J.; Strak, J.; Koper, O.; Mohs, C.; Park, D.; Decker, S.; Jiang, Y.; Lagadic, I.; Zhang. D. Nanocrystals as stoichiometric reagents with unique surface chemistry. *The Journal of Physical Chemistry.* 100, 1996, 12141.
88. McNamara, L.E.; Sjostrom, T.; Burgess, K.E.; Kim, J.J.; Liu, E.: Gordonov, S.; Moghe, P.V.; Meek, R.M.; Oreffo, R.O.; Su, B.; Dalby, M.J. Skeletal stem cell physiology on functionally distinct titaniananotopographies. *Biomaterials.* 32, 2011, 7403–7410.
89. Yim, E.K.; Darling, E.M.; Kulangara, K.; Guilak, F.; Leong, K.W. Nanotopography-induced changes in focal adhesions, cytoskeletal organization, and mechanical properties

of human mesenchymal stem cells. *Biomaterials.* 31, 2010, 1299–1306.
90. Mendonça, G.; Mendonca, D.B.; Simoes, L.G.; Araujo, A.L.; Leite, E.R.; Aragao, F.J. Cooper, L.F. The effects of implant surface nanoscale features on osteoblast-specific gene expression. *Biomaterials.* 30, 2009, 4053–4062.
91. Mendonça, G.; Mendonca, D.B.; Aragao, F.J.; Cooper, L.F. Advancing dental implant surface technology From micron to nanotopography. *Biomaterials.* 29, 2008, 3822–3835.
92. Dalby, M.J.; Riehle, M.O.; Johnstone H.; Affrossman, S.; Curtis, A.S.G. In vitro reaction of endothelial cells to polymer demixednanotopography. *Biomaterials.* 23(14), 2002, 2945–2954.
93. Dalby, M.J.; Mccloy, D.; Robertson, M.; Agheli, H.; Sutherland, D.; Affrossman, S. Osteoprogenitor response to semi-ordered and random nanotopographies. *Biomaterials.* 27(15), 2006, 2980–7.
94. Ogura N.; Kawada, M.; Chang, W.J.; Zhang, Q.; Lee, S.Y.; Kondoh, T; Abiko, Y. Differentiation of the human mesenchymal stem cells derived from bone marrow and enhancement of cell attachment by fibronectin. *Journal of Oral Science.* 46(4), 2004, 207–13.
95. Zaidel-Bar, R.; Cohen, M.; Addadi, L.; Geiger, B. Hierarchical assembly of cell matrix adhesion complexes. *Biochemical Society Transactions.* 32, 2004, 416–20.
96. Kulangara, K. Yang, Y.; Yang, J.; Leong, K.W. Nanotopography as modulator of human mesenchymal stem cell function. *Biomaterials.* 33, 2012, 4998–5003
97. Chrzanowska-Wodnicka, M.; Burridge, K. Rho-stimulated contractility drives the formation of stress fibers and focal adhesions. *Journal of Cell Biology.* 133, 1996, 1403–1415.
98. Ridley, A.J.; Hall, A. The small GTP-binding protein rho regulates the assembly of focal adhesions and actin stress fibers in response to growth factors. *Cell.* 70, 1992, 389–399.
99. Yang, W.E.; Huang, H.H. Improving the biocompatibility of titanium surface through formation of a TiO_2nano-mesh layer. *Thin Solid Films.* 518, 2010, 7545–7550.
100. Oh, S.; Daraio, C.; Chen, L.H.; Pisanic, T.R.; Finones, R.R; Jin. S. Significantly accelerated osteoblast cell growth on aligned TiO_2nano tubes. *Journal of Biomedical Materials Research A.* 78, 2006, 97–103.
101. Yu, W.Q.; Zhang, Y.L.; Jiang, X.Q.; Zhang, F.Q. In vitro behavior of MC3T3-E1 preosteoblast with different annealing temperature titania nanotubes. *Oral Diseases.* 16, 2010, 624–630.
102. Zhao, X; Wang, T.; Qian, S.; Liu, X.; Sun, J. Li, B. Silicon-Doped Titanium Dioxide Nanotubes Promoted Bone Formation on Titanium Implants. *International Journal of Molecular Science.* 17(3), 2016, 292.
103. Ellingsen, J.E.; Johansson, C.B.; Wennerberg, A.; Holmen A. Improved retention and bone-to-implant contact with fluoride-modified titanium implants. *The International Journal of Oral and Maxillofacial Implants.* 19, 2004, 659–66.
104. Berglundh, T.; Abrahamsson, I.; Albouy, J.P.; Lindhe. J. Bone healing at implants with a fluoride-modified surface: An experimental study in dogs. *Clinical Oral Implants Research.* 18, 2007, 147–52.
105. Reilly, G.C.; Engler; A.J. Intrinsic extracellular matrix properties regulate stem cell differentiation. *Journal of Biomechanics.* 43, 2010, 55–62.
106. Ramires, P.A. The influence of titania/hydroxyapatite composite coatings

on in vitro osteoblasts behavior. *Biomaterials*. 22, 2001, 1467–1474.
107. Cosentino, F.; Licciulli, A.; Massaro, C.; Milella, E.; Preparation and characterization of titania/hydroxyapatite composite coatings obtained by sol-gel process. *Biomaterials*. 22(11), 2001, 1425–31.
108. Han, J.Y.; Yu, Z.T.; Zhou, L. Hydroxyapatite/titania composite bioactivity coating processed by the sol-gel method. *Biomedical Materials*. 3(4), 2008, 044109.
109. Li, P.; De, K.; Groot; Kokubo, T. Bioactive $Ca_{10}(PO_4)_6(OH)_2$-TiO_2 composite coating prepared by sol-gel process. *Journal of Sol-Gel Science and Technology*. 7, 1996, 27–34.
110. Jokinen, M.; Paktsi, M.; Rahiala, H.; Peltola, T.; Ritala, M.; Rosenholm, J.B. Influence of sol and surface properties on in vitro bioactivity of sol-gel-derived TiO_2 and TiO_2-SiO_2 films deposited by dip-coating method. *Journal of Biomedical Materials Research*. 42, 1998, 295–302.
111. Li, P.; Ohtsuki, C.; Kokubo, T.; Nakanish, K.; Soga, N.; Groot, K.D. A role of hydrated silica, titania and alumina in forming biologically active bone-like apatite on implant. *Journal of Biomedical Materials Research*. 28, 1994, 7–15.
112. Li, P.; Groot, K. D.; Kokubo, T. Bonelike hydroxyapatite induction by sol/gel derived titania coating on a titanium substrate. *Journal of the American Ceramic Society*. 77, 1994, 1307–1315.
113. Belyakov, A.; Tsuzaki, K.; Hiura, H.; Sakai, T. Effect of initial microstructures on grain refinement in a stainless steel by large strain deformation. *Acta Materialia*. 51, 2003, 847–861.
114. Bacon, D.J.; Martin, J.W. The atomic structure of dislocations in h.c.p metals I.Potentials and unstressed crystals. *Philosophical Magazine*. 43A, 1981, 883–900.
115. Ohtsuki, C.; Iida, H.; Hayakawa, S.; Osaka, A. Bioactivity of titanium treated with hydrogen peroxide solutions containing metal chlorides. *Journal of Biomedical Materials Research Part B*. 35, 1997, 39.
116. Wen, H.B.; Liu, Q.; DeWijn, J.R.; Groot, K.De; Cui, F.Z. Preparation of bioactive microporous titanium surface by a new two-step chemical treatment. *Journal of Materials Science: Materials in Medicine*. 9, 1998, 121.
117. Kokubo, T.; Miyaji, F.; Kim, H.M.; Nakamura, T. spontaneous formation of bonelike apatite layer on chemically treated titanium metals. *Journal of the American Ceramic Society*. 79, 1996, 1127–1129.
118. Kim, H. M.; Miyaji, F.; Kokubo, T.; Nakamura, T. Bonding strength of bonelike apatite layer to Ti metal substrate. *Journal of Biomedical Materials Research*. 38, 1997, 121.
119. Kim, H. M.; Miyaji, F.; Kokubo, T.; Nishiguchi, S.; Nakamura, T. Graded surface structure of bioactive titanium prepared by chemical treatment. *Journal of Biomedical Materials Research*. 45, 1999, 100–107.
120. Kim, H. M.; Takadama, H.; Kokubo, T.; Nishiguchi, S.; Nakamura, T. Formation of a bioactive graded surface structure on Ti-15Mo-5Zr-3Al alloy by chemical treatment. *Biomaterials*. 21, 2000, 353–358.
121. Kim, H. M.; Miyaji, F.; Kokubo, T.; Nakamura, T. Preparation of bioactive Ti and its alloys via simple chemical surface treatment. *Journal of Biomedical Materials Research*. 32, 1996, 409–417.
122. Yan, W.Q.; Nakamura, T.; Kobayashi, M.; Kim, H.M.; Miyaji, F.; Kokubo, T. Bonding of chemically treated titanium implants to bone. *Journal of*

Biomedical Materials Research. 37, 1997. 267–275.

123. Nishiguchi, S.; Nakamura, T.; Kobayashi, M.; Kim, H. M.; Miyaji, F.; Kokubo, T. The effect of heat treatment on bone-bonding ability of alkali-treated titanium. *Biomaterials.* 20, 1999, 491–500.

CHAPTER 18

Perspectives on Translational Outcomes Research Using Nanobiotechnological Agents

SHINJINI MITRA, NANDITA GHOSH, PRAMATHADHIP PAUL, PAYAL PAL, RANITA BOSE, ADETI MUNMUN SENGUPTA, and ENA RAY BANERJEE*

Immunobiology and Regenerative Medicine Research Laboratory, Department of Zoology, University of Calcutta, Kolkata, West Bengal, India

*Corresponding author.
E-mail: enaraybanerjee@gmail.com; erb@caluniv.ac.in*

ABSTRACT

Inflammation and degeneration often go hand-in-hand in a number of diseases. In this chapter, we have attempted to highlight the mechanisms involved in inflammation and degeneration, with special focus on some common, and often fatal, inflammatory and/or degenerative diseases. All the diseases mentioned in this chapter, including asthma, idiopathic pulmonary fibrosis, peritonitis, and atopic dermatitis, have lacunae in the therapies available currently. Here, we have proposed some therapeutic strategies that are different from the current strategies, in the fact that we have used nanobiotechnological agents and the principle of drug repurposing. Nanobiotechnological strategies, like use of nanodrugs, nanovehicles, scaffolds, and nanoantibodies, are advantageous because of their greater penetrability into tissues and lower saturation kinetics. The strategies outlined in this chapter may help in reducing the disease burden, since they are most likely to be easily available and cost-effective.

18.1 INTRODUCTION

18.1.1 INFLAMMATION

The observation of the phenomena of inflammation was first recorded by the Roman writer Aulus Cornelius Celsius in his only surviving work *De Medicina*, in which he described redness, warmth, swelling, and pain to be the four characteristic signs of inflammation [1, 24, 27, 42]. A few centuries later, Galen added "loss of function" as the fifth manifestation of inflammation and considered inflammation to be a beneficial response in case of injury [24, 27, 42]. Today, the characteristic signs of inflammation can be explained by vasodilation and increased blood flow, increased cellular metabolism, the release of mediators like cytokines and chemokines, and infiltration of cells [24].

Inflammation can be considered to be a part of the biological response of the body tissues against harmful stimuli which may be endogenous or exogenous [17, 47, 24]. The stimuli may range from infections and pathogens that have entered the body to damaged or altered cells and even nonliving entities such as splinters or other irritants that may have breached the epithelial barrier. Inflammation is a part of nonspecific response of the immune system against any irritation and is thus considered to be a part of innate immunity, in contrast to adaptive immunity which is specific for each pathogen [24].

Inflammation is categorized to be either *acute*, the initial response by the body to invading stimuli, or *chronic*, a prolonged mode of inflammation, characterized by a gradual change in the type of the participating cells [17, 24, 47]. Acute inflammation is primarily mediated by cytokines like interleukin (IL)-1β, secreted by tissue-resident and blood macrophages which act as the first responders to the site of infection [1, 27]. The immune response is stimulated by recognition of the antigen by receptors of the immune system, like the toll-like receptors (TLRs), followed by the production of inflammatory mediators [17, 47]. This phase is also marked by infiltration to the site of inflammation by granulocytes, especially neutrophils, and vasodilation of the blood vessels. All these changes result in the production of heat and swelling caused by the accumulation of leukocytes. These infiltrating neutrophils try to kill the antigens by releasing enzymes, reactive oxygen species (ROS), and reactive nitrogen species from their granules [47]. The final stage in a successful inflammatory response is the elimination of the antigen, resulting in resolution of the inflammation and repair of the tissue, which is mediated mainly by macrophages [47]. The change from inflammation to resolution involves a switch in lipid mediators, from proinflammatory prostaglandins to anti-inflammatory lipoxins [47].

If the acute inflammatory response is unable to remove the invading pathogen, the inflammatory process takes the form of chronic inflammation [17, 24, 47]. All these principles also underlie the event of chronic inflammation; the phase is marked by a progressive shift from multinuclear neutrophils and eosinophils to mononuclear macrophages and lymphocytes [24, 47]. Macrophages and monocytes play an important role in the ultimate manifestation of chronic inflammation, that is, loss of tissue function because of fibrosis [24, 47].

The most common pathways usually activated during an inflammatory response are the NFκB pathway, the mitogen-activated protein kinase pathway, and the JAK-STAT pathway [17]. The interleukins (ILs), interferons (IFNs), tumor necrosis factors (TNFs), transforming growth factors (TGFs), C reactive proteins, superoxide dismutase, and inducible nitric oxide synthase (iNOS) are all inflammatory markers [17].

Though inflammation is useful in combating the infection initially, unregulated inflammation can be detrimental [47]. Prompted by the escape of the pathogen into the systemic circulation or by a severe infection, or even by internal hemorrhage, inflammation may spread throughout the entire body and result in what is referred to as systemic inflammatory response syndrome, which may result in septic shock or even death [47].

In this chapter, we will take an in-depth look at the various mechanisms that underlie the process of inflammation and how it is regulated by various bodily factors to maintain homeostasis and thus serve as the first line of defense against any irritating stimuli. We will also look at how acute inflammation transitions into chronic inflammation which forms the basis of our knowledge to cure inflammatory disorders, such as rheumatoid arthritis and asthma.

18.1.2 DEGENERATION

Tissue degeneration is the functional deterioration and structural damage of tissue. Tissues are made of different types of cells, which need to be maintained at correct ratios for optimal function of the tissue [2]. Any change in, or loss of, this tissue homeostasis, may lead to diseases. Degenerative diseases are caused by the loss of a particular type of resident cells and worsens over time [2]. On the other hand, fibrosis is the result of hyperactivity of a cell type [2].

The process of degeneration involves the following mechanisms:

(1) Cessation of ATP production
(2) Lysis of cell membrane
(3) Absence of vital proteins
(4) Failure of vital metabolic reactions.

The production and removal of ROS influence the cellular redox status, that is, the oxidation and reduction environment, which is critical in the regulation of several physiological processes [5]. Acute and chronic degenerative diseases are often characterized by unbalanced ROS levels, which in turn, cause inflammatory and oxidative stresses [5].

18.1.3 SOME INFLAMMATORY DISEASES

Inflammatory diseases, as the name suggests, are a range of disorders that involve inflammation and include allergy, asthma, inflammatory bowel disease, atopic dermatitis (AD) rheumatoid arthritis, hepatitis, glomerulonephritis, coeliac disease, reperfusion injury, and transplant rejection. In this chapter, we have focused on the etiology, preclinical mouse models, and unmet needs of the few inflammatory diseases which we work with in our laboratory.

18.1.3.1 ASTHMA

18.1.3.1.1 Etiology of Asthma

Allergic asthma is considered a type I hypersensitivity reaction, involving the reaction of allergen-specific IgE with receptors on mast cells and basophils [3]. Exposure of an asthmatic patient to an allergen can lead to two types of responses—the early response, which occurs within 15–30 min of exposure and involves bronchospasms, and the late-phase response, which occurs 4–6 h after exposure and reduces over time [21]. The key event in asthma is considered to be IgE-mediated degranulation of mast cells and/or basophils, leading to the release of mediators like histamine, leukotrienes, cytokines (like IL-4 and IL-5), and prostaglandins [3, 21]. These mediators play a role in the contraction of the airway smooth muscles, induction of mucus overproduction, and subsequent constriction of airways [3]. This airway constriction is the cause of the wheezing and shortness of breath seen in asthmatic patients. Asthma leads to a Th2 cell-mediated inflammatory response in the airways [21]. The secreted mediators attract eosinophils, macrophages, neutrophils, and T cells, which release toxic granules and cause damage to the epithelial layers [3]. This damage, along with the profibrotic cytokines released, leads to structural changes including thickening of the airway walls, fibrosis in the airways, mucus metaplasia, and hypertrophy and hyperplasia of myocytes [3, 21].

18.1.3.1.2 Preclinical Mouse Model of Asthma

Asthma, be it acute or chronic, can be induced in mice, using ovalbumin

(Ova), a 42.7 kDa protein containing 385 amino acids and several modifications like acetylation, glycosylation, and phosphorylation [54]. Ova simulates the characteristics of eosinophilic asthma [73].

To induce acute asthma, 6–8-week-old female BALB/c mice are used, as it has been observed that female mice show higher susceptibility than male mice [73]. The mice are sensitized with 100 μg Ova intraperitoneally on day 0, followed by intratracheal challenges, with 250 μg Ova on day 8, and with 125 μg Ova on days 15, 18, and 21 [48]. The mice are then sacrificed on day 22, 24 h after the last intratracheal treatment, tissues collected and various assays done to assess the changes in disease phenotype with Ova and with therapy. Our results show that treatment with Ova increases the total cellular infiltration, as well as infiltration of eosinophils, $CD3^+$ T cells, $CD4^+$ T_H cells, and $CD8^+$ T_C cells, into the lung and bronchoalveolar lavage fluid (BALF). Ova-induced asthma causes an increase in the concentration of serum IgE and in the concentration of nitric oxide (NO) in the lung [48]. Ova increases the expression of proinflammatory cytokines like IL4, IL5, IL13, TNFα, and IFNγ [48]. It also reduces the clonogenic potential of lung cells, which is a measure of the ability of cells to regenerate.

Chronic asthma is induced by sensitizing 6–8-week-old C57BL/6J mice with 100 μg Ova intraperitoneally on day 0. They are then challenged, intratracheally, with 250 μg Ova on day 8, and with 125 μg Ova every three days from day 15 to 54 [9, 61, 62]. The mice are then sacrificed on day 55, tissues collected and various assays done to assess the changes in disease phenotype with Ova and with therapy. Our results show that treatment with Ova increases the total cellular infiltration, as well as infiltration of eosinophils, $CD3^+$ T cells, $CD4^+$ T_H cells, and $CD8^+$ T_C cells, into the lung and BALF [61, 62]. Ova-induced asthma causes an increase in the concentration of serum IgE [61, 62]. There is an increase in the collagen content in the lungs, which is measured by estimating the concentration of hydroxyproline [61, 62]. Ova increases the expression of proinflammatory cytokines like NFκB, IL4, IL5, IL13, TNFα, IFNγ, iNOS, TGFβ, and STAT6 [61, 62]. There is also a reduction in the clonogenic potential of lung cells [61, 62].

18.1.3.1.3 Unmet Needs for New Therapeutic Strategies for Asthma

Asthma is a global disease, affecting a large portion of the world's population. Till date, there is no permanent cure for asthma [61, 62]. Inhaled corticosteroids and β-antagonists are used to control the symptoms of asthma temporarily [61, 62]. However, these drugs have severe side effects and often patients stop responding to them after a period of time. Limited access to medications and health services are also leading causes of the rise in asthma occurrence [61, 62].

18.1.3.2 IDIOPATHIC PULMONARY FIBROSIS

18.1.3.2.1 Etiology of Idiopathic Pulmonary Fibrosis

Pulmonary fibrosis is a chronic lung disease, which, if not successfully managed, can turn to fatal interstitial lung disease. It may be caused by a variety of factors, including other chronic inflammations, infections, systemic autoimmune diseases, environmental agents, and exposure to ionizing radiations or certain medications. Idiopathic pulmonary fibrosis (IPF) is pulmonary fibrosis with unknown etiology and is associated with progressive loss of lung function [32, 41, 46]. The prevalence of IPF increases with age, with males being more prone to it, and the occurrence of IPF is on the rise [19, 20, 41, 64].

IPF is associated with the expansion of fibroblasts or myofibroblasts, leading to excessive assembly and deposition of extracellular matrix (ECM) with the formation of fibrotic foci, which are the hallmark lesions of IPF located immediately adjacent to areas of alveolar epithelial (AE) cell injury and repair [56]. Repetitive injury and subsequent repair of AE cells lead to impairment of the AE-mesenchymal crosstalk which plays a key role in the pathobiology of IPF. Type I alveolar epithelial (AE-I) cells cover most of the alveolar surface area, which is the site of gas exchange. Under normal circumstances, AE-I cells regulate the secretion of surfactant by type II alveolar epithelial (AE-II) cells in response to stretch. In the initial injuries associated with the death of AE-I cells, AE–II cells replicate and become hypertrophic to cover the exposed basement membranes. Within normal tissue repair, these events are accomplished by apoptosis of the hyperplastic AE-II cells and differentiation of remaining cells to AE-I cells [18]. However, continuous disruption of the AE cell layer leads to the early death of AE-II cells and accumulation of resident intrapulmonary fibroblasts in these areas of damage. These events occur in response to a variety of profibrogenic factors such as TGFβ, interleukin-17A (IL-17A), platelet-derived growth factor, or Wnt. Fibroblasts differentiate into a smooth muscle actin expressing myofibroblasts, leading to deposition of collagen and other ECM proteins, and formation of a scar as a consequence [23, 44].

18.1.3.2.2 Preclinical Mouse Model of IPF

IPF can be induced in mice using bleomycin (Bleo). Bleo is a small peptide, with a molecular weight of approximately 1.5 kDa. It is used as an anticancer drug, but it causes several side effects with prolonged use, the most common of which is lung toxicity [66]. Due to this property of Bleo, it is used to induce IPF in preclinical mice models, where it causes oxidative damage and subsequent inflammation and fibrosis [66]. Studies show that Bleo causes strand breaks in DNA due to the overproduction of ROS, which ultimately leads to inflammatory response, pulmonary toxicity, and subsequent fibrosis [52]. Tissues contain Bleo hydrolase, which inactivates the drug. However, the low levels of the enzyme in the lungs make them susceptible to the action of Bleo [52].

To induce IPF, BALB/c mice are administered a single dose of 0.075 U/mL Bleo in 0.9% saline both intratracheally and intranasally on day 0 [8, 10, 36, 37]. The mice are sacrificed on day 28, and tissues are collected for various assays to assess the changes in the disease phenotype. Our results show that Bleo treatment elevates the total cell count in the peripheral blood, lung, and BALF [36, 37]. There is an increase in the concentration of superoxide radicals and in the concentration of the NO radical in the blood, lung, and BALF, as seen by the NBT assay and with Griess reagent, respectively [36, 37]. The clonogenic potential of blood, lung, and BALF reduces with Bleo administration, as does the cell viability of the tissues, as seen by the MTS assay [36, 37]. Fibrosis caused by Bleo leads to the deposition of collagen in the lungs, which is measured by estimating the concentration of hydroxyproline in the lung [36, 37]. Histological assessment and subsequent Masson's Trichrome staining also indicates the establishment of fibrosis [36, 37], with the presence of intra-alveolar buds, loss of alveolar spaces, and deposition of collagen [52].

Expression of proinflammatory cytokines like IL1, IL6, TNFα, and IFNγ increases in the initial stages of the disease, followed by an increase in the profibrotic markers, like TGFβ1, procollagen-1, and fibronectin [52]. The highest level of expression is seen at around day 14, and the transition from inflammation to fibrosis is seen approximately 9 days after Bleo administration [52].

18.1.3.2.3 Unmet Needs for New Therapeutic Strategies

Current therapies for IPF are ineffective, resulting in the death from respiratory insufficiency within 2.5–3.5 years after diagnosis [20, 41]. Studies have demonstrated that IPF patients treated with the combination of prednisone, azathioprine, and *N*-acetylcysteine (known as standard therapy for IPF) have an increased rate of death compared to placebo [30, 44]. Two novel compounds Nintedanib, a tyrosine-kinase inhibitor targeting growth factor receptors, and Pirfenidone, a TGFβ inhibitor have been found to reduce the rate of decline in vital capacity in IPF patients, but the disease continues to progress, and these drugs offer no clear advantage to patients [16]. At present, lung transplantation is the only intervention for improving survival in IPF patients. Five-year survival after lung transplantation for patients with IPF is about 44% [69].

18.1.3.3 PERITONITIS

18.1.3.3.1 Etiology of Peritonitis

Peritonitis is mainly caused due to the release of proinflammatory cytokines and mediators at the site of injury within the peritoneum cavity, be it due to infection or surgical trauma [7]. These types of injuries can damage the integrity of the peritoneal membrane, leading to tissue damage, fibrosis and, in severe cases, peritoneal sclerosis [7]. The initial effector cells of peritonitis are the resident macrophages, mesothelial cells, mast cells, and lymphocytes, which secrete proinflammatory cytokines and chemokines [13].

18.1.3.3.2 Preclinical Mouse Model of Peritonitis

Aseptic peritonitis, a systemic inflammation of the peritoneum, can be induced in mice using thioglycollate (TG). In this model, the neutrophil count increases between 4 and 24 h after treatment, followed by a rise in the number of macrophages after 3–4 days [43, 70]. Injection of TG intraperitoneally leads to infiltration of immune cells into the peritoneal cavity, including T cells, B cells, NK cells, and dendritic cells (DCs) [70]. TG causes apoptosis of peritoneal cells, and upregulation of cytokines like IL-6, TNFα, and IL-12p40 within 2–4 h of treatment [70].

In our lab, we have used 3% TG to induce peritonitis in BALB/c mice [49] as well as in C57Bl/6J [50] mice. We have found that TG administration leads to an influx of total cells into the peritoneum, as well as of $CD3^+$ T cells, $B220^+$ B cells, $GR1^+$ neutrophils, and $F4/80^+$

macrophages [49, 50]. We also found that TG leads to a decrease in the clonogenic potential and the viability of the peritoneal fluid (PF) cells [49, 50]

18.1.3.3.3 Unmet Needs for New Therapeutic Strategies for Peritonitis

Current therapy for peritonitis includes antibiotics, IV fluid, and surgery. If left untreated, peritonitis can be fatal.

18.1.3.4 ATOPIC DERMATITIS

18.1.3.4.1 Etiology of AD

AD is a chronic skin disorder that involves inflammation of the skin and defective skin barrier function. Redness and itching are the main symptoms of AD. The defective skin barrier function, immune dysregulation, environmental, and infectious agents are associated with the pathogenesis of AD [35]. The genetic factor involved in the pathogenesis of AD involves mutation within the filaggrin gene and impaired expression of its associated structural protein, resulting in defective skin barrier function. The AD skin becomes deficient in lipid molecules (ceramides) and antimicrobial peptides (cathelicidins), which makes it susceptible to cutaneous secondary bacterial, fungal, as well as viral infection [65]. The skin barrier mainly begins with the stratum corneum (SC) which makes a permeable barrier that prevents water loss, provides an antimicrobial barrier, and encourages the colonization by nonpathogenic bacteria flora. The lamellar sheets stand as the supportive hydrophobic matrix for SC. The antimicrobial molecules and other enzymes are regulated by lamellar bodies and delivered to SC intercellular matrix. The disruption in these components of skin results in the thinning of SC causing disrupted skin barrier followed by skin inflammation. The pH of skin and calcium gradient sometimes alter the expression of proteins and enzymes required for proper barrier function [14]. The immune dysregulation in AD involves the production of proinflammatory cytokines (IL-4, IL-5, IL-13, IL-31, CCL-18, IL-1β, and TNFα) on the induction of any mechanical trauma (scratching) and disrupted skin. These cytokines then facilitate the production of chemokines (CCL11/eotaxin-1, CCL24/eotaxin-2, and CCL3/eotaxin-3) from epidermal keratinocytes which signal the infiltration of leucocytes to the injury site. Mainly the activated epidermal langerhans cells and DCs are involved in the release of these chemokines [40]. Eotaxin binds to CCR-3 receptors on eosinophils and directs these cells to the inflammatory site. AD is characterized by Th2 type mediated cytokines in both lesional and nonlesional skin [11]. These cytokines act as mediators to downregulate the expression of multiple genes in the epidermal differentiation complex and tight junction proteins, such as claudins, and thus leads to defective skin barrier in AD [15]. The Th2 cytokines also downregulate the expression of TLRs which limit their role in recognition of pathogen-associated molecular patterns and antimicrobial defenses, making skin more susceptible to infections [40]. The recruitment of leukocytes such as eosinophils to the inflammatory site results in the phosphorylation of STAT6 by IL-4, thereby activating STAT6. Thus JAK-STAT pathway plays a major role in Th2 cell proliferation and its associated cytokines. The Th2 upregulation triggers B cell maturation and IgE hypersecretion which binds to mast cells and causes the release of histamines for the disease progression [11]. Thus, both tissue-specific cells (site of injury) and systemic cells aid to exacerbate AD.

18.1.3.4.2 Preclinical Mouse Model of AD

AD is induced in BALB/c mice using Oxazolone (Oxa) (4-ethoxymethylene-2-phenyl-2-oxazoline-5-one), a compound with a chemical formula of $C_3H_3NO_2$, which acts as a hapten [34]. Oxa mimics the symptoms of AD. Repeated treatment with Oxa induces a Th2-like response, characterized by infiltration of Th2 lymphocytes, mast cells and eosinophils, overexpression of IL-4, and overproduction of IgE [34]. It also causes skin hyperplasia and dehydration of the skin [34].

Mice are sensitized topically, on both ears, with 0.8% Oxa, dissolved in acetone, on day 0 [25]. They are then challenged topically, on both ears, with 0.4% Oxa, dissolved in acetone, on days 7, 10, 12, 14, 17, 19, 21, 23, 25, and 27 [25]. The mice are sacrificed on day 28 for assays [25]. The main symptom of AD is the swelling and reddening of the site of inflammation, in this case, the ears. As we have seen, the skin around the ears becomes very dry, the fur in that area sheds and the ears themselves become red and swollen. The total cellular infiltration to the skin increases significantly in AD [25, 33], as does the serum IgE [25, 33] and expression of IL-4 genes [33]. The count of eosinophils in the blood increases [25]. The $CD3^+$ T cells, $CD4^+$ TH cells, and $GR1^+$ neutrophils increase in the blood as well as the skin [25]. Hematoxylin-eosin staining of histological sections of the ear shows a thickened epidermis, with an increase in the eosinophils and mast cells [25]. There is also a decrease in the clonogenic potential of the cells of the blood and skin, showing the degenerative nature of Oxa [25].

18.1.3.4.3 Unmet Needs for New Therapeutic Strategies for AD

There is no permanent cure for AD. The therapies are used only to reduce the symptoms temporarily. Lotions and emollients are prescribed to reduce the itching and dryness [25]. Topical corticosteroids and topical calcineurin inhibitors inhibit the proinflammatory cytokines, but these drugs are associated with adverse effects like skin atrophy and malignancies [25]. Antibiotics are used to treat the secondary bacterial infections associated with AD [25]. Newer therapeutic strategies, like phototherapy and antibody therapy, have been developed, but the effects of natural compounds have not been fully explored [25].

18.1.4 DRUG REPURPOSING

Drug repurposing or drug repositioning is the investigation and use of an existing drug as a therapy for other diseases [6, 68]. For example, thalidomide has been successfully used for leprosy and multiple myeloma [6]. High attrition rates, high costs, and slow speed of drug discovery and development have led to the repurposing of "old" drugs [63]. Drug repurposing has the added benefit of having lower costs and lesser development time, thereby increasing productivity [6, 63].

Our lab has used several compounds in different in-vivo preclinical models of inflammatory diseases, showing the possibility of drug repurposing.

18.2 NANOBIOTECHNOLOGICAL AGENTS

Nanoparticles (NPs) are particles or small objects, usually between 1 and 100 nm in size, which behaves as an individual unit, with unique properties. NPs typically have a surrounding interfacial layer, which influences the properties of the NPs. The ISO Technical Specification 80004 defines an NP as "a nanoobject with all three external dimensions in the nanoscale, whose longest and shortest axes do not differ significantly, with a significant difference typically being a factor of at least 3" [31]. The physical and optical properties of NPs are unique and different from the material that they are made from [29], and these properties depend on the size of the NP [39]. NPs can be functionalized by coating them with organic molecules or polymers, whereby the properties of the NP can be changed [67]. For biological uses, NPs are usually coated with polar molecules, especially polysaccharides [59], to bestow high water-solubility, to improve biocompatibility, and to prevent unwanted aggregation [45]. They can be conjugated to biological molecules and used to target the NP to specific sites in the body or cell [4], or for imaging purposes [59]. The penetrability of NPs into cells and tissues can be further enhanced by functionalizing them with lipid or phospholipid molecules [45].

The rationale behind using NPs as anti-inflammatory nanodrugs and/or nanovehicles for targeting in anti-inflammatory or prophylaxis thereof is as follows:

(1) Greater penetrability in tissue.
(2) Lowering saturation kinetics, that is, not only enhancing drug efficacy at lower doses but also identifying targets of such NPs with higher affinity in the target tissue/organ.
(3) Developing the same as a probe and as a scaffold for tissue engineering (in-vitro or ex-vivo) and engraftment (in-vivo).

The rationale behind the choice of NPs was in the structure of the same, that is, for fishing out complementary epitopes and "lighting up" the same with an autofluorescent—or a complexed—fluorophore. On the one hand, an NP is a reliable probe in complex signaling networks, and on the other hand, the property of self-recognition allows greater permeability through biomembranes, bypassing the off-target side effects/contraindications of nonnanotized particles.

In our choice of NPs, we have used carbon and hydrocarbon, peptides, and sugar to identify, characterize, and validate:

(1) Targeting by NPs.
(2) Target-enhancement and downstream activity of NPs.
(3) Tracing the path of such activity by the aforementioned NP-photoreactivity.
(4) Using nanoporous scaffolds to develop an ideal stroma for culture, characterization, and translation in tissue engineering.

18.2.1 NANODRUGS

18.2.1.1 GUAR GUM NP AS NANODRUG

Guar gum (GG) is a nonionic polysaccharide gum found in the endosperm of the seed of the guar plant, *Cyamopsis tetragonoloba* [25, 26]. GG is a galactomannan, containing a mannose backbone with galactose side chains, with mannose and galactose usually present in a 2:1 ratio (Figure 18.1A). GG is already used in the food industry, as well as in the pharmaceutical industry. We have used a novel NP synthesized with the GG [25, 26]. The *guar gum nanoparticle* (GN), prepared by the acid hydrolysis of the GG, is a spherical molecule with a size ranging from 30 to 80 nm (Figure 18.1B). The mannose: galactose ratio has also been modified, from 2:1 in GG to 4:1 in GN, allowing it to be taken in by cells containing mannose receptors on their surfaces [26].

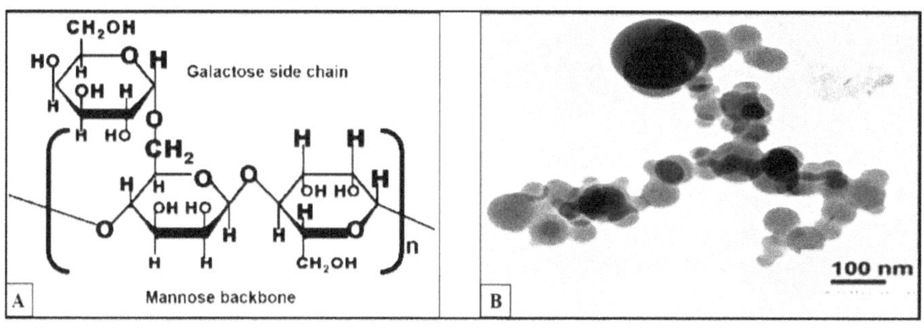

FIGURE 18.1 (A) Guar gum (GG) is a galactomannan, with mannose and galactose present in a 2:1 ratio. (B) GG nanoparticle (GN) is a spherical molecule with a size of 30–80 nm.

GN was found to have anti-inflammatory and proregenerative effects both in-vitro and in-vivo. We administered 40 μg/mL GN to lipopolysaccharide (LPS)-treated RAW 264.7 macrophage cell lines and assessed its anti-inflammatory effect using Griess reagent, and its proregenerative effect using MTS assay [26]. While it reduced the NO concentration of the cells, it also increased cellular viability. We hypothesized that GN works by being taken up the cells via their mannose receptors, which are abundantly present on macrophages [26]. We also found that GN was successful in healing a wound, induced by a scratch in-vitro in NIH3T3 fibroblast cell line, within a very short period of time [25].

18.2.1.2 REPURPOSING OF GUAR GN IN PERITONITIS & AD

18.2.1.2.1 In peritonitis [26]

One hour after the TG treatment, mice were given 200 μg GN intra-peritoneally. Twenty-four hours after GN treatment, the mice were sacrificed and tissues collected. GN was found to reduce the total cellular infiltration into the blood and PF, as well as the count of monocytes and neutrophils. The NO content of the PF decreased. While the number of $CD3^+$ T cells, $GR1^+$ neutrophils, and the $F4/80^+$ macrophages decreased in the blood, in the PF the $B220^+$ B cells and $F4/80^+$ macrophages decreased. These observations indicated that GN shows an anti-inflammatory effect in-vivo. GN also successfully restored the clonogenic potential and the viability of the PF cells, indicating a proregenerative effect as well.

18.2.1.2.2 In AD [25]

Around 100 μg GN was administered topically to each ear of the mice on days 14, 17, 19, 21, 23, 25, and 27. With Oxa treatment, the ears of the mice became red, dry, and swollen, and the fur around the ears was lost. GN not only reduced the swelling in the ears that were caused by Oxa treatment but also restored the lost fur to some extent. GN successfully inhibited the infiltration of cells into the blood and skin, including eosinophils, $CD3^+$ T cells, $CD4^+$ T_H cells, and $GR1^+$ neutrophils. The serum IgE level also decreased with GN treatment, as did the thickness of the epidermis as seen in histological sections. GN also shows a proregenerative effect, as indicated by the restoration of the clonogenic potential of both blood and skin.

18.2.1.3 Antibody Therapy

Antibody therapy is a type of immunotherapy, where monoclonal antibodies (mAbs) are used to bind specifically to harmful cells or protein, with the intent to elicit the patient's immune response. It is possible to design mAbs against any extracellular or cell surface targets. With this aspect of antibodies in mind, we have developed a technology whereby we use camelid antibodies as therapeutic agents against inflammatory diseases.

Traditional antibodies are large, Y-shaped glycoprotein molecules, of approximately 150 kDa, produced by the body's immune cells to fight pathogens. A unique molecule on the pathogen (antigen) is recognized by the antibody's Fab, or antibody binding, region. This allows the immune cells to recognize the antigen-antibody complex, and destroy the antigen. The typical antibody (Figure 18.2A) has two identical polypeptide heavy chains, connected to two polypeptide light chains via disulphide bonds. Each heavy chain has a variable (Fv) domain, and a constant (Fc) region made of three domains (CH1, CH2, and CH3) [60]. The antigen-binding region is made of the Fv and CH1 domains. Camelids, members of the family Camelidae, like camels and llamas, have another type of antibodies (Figure 18.2B). These camelid antibodies do not contain the light chain or the CH1 domain, but only the variable (VHH), CH2 and CH3 domains of the heavy chain [28, 60]. Camelid antibodies are lighter than normal antibodies, with an approximate molecular weight of 80 kDa.

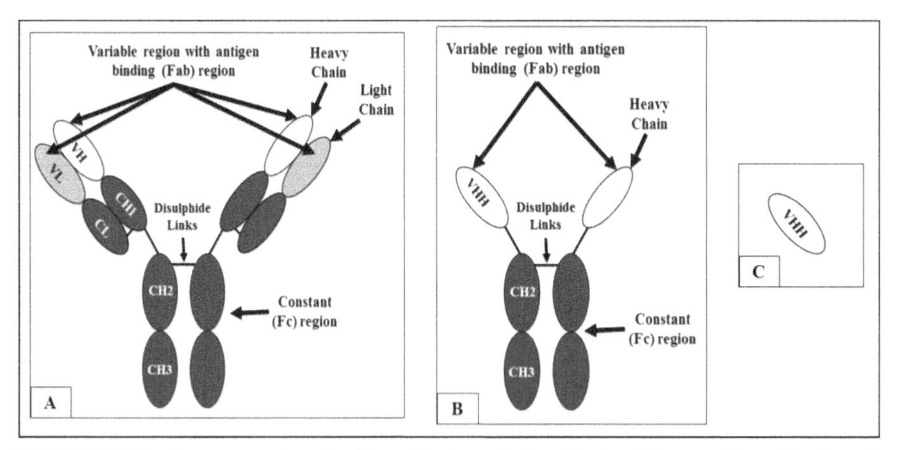

FIGURE 18.2 Structure of antibodies. (A) A typical antibody; (B) camelid antibody; (C) single-domain antibody.

Single-domain antibodies comprise of only the very heavy VHH domain (around 15 kDa in size) (Figure 18.2C), can be raised against any protein present in diseased conditions, and, due to their small size, have the ability to reach such sites on an antigen that classical antibodies may not be able to reach [60]. They can also be expressed in any expression system, which is very stable and has high solubility [28]. In our work, we have cloned the 400 bp gene for the VHH domain of camels (*Camelus dromedarius*) and prepared libraries of antibodies against specific antigens. To obtain antibodies against alkaline phosphatase (AP) and α-amylase (AA), commercially available AP and AA are injected into camels. For asthma-specific IgE, chronic asthma is the first induced in mice using Ova, then their blood, containing anti-Ova IgE, is collected and injected into camels. One hundred five days after injection, camel blood is collected, RNA isolated, and cDNA prepared from it. Then, a nested PCR was done to obtain the 700 bp gene encoding the VHH+CH2 domains, followed by a second nested PCR to get the 400 bp gene for the VHH domain. This 400 bp gene is ligated into a plasmid and transformed into a bacterial vector, which is then infected with a phage so that the protein library translated from the 400 bp gene gets displayed on the phages. Biopanning is then done to enrich the specific antibody, which can then be used as therapy.

18.2.2 NANOVEHICLES

Mesoporous carbon nanoparticle (MCN) is a porous NP with a size ranging between 100 and 200 nm

[49]. Its high surface area, large pore volume, chemical inertness, biocompatibility, and solubility in water make it useful as an NP [49]. It has a uniform pore size of 3 nm, and the pores can be loaded with large amounts of a drug [49]. In our lab, we have loaded the pores of MCN with fisetin, a flavonoid with known anti-oxidant, anti-inflammatory, anticarcinogenic, and immuno-suppressing effects [57, 72]. We have used fisetin loaded on MCN in preclinical models of several inflammatory diseases, like acute asthma [48], peritonitis [49], and IPF [36, 37].

18.2.2.1 REPURPOSING OF MCN AS NANOVEHICLES FOR FISETIN IN INFLAMMATORY DISEASES

18.2.2.1.1 In Acute Asthma [48]

In our study, 50 nM (~2 μM/kg) fisetin was mixed with an equal amount of 5.1 mg/mL MCN and administered intratracheally on days 8, 15, 18, and 21, 1 h before each intratracheal treatment of Ova. Fisetin + MCN decreased the total cellular infiltration, as well as the infiltration of eosinophils, in the blood, lung, and BALF. It significantly decreased the concentration of NO in the lung. It decreased the expression of IL2, IL4, IL5, TNFα, and IFNγ. The significant reduction in the inflammatory markers indicates that fisetin+MCN is an effective anti-inflammatory agent in acute asthma. However, the anti-inflammatory effect of fisetin+MCN was less than that of fisetin alone, showing that MCN did not impart any extra beneficial effect to fisetin.

18.2.2.1.2 In Idiopathic Pulmonary Fibrosis [36, 37]

In our study, 50 nM (~2 μM/kg) fisetin was mixed with an equal amount of 5.1 mg/mL MCN and administered intratracheally on days 7, 14, and 21. Fisetin+MCN significantly reduced the symptoms of IPF induced by Bleo. The total cell count of blood, lung, and BALF decreased significantly, as did the concentrations of superoxide radical and NO radicals. The clonogenic potential and the cell viability were restored, showing a proregenerative effect of fisetin+MCN. The reduction in the hydroxyproline content of the lung, and the decrease in the collagen deposition as seen in the histological sections, indicate an antifibrotic effect as well. In the case of IPF, however, MCN appeared to enhance the effects of fisetin, and thus, worked well as a nanovehicle.

18.2.2.1.3 In Peritonitis [49]

We have used fisetin in peritonitis. As a therapeutic, we have administered

50 nM (~2 μM/kg) fisetin, mixed with an equal amount of 5.1 mg/mL MCN, orally, 1 h after each TG treatment. Fisetin+MCN reduced the inflammation caused by TG, as indicated by the reduction in the total cell count, differential cell count, and NO concentration in the blood and PF. It also restored the clonogenic potential and viability of the PF cells. In the case of peritonitis, MCN has enhanced the anti-inflammatory effects of fisetin till up to 48 h, after which there is not much difference with fisetin. However, it does not show any added effect to fisetin's regenerative activity.

18.2.3 SCAFFOLDS

Scaffolds are porous biomaterials, which act as templates or supports for the growth of new tissues [55]. Polymeric scaffolds, usually designed as meshes, sponges, or fibers, create a microenvironment for cells to grow into tissues and can act as a substitute for the ECM of cells [51]. The porous nature and large surface area of scaffolds allow the cells and nutrients to be distributed evenly [51]. Scaffolds can either be used as physical supports on which the cells can grow, or they can be used to deliver growth factors to the cells [51].

In our lab, we used a scaffold designed by GG. The GG was modified and functionalized with a carboxymethyl group, forming the carboxymethyl GG scaffold, the nanofibers of which form pores with an average diameter of 10 μm [51]. We have grown RAW 264.7 murine macrophage cells, NIH3T3 murine fibroblast cells, and murine BMMSCs on the scaffold, to assess whether it could support their growth, and to determine whether it has any protective effect on the cells under inflammatory and hypoxic conditions [51].

We found that cells grown on the scaffold had better viability than the cells grown without the scaffold. While the RAW 264.7 and NIH3T3 cells maintained their viability for up to 72 h, the BMMSCs lost their viability after 48 h [51]. We also found that the scaffold exerted a protective effect on the cells in the presence of LPS (to induce inflammation) and $CoCl_2$ (to induce hypoxia), as the cells in the scaffold had better viability than without [51].

18.3 DISCUSSION

Bioprospecting is a process of searching for plants and animals which can serve as sources of medicinal compounds and can act as bioresources [12]. All types of organisms (like bacteria, fungi, plants, invertebrates, and vertebrates) can be used in bioprospecting.

Natural compounds and their extracts are gathering importance in the field of drug discovery with the increase in antibiotic resistance. Natural products or bioactive compounds that are derived from natural sources such as plants, animals, or micro-organisms have been in use for thousands of years [22]. The importance of natural products for medicine and health has been enormous. Unmet needs in medicine and unknown phenomena prevailing in the mechanism of disease onset remain. In traditional medicine that has provided the solution to prevailing health issues at a global level, medicinal plants continue to provide valuable therapeutic agents. To avoid and mitigate various side effects and complications of modern medicine, and to address unmet needs of diseases, especially in the context of emerging complex etiopathophysiological pathways, traditional medicine is gaining importance and is now being studied systematically using biotechnological tools, to find the scientific basis of their therapeutic actions. Since early times, when people chewed on certain herbs to relieve pain, or wrapped leaves around wounds to improve healing, natural products have often been the sole means to treat diseases and injuries. Different remote communities in India claim that the medicines obtained from these natural products are cheaper and more effective than modern medicines. However, natural products have taken a secondary role in drug discovery and drug development, after the advent of molecular biology and combinatorial chemistry made possible the rational design of chemical compounds to target specific molecules. Different varieties of *Piper betel* leaf extracts are known for their antioxidative, anti-inflammatory, and antigeriatric activities. Betel leaf is known to be useful for the treatment of bad breath, abscesses, conjunctivitis, constipation, headache, itches, leucorrhoea and otorrhoea, swelling of gum, rheumatism, and injuries. The leaves possess antibacterial, antiprotozoan, and antifungal properties, and the oil is used as industrial raw material for the manufacture of medicines, perfumes, mouth fresheners, tonics, and food additives. Betel vine and date palm have a high content of various phenols, flavonoids, and alkaloids, as well as other bioactive compounds, such as fisetin, curcumin, β-carotene, α-tocopherol, hydroxychavicol, and eugenol [53, 58, 74]. These compounds have anticarcinogenic, antimutagenic, antidiabetic, antimicrobial, and antiulcer activities. Betel vine and date palm also may contribute as a nutraceutical substance and a prophylactic or therapeutic agent in some tissue-specific inflammatory diseases [58].

Recently the use of natural dietary substances, found in fruits, vegetables, and herbs, has received considerable attention as chemopreventive and chemotherapeutic agents worldwide. Fisetin is a bioactive flavonol found in fruits and vegetables such as strawberries, apples, persimmon, grape, onion, and cucumber [38]. It has been shown to possess both direct intrinsic antioxidant and indirect antioxidant effects. Fisetin shows multiple beneficial pharmacological activities such as anti-inflammatory activity, antioxidant activity, and anti-cancer [38]. The highest concentration of fisetin is found in strawberries (160 μg/g), followed by apples (26.9 μg/g) [38].

The consumption of antioxidant-rich food helps neutralize the ROS released in inflammation, thus preventing oxidative damage. It has been proven that antioxidants could reduce the chances of cardiovascular diseases and other chronic diseases, like AD, rheumatoid arthritis, and allergic asthma.

It is well established that all camelids have unique antibodies circulating in their blood. Camelids produce functional antibodies devoid of light chains. The single N-terminal domain of the light chain is fully capable of antigen-binding. These VHHs or nanobodies (single-domain antibody fragments) have received a progressive interest in the biotechnology industries. Nanobodies have been an ideal research tool for the development of sophisticated nanodiagnostic, nanoprophylactic, and nanotherapeutic technologies because nanobodies have some peculiar properties—nanoscale size (MW ~15 kDa), reversible refolding, robust structure, stability and solubility in aqueous solution, superior cryptic cleft accessibility, high affinity and specificity for only one cognate target, and deep tissue penetration [71]. For their use in targeting drugs across the blood–brain barrier (BBB) into the brain, VHHs were selected that transmigrate the human BBB in an in vitro model and accumulate in the brain after intravenous injection into mice [28]. These could be used for the treatment of neurological disorders. VHHs have many advantages for biotechnological applications. They can be economically produced in microorganisms and have high stability. Furthermore, they are highly suited for expression as multivalent, including bispecific, formats, or as enzyme fusions. This permits a plug-and-play approach, where, depending on the target, biological potency can be increased by multivalent constructs or bispecific VHH recognizing two different targets can be made [28]. Stem cell-based therapies also hold tremendous promise for the treatment of degenerative diseases, other serious diseases, and injuries.

18.4 CONCLUSION

Inflammatory diseases are extremely common in today's environment and lifestyle. Often, these diseases remain untreated due to a variety of reasons, the most common of which is the unavailability of permanent cure and the inability of many people to afford treatment. The strategies outlined in this chapter may go a long way in reducing this disease burden, by using easily available drug sources and cost-effective strategies.

KEYWORDS

- **inflammation**
- **degeneration**
- **nanotechnology**
- **drug repurposing**

REFERENCES

1. Abbas AB, Lichtman AH. "Ch.2: Innate immunity". In: Basic Immunology: Functions and Disorders of the Immune System, 3rd ed. Saunders: London, 2009. ISBN 978–1-4160–4688-2.
2. Adler M, Mayo A, Zhou Z, Franklin RA, Jacox JB, Medzhitov R, Alon U. Endocytosis as a stabilizing mechanism for tissue homeostasis. PNAS. 2018. 115 (8): E1926–E1935.
3. Agrawal DK, Shao Z. Pathogenesis of allergic airway inflammation. Curr Allergy Asthma Rep. 2010. 10(1): 39–48.
4. Åkerman ME, Chan WCW, Laakkonen P, Bhatia SN, Ruoslahti E. Nanocrystal targeting in vivo. Proc Natl Acad Sci USA. 2002. 99(20): 12617–12621.
5. Angeloni C, Maraldi T, Vauzour D. Redox signalling in degenerative diseases: from molecular mechanisms to health implications. BioMed Res Int. 2014. 2014, 245761.
6. Ashburn TT, Thor KB. Drug repositioning: identifying and developing new uses for existing drugs. Nat Rev Drug Discov. 2004. 3(8): 673–83.
7. Aufricht C, Neuhofer W, Topley N, Wörnle M. Peritoneal infection and inflammation. Mediators Inflamm. 2012. 2012. 456985.
8. Banerjee ER, Henderson WR Jr. Characterization of lung stem cell niches in a mouse model of bleomycin-induced fibrosis. Stem Cell Res Ther. 2012. 3(3): 21–42.
9. Banerjee ER, Jiang Y, Henderson WR Jr, Latchman Y, Papayannopoulou T. Absence of alpha 4 but not beta 2 integrins restrains development of chronic allergic asthma using mouse genetic models. Exp Hematol. 2009. 37(6): 715–727.
10. Banerjee ER, LaFlamme MA, Papayannopoulou T, Kahn M, Murry CE, Henderson WR Jr. Human embryonic stem cells differentiated to lung lineage-specific cells ameliorate pulmonary fibrosis in a xenograft transplant mouse model. PLoS One. 2012. 7(3): e33165:1–15.
11. Bao L, Zhang H, Chan LS. The involvement of the JAK-STAT signaling pathway in chronic inflammatory skin disease atopic dermatitis. JAK-STAT. 2013. 2(3): e24137.
12. Beattie AJ, Hay M, Magnusson B, de Nys R, Smeathers J, Vincent JFV. Ecology and bioprospecting. Austral Ecol. 2011. 36(3): 341–356.
13. Beyer K, Menges P, Keßler W, Heidecke CD. Pathophysiology

of peritonitis [Article in German]. Chirurg. 2016. 87(1): 5–12.
14. Boothe WD, Tarbox JA, Tarbox MB. "Ch. 3: Atopic dermatitis: pathophysiology". In: Management of Atopic Dermatitis, Advances in experimental medicine and biology springer, Fortson et al. (eds.), International Publishing AG: Basel. 2017, 21 E.A, 21–37.
15. Brunner PM, Yassky EG, Leung DYM. The immunology of atopic dermatitis and its reversibility with broad-spectrum and targeted therapies. J Allergy Clin Immunol. 2017. 139(4S): S65–S76.
16. Canestaro WJ, Forrester SH, Raghu G, Ho L, Devine BE. Drug treatment of idiopathic pulmonary fibrosis: systematic review and network meta-analysis. Chest. 2016. 149(3): 756–766.
17. Chen L, Deng H, Cui H, Fang J, Zuo Z, Deng J, Li Y, Wang X, Zhao L. Inflammatory responses and inflammation-associated diseases in organs. Oncotarget. 2018. 9(6): 7204–7218.
18. Cool CD, Groshong SD, Rai PR, Henson PM, Stewart JS, Brown KK. Fibroblast foci are not discrete sites of lung injury or repair: the fibroblast reticulum. Am J Respir Crit Care Med. 2006. 174(6): 654–658.
19. Coultas DB, Zumwalt RE, Black WC, Sobonya RE. The epidemiology of interstitial lung diseases. Am J Respir Crit Care Med. 1994. 150(4): 967–72.
20. Demedts M, Wells AU, Antó JM, Costabel U, Hubbard R, Cullinan P, Slabbynck H, Rizzato G, Poletti V, Verbeken EK, Thomeer MJ, Kokkarinen J, Dalphin JC, Taylor AN. Interstitial lung diseases: an epidemiological overview. Eur Respir J Suppl. 2001. 32, 2s–16s.
21. Elias JA, Lee CG, Zheng T, Ma B, Homer RJ, Zhu Z. New insights into the pathogenesis of asthma. J Clin Invest. 2003. 111(3): 291–297.
22. Fabricant DS, Farnsworth NR. The value of plants used in traditional medicine for drug discovery. Environ Health Perspect. 2001. 109(Suppl 1): 69–75.
23. Fernandez IE, Eickelberg O. New cellular and molecular mechanisms of lung injury and fibrosis in idiopathic pulmonary fibrosis. The Lancet. 2012. 380(9842): 680–688.
24. Ferrero-Miliani L, Nielsen OH, Andersen PS, Girardin SE. Chronic inflammation: importance of NOD2 and NALP3 in interleukin-1β generation. Clin Exp Immunol. 2007. 147(2): 227–235.
25. Ghosh N, Mitra S, Banerjee ER. Therapeutic effects of topically-administered guar gum nanoparticles in oxazolone-induced atopic dermatitis in mice. Biomed Res Ther. 2018a. 5(5): 2305–2325.
26. Ghosh N, Mitra S, Biswas S, Banerjee ER. Mannose-rich guar gum nanoparticle as a novel therapeutic drug against inflammatory diseases. J Mol Biochem. 2018b. 7, 14–27.
27. Granger DN, Senchenkova E. Inflammation and the Microcirculation. Morgan & Claypool Life Sciences: San Rafael, CA, 2010.
28. Harmsen MM, De Haard HJ. Properties, production, and applications of camelid single-domain antibody fragments. Appl Microbiol Biotechnol. 2007. 77(1): 13–22.
29. Hewakuruppu YL, Dombrovsky LA, Chen C, Timchenko V, Jiang X, Baek S, Taylor RA. Plasmonic "pump–probe" method to study semi-transparent nanofluids. Appl Optics. 2013. 52(24): 6041–6050.
30. IPFCRN Idiopathic Pulmonary Fibrosis Clinical Research Network. Raghu G, Anstrom KJ, King TE Jr, Lasky JA,

Martinez FJ. Prednisone, azathioprine, and N-acetylcysteine for pulmonary fibrosis. N Engl J Med. 2012. 366(21): 1968–1977

31. ISO/TS 80004–2. Nanotechnologies—vocabulary—Part 2: nano-objects. 2015. International Organization for Standardization, Geneva.
32. Javad-Mousavi SA, Hemmati AA, Mehrzadi S, Hosseinzadeh A, Houshmand G, Nooshabadi MRR, Mehrabani M, Goudarzi M. Protective effect of Berberis vulgaris fruit extract against Paraquat-induced pulmonary fibrosis in rats. Biomed Pharmacother. 2016. 81: 329–336.
33. Jegal J, Park NJ, Bong SK, Jegal H, Kim SN, Yang MH. Dioscorea quinqueloba ameliorates oxazolone- and 2,4-dinitrochlorobenzene-induced atopic dermatitis symptoms in murine models. Nutrients. 2017. 9(12): E1324.
34. Jin H, He R, Oyoshi M, Geha R. Animal models of atopic dermatitis. J Invest Dermatol. 2009. 129(1): 31–40.
35. Kapur S, Watson W, Carr S. Atopic dermatitis. Allergy Asthma Clin Immunol. 2018. 14(Suppl 2): 43–52.
36. Kar S, Konsam S, Hore G, Mitra S, Biswas S, Sinha A, Jana NR, Banerjee ER. Therapeutic use of fisetin, curcumin, and mesoporous carbon nanoparticle loaded fisetin in bleomycin-induced idiopathic pulmonary fibrosis. Biomed Res Ther. 2015. 2(4): 250–262.
37. Kar S, Biswas S, Banerjee ER. Evaluating the ameliorative potential of plant flavonoids and their nanocomposites in bleomycin induced idiopathic pulmonary fibrosis. Biomed Res Ther. 2016. 3(7): 707–722.
38. Khan N, Syed DN, Ahmad N, Mukhtar H. Fisetin: a dietary antioxidant for health promotion. Antioxid Redox Signal. 2013. 19(2): 151–162.
39. Khan I, Saeed K, Khan I. Nanoparticles: properties, applications and toxicities. Arab J Chem. 2017. https://doi.org/10.1016/j.arabjc.2017.05.011.
40. Kim JE, Kim JS, Cho DH, Park HJ. Molecular mechanisms of cutaneous inflammatory disorder: atopic dermatitis. Int J Mol Sci. 2016. 17(8): E1234.
41. King TE Jr, Pardo A, Selman M. Idiopathic pulmonary fibrosis. The Lancet. 2011. 378(9807): 1949–1961
42. Kuprash DV, Nedospasov SA. Molecular and cellular mechanisms of inflammation. Biochemistry (Mosc). 2016. 81(11): 1237–1239.
43. Lam D, Harris D, Qin Z. Inflammatory mediator profiling reveals immune properties of chemotactic gradients and macrophage mediator production inhibition during thioglycollate elicited peritoneal inflammation. Mediators Inflamm. 2013. 2013, 931562.
44. Loomis-King H, Flaherty KR, Moore BB. Pathogenesis, current treatments and future directions for idiopathic pulmonary fibrosis. Curr Opin Pharmacol. 2013. 13(3): 377–385.
45. Luchini A, Vitiello G. Understanding the nano-bio Interfaces: lipid-coatings for inorganic nanoparticles as promising strategy for biomedical applications. Front Chem. 2019. 7, 343.
46. Luzina IG, Todd NW, Iacono AT, Atamas SP. Roles of T lymphocytes in pulmonary fibrosis. J Leukoc Biol. 2008. 83(2): 237–244.
47. Medzhitov R. Origin and physiological roles of inflammation. Nature. 2008. 454(7203): 428–435.
48. Mitra S, Paul P, Mukherjee K, Biswas S, Jain M, Sinha A, Jana NR, Banerjee ER. Mesoporous nano-carbon particle loaded fisetin has a positive therapeutic effect in a murine preclinical model of ovalbumin induced acute allergic

asthma. J Nanomedine Biotherapeutic Discov. 2015a. 5: 132.
49. Mitra S, Biswas S, Sinha A, Jana NR, Banerjee ER. Therapeutic use of fisetin and fisetin loaded on mesoporous carbon nanoparticle (MCN) in thioglycollate-INDUCED peritonitis. J Nanomed Nanotechnol. 2015b. 6(6): 332.
50. Mitra S, Mukherjee K, Biswas S, Banerjee ER. Prophylactic use of fisetin in thioglycollate-induced peritonitis in mice. Biol Syst Open Access. 2015c. 4: 144.
51. Mitra S, Ghosh N, Banerjee ER. Carboxymethyl guar gum nanoscaffold as matrix for cell growth in vitro. J Lung Pulm Respir Res. 2018. 5(1): 00156.
52. Moeller A, Ask K, Warburton D, Gauldie J, Kolb M. The bleomycin animal model: a useful tool to investigate treatment options for idiopathic pulmonary fibrosis? Int J Biochem Cell Biol. 2008. 40(3): 362–382.
53. Mukherjee K, Paul P, Banerjee ER. Free radical scavenging activities of date palm (Pheonix sylvestris) fruit extracts. Nat Prod Chem Res. 2014. 2(6): 151.
54. Nisbet AD, Saundry RH, Moir AJG, Fothergill LA, Fothergill JE. The complete amino-acid sequence of hen ovalbumin. Eur J Biochem. 1981. 115(2): 335–345.
55. O'Brien FJ. Biomaterials & scaffolds for tissue engineering. Mater Today. 2011. 14(3): 88–95.
56. Oruqaj G, Karnati S, Vijayan V, Kotarkonda LK, Boateng E, Zhang W, Ruppert C, Günther A, Shi W, Baumgart-Vogt E. Compromised peroxisomes in idiopathic pulmonary fibrosis, a vicious cycle inducing a higher fibrotic response via TGF-β signaling. Proc Natl Acad Sci USA. 2015. 112(16): E2048– E2057.
57. Pal HC, Pearlman RL, Afaq F. Fisetin and its role in chronic diseases. Adv Exp Med Biol. 2016. 928, 213–244.
58. Patel N, Mohan JSS. Isolation and characterization of potential bioactive compounds from piper betle varieties Banarasi and Bengali leaf extract. Int J Herb Med. 2017. 5(5): 182–191.
59. Patitsa M, Karathanou K, Kanaki Z, Tzioga L, Pippa N, Demetzos C, Verganelakis DA, Cournia Z, Klinakis A. Magnetic nanoparticles coated with polyarabic acid demonstrate enhanced drug delivery and imaging properties for cancer theranostic applications. Sci Rep. 2017. 7(1): 775.
60. Paul P, Biswas S, Banerjee ER. Development of a novel format of stable single chain antibodies against alkaline phosphatase as therapeutic agents against immune diseases. J Nanomed Biother Discov. 2016. 6, 144.
61. Paul P, Majhi S, Mitra S, Banerjee ER. Immuno-modulatory and therapeutic effect of curcumin in an allergen-sensitized murine model of chronic asthma. J Clin Cell Immunol. 2018. 9, 551.
62. Paul P, Majhi S, Mitra S, Banerjee ER. Orally administered fisetin as an immunomodulatory and therapeutic agent in a mouse model of chronic allergic airway disease. Biomed Res Ther. 2019. 6(7): 3262–3273.
63. Pushpakom S, Iorio F, Eyers PA, Escott KJ, Hopper S, Wells A, Doig A, Guilliams T, Latimer J, McNamee C, Norris A, Sanseau P, Cavalla D, Pirmohamed M. Drug repurposing: progress, challenges and recommendations. Nat Rev Drug Discov. 2019. 18(1): 41–58.
64. Raghu G, Collard HR, Egan JJ, Martinez FJ, Behr J, Brown KK, Colby TV, Cordier JF, Flaherty

KR, Lasky JA, Lynch DA, Ryu JH, Swigris JJ, Wells AU, Ancochea J, Bouros D, Carvalho C, Costabel U, Ebina M, Hansell DM, Johkoh T, Kim DS, King TE Jr, Kondoh Y, Myers J, Müller NL, Nicholson AG, Richeldi L, Selman M, Dudden RF, Griss BS, Protzko SL, Schünemann HJ, ATS/ERS/JRS/ALAT Committee on Idiopathic Pulmonary Fibrosis. An official ATS/ERS/JRS/ALAT statement: idiopathic pulmonary fibrosis: evidence-based guidelines for diagnosis and management. Am J Respir Crit Care Med. 2011. 183(6): 788–824

65. Rahman S, Collins M, Williams CMM, Ma HL. The pathology and immunology of atopic dermatitis. Inflamm Allergy Drug Targets. 2011. 10(6): 486–496.
66. Reinert T, da Rocha Baldotto CS, Nunes FAP, de Souza Scheliga AA. Bleomycin-induced lung injury. J Cancer Res. 2013. 2013, 480608.
67. Sadri R, Hosseini M, Kazi SN, Bagheri S, Ahmed SM, Ahmadi G, Zubir N, Sayuti M, Dahari M. Study of environmentally friendly and facile functionalization of graphene nanoplatelet and its application in convective heat transfer. Energ Convers Manage. 2017. 150, 26–36.
68. Sleigh SH, Barton CL. Repurposing strategies for therapeutics. Pharm Med. 2010. 24(3): 151– 159.
69. Thabut G, Christie JD, Ravaud P, Castier Y, Dauriat G, Jebrak G, Fournier M, Lesèche G, Porcher R, Mal H. Survival after bilateral versus single-lung transplantation for idiopathic pulmonary fibrosis. Ann Intern Med. 2009. 151(11): 767–74.
70. Wan H, Coppens JM, van Helden-Meeuwsen CG, Leenen PJ, van Rooijen N, Khan NA, Kiekens RC, Benner R, Versnel MA. Chorionic gonadotropin alleviates thioglycollate-induced peritonitis by affecting macrophage function. J Leukoc Biol. 2009. 86(2): 361–370.
71. Wang Y, Fan Z, Shao L, Kong X, Hou X, Tian D, Sun Y, Xiao Y, Yu L. Nanobody-derived nanobiotechnology tool kits for diverse biomedical and biotechnology applications. Int J Nanomed. 2016. 11, 3287–3303.
72. Wu MY, Hung SK, Fu SL. Immunosuppressive effects of fisetin in ovalbumin-induced asthma through inhibition of NF-κB activity. J Agric Food Chem. 2011. 59(19): 10496–10504.
73. Yu QL, Chen Z. Establishment of different experimental asthma models in mice. Exp Ther Med. 2018. 15(3): 2492–2498.
74. Zahira A, Thamilmani K. Evaluation of bioactive compounds present in piper betle Linn. by elution chromatography coupling technique. World J Pharm Pharm Sci. 2016. 5(5): 1405–1413.

Index

C

I

N

O

P

Q

R

S

T

U

V